MARINE MICROBIOLOGY

ECOLOGY AND APPLICATIONS

Second Edition

MARINE MICROBIOLOGY

ECOLOGY AND APPLICATIONS

Second Edition

Colin Munn

With Foreword by Farooq Azam

Garland Science
Taylor & Francis Group
NEW YORK AND LONDON

Garland Science
Vice President: Denise Schanck
Senior Editor: Gina Almond
Assistant Editor: Dave Borrowdale
Senior Production Editor: Georgina Lucas
Illustrator: Oxford Designers and Illustrators
Cover Design: Andrew Magee
Copyeditor: Susannah Lord
Typesetting: EJ Publishing Services
Proofreader: Sally Huish
Indexer: Liza Furnival

Front cover image. Atomic force microscopy of pelagic bacteria surrounded by an apparent network of gel and nanometer-sized particles, illustrating microscale structuring of ocean ecosystems. (Image courtesy of Francesca Malfatti and Farooq Azam, Scripps Institution of Oceanography.)

Back cover images. Scanning electron micrographs of dinoflagellates. From top to bottom, zoospore of *Pfiesteria piscicida*, with its feeding peduncle extended (courtesy of Howard Glasgow, North Carolina State University Center for Applied Aquatic Ecology); *Protoperidinium incognitum*; *Protoperidinium antarcticum*; *Gonyaulax striata* (all courtesy of Rick van den Enden, Electron Microscopy Unit, Australian Antarctic Division, © Commonwealth of Australia).

Colin B. Munn obtained a BSc (Hons) degree from University College London and a PhD from the University of Birmingham. He is currently Associate Professor of Microbiology and Admissions Tutor for Marine Biology programmes at the University of Plymouth, England. He has held former positions as Visiting Professor at the University of Victoria, Central University of Venezuela and St Georges University, Grenada and a visiting researcher (Leverhulme Fellow) at James Cook University and the Australian Institute of Marine Science. He has served as external examiner for Bachelor's, Master's and PhD degrees in many countries and as a special assessor for molecular and organismal biology for the UK Quality Assurance Agency for Higher Education.

ISBN 978-0-8153-6517-4

Library of Congress Cataloging-in-Publication Data

Munn, C. B. (Colin B.)
  Marine microbiology : ecology and applications / Colin Munn. -- 2nd ed.
    p. cm.
  IS BN 978-0-8153-6517-4 (alk. paper)
  1. Marine microbiology. I. Title.
    QR106.M86 2011
    579'.177--dc22

                                  2011000917

Published by Garland Science, Taylor & Francis Group, LLC, an informa business, 270 Madison Avenue, New York NY 10016, USA, and 2 Park Square, Milton Park, Abingdon, OX14 4RN, UK.

Printed in the United States of America

15 14 13 12 11 10 9 8 7 6 5 4 3 2 1

Garland Science
Taylor & Francis Group

Visit our website at http://www.garlandscience.com

FSC
www.fsc.org

MIX
Paper from responsible sources
FSC® C005928

7459104

Foreword

When we think of the ocean we might think of whales, waves, dolphins, fish, the smell of the sea, its blue color, and its vastness; most of us would not look out at the sea and think of marine microbes, nor did marine scientists for over a century. They sailed the seas and strenuously dragged plankton nets through the ocean's pelagic zone to capture what they judged would represent the marine biota. But they were unaware that the great majority of the biota, perhaps 99 percent, easily streamed through the holes of their nets; the holes were simply too big to capture these microbes. Even when membrane filters and microscopy were used, and they revealed great diversity of microplankton and nanoplankton, most microbes evaded detection. The view of the pelagic web of life that emerged, and became entrenched for a century, was then based on a tiny fraction of marine biota. As a result, fisheries scientists used models that did not include marine microbes, as did marine chemists and geochemists who studied how biological forces influenced the grand cycles of elements in the ocean. Much had to be revised as the major roles of the microbes were discovered, following the development of new concepts, incisive imaging, and molecular methods to observe and study marine plankton.

Munn enthusiastically and persuasively tells the story of the dramatic transformation of marine microbiology since the mid-1970s ("one of the most important advances in modern science"). The essence of it is that literally a billion (per liter of seawater) previously unsuspected picoplankton and ten-times more abundant and diverse femtoplankton (viruses) account for much of the marine biodiversity, abundance, and metabolism. It is not just the enormous numbers that are remarkable (a billion times more microbes in the sea than all the stars in the known universe, quotes Munn) but the microbes' metabolic capabilities. For instance, in 1988 a 0.6 μm marine photosynthetic bacterium *Prochlorococcus marinus* was discovered. We now know this is the most abundant photosynthetic organism on Earth, and is responsible for a large fraction of marine, and indeed planetary, photosynthesis. In pelagic ecosystems a microbial web of life forms the fine fabric on which macrobial life is visibly embroidered. The story of marine microbiology is truly exciting, and it is still unfolding. Munn captures the excitement of this dynamic field of marine microbiology and conveys it lucidly, insightfully, and in an engaging and accessible manner.

A distinct strength of the book is that it conveys the basic information on marine microbes and their ecophysiology in a highly accessible narrative; detailed and well-referenced discussions are placed in boxes that read like mini-reviews (and they are remarkably up to date and well-referenced including some 2010 citations.) Bacterial physiology and biochemistry is streamlined and marine examples are used, so the relevance to ocean processes is always maintained. Many currently topical research questions are also presented and discussed, including some hypotheses ripe for testing. A prominent feature of marine microbiology is that it is expanding rapidly to create new interdisciplinary interfaces. As we face complex environmental issues, so marine microbiologists need to work with ecologists, geochemists, fisheries and conservation biologists, climate scientists, biomedicine researchers, and epidemiologists to incorporate a microbial perspective into studies of ocean and Earth systems. Munn's book should be of much value not only to students of marine microbiology but also to scientists from related disciplines, and could foster interdisciplinary research.

Munn reflects the times and the environmental concerns, returning over and over again to the theme of global climate change. He is passionate and creative as he debates the geoengineering proposals to sequester atmospheric CO_2 (he expresses strong reservations on geoengineering) and the environmental threats to coral reefs. In the "good old days" marine microbes (and microbiologists) were not included in such weighty issues. So, this is a relatively new arena for marine microbiologists. We have made impressive progress in uncovering microbial diversity, spatial and temporal dynamics, and have predicted metabolic capabilities for interactions with ocean systems. However, we are only beginning to understand how marine microbes influence system *variability* in an ecosystem context. For example, what regulatory mechanisms underlie the variability in the microbial carbon cycling in the ocean? How would they respond, for example, to climate change or geoengineering? These questions involve regulatory interactions among diverse microbes, from molecular and microspatial to an ocean-basin scale. Munn pays considerable and deep attention to the microspatial context for microbial biogeochemistry. After all, *this* is the realm of the microbe and the scale at which the individual microbes minutely structure the ecosystems and respond to global change. Their strength lies in their astronomical numbers, incredible diversity, and intense activity, which still evade the casual observer of the sea.

Farooq Azam
Distinguished Professor
Scripps Institution of Oceanography
University of California, San Diego

Preface

Marine microbiology has emerged as one of the most important areas of modern science, and I hope that readers will share a sense of my excitement of learning about new discoveries in this fascinating and fast-moving subject. This book is intended primarily for upper-level undergraduates and graduate students, but I anticipate that it will also prove useful to researchers who are interested in some of the broader aspects and applications outside of their specific area of investigation. University courses often include some element of marine microbiology as a specialist option for oceanography or marine biology majors who have little previous knowledge of microbiology, but marine microbiology is lightly covered in most marine biology textbooks. I hope that this book may play some part in rectifying this deficiency. I also hope that the book will be useful to microbiology majors studying courses in environmental microbiology, who may have little knowledge about ocean processes or the applications of the study of marine microorganisms. I have attempted to make the book sufficiently self-contained to satisfy all of the various potential audiences. Above all, I wanted to create a book that is enjoyable to read, with the overall aim of bringing together an understanding of microbial biology and ecology with consideration of the applications for environmental management, human welfare, health, and economic activity.

As will become evident, many common themes and recurring concepts link the activities, diversity, ecology, and applications of marine microbes. I have attempted to summarize the current state of knowledge about each aspect with extensive cross-linking to other sections. To improve readability, I rarely use references in the main text, but each chapter contains special interest boxes, which contain references to recent research. Short boxes marked with the symbols (?) or (i) highlight important questions or interesting facts that supplement the main text. The choice of these topics is entirely my personal whim—they represent subjects that I think are particularly intriguing, exciting, or controversial. The longer *Research Focus Boxes* are intended to be relatively self-contained "mini-essays," which explore in more detail some topical areas of investigation. Space does not permit exhaustive reviews of these topics, but they may serve as a stimulus for students to follow up some of the original research papers suggested and use these as a starting point for further enquiry. Also included is a glossary of some of the most important key terms.

The great advances in marine microbiology have occurred because of the development and application of innovative new techniques. A basic understanding of the main principles of methodology is essential if the student is to make sense of research findings. Chapter 2 is written in a style that students with limited practical experience can understand, by concentrating on the principles and avoiding too much technical detail, so I implore students to read this chapter right through at an early stage and refer back to it whenever a particular technique is mentioned later.

What's new in this edition? The general aims and structure of the book are similar, but all material has been updated and most sections have been completely rewritten to take account of the many new discoveries in this field since the first edition was published in 2003. I have also tried to incorporate the numerous helpful comments received from students, course leaders, and reviewers. There is more emphasis on the role of microbes in ocean processes and nutrient cycles, new information about the importance of viruses, and expanded coverage of the beneficial role of marine microbes in biotechnology. Despite an exhaustive review process, astute readers may spot some errors and omissions or have suggestions for improvement. I welcome comments from students and instructors—please e-mail me at c.munn@plymouth.ac.uk.

Acknowledgments

This book would not have been possible without the continued stimulation of ideas that I receive from the students who take my courses. I am indebted to the many scientific colleagues listed below, who gave up valuable time to review sections of the text. I also thank all those colleagues who provided images. I am particularly grateful to Gina Almond, Senior Editor at Garland Science, for her enthusiasm and encouragement to produce a second edition and developing the approach and new look for the series. I also thank the editorial team at Garland Science, especially Sarah Holland, David Borrowdale, and Georgina Lucas for their patience and help. Finally, to Sheila—thank you for your constant love and support.

Colin Munn
Plymouth, UK

Reviewers

Roger Anderson (Columbia University, USA); Esther Angert (Cornell University, USA); Josefa Anton Botella (University of Alicante, Spain); Brian Austin (University of Stirling, UK); Harry Birkbeck (University of Glasgow, UK); Katherine Boettcher (University of Maine, USA); David Bourne (Australian Institute of Marine Science, Australia); Paul Broady (University of Canterbury, New Zealand); Vivia Bruni (University of Messina, Italy); Sergey V. Dobretsov (Hong Kong University of Science and Technology, China); Stuart Donachie (University of Hawaii, USA); Gail Ferguson (University of Aberdeen, UK); Shana Goffredi (California Institute of Technology, USA); Lone Høj (Australian Institute of Marine Science, Australia); Robert Hall (Department of Health & Human Services, National Institutes of Health, USA); Fiona Hannah (University of Glasgow, UK); Dennis Hansell (University of Miami, USA); Tilmann Harder (University of New South Wales, Australia); Terry Hazen (Lawrence Berkeley National Laboratory, USA); Russell T. Hill (University of Maryland, USA); Zackary Johnson (Duke University, USA); Andrew Johnston (University of East Anglia, UK); Markus Karner (Karner Consulting, Singapore); David Kirchman (University of Delaware, USA); Angela Knapp (University of Miami, USA); Phyllis Lam (Max Planck Institute of Marine Microbiology, Germany); Manual Lemos (University of Santiago de Compostela, Spain); Peter McCarthy (Florida Atlantic University, USA); Diane McDougald (University of New South Wales, Australia); Terry McGenity (University of Essex, UK); Balbina Nogales (University of the Balearic Islands, Spain); James Oliver (University of North Carolina, USA); Mark Osborn (University of Hull, UK); Leigh Owens (James Cook University, Australia); Fernando Perestelo Rodriquez (University of La Laguna, Spain); Kenneth Peterson (Louisiana State University, USA); Dann Rittschof (Duke University, USA); Liz Robertson (Max Planck Institute of Marine Microbiology, Germany); Eugene Rosenberg (Tel Aviv University, Israel); Ned Ruby (University of Wisconsin–Madison, USA); Hendrik Schafer (University of Warwick, UK); Heide Schulz-Vogt (Max-Planck Institute for Marine Microbiology, Germany); Steven M. Short (University of Toronto, Canada); Val Smith (University of St Andrews, UK); Peter Steinberg (University of New South Wales, Australia); Frank Stewart (Massachusetts Institute of Technology, USA); Mike Taylor (University of Auckland, New Zealand); Fabiano Thompson (University of Rio de Janeiro, Brasil); Karen Visick (Loyola University, USA); Bess Ward (Princeton University, USA); Rui Zhang (Xiamen University, China).

Contents

Chapter 3 Metabolic Diversity and Ecophysiology 59

Chapter 11 Microbial Diseases of Marine Organisms   223

List of Research Focus Boxes

Chapter 1

Microbes in the Marine Environment

Viewed from space, it is clear why our planet would be better named "Ocean" than "Earth." More than 70% of the planet's surface is covered by interconnected bodies of water. Life originated in the oceans about 3.5 billion years ago and microbes were the only form of life for two thirds of the planet's existence. The development and maintenance of all other forms of life depend absolutely on the past and present activities of marine microbes. Yet the vast majority of humans—including many marine scientists—live their lives completely unaware of the diversity and importance of marine microbes. Such understanding is vital as we now live in a period of rapid global change. This chapter introduces the scope of marine microbiology, the different types of marine microbe (viruses, bacteria, archaea, fungi, and protists), and their place in the living world. The role of microbes in the many diverse habitats found in the marine environment is explored.

Key Concepts

- Modern methods have led to new ideas about the evolution of microbial life.
- Marine microbes exist in huge numbers and form a major component of biomass on Earth.
- Although there is a wide range of sizes, most marine microbes are exceptionally small.
- A wide range of physical and chemical conditions provide diverse specialized habitats.
- Microbes are major components of plankton and marine snow.
- Microbes are important in sediment formation and there is abundant life below the seafloor.
- Microbes colonize the surfaces of inanimate objects and other living organisms by the formation of biofilms.

Marine microbiology is one of the most exciting and important areas of modern science

Ever since a detailed study of the microbial world began at the end of the nine-teenth century, microbiologists have asked questions about the diversity of microbial life in the sea, its role in ocean processes, its interactions with other marine life, and its importance to humans. However, despite excellent work by pioneering scientists, progress in understanding these issues was often slow and most microbiologists remained unaware of this field of study until recently. Toward the end of the twentieth century, a number of factors conspired to pro-pel marine microbiology to the forefront of "mainstream" science, and the sub-sequent application of new technology means that it is now one of the most exciting and fast-moving areas of investigation. Powerful new tools in molecular biology, remote sensing, and deep-sea exploration have led to astonishing dis-coveries of the abundance and diversity of marine microbial life and its role in global ecology. Continuing new discoveries in marine microbiology necessitate radical rethinking of our understanding of ocean processes. We now realize the vital role that marine microbes play in the maintenance of our planet, a fact that will have great bearing on our ability to respond to problems such as the increase in human population, overexploitation of fisheries, climate change, ocean acidi-fication, and marine pollution. Study of the interactions of marine microbes with other organisms is providing intriguing insights into the phenomena of food webs, symbiosis, and pathogenicity. Since some marine microbes produce dis-ease or damage, we need to study these processes and develop ways to overcome them. Finally, marine microbes have beneficial properties such as the manufac-ture of new products and development of new processes in the growing field of marine biotechnology. This chapter sets the scene for the discussion of all these topics in this book.

Marine microbiology encompasses all microscopic organisms and viruses

Defining the terms "microbiology" and "microorganism" is surprisingly difficult! Microbiology is the study of very small organisms that are too small to be seen clearly with the naked eye (i.e. less than about 1 mm diameter), but most micro-biologists are concerned with the activities or molecular properties of microbial communities rather than viewing individual cells with a microscope. The term "microorganism" simply refers to a form of life that falls within the microscopic size range, but there is a huge spectrum of diversity concealed by this all-encom-passing term. Indeed, some "microorganisms" are large enough to see without using a microscope, so this is not entirely satisfactory either. Some scientists would argue that the distinguishing features of microorganisms are small size, unicellular organization, and osmotrophy (feeding by absorption of nutrients). The osmotrophic characteristic is important because diffusion processes are a major limitation to cell size, as discussed in the next section. However, this char-acteristic would exclude many microscopic unicellular protists (a loose group-ing of simple eukaryotes), many of which feed by phagotrophy (engulfment of particles). These "plant-like" or "animal-like" groups are most commonly studied by specialists who traditionally have a background in botany or zoology. Indeed, the study of marine protists and recognition that they are microbes with a major role in ocean processes has lagged behind the study of bacteria. Many marine protists are mixotrophic and can switch from photosynthesis to phagotrophic feeding, so the plant or animal similarity is meaningless. Additionally, viruses are microscopic and are obviously included in the remit of microbiologists, but they are not cellular and it can be argued that they are not living organisms either (this question is explored in depth in Chapter 7). There is a huge diversity of inter-connected microbial life forms in the marine environment, and worrying about such artificial divisions is not going to be helpful; so in this book I use the term "microbe" as a generic descriptor for all microscopic organisms (i.e. the bacteria, archaea, fungi, and unicellular protists), together with the viruses.

Marine microbes are found in all three domains of cellular life

Biologists usually rely on the study of morphology and physiological properties to classify living organisms, but these characteristics have always proved frustratingly unhelpful when dealing with microbes. Phylogenetic systems of classification depend on comparisons of the information content of their macromolecules, especially nucleic acids and proteins. If two organisms are very closely related, we expect the sequence of the individual units in a macromolecule to be more similar than they would be in two unrelated organisms. In the 1970s, Carl Woese and colleagues pioneered the use of ribosomal RNA (rRNA) sequencing in order to develop a better view of microbial diversity. Our view of the living world has since been revolutionized by advances in this approach, made possible because of the parallel advances in molecular biological techniques and computer processing of the large amounts of information generated. Because the secondary structure of rRNA is so important in the ribosome and the vital cell function of protein synthesis, base sequence changes in the rRNA molecule occur quite slowly in evolution. In fact, some parts of rRNA are highly conserved and sequence comparisons can be used to ascertain the similarity of organisms on a broad basis. The methods and applications of this major technique are described in Chapter 2.

Woese identified three distinct lineages of cellular life, referred to as the domains *Archaea*, *Bacteria*, and *Eukarya*. A phylogenetic "tree of life" based on rRNA sequences envisaged divergence of the three domains from an original "universal ancestor" (*Figure 1.1A*). The three-domain system of classification is used by most microbiologists, although it has never been universally adopted by other biologists and remains a topic of fierce controversy. Apart from our preference for a phylogenetic system, it allows microbiologists to say that we study two entire domains of life, and a significant proportion of the third! The most important consequence of the three-domain tree of life is that we now realize that the *Archaea* are not a peculiar, specialized group of bacteria as originally thought (for many years they were called the archaebacteria), but are in fact a completely separate group with closer phylogenetic relationships to the *Eukarya* than to the *Bacteria*.

Figure 1.1 Representations of the three domains of life. (A) Simple tree based on ribosomal RNA sequencing. In this model, the root of the tree is envisaged as a hypothetical universal ancestor from which all life evolved. (Adapted from Martin and Embley [2004] by permission of Macmillan Publishers Ltd.) (B) A three-domain tree based on evidence of extensive lateral gene transfer, revealed by studies of other genes. (Image courtesy of Gary J. Olsen, University of Illinois; based on concept of W. Ford Doolittle.)

(A)

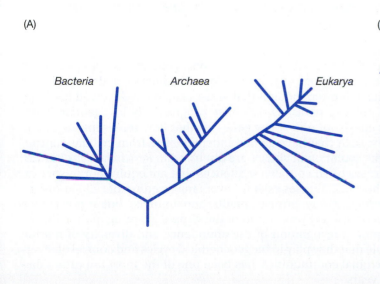

(B)

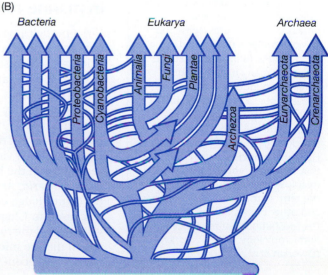

Eukaryotes are distinguished by a membrane-bound nucleus and organelles with specific functions. Mitochondria occur in all eukaryotic cells, with the exception of a few anaerobic protozoa, and carry out the processes of respiratory electron transport. In photosynthetic eukaryotes, chloroplasts carry out reactions for the transfer of light energy for cellular metabolism. Various lines of evidence (especially the molecular analysis of the nucleic acids and proteins) support the hypothesis that the organelles of eukaryotic cells have evolved by a process of endosymbiosis, in which one cell lives inside another cell for mutual benefit. This hypothesis proposes that the original source of mitochondria in eukaryotic cells occurred when primitive cells acquired respiratory bacteria (most closely related to proteobacteria) and that the chloroplasts evolved from endosymbiosis with cyanobacteria. Such interactions between different types of cell have continued throughout evolution, and Chapter 10 contains many examples of endosymbiosis involving microbes.

Horizontal gene transfer confounds our understanding of evolution

Since it became possible to study the whole genome sequences of organisms, we have found increasing evidence of extensive lateral gene transfer (LGT; also known as horizontal gene transfer, HGT) between microbes. Such transfer occurs most commonly between related organisms, but transfer across bigger genetic distances also occurs—even between domains. Members of the *Bacteria* and *Archaea* contain some genes with very similar sequences, and members of the *Eukarya* contain genes from both of the other domains. Some members of the domain *Bacteria* have even been shown to contain eukaryotic genes. Previously, evolution was explained only by the processes of mutation and sexual recombination, but we now know that the pace of evolution is accelerated by the transfer and acquisition of modules of genetic information. This phenomenon is widespread in modern members of the *Bacteria* and *Archaea* and can occur via three processes. During the process known as transformation, cells may take up and express naked DNA; whilst conjugation relies on cell–cell contact mediated by pili. The most important source of LGT is the process of transduction by phages (viruses infecting bacteria); this is explored in detail in Chapter 7. The enormous diversity of marine viruses and the identification of a viral origin of genes in many marine organisms indicate how important this process has been throughout evolution.

Viruses are noncellular entities with great importance in marine ecosystems

Virus particles (virions) consist of a protein capsid containing the viral genome composed of either RNA or DNA. Because they only contain one type of nucleic acid, viruses must infect living cells and take over the host's cellular machinery in order to replicate. It is often thought that viruses could have evolved (perhaps from bacteria) as obligate parasites that have progressively lost genetic information until they consist of only a few genes, or that they represent fragments of host-cell RNA or DNA that have gained independence from cellular control. New ideas about the evolution of viruses are discussed in *Research Focus Box 7.2*. The genome of viruses often contains sequences that are equivalent to specific sequences in the host cell. Viruses exist for every major group of cellular organisms (*Bacteria*, *Archaea*, *Fungi*, protists, plants, and animals), but at present we have knowledge of only a tiny proportion of the viruses infecting marine life. As discussed in Chapter 7, recognition of the abundance and diversity of marine viruses, and the role that they play in biogeochemical cycles and control of diversity in marine microbial communities, has been one of the most important discoveries of recent years.

? WHAT HAPPENED TO THE PROKARYOTES?

Traditionally, members of the domains *Bacteria* and *Archaea* have been grouped together as "prokaryotes," because they share a simple internal cellular structure, with their genetic material not bound by a nuclear membrane. However, this division of life is not supported by modern studies showing that *Bacteria* and *Archaea* are completely different phylogenetic groups. As pointed out by Pace (2009): "No-one can tell you what a prokaryote is, they can only tell you what it is not." He argues that the prokaryote–eukaryote model is scientifically illogical and wrongly implies that prokaryotes gave rise to eukaryotes. Marine microbiology provides numerous examples that emphasize the fact that the prokaryotic designation is no longer appropriate—some marine bacteria are much larger than eukaryotic cells, some have complex intracellular structures, some show obvious multicellularity, and some differentiate during their lifecycle. Although many microbiologists defend continued use of the term "prokaryote" and reinterpretation of the concept in modern terms (see Whitman, 2009), I have decided not to use it in this book. However, readers should remember that the prokaryote concept is still deeply embedded, and you will find the term in many research papers and other books. Indeed, one of the most important reference works in microbiology is *The Prokaryotes* (Dworkin et al., 2007).

Microbial processes shape the living world

Probably the most important overriding features of microbes are their exceptional diversity and ability to occupy every conceivable habitat for life. Indeed, what we consider "conceivable" is challenged constantly by the discovery of new microbial communities in habitats previously thought of as inhospitable, or carrying out processes that we had no idea were microbial in nature. Bacteria and archaea have shaped the subsequent development of life on Earth ever since their first appearance—the metabolic processes that they carry out in the transformation of elements, degradation of organic matter, and recycling of nutrients play a central role in innumerable activities that affect the support and maintenance of all other forms of life. Microbial life and the Earth have evolved together and the activities of microbes have affected the physical and geochemical properties of the planet. Indeed, they are actually the driving forces responsible for major planetary processes like changes in the composition of the atmosphere, oceans, soil, and rocks. This is especially relevant to our consideration of the marine environment, in view of the huge proportion of the biosphere that this constitutes. Despite the preponderance of microbes and the importance of their activities, they are unseen in everyday human experience. Microbes live and grow almost everywhere, using a huge range of resources, whereas plants and animals occupy only a small subset of possible environments and have a comparatively narrow range of lifestyles.

Marine microbes show great variation in size

Table 1.1 shows the range of dimensions and volumes of some representative marine microbes. Even by the usual standards of microbiology, the most abundant microbes found in seawater are *exceptionally* small. Their very small size is the main reason appreciation of their abundance eluded us until quite recently. As described in Chapter 2, recognition of the abundance of marine microbes depended on the development of fine-pore filters and direct counting methods using epifluorescence microscopy and flow cytometry. Small cell size has great significance in terms of the physical processes that affect life. At the microscale, the rate of molecular diffusion becomes the most important mechanism for transport of substances into and out of the cell. Small cells feeding by absorption (osmotrophy) can take up nutrients more efficiently than larger cells. The surface area:volume ratio (SA/V) is the critical factor because as cell size increases, volume (V) increases more quickly than surface area (SA), as shown in *Figure 1.2*.

The most abundant ocean bacteria and archaea have very small cell volumes and large SA/V ratios. The majority are smaller than about 0.6 µm in their largest dimension, and many are less than 0.3 µm, with cell volumes as low as 0.003 µm³. If nutrients are severely limiting, as they are in most of the oceans, selection will favor small cells. Since the first description of such small cells, termed "ultramicrobacteria," their size has provoked considerable controversy. Such extremely small cells could result from a genetically fixed phenotype maintained throughout the cell cycle or because of physiological changes associated with starvation. The latter explanation is supported by the fact that some cultured bacteria become much smaller when starved. Most naturally occurring bacteria have been impossible to grow in culture—this is a central problem in marine microbiology, which we shall return to on several occasions in future chapters. Because of this, it has been difficult to determine whether small size is a genotypically determined condition for marine bacteria. However, studies with some recently cultured marine bacteria from low-nutrient (oligotrophic) ocean environments show that addition of nutrients does not cause an increase in cell size. Small cell size also has important implications for mechanisms of active motility and chemotaxis, because of the microscale effects of Brownian movement (bombardment by water molecules) and shear forces. Small marine bacteria have

? **IS IT TIME TO CHOP DOWN THE "TREE OF LIFE"?**

The idea that relationships between all living organisms can be represented as a tree of life helped to shape Darwin's theory of evolution by natural selection and has been deeply embedded in the philosophy of biology for more than 150 years. As the importance of endosymbiosis and LGT became better understood, some evolutionary scientists began to question the validity of the "tree of life" concept. A seminal paper by Doolittle (1999) argued that "Molecular phylogeneticists will have failed to find the 'true tree', not because their methods are inadequate or because they have chosen the wrong genes, but because the history of life cannot properly be represented as a tree." Relationships are now envisaged as complex intertwined branches, more like a web (see *Figure 1.1B*) or network of genomes (Dagan and Martin, 2009). However, this remains a controversial topic, and some have argued that analysis of genome sequences for "core genes" still supports the idea of a common ancestor and branching tree (Ciccarelli et al., 2006).

Figure 1.2 Diagrammatic representation of three spherical cells showing a reduction in the ratio of surface area (SA) to volume (V) as size increases. V is a function of the cube of the radius ($V = \frac{4}{3}\pi r^3$), whereas SA is a function of the square of the radius ($SA = 4\pi r^2$). Cells with large SA/V ratios are more efficient at obtaining scarce nutrients by absorption across the membrane.

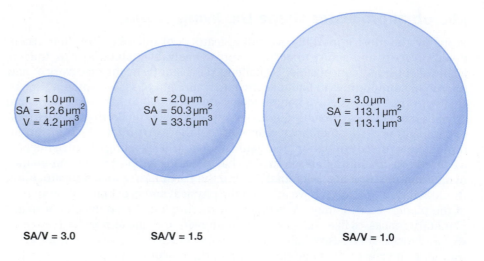

r = 1.0 µm
SA = 12.6 µm²
V = 4.2 µm³

r = 2.0 µm
SA = 50.3 µm²
V = 33.5 µm³

r = 3.0 µm
SA = 113.1 µm²
V = 113.1 µm³

SA/V = 3.0   SA/V = 1.5   SA/V = 1.0

mechanisms of motility and chemotaxis quite unlike those with which we are familiar in "conventional" microbiology. Cells use various strategies to increase the SA/V ratio and thus improve efficiency of diffusion and transport. In fact, spherical cells are the least efficient shape for nutrient uptake, and many marine bacteria and archaea are thin rods or filaments or may have appendages such as stalks or buds. *Figure 1.3* shows examples of the diverse morphology of marine bacteria. Many of the larger organisms overcome the problems of diffusion by having extensive invaginations of the cytoplasmic membrane or large intracellular vacuoles, increasing the SA.

As shown in *Table 1.1*, marine eukaryotic microbes also show a considerable variation in size. Many flagellates are in the 1–2 µm range, cellular dimensions that are more typical of many familiar bacteria. The smallest known eukaryote is *Ostreococcus tauri,* which is only about 0.8 µm diameter (*Figure 1.4B*). Again, the realization that such small cells (now referred to as "picoeukaryotes") play a key role in ocean processes escaped attention until quite recently. Many small protists seem capable of engulfing bacteria of almost the same size as themselves or can prey on much larger organisms. Many groups of the flagellates, ciliates, diatoms, and dinoflagellates are somewhat larger, reaching sizes up to 200 µm, and

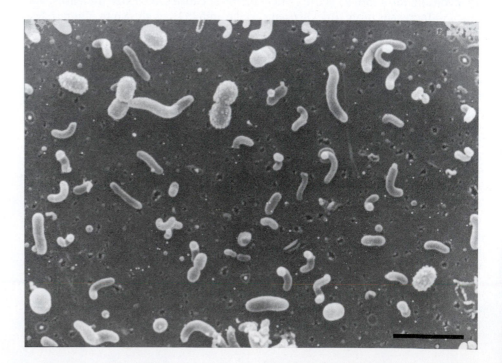

Figure 1.3 Scanning electron micrograph of picoplankton, showing various cell morphologies of marine bacteria. Bar represents ~ 1 µm. (Image courtesy of Ed DeLong, Massachusetts Institute of Technology.)

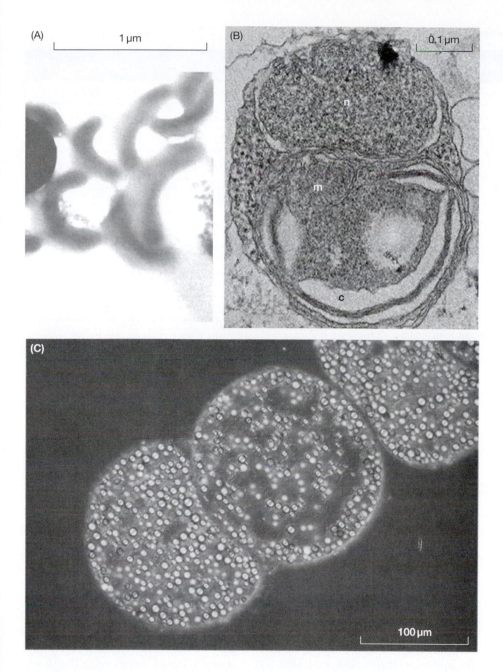

Figure 1.4 Extremes of size in marine microbes (note different scale bars). (A) Electron micrograph of crescent-shaped cells of *Candidatus* Pelagibacter ubique, one of the smallest bacteria known. (B) Electron micrograph of section of *Ostreococcus tauri*, the smallest known eukaryote. n = nucleus, m = mitochondrion, c = chloroplast. (C) Light micrograph of *Thiomargarita namibiensis*, the largest known bacterium, showing sulfur granules. (Images courtesy of (A) Oregon State University Laboratory for the Isolation of Novel Species; (B) Henderson et al. [2007], Copyright Public Library of Science; (C) Heide Schulz-Vogt, Max Planck Institute for Marine Microbiology, Bremen, Germany.)

amoeboid types (radiolarians and foraminifera) can be even larger and visible to the naked eye. Finally, a few types of bacteria are bigger than many protists, the largest of which is *Thiomargarita namibiensis* (*Figure 1.4C*).

The world's oceans and seas form an interconnected water system

The oceans cover 3.6×10^8 km^2 (71% of the Earth's surface) and contain 1.4×10^{21} liters of water (97% of the total on Earth). The average depth of the oceans is 3.8 km, with a number of deep-sea trenches, the deepest of which is the Marianas Trench in the Pacific (11 km). The ocean floor contains large mountain ranges and is the site of almost all the volcanic activity on Earth. More than 80% of the area and 90% of the volume of the oceans constitutes the deep sea, most of which remains unexplored. It is usual to recognize five major ocean basins, although they actually form one interconnected water system, as illustrated in *Figure 1.5*.

Table 1.1 Size range of some representative marine microbes

Organism	Characteristics	Size (μm)[a]	Volume (μm³)[b]
Parvovirus	Icosahedral DNA virus infecting shrimp	0.02	0.000004
Coccolithovirus	Icosahedral DNA virus infecting *Emiliania huxleyi*	0.17	0.003
Thermodiscus	Disc-shaped. Hyperthermophilic archaeon	0.2 × 0.08	0.003
Pelagibacter ubique[c]	Crescent-shaped *Bacteria* ubiquitous in ocean plankton (cultured example of SAR11 clade)	0.1 × 0.9	0.01
Prochlorococcus	Cocci. Dominant photosynthetic ocean bacterium	0.6	0.1
Ostreococcus	Cocci. Prasinophyte alga. Smallest known eukaryote	0.8	0.3
Vibrio	Curved rods. Bacteria common in coastal environments and associated with animals	1 × 2	2
Pelagomonas calceolata	Photosynthetic flagellate adapted to low light	2	24
Pseudo-nitzschia	Pennate diatom which produces toxic domoic acid	5 × 80	1600
Staphylothermus marinus	Cocci. Hyperthermophilic archaeon	15	1800
Thiploca auracae	Filamentous. Sulfur bacterium	30 × 43	30000
Lingulodinium polyedrum	Bioluminescent bloom-forming dinoflagellate	50	65000
Beggiatoa	Filamentous. Sulfur bacterium	50 × 160	314000
Epulopiscium fishelsoni	Rods. Bacteria symbiotic in fish gut	80 × 600	3000000
Thiomargarita namibiensis	Cocci. Sulfur bacterium	300[d]	14137100

[a]Approximate diameter × length; where one value is given, this is the diameter of spherical virus particles or cells. [b]Approximate values, calculated assuming spherical or cylindrical shapes. [c]*Candidatus*; provisional taxonomic name—see Glossary. [d]Cells up to 750 μm have been recorded.

The Pacific is the deepest and largest ocean, almost as large as all the others combined. This single body of water has an area of 1.6×10^8 km² and covers about 28% of the Earth's surface, more than the total land area. The ocean floor in the eastern Pacific is dominated by the East Pacific Rise, while the western Pacific is dissected by deep trenches. The Atlantic Ocean is the second largest with an area of 7.7×10^7 km² lying between Africa, Europe, the Southern Ocean, and the Americas. The mid-Atlantic Ridge is an underwater mountain range stretching down the entire Atlantic basin. The deepest point is the Puerto Rico Trench (8.1 km). The Indian Ocean has an area of 6.9×10^7 km² and lies between Africa, the Southern Ocean, Asia, and Australia. A series of ocean ridges cross the basin, and the deepest point is the Java Trench (7.3 km). The Southern Ocean is the body of water between 60°S and Antarctica. It covers 2.0×10^7 km² and has a fairly constant depth of 4–5 km, with a continual eastward water movement called the Atlantic Circumpolar Current. The Arctic Ocean, lying north of the Arctic Circle, is the smallest ocean, with an area of 1.4×10^7 km². About half of the ocean floor is continental shelf, with the remainder being a central basin interrupted by three submarine ridges. As well as the major oceans, there are marginal seas, including the Mediterranean, Caribbean, Baltic, Bering, South China Seas, and many others.

At the margins of major landmasses, the ocean is shallow and lies over an extension of the land called the continental shelf. This extends offshore for a distance ranging from a few kilometers to several hundred kilometers and slopes gently to a depth of about 100–200 m, before there is a steep drop-off to become the continental slope. The abyssal plain covers much of the ocean floor; this is a mostly flat surface with few features, but is broken in various places by ocean ridges, deep-sea trenches, undersea mountains, and volcanic sites.

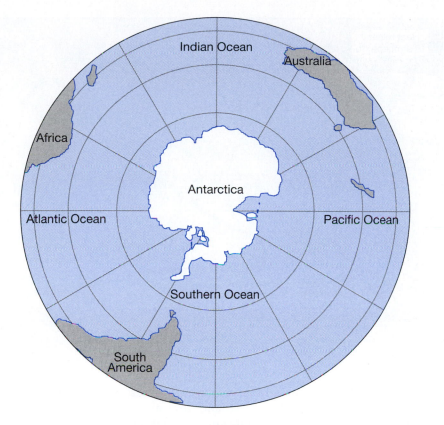

Figure 1.5 Earth as seen from a south polar view, showing that the world's oceans form one interconnected system. The Arctic Ocean is not shown.

The upper surface of the ocean is in constant motion owing to winds

Wind belts created by differential heating of air masses generate the major surface current systems, shown in *Figure 1.6*. Rotation of the Earth deflects moving water in a phenomenon known as the Coriolis Effect, which results in boundary currents between large water masses. This leads to large circular gyres that move clockwise in the northern hemisphere and anticlockwise in the southern hemisphere. Such gyres and currents affect the distribution of nutrients and marine organisms. On the basis of surface ocean temperatures, the marine ecosystem

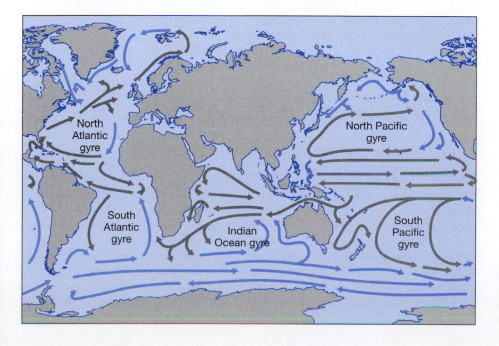

Figure 1.6 Major ocean currents of the world. Warm currents are shown as gray arrows and cold currents as blue arrows.

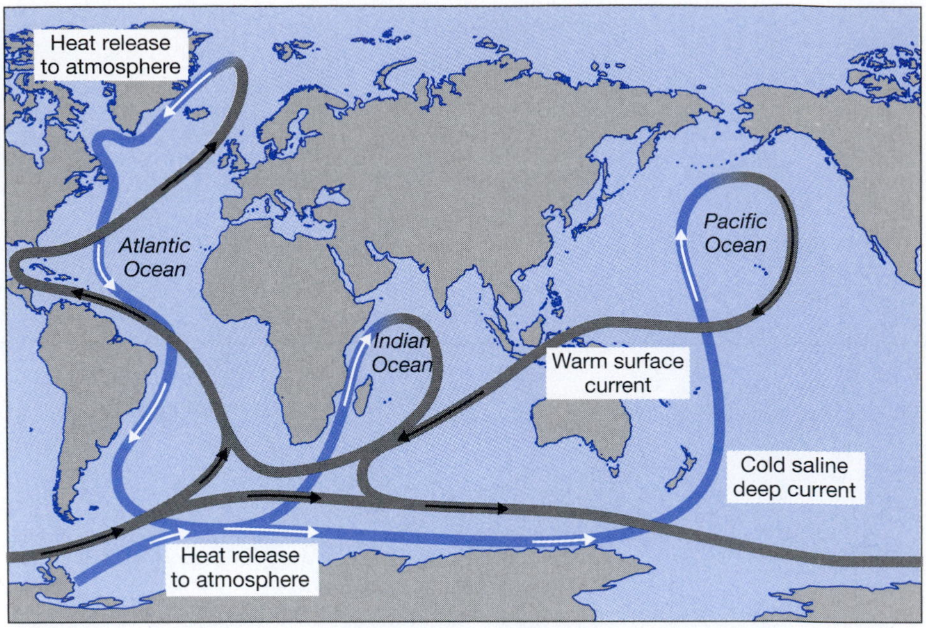

Figure 1.7 Schematic representation of the thermohaline circulation system ("global ocean conveyor belt"). Currents of warmer water are indicated in gray; cold saline deep currents are indicated in blue.

can be divided into four major biogeographical zones, namely polar, cold temperate, warm temperate (subtropical), and tropical (equatorial). The boundaries between these zones are not absolute and vary with season.

Deep-water circulation systems transport water between the ocean basins

Below a depth of about 200 m, ocean water is not affected by mixing and wind-generated currents. However, a system of vast undersea rivers transports water around the globe and has a major influence on the distribution of nutrients and classic processes (*Figures 1.7* and *8.1*). This thermohaline circulation system—often referred to as the "global conveyor belt"—is formed by the effects of temperature and salinity causing differences in the density of water. Surface water in the North Atlantic flows toward the pole as the Gulf Stream. Water is removed to the atmosphere as it cools through evaporation, resulting in higher salinity. Water is also removed during the formation of sea ice. Thus, the water becomes denser and sinks to form a deep pool, which then flows toward Antarctica, where more cold, dense water is added. The current then splits, with one branch going toward the Indian Ocean and the other to the Pacific Ocean. As the current nears the equator it warms and becomes less dense, so upwelling occurs. The warmer waters loop back to the Atlantic Ocean, where they start the cycle again.

Seawater is a complex mixture of inorganic and organic compounds

Seawater is a slightly alkaline (pH 7.5–8.4) aqueous solution—a complex mixture of more than 80 solid elements, gases, and dissolved organic substances. The concentration of these varies considerably according to geographic and physical factors, and it is customary to refer to the salinity of seawater in parts per thousand (‰) to indicate the concentration of dissolved substances. The open ocean has a constant salinity in the range 34–37‰, with differences due to dilution by rainfall and evaporation. Oceans in subtropical latitudes have the highest salinity as a result of higher temperatures, whilst temperate oceans have lower salinity as a result of less evaporation. In coastal regions, seawater is diluted considerably

? WHY IS THE SEA SALTY?

The constant percolation of rainwater through soil and rocks leads to weathering, in which some of the minerals are dissolved. Ground water has very low levels of salts and we cannot taste it in the water we drink. The addition of salts to the oceans from rivers is thus a very slow process, but evaporation of water from the oceans to form clouds means that the salt concentration has increased to its present level over hundreds of millions of years. Seawater also percolates into the ocean crust where it becomes heated and dissolves minerals, emerging at hydrothermal vents (see *Figure 1.10*). Submarine volcanoes also result in reactions between seawater and hot rock, resulting in the release of salts. The salt concentration in the oceans appears to be stable, with deposition of salts in sediments balancing the inputs from weathering, hydrothermal vents and volcanic activity.

by freshwater from rivers and terrestrial runoff and is in the range 10–32‰. Conversely, in enclosed areas such as the Red Sea and Arabian Gulf, the salinity may be as high as 44‰. In polar areas, the removal of freshwater by the formation of ice also leads to increased salinity. The major ionic components of seawater are sodium (55% w/v), chloride (31%), sulfate (8%), magnesium (4%), calcium (1%), and potassium (1%). Together, these constitute more than 99% of the weight of salts. There are four minor ions–namely bicarbonate, bromide, borate, and strontium–which together make up just less than 1% of seawater. Many other elements are present in trace amounts (<0.01%), including key nutrients such as nitrate, phosphate, silicate and iron. The concentration of these is crucial in determining the growth of marine microbes and the net productivity of marine systems, as discussed in Chapter 9.

The concentration of salts has a marked effect on the physical properties of seawater. The freezing point of seawater at 35‰ is −1.9°C, and seawater increases in density up to this point. As noted above, this results in the formation of masses of cold, dense water in polar regions, which sink to the bottom of the ocean basins and are dispersed by deep-water circulation currents. Differences in the density of seawater create a discontinuity called the pycnocline, which separates the top few hundred meters of the water column from deeper water. This has great significance, because the gases oxygen and carbon dioxide are more soluble in cold water.

Oxygen is at its highest concentrations in the top 10–20 m of water, owing to exchange with the atmosphere and production of oxygen by photosynthesis. Concentration decreases with distance from the surface until it reaches a minimum between 200 and 1000 m, and bacterial decomposition of organic matter may create conditions that are almost anoxic. Below this, the oxygen content increases again as a result of the presence of dense water (with increased solubility at lower temperature) that has sunk from polar regions and been transported on the thermohaline circulation system. This oxygen gradient varies greatly in different regions, and there are several regions where large bodies of hypoxic water occur at relatively shallow depths—these are the oxygen minimum zones (see *Plate 9.1*).

The solubility of carbon dioxide is an important factor in controlling the exchange of carbon between the atmosphere and the oceans and therefore is of huge significance in understanding climatic processes, as discussed in Chapter 8. Only a very small proportion of dissolved inorganic carbon (DIC) is present in the form of dissolved CO_2 gas. Carbon dioxide reacts with water to form carbonic acid, which rapidly dissociates to form bicarbonate, hydrogen ions, and carbonate in the reactions:

$$CO_2 \text{ (gas)} + H_2O \rightleftharpoons H_2CO_3 \rightleftharpoons H^+ + HCO_3^- \rightleftharpoons 2H^+ + CO_3^{2-}$$

These reactions tend to stay in equilibrium, buffering the pH of seawater within a narrow range and constraining the amount of CO_2 taken up from the atmosphere. However, the accelerating atmospheric concentration of high levels of CO_2 since the industrial revolution is shifting the equilibrium and causing the pH to fall because of increased levels of H^+ ions. This phenomenon is called ocean acidification and some of its possible consequences for microbial life are discussed in *Research Focus Box 1.1*.

Light and temperature have important effects on microbial processes

Light is of fundamental importance in the ecology of microbes that use light energy for photosynthesis and other functions (see *Research Focus Box 3.1*), thus affecting primary productivity. The extent to which light of different wavelengths

BOX 1.1   **RESEARCH FOCUS**

Ocean acidification—"the other CO₂ problem"
How will microbes respond to rapid changes in ocean chemistry?

The effects of high levels of CO_2 (together with other gases such as methane and nitrous oxide) in promoting global warming via the "greenhouse effect" are well known and a major topic of current sociopolitical concern. However, a less well-known effect of atmospheric CO_2 is the process of ocean acidification (OA), which has been termed "the other CO_2 problem" to bring it to the attention of the public (see Mitchell, 2009). When CO_2 is absorbed from the atmosphere, it leads to an increase in the concentration of hydrogen ions and hence a fall in pH (see equation on p. 11). Approximately half of the CO_2 produced by human activities in the past 200 years has been absorbed by the oceans, leading to a fall in the average pH of ocean surface water from about 8.21 to 8.10. Although this seems small, it represents a 30% increase in the hydrogen ion concentration because pH is a logarithmic scale. Some models predict that the average ocean pH of the oceans could fall to pH 7.9 by 2100—at which point hydrogen ions will be three times the current levels—unless CO_2 emissions are drastically reduced. This level is lower than has occurred for hundreds of millions of years; more importantly, this *rate* of change has probably never occurred in the history of the planet (Royal Society, 2005). Even if the world succeeds in controlling CO_2 emissions, much of this change will be irreversible, because of the long time needed for mixing of deep waters and natural buffering processes to take effect.

One of the main effects of the altered seawater chemistry as a result of OA is likely to be on calcifying organisms that use calcium carbonate ($CaCO_3$) to construct a skeleton or shell. As well as animals such as corals, crustaceans and mollusks, there are some very important planktonic microbes that carry out calcification, especially the coccolithophores (see Chapter 6). Orr et al. (2005) concluded that such calcifying organisms could begin to experience difficulties in maintaining their $CaCO_3$ skeletons as early as 2050, especially in polar oceans, where the state of $CaCO_3$ saturation is lower. When considering the effect of OA on microbes, some examples of recent research given here illustrate how difficult it is to reach clear conclusions about the effect of OA on microbial processes.

When coccolithophores make the plates of calcite that surround the cell (see *Figure 6.5B*), they release CO_2, but they also fix CO_2 during photosynthesis. Thus, the balance between these processes is very important in the global carbon cycle. Research has indicated that there is great variation in the responses of different species (and strains within species) to changes in ocean pH and CO_2 levels. Riebesell (2004) concluded that increased CO_2 levels enhance photosynthetic carbon fixation of some phytoplankton groups and predicted that calcifying plankton might benefit at the expense of some other groups. Mesocosm studies (see *Figure 2.2*) of the response of blooms of the coccolithophore *Emiliania huxleyi* showed a reduction in calcification rates when CO_2 levels simulating end of the century conditions were applied (Delille et al., 2005; Engel et al., 2005). However, Iglesias-Rodriguez et al. (2008a) developed observations from sediment cores from a site in the North Atlantic, indicating that coccolithophores have increased their calcification rates in response to rising CO_2 levels—the average coccolith mass has increased by ~40% over the past 220 years. In laboratory cultures, a strain of *E. huxleyi* increased photosynthesis and calcification rates by 100–150% at high CO_2 levels, a level much higher than that observed in previous studies. This paper has attracted a lot of interest because of the important conclusion that coccolithophores are adapting to a high CO_2 ocean and the impact that this will have on predicting future trends. Following publication, Riebesell et al. (2007) were highly critical of the experimental setup, but Iglesias-Rodriguez et al. (2008b) have defended their conclusions. In another mesocosm study, Riebesell et al. (2007) showed that CO_2 uptake by phytoplankton was 27% and 39% higher, respectively, when CO_2 was at two or three times present-day levels. These authors also found that the ratio of carbon to nitrogen uptake increased at higher CO_2 concentrations, whereas the carbon–nitrogen ratio within the cells of the phytoplankton was unaltered. It seems that extra carbon incorporated through photosynthesis was rapidly lost from the cells to form transparent exopolymer particles that contribute to the formation of marine snow and the flux of organic carbon from the surface to the deep sea (see *Figure 1.9*). Thus, increased productivity might provide a partial negative feedback mechanism by which some of the increased CO_2 dissolved in seawater is removed. Further evidence for a possible negative feedback mechanism comes from the observation by Ramos et al. (2007) that the cyanobacterium *Trichodesmium* increases its rate of nitrogen fixation very markedly at high CO_2 levels. The authors suggest that this could enhance the productivity of oligotrophic oceans that are currently limited by nitrogen limitation and increase the flux of carbon in the biological pump.

From the few examples of conflicting results presented here, it is clear that our knowledge of the effects of OA on the physiology and diversity of marine microbes is very poorly understood. Experts have recently been coordinating efforts to highlight the challenges for future research in this area and the need for coordinated methodological approaches (C-MORE, 2009; EPOCA, 2009). Microbes have adapted to previous changes in ocean chemistry that have occurred periodically over the past 3 billion years. How will they adapt to the extreme changes currently occurring?

penetrates seawater depends on a number of factors, notably cloud cover, the polar ice caps, dust in the atmosphere, and variation of the incident angle of solar radiation according to season and location on the Earth's surface. Light is absorbed or scattered by organisms and suspended particles. Even in the clearest ocean water, photosynthesis is restricted by the availability of light of sufficient intensity to the upper 150–200 m. This is termed the photic or euphotic zone (from the Greek, "well lit"). Blue light has the deepest penetration, and photosynthetic microbes at the lower part of the photic zone have light-harvesting systems that are tuned to collect blue light most efficiently (see p. 114). In turbid coastal waters, during seasonal plankton blooms, the euphotic zone may be only a few meters deep.

Solar radiation also leads to thermal stratification of seawater. In tropical seas, the continual input of energy from sunlight leads to warming of the surface waters to 25–30°C, causing a considerable difference in density from that of deeper waters. Thus, throughout the year, there is a marked thermocline at about 100–150 m, below which there is a sudden reduction in temperature to 10°C or less. Little mixing occurs between these layers. In polar seas, the water is permanently cold except for a brief period in the summer, when a slight thermocline results. Apart from this period, turbulence created by surface winds generates mixing of the water to considerable depths. Temperate seas show the greatest seasonal variation in the thermocline, with strong winds and low temperatures leading to extensive mixing in the winter. The thermocline develops in the spring, leading to a marked shallow surface layer of warmer water in summer. As the sea cools and wind increases, the thermocline breaks down again in the autumn. Combined with seasonal variations in light intensity, these temperature stratification effects and vertical mixing have a great impact on rates of photosynthesis and other microbial activities.

Marine microbes form a major component of the plankton

Microbes occur in all the varied habitats found in the oceans, shown in *Figure 1.8*. The surface interface (neuston) between water and atmosphere is rich in organic matter and often contains high numbers of diverse microbes. From the limited studies conducted in coastal environments, there does not seem to be much evidence of a stable and unique neuston community, although it might be expected that there are some specialized bacteria that are permanent members of this interface, adapted to high levels of solar radiation.

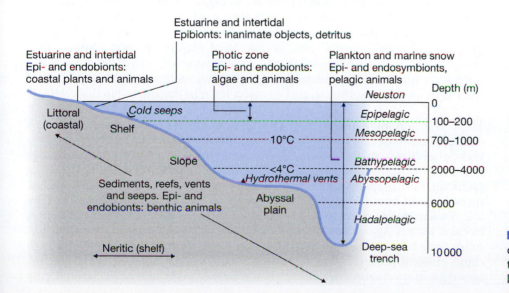

Figure 1.8 Schematic representation of the major ecological zones of the oceans and marine microbial habitats. (Not to scale.)

Table 1.2 Classification of plankton by size

Size category	Size range (μm)	Microbial groups
Femtoplankton	0.01–0.2	Viruses
Picoplankton	0.2–2	Bacteria[a], archaea, some flagellates
Nanoplankton	2–20	Flagellates, diatoms, dinoflagellates
Microplankton	20–200	Ciliates, diatoms, dinoflagellates, other algae

SIZE ↓   ABUNDANCE ↑

[a]Some filamentous cyanobacteria and sulfur-oxidizing bacteria occur in larger size classes (see *Table 1.1*).

Plankton is a general term in marine biology referring to organisms suspended in the water column that do not have sufficient power of locomotion to resist large-scale water currents (in contrast to the nekton, which are strong-swimming animals). Traditionally, biologists refer to the phytoplankton (plants) and zooplankton (animals). Using this approach, we could add the terms bacterioplankton for bacteria and virioplankton for viruses. Traditional concepts of "plant" and "animal" are unsatisfactory, and the term phytoplankton now refers to all photosynthetic microbes, including cyanobacteria as well as eukaryotic protists. Another approach to classifying the plankton is in terms of size classes, for which a logarithmic scale ranging from megaplankton (>20 mm) to nanoplankton (<0.2 μm) has been devised. *Table 1.2* shows the size classes that encompass marine microbes. Thus, the viruses constitute the femtoplankton, bacteria mainly occur in the picoplankton, whilst protists occur in the picoplankton, nanoplankton, and microplankton.

Microbes, particles, and dissolved nutrients are not evenly distributed in seawater

It is tempting to think of seawater as a homogeneous fluid, with planktonic microbes and nutrients evenly distributed within it. However, a growing body of evidence indicates that there is microscale heterogeneity in the distribution of nutrients around organisms and particles of organic matter. Large quantities of transparent colloidal polymers structure seawater into a complex gel-like matrix (see *Plate 2.4B*). Large-scale processes like productivity, nutrient recycling, and geochemical cycles are the result of microbial activity. In turn, physical processes like turbulence, photon flux, and gas exchange are translated down to the microscale level, affecting microbial behavior and metabolism. Physical factors such as diffusion, shear forces, and viscosity must be considered in this context.

There is a continuous shower of clumps and strings of material which falls through the water column—this is termed "marine snow" because of its resemblance to falling snowflakes when illuminated underwater. Marine snow consists of aggregates of inorganic particles, plankton cells, detritus from dead or dying plankton, and zooplankton fecal material, glued together by a matrix of polymers released from phytoplankton and bacteria (*Figure 1.9*). Most particles are 0.5 to a few micrometers in diameter, but they can grow to several centimeters in calm waters. Aggregates form as a result of collision and coagulation of primary particles, and they increase in size as they acquire more material through these physical processes. The nucleus for snow formation is often the mucus-based feeding structures used by salps and larvaceans in the zooplankton. Dying

WHAT DOES "DISSOLVED ORGANIC MATTER" REALLY MEAN?

Oceanographers have traditionally talked about "dissolved" and "particulate" organic matter (DOM and POM, respectively). Measurements of concentrations and fluxes of DOM and its constituent elements carbon (DOC), nitrogen (DON) and phosphorus (DOP) are among the most important factors in the study of ocean processes. It is important to remember that the difference between "dissolved" and "particulate" is a purely empirical distinction and simply reflects the size of filters used in sample preparation. There is no absolute definition, but most filters used in studies of DOM and POM have pore sizes from about 0.45 to 1.0 μm. Many bacteria and almost all viruses would pass through such filters and therefore appear in the DOM fraction. Colloidal material and polymers aggregate to form particles, and it is only low-molecular-weight compounds like sugars and amino acids that are truly dissolved. Thus, DOM and POM form a continuum, with microbes spanning both fractions.

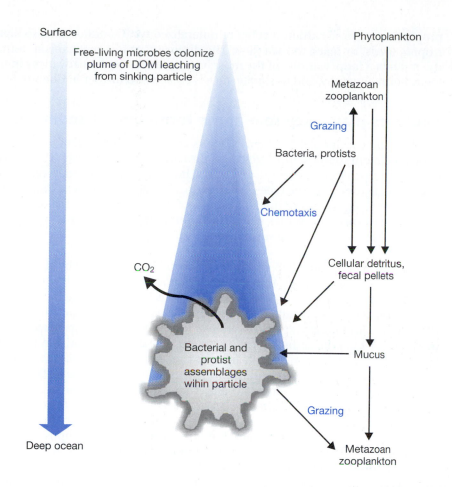

Figure 1.9 Schematic diagram showing the microbial processes occurring in the formation and fate of a marine snow particle as it falls through the water column. The action of extracellular enzymes and viral lysis leads to the release of dissolved organic material (DOM).

diatoms, at the end of a bloom, often precipitate large-scale snow formation owing to the production of large amounts of mucopolysaccharides in their cell walls. The generation of water currents during feeding by flagellates and ciliates colonizing the aggregate also collects particles from the surrounding water and leads to growth of the snow particle.

Marine snow is mainly produced in the upper 100–200 m of the water column, and large particles can sink up to 100 m per day, allowing them to travel from the surface to the ocean deep within a matter of days. This is the main mechanism by which a proportion of the photosynthetically fixed carbon is transported from the surface layers of the ocean to deeper waters and the seafloor. However, aggregates also contain active complex assemblages of bacteria and protists that graze on them. Levels of microorganisms in marine snow are typically 10^8–10^9 ml^{-1}, which are about 100–10000-fold higher than in the bulk water column. As particles sink, organic material is degraded by extracellular enzymes produced by the resident microbial population. Microbial respiration creates anoxic conditions, so that diverse aerobic and anaerobic microbes colonize different niches within the snow particle. The rate of solubilization exceeds the rate of remineralization, so dissolved material leaks from snow particles, leaving a plume of nutrients in its wake as it spreads by diffusion and advection. This may send chemical signals that attract small zooplankton to consume the particle as food. The trailing plume also provides a concentrated nutrient source for suspended planktonic bacteria, which may show chemotactic behavior in order to remain within favorable concentrations. Thus, much of the carbon is recycled during its descent, but some material reaches the ocean bottom, where it is consumed by benthic organisms or leads to the formation of sediments. Photosynthesis by algae and bacteria leads to the formation of organic material through CO_2 fixation, but viruses, heterotrophic bacteria, and protists all play a part in the fate of this fixed carbon. The balance of their activities throughout the water column determines the

? HOW DID CHERNOBYL FALLOUT HELP IN THE STUDY OF MARINE SNOW?

In 1986, a major environmental catastrophe occurred when the nuclear reactor in Chernobyl, Ukraine, exploded, releasing hundreds of tons of radioactive particles. These were carried by winds to many distant regions and settled with rain on land and sea. Fowler et al. (1987) were able to extract some benefit from this tragedy. They had already set up time-series sediment traps to measure vertical transport in the Mediterranean Sea. Following the deposition of Chernobyl fallout, they found that the pulse of radioactivity—in particular the rare earth nuclides ^{141}Ce and ^{144}Ce— was rapidly removed from surface waters and transported to depths of over 200 m within a few days at an average settling rate of 29 m per day. Physical processes alone could not account for such rapid settlement. Fowler and colleagues identified the fecal pellets of zooplankton, leading to the conclusion that zooplankton were ingesting particles adsorbed to their food source and repackaging them as larger, denser particles that aggregated with marine snow to sink at a high rate. Subsequent studies have shown that the fecal pellets of different zooplankton species vary greatly in their density and sinking rate, and this is affected by many factors.

proportions of fixed carbon that are remineralized to CO_2, transferred to higher trophic levels, or reach the sea floor. The discovery of this mechanism, termed the microbial loop, was one of the most important conceptual advances in biological oceanography, and its significance is considered further in Chapter 8.

Microbes play a key role in the formation of sediments

Much of the continental shelf and slope is covered with terrigenous or lithogenous sediments derived from erosion of the continents and transported into the ocean as particles of mud, sand, or gravel. The mineral composition reflects the nature of the rocks and the type of weathering. Large rivers such as the Amazon, Orinoco or Ganges transfer millions of tons of fine sediments to the ocean each year. Most of this mud settles along the continental margins or is funneled by submarine currents as dilute suspensions.

In the deep ocean, about 75% of the deep ocean floor is covered by abyssal clays and oozes. Abyssal clays are formed by the deposition of wind-blown terrestrial dust from the continents, mixed with volcanic ash and cosmogenic dust from meteor impact. These accumulate very slowly—less than 1 mm per 1000 years—whilst oozes accumulate at up to 4 cm per 1000 years. Biogenous oozes contain over 30% of material of biological origin, mainly shells of protistan plankton, mixed with clay. Oozes are usually insignificant in the shallow waters near continents. Calcareous oozes or muds cover nearly 50% of the ocean floor, especially in the Indian and Atlantic Oceans. They are formed by the deposition of the calcium carbonate shells (tests) of two main types of protist: the coccolithophorids and the foraminifera (see Chapter 6). Siliceous oozes are formed from the shells (frustules) of diatoms and radiolarians, which are composed of opaline silica ($SiO_2.nH_2O$). The rate of accumulation of biogenous oozes depends on the rate of production of organisms in the plankton, the rate of destruction during descent to the seafloor, and the extent to which they are diluted by mixing with other sediments. In the case of coccolithophorids and foraminifera, depth has an important effect on dissolution of the tests. At relatively high temperatures near the surface, seawater is saturated with $CaCO_3$. As calcareous shells sink, $CaCO_3$ becomes more soluble as a result of the increased content of CO_2 in water at lower temperatures and higher pressures. The carbonate compensation depth is the depth at which carbonate input from the surface waters is balanced by dissolution in deep waters; this varies between 3000 m in polar waters and 5000 m in tropical waters. For this reason, calcareous oozes tend not to form in waters more than 5000 m deep. Similarly, not all of the silica in the frustules of diatoms reaches the ocean floor because bacterial action has been shown to play a large part in the dissolution of diatom shells during their descent (see p. 145). The rate of deposition of protist remains to the seabed is much more rapid than would be assumed from their small size. This is because they are aggregated into larger particles through egestion as fecal pellets after grazing by zooplankton and through the formation of marine snow as described above. In shallower waters near the continental shelf, the high input of terrigenous sediments mixes with and dilutes sediments of biogenous origin.

Remineralization of readily degradable organic matter in the water column through microbial action means that only a small fraction of fixed carbon—probably less than 1% of primary production—reaches the deep ocean floor. Oxygen only penetrates the top few millimeters of sediments, and below this the sediments are anoxic. Indeed, the overlying water column may be completely anoxic in some regions where high nutrient concentrations promote high oxygen consumption by microbes. For example, the Black Sea contains no free oxygen from a depth of 150 m to the bottom, at 2000 m, owing to sulfate reduction. There are also large anoxic basins off the coasts of Venezuela and Mexico and in the Eastern

Mediterranean (see *Research Focus Box 1.2*). As organic material settles deeper into anoxic sediments, it accumulates more rapidly than it is degraded and thus joins the geological cycle, reemerging millions of years later when uplifted in continental rocks through tectonic processes.

There is increasing recognition of the importance of microbial activities in the sediment–water interface (SWI) and deep-sea benthic boundary layer (BBL), which is a layer of homogeneous water 10 m or more thick, adjacent to the sediment surface. The SWI includes high concentrations of particulate organic debris and dissolved organic compounds that become absorbed onto mineral particles. The structure and composition of the microbial habitat is modified by benthic "storms" and the action of animals such as worms and burrowing shrimps, which move and resuspend sediments, transporting oxygen into deeper layers.

As well as the constant "snowfall" of plankton-derived material, concentrated nutrient inputs reach the seabed in the form of large animal carcasses. For example, time-lapse photography has shown how quickly fallen whale carcasses attract colonies of animals, and microbiological studies accompanying these investigations have yielded novel bacteria, some with biotechnological applications (see p. 310). The microbial communities and symbioses that develop are similar to those found at hydrothermal vents and cold seeps. Other types of sediment that provide special habitats for microbes include those in salt marshes, mangroves, and coral reefs.

Studies of the extent to which carbon fixed in the photic zone finds its way to the seabed, and its fate in sediments, are important in understanding the role of the oceans in the planetary carbon cycle. Microbial processes such as production and oxidation of methane and oxidation and reduction of sulfur compounds are of special interest. Studies of the diversity and activity of microbial life in the various types of sediment are yielding many new insights, mainly because of the application of molecular techniques, and are described in subsequent chapters.

Microbes colonize surfaces through formation of biofilms

In the last two decades, the special phenomena that govern the colonization of surfaces by microbes have come under intense scrutiny, with the growing recognition that such biofilm formation involves complex physicochemical processes and community interactions. Biofilms consist of a collection of microbes bound to a solid surface by their extracellular products, which trap organic and inorganic components. In the marine environment, all kinds of surfaces including other microbes, plants, animals, rocks, and fabricated structures may be colonized by biofilms. As a result of metabolic processes, ecological succession results in the development of microenvironments and colonization by mixed communities of bacteria and protists to form layered structures known as microbial mats, which can be several millimeters thick. These are particularly important in shallow and intertidal waters. The composition of microbial mats is affected by physical factors such as light, temperature, water content, and flow rate; and by chemical factors such as pH, redox potential, the concentration of molecular oxygen and other chemicals (especially sulfide, nitrate, and iron), and dissolved organic compounds. Phototrophic bacteria and diatoms are major components of stratified microbial mats, and the species composition and zonation are determined by the intensity and wavelength of light penetration into the mat. Light normally only penetrates about 1 mm into the mat and anoxic conditions develop below this. The formation and diurnal variations of oxygen and sulfide gradients has a major effect on the distribution of organisms in the mat. Biofilm formation is considered in more detail in Chapter 3, and the economic importance in biofouling is discussed in Chapter 13.

BOX 1.2   RESEARCH FOCUS

Living on the edge—an insight into extraterrestrial life?

Microbes are abundant at the interface of brine lakes in the deep sea

An international consortium of scientists is studying the microbiology of deep hypersaline anoxic basins (DHABs) as part of the European Commission Biodeep project. There are at least four known DHABs (called Bannock, L'Atalante, Discovery, Urania), occurring at depths of 3.2–3.6 km in the Eastern Mediterranean. It is thought that they were formed when the Mediterranean Sea evaporated about 250 million years ago, leaving a layer of rock salt, which then became covered by sediments. Movement of the tectonic plates has exposed the salt deposits in a few areas. Here, the salts have dissolved to form dense pockets of highly concentrated salt solutions, separated from the overlying seawater by extremely sharp environmental interfaces (chemoclines). These undersea brine lakes have extreme physical and chemical conditions, notably the complete lack of oxygen and near-saturated solutions of salts (up to 10 times the salinity of seawater).

The interface is just a few meters thick, and accurate sampling within this chemocline at such a great depth requires great ingenuity. Conventional remotely operated vehicles (ROVs) and manned submersibles cannot be used because of the damaging properties of the brines. Daffonchio et al. (2006) describe the use of a specially adapted sampling module with a CTD probe (see p. 28) containing a sensor to record the exact pressure at which sampling bottles are closed. Operators on the research vessel can observe entry into the narrow band of water at the interface with a camera. Daffonchio and colleagues found that the interface at the Bannock basin contained about 10^6 cells ml^{-1}, compared with about 10^4 cells ml^{-1} in both the overlying water and underlying brine. As in previous studies of the deep sea (Karner et al., 2001), the proportion of organisms belonging to the *Bacteria* and *Archaea* in overlying seawater was about the same, but within the interface the proportion of *Bacteria* was much higher. The interface also showed significantly higher metabolic activity than the overlying or underlying layers. By careful fractionation of the samples from different depths, Daffonchio et al. (2006) found two peaks in biomass and biodiversity, occurring at about 8% and 15–22% salinity. Because salt creates a high osmotic potential, organisms growing in such habitats require special adaptations, and it might be expected that the diversity of bacterial types would be low. However, through the use of DNA-based identification methods, many previously unknown groups of *Bacteria* were identified, occupying narrow niches in the highly stratified interface.

Borin et al. (2009) also used the fine-scale sampling to study microbial processes in the Urania basin, which has lower salt concentrations but exceptionally high levels of sulfide. Two chemoclines occurred, indicating two overlying lakes of brine with very little mixing. The first interface (about 2 m thick), is a very sharp barrier between oxic and anoxic conditions, with differences in temperature and salinity and the presence of different electron acceptors. Again, bacterial abundance and diversity was higher along the chemoclines than in the seawater or in the uniform brines. The main group of bacteria found belonged to the *Delta-proteobacteria* (sulfate reducers) and *Epsilonproteobacteria* (sulfide oxidizers), whilst the deepest, most saline parts of the basin were dominated by archaea responsible for very high levels of methane production. What sustains this dense microbial community? At such a depth, photosynthesis is obviously impossible, and the chemocline acts as powerful density barrier to nutrient particles settling through the water column. Could chemosynthesis be responsible? Yakimov et al. (2007) measured carbon fixation through uptake of radioactively labeled bicarbonate and showed that this was much higher within the brine lake than in the overlying water. By using gene probes to detect genes involved in the anaerobic oxidation of ammonia (anammox, see p. 69), it was shown that *Crenarchaeota* dominate the brine lake.

In the Discovery Basin, the brine is an almost saturated solution of $MgCl_2$ rather than NaCl. There is a steep gradient of $MgCl_2$ from 55 mM at the seawater face to 5.05 M at the lower face. Although Mg is essential in low concentrations, life would be inhibited by such high levels of $MgCl_2$ because it is a highly chaotropic salt (i.e. it weakens electrostatic interactions that help to stabilize biological molecules). Hallsworth et al. (2007) isolated and cultured some bacteria from the upper parts of the interface, but laboratory tests showed that these would not grow—even after 18 months—above 1.26 M $MgCl_2$. DNA corresponding to sulfate-reducing bacteria and methanogenic archaea was found throughout the interface, but mRNA (a reliable indicator of active microbes because it only has a short life) was only found in the uppermost layer, with $MgCl_2$ concentrations below 2.3 M, suggesting that this is the limit for life.

As well as bacteria and archaea, Edgcomb et al. (2009) recently showed that the interface in Bannock and Discovery Basins harbors an astonishing diversity of protists. They obtained more than 1500 protistan gene sequences, most of which have never been described before.

Clearly, microbes thrive in environments that we previously thought completely inhospitable to life. Such studies are important for our understanding of life and biogeochemical processes in the deep sea, as well as having biotechnological potential through discovery of new genes and compounds. They will also inform the search for life on other planets or moons, for example in the planned missions to Mars and Europa.

Microbes in sea ice form an important part of the food chain in polar regions

At the poles, the temperature is so low during the winter that large areas of sea-water freeze to form sea ice, some of which forms adjacent to the coastal shore-line and some of which forms floating masses of pack ice. Sea ice forms when the temperature is less than –1.9°C, the freezing point of water at 35‰ salinity. The first stage in sea-ice formation is the accumulation of minute crystals of frazil ice on the surface, which are driven by wind and wave action into aggregated clumps called grease ice. These turn into pancake-shaped ice floes that freeze together and form a solid ice cover. At the winter maxima, the combined cover-age by sea ice at the north and south polar regions is almost 10% of the Earth's surface (1.8×10^7 km^2 in the Antarctic and 1.5×10^7 km^2 in the Arctic). During the formation of frazil ice, planktonic microbes become trapped between the ice crystals and wave motion transports more organisms into the grease ice during its formation. Near the ice–air interface, temperatures may be as low as –20°C during the depths of the polar winter, whilst the temperature at the ice–water interface remains fairly constant at about –2°C. When seawater freezes, it forms a crystalline lattice of pure water, excluding salts from the crystal structure. The salinity of the liquid phase increases and its freezing point drops still further. This very cold, high-density, high-salinity (up to 150‰) water forms brine pockets or channels within the ice, which can remain liquid to –35°C. The ice becomes less dense than seawater and rises above sea level, with the channels draining brine through the ice to the underlying seawater. Thus, sea ice is very different from freshwater glacial ice.

The structure of sea ice provides a labyrinth of different microhabitats for microbes, with variations in temperature, salinity, nutrient concentration, and light penetration. This enables colonization and active metabolism by distinc-tive mixed communities of cold-adapted (psychrophilic) photosynthetic and heterotrophic protists and bacteria, as well as viruses. Microbial activities also alter the physicochemical conditions, mainly owing to the production of large amounts of cryoprotectant compounds and extracellular polymers, leading to the creation of additional microenvironments. The dominant photosynthetic organisms near the ice–sea interface are pennate diatoms and small dinoflagel-lates. The density of diatoms in sea ice may be up to 1000 times that in surface waters. Through photosynthesis, the microalgae make a small, but significant, contribution to primary productivity in the polar regions. For example, the con-tribution of sea ice to primary productivity in the Southern Ocean is only about 5% of the total, but it extends the short summer period of primary production and provides a concentrated food source that sustains the food web during the winter. During the Antarctic winter, microalgae on the undersurface of sea ice are the main source of food for grazing krill, shrimp-like crustaceans that are the main diet of fish, birds, and mammals in the Southern Ocean. A wide range of protists and heterotrophic bacteria have been found in sea ice, including new species with biotechnological potential. Some microbes remain active—albeit at a much reduced metabolic rate—even in the coldest parts of the ice and in "frost flowers" formed on the surface of the ice, where they are trapped in pockets of very low temperatures and high salinity.

Microbial activity at hydrothermal vents provides an oasis of life in the deep sea

Hydrothermal vents form a specialized and highly significant habitat for microbes. They occur mainly at the mid-ocean ridges at the boundary of the Earth's tectonic plates, where seafloor spreading and formation of new ocean crust is occurring. Numerous such sites have been studied in the Pacific and Atlantic Oceans. Sea-water permeates through cracks and fissures in the crust and interacts with the

WHAT WILL HAPPEN WHEN THE SEA-ICE MELTS?

Global warming is affecting polar regions more rapidly than predicted by previous models. Although there is a large natural variation in the Arctic climate, since the 1970s, the extent of summer sea ice in the Arctic has declined by an average 8% per decade and ice packs have become thinner. The loss of cover in 2007 was especially severe and led to fears that a "tipping point" had been reached, but the loss has not been so severe in recent years. The loss of ice cover results in changes to wind patterns and warmer and less saline waters in the upper ocean. Most significantly, loss of summer ice also reduces the albedo effect, resulting in a positive feedback, which increases the absorption of solar energy by seawater rather than reflecting sunlight, further hastening the rise in temperature. An intensive research program to compare the microbiology of current and archived samples of deep-sea sediment has recently been initiated by Antje Boetius and colleagues at the Max Planck Institute of Marine Microbiology, Bremen. This is revealing important information about changes in microbial diversity and how the flux of particulate organic matter is affected by the loss of permanent ice cover. Besides producing major changes in marine ecology, the melting of summer sea ice means that shipping can now pass through the ice-free North West Passage and is enabling exploration for oil and gas in previously inaccessible parts of the Arctic Ocean.

heated underlying rocks, thereby changing the chemical and physical character-istics of both the seawater and the rock. The permeability structure of the ocean crust and the location of the heat source determine the circulation patterns of hydrothermal fluids. As cold seawater penetrates into the ocean crust, it is gradu-ally heated along its flow path, leading to the removal of magnesium from the fluid into the rock, with production of acid during the process. This leads to the leaching of other major elements and transition metals from the rock into the hydrothermal fluid, and sulfate in the seawater is removed by precipitation and reduction to hydrogen sulfide. As the percolating fluids reach the proximity of the magma heat source, extensive chemical reactions occur within the rock and the pressurized fluids are heated to over 350°C, becoming buoyant and rising toward the ocean floor. As they rise, decompression causes the fluids to cool slightly, and precipitation of metal sulfides and other compounds occurs en route. The hydro-thermal fluid is injected into the ocean as plumes of mineral-rich superheated water. The hottest plumes (up to 350°C) are generally black, because of the high content of metal sulfide and sulfate particles, and precipitation occurs as the hot plume mixes with the cold seawater. Some of these precipitates form chimney structures called "black smokers," whilst others are dispersed through the water and form sediments in the vicinity. In other parts of the vent field, the circula-tion of hydrothermal fluid may be shallower, leading to diffuse plumes of water heated to 6–23°C (*Figure 1.10*). The gradients of temperature and nutrients that exist at hydrothermal systems provide a great diversity of habitats for microbes suspended in the surrounding heated waters, in sediments, and attached to surfaces of the chimneys. Many of these are hyperthermophilic bacteria and archaea, which can grow at temperatures up to 121°C, whilst others grow at lower temperatures further from the fluid emissions. Molecular studies are revealing an astonishing diversity of such organisms, many of which have biotechnologi-cal applications. The microbiology of the deep subsurface rocks beneath vents is also now under investigation, and many novel microbes and metabolic processes are being discovered. Microbial activity in the deep subsurface contributes to the chemical changes in composition during circulation of the hydrothermal fluids.

Hydrothermal vent systems were first described in 1977, when scientists aboard the submersible *Alvin*, from Woods Hole Oceanographic Institution, were explor-ing the seabed about 2500 m deep near the Galapagos Islands. The discovery of life around the vents was totally unexpected. The *Alvin* scientists observed dense communities of previously unknown animals, including tubeworms, clams, anemones, crabs, and many others (see *Plate 10.3*). Subsequent research showed that the warm waters near hydrothermal vents contain large populations of che-mosynthetic bacteria and archaea, which fix CO_2 using energy from the oxidation of sulfides in the vent fluids. This metabolism supports a food chain with many trophic levels that is independent of photosynthesis. We now know that many of the animals at vent sites contain chemosynthetic bacteria as symbionts within their tissues or on their surfaces—these relationships are discussed in Chapter 10. In addition, bacterial populations directly support the growth of filter-feeding animals, such as clams and mussels, or shrimps, which graze on microbial mats. Thus, hydrothermal vents are an oasis of life in the deep sea. Previously, life was thought always to rely ultimately on the fixation of CO_2 by photosynthesis, but the vent communities function without the input of material derived from the use of light energy. However, note that sulfide oxidation depends on dissolved oxygen in the water, and the origin of this is photosynthetic.

Cold seeps also support diverse life

Cold seeps are abundant along the continental shelf and slope, where the upwards percolation of fluids through the sediments is influenced by plate tectonics and other geological processes. At these sites, high concentrations of methane and sulfide support prolific chemosynthetic communities consisting of free-living

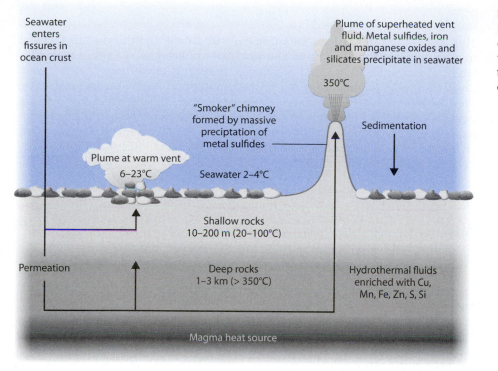

Seawater enters fissures in ocean crust

Plume of superheated vent fluid. Metal sulfides, iron and manganese oxides and silicates precipitate in seawater

350°C

"Smoker" chimney formed by massive precipitation of metal sulfides

Sedimentation

Plume at warm vent 6–23°C

Seawater 2–4°C

Shallow rocks 10–200 m (20–100°C)

Permeation

Deep rocks 1–3 km (> 350°C)

Hydrothermal fluids enriched with Cu, Mn, Fe, Zn, S, Si

Magma heat source

Figure 1.10 Processes occurring at a hydrothermal vent system. Gradients of minerals and temperature around the vents create a variety of habitats for diverse microbial and animal communities.

bacteria and archaea, as well as those living symbiotically with invertebrates (Chapter 10). Extensive populations of foraminifera also occur at cold seeps. Some sites are associated with seeps of hypersaline brines or hydrocarbons.

Living organisms are the habitats of many microbes

Microbial biofilms also form on the surfaces of all kinds of animals, algae, and coastal plants, which provide a highly nutritive environment through secretion or leaching of organic compounds. Many organisms seem selectively to enhance surface colonization by certain microbes and discourage colonization by others. This may occur by the production of specific compounds that inhibit microbial growth or interfere with microbial attachment processes. Once established, particular microbes may themselves influence colonization by other types. These processes offer obvious applications in the control of biofouling (see p. 284). As well as surface (epibiotic) associations, microorganisms can form endosymbiotic associations within the body cavities, tissue, or cells of living organisms.

Many microalgae such as diatoms and dinoflagellates harbor bacteria on their surfaces, or as endosymbionts within their cells. Intimate associations between bacterial and archaeal cells are also being revealed by new imaging techniques (see *Plate 5.1*). Seaweeds and seagrasses have dense populations of bacteria (up to 10^6 per cm^2) on their surfaces, although this varies considerably with species, geographic location and climatic conditions.

The external surfaces and intestinal content of animals provide a variety of habitats to a wide diversity of microbes. Such associations may be neutral in their effects, but commonly lead to some mutual benefit for host and microbe (symbiosis). Examples of symbiotic interactions between animals and microbes are considered in Chapter 10, whilst diseases of marine organisms are discussed in Chapter 11.

Conclusions

This chapter has introduced the marine environment and the various types of marine microbes that are found. The adaptations of marine microbes to the different physical, chemical, and biological conditions encountered in the wide range of marine habitats have led to the evolution of highly diverse microbes. The discovery that these tiny microbes are present in such large numbers and are largely responsible for the biogeochemical processes that shape our planet can be viewed as one of the most important advances of modern science. Subsequent chapters build on this introduction by exploring the mechanisms underlying this diversity of form and function.

References

Borin, S., Brusetti, L., Mapelli, F., et al. (2009) Sulfur cycling and methanogenesis primarily drive microbial colonization of the highly sulfidic Urania deep hypersaline basin. *Proc. Natl Acad. Sci. USA* 106: 9151–9156.

Ciccarelli, F. D., Doerks, T., von Mering, C., Creevey, C. J., Snel, B. and Bork, P. (2006) Toward automatic reconstruction of a highly resolved tree of life. *Science* 311: 1283–1287.

C-MORE (2009) Rising CO$_2$, ocean acidification, and their impacts on marine microbes. Center for Microbial Oceanography: Research and Education, http://cmore.soest.hawaii.edu/oceanacidification/ [Accessed 10 December 2010].

Daffonchio, D., Borin, S., Brusa, T., et al. (2006) Stratified prokaryote network in the oxic/anoxic transition of a deep-sea halocline. *Nature* 440: 203–207.

Dagan, T. and Martin, W. (2009) Getting a better picture of microbial evolution en route to a network of genomes. *Philos. Trans. R. Soc. Lond. B Biol. Sci.* 364: 2187–2196.

Danovaro, R., Dell'Anno, A., Corinaldesi, C., Magagnini, M., Noble, R., Tamburini, C. and Weinbauer, M. (2008) Major viral impact on the functioning of benthic deep-sea ecosystems. *Nature* 454: 1084–1087.

Delille, B., Harlay, J., Zondervan, I., et al. (2005) Response of primary production and calcification to changes of pCO$_2$ during experimental blooms of the coccolithophorid *Emiliania huxleyi*. *Global Biogeochem. Cycles* 19: GB2023.

Doolittle, W. F. (1999) Phylogenetic classification and the universal tree. *Science* 284: 2124–2128.

Dworkin, M., Falkow, S., Rosenberg, E., Schleifer, K. H. and Stackebrandt, E. (2007) *The Prokaryotes*. Springer, New York.

Edgcomb, V., Orsi, W., Leslin, C., et al. (2009) Protistan community patterns within the brine and halocline of deep hypersaline anoxic basins in the eastern Mediterranean Sea. *Extremophiles* 13: 151–167.

Engel, A., Zondervan, I., Aerts, K., et al. (2005) Testing the direct effect of CO$_2$ concentration on a bloom of the coccolithophorid *Emiliania huxleyi* in mesocosm experiments. *Limnol. Oceanogr.* 50: 493–507.

EPOCA (2009) European Project on Ocean Acidification website, accessed August 28, 2010, http://www.epoca-project.eu/.

Fowler, S. W., Buat-Menard, P., Yokoyama, Y., Ballestra, S., Holm, S. and Van Nguyen, H. (1987) Rapid removal of Chernobyl fallout from Mediterranean surface waters by biological activity. *Nature* 329: 56–58.

Hallsworth, J. E., Yakimov, M. M., Golyshin, P. N., et al. (2007) Limits of life in MgCl$_2$-containing environments: chaotropicity defines the window. *Environ. Microbiol.* 9: 801–813.

Henderson, G. P., Gan, L. and Jensen, G. J. (2007) 3-D ultrastructure of *Ostreococcus tauri*: electron cryotomography of an entire eukaryotic cell. *PLoS ONE* 2(8): e749.

Iglesias-Rodriguez, M. D., Halloran, P. R., Rickaby, R. E. M., et al. (2008a) Phytoplankton calcification in a high-CO$_2$ world. *Science* 320: 336–340.

Iglesias-Rodriguez, M. D., Buitenhuis, E. T., Raven, J. A., et al. (2008b) Response to comment on "Phytoplankton calcification in a high-CO$_2$ world". *Science* 322: 1466.

Karner, M. B., De Long, E. F. and Karl, D. M. (2001) Archaeal dominance in the mesopelagic zone of the Pacific Ocean. *Nature* 409: 507–510.

Lipp, J. S., Morono, Y., Inagaki, F. and Hinrichs, K. U. (2008) Significant contribution of Archaea to extant biomass in marine subsurface sediments. *Nature* 454: 991–994.

Martin, W. and Embley, T. M. (2004) Evolutionary biology: early evolution comes full circle. *Nature* 431: 134–137.

Mitchell, A. (2009) *Sea Sick. The Hidden Ecological Crisis of the Global Ocean*. One World Publications, Oxford.

Orr, J. C., Fabry, V. J., Aumont, O., et al. (2005) Anthropogenic ocean acidification over the twenty-first century and its impact on calcifying organisms. *Nature* 437: 681–686.

Pace, N. R. (2009) It's time to retire the prokaryote. *Microbiol. Today* 36: 85–87.

Ramos, J. B. E., Biswas, H., Schulz, K. G., LaRoche, J. and Riebesell, U. (2007) Effect of rising atmospheric carbon dioxide on the marine nitrogen fixer *Trichodesmium*. *Global Biogeochem. Cycles* 21: GB2028; doi: 10.1029/2006GB002898.

Riebesell, U. (2004) Effects of CO$_2$ enrichment on marine phytoplankton. *J. Oceanogr.* 60: 719–729.

Riebesell, U., Schulz, K. G., Bellerby, R. G. J., et al. (2007) Enhanced biological carbon consumption in a high CO$_2$ ocean. *Nature* 450: 545–548.

Riebesell, U., Schulz, K. G., Bellerby, R. G. J., Botros, M., Fritsche, P., Meyerhofer, M., et al. (2007) Enhanced biological carbon consumption in a high CO$_2$ ocean. *Nature* 450: 545–548.

Royal Society (2005) Ocean acidification due to increasing atmospheric carbon dioxide. Policy document 12/05. The Royal Society, London, 2005, accessed August 28, 2010, http://royalsociety.org/.

Suttle, C. A. (2005) Viruses in the sea. *Nature* 437: 359–361.

Whitman, W. B. (2009) The modern concept of the prokaryote. *J. Bacteriol.* 191: 2000–2005.

Whitman, W. B., Coleman, D. C. and Wiebe, W. J. (1998) Prokaryotes: the unseen majority. *Proc. Natl Acad. Sci. USA* 95: 6578–6583.

Yakimov, M. M., La Cono, V., Denaro, R., D'Auria, G., Decembrini, F., Timmis, K. N., Golyshin, P. N. and Giuliano, L., et al. (2007) Primary producing prokaryotic communities of brine, interface and seawater above the halocline of deep anoxic lake L'Atalante, Eastern Mediterranean Sea. *ISME J.* 1: 743–755.

Further reading

Historical perspectives

Karl, D. M. and Proctor, L. M. (2008) Foundations of microbial oceanography. *Oceanography* 20: 16–27.

Yayanos, A. A. (2003) Marine microbiology at Scripps. *Oceanography* 16: 63–75.

Microbial diversity and evolution

Caron, D. A., Worden, A. Z., Countway, P. D., Demir, E. and Heidelberg, K. B. (2008) Protists are microbes too: a perspective. *ISME J.* 3: 4–12.

Lawton, G. (2009) Why Darwin was wrong about the tree of life. *New Scientist* 2692: 34–39.

Schulz, H. N. and Jørgensen, B. B. (2001) Big bacteria. *Annu. Rev. Microbiol.* 55: 105–137.

Sherr, B. F., Sherr, E. B., Caron, D. A., Vaulot, D. and Worden, A. Z. (2007) Oceanic protists. *Oceanography* 20: 130–134.

Microscale processes, marine snow, and dissolved organic matter

Azam, F. and Long, R. A. (2001) Sea snow microcosms. *Nature* 414: 495–498.

Azam, F. and Malfatti, F. (2007) Microbial structuring of marine ecosystems. *Nat. Rev. Microbiol.* 5: 782–791.

Grossart, H. P., Kiørboe, T., Tang, K. W., Allgaier, M., Yam, E. M. and Ploug, H. (2006) Interactions between marine snow and heterotrophic bacteria: aggregate formation and microbial dynamics. *Aquat. Microb. Ecol.* 42: 19–26.

Kjorbøe, T. (2001) Formation and fate of marine snow: small-scale processes with large-scale implications. *Sci. Mar.* 65 (S2): 57–71.

Verdugo, P., Alldredge, A. L., Azam, F., Kirchman, D. L., Passow, U. and Santschi, P. H. (2004) The oceanic gel phase: a bridge in the DOM-POM continuum. *Marine Chem.* 92: 67–85.

Sea-ice microbiology and polar oceans

Arrigo, K. R. and Thomas, D. N. (2004) Large scale importance of sea ice biology in the Southern Ocean. *Antarctic Sci.* 16: 471–486.

Kirchman, D. L., Moran, X. A. G. and Ducklow, H. (2009) Microbial growth in the polar oceans—role of temperature and potential impact of climate change. *Nat. Rev. Microbiol.* 7: 451–459.

Mock, T. and Thomas, D. N. (2005) Recent advances in sea-ice microbiology. *Environ. Microbiol.* 7: 605–619.

Sediments and subsurface microbiology

Jørgensen, B. B. and Boetius, A. (2007) Feast and famine—microbial life in the deep-sea bed. *Nat. Rev. Microbiol.* 5: 770–781.

Hydrothermal vents

Martin, W., Baross, J., Kelley, D. and Russell, M. J. (2008) Hydrothermal vents and the origin of life. *Nat. Rev. Microbiol.* 6: 805–814.

Reysenbach, A.-L., Cady, S. L. (2001) Microbiology of ancient and modern hydrothermal systems. *Trends Microbiol.* 9: 79–86.

Chapter 2

Methods in Marine Microbiology

The application of new techniques, especially those based on molecular biology, has resulted in a huge recent expansion of our knowledge of marine microbes. Although knowledge of technical details is not necessary unless directly involved in research, it is essential for the student to appreciate the underlying principles of these methods in order to understand research articles. Whilst it can be tempting to concentrate only on these latest innovations, this chapter will show that "traditional" microbiological methods based on culture and microscopic observation remain absolutely essential. The chapter is divided into four sections: the first part describes sampling and general experimental protocols; the second part considers the use of microscopy and flow cytometry; the third part describes culture-based methods of microbiology; and the final section considers the use of molecular methods based on analysis of nucleic acids, for studying both isolated microbes and their diversity and activities in the marine environment.

Key Concepts

- Sampling and experimental observation of marine microbes requires special techniques.

- Epifluorescence microscopy and flow cytometry led to recognition of the abundance and diversity of marine microbes in the oceans.

- The most significant recent advances in marine microbiology have occurred as a result of development of methods for study of nucleic acids isolated directly from the environment.

- Full understanding of marine microbial diversity, physiology, and ecology requires the combination of culture-based and molecular methods and awareness of the bias inherent in different methods.

- New sequencing technologies and bioinformatics tools are resulting in massive expansion of knowledge through metagenomics, metatranscriptomics, and metaproteomics.

SAMPLING, GENERAL EXPERIMENTAL PROCEDURES, AND REMOTE SENSING

The aim of microbial ecology is the study of the diversity and activities of microbes *in situ*

Methods in microbial ecology are designed to answer two fundamental questions about microbes in the environment: "who's there?" and "what are they doing?" Measurements of the overall activities within a microbial community can be combined with modern knowledge and tools to investigate diversity, so that we can begin to build a picture of the activities of individual community members. The importance of scale cannot be overemphasized—*micro*organisms carry out their activities in *micro*environments. Measuring techniques must be designed to cause the minimum disturbance of the microenvironment that they are probing, because microbial communities may exist in a delicate three-dimensional structured organization, which is easily disrupted. At the micrometer scale, physicochemical gradients are very steep. For example, anaerobic, microaerophilic, and aerobic conditions may occur in a sediment particle or a microbial mat just a few hundred micrometers thick. Other conditions such as boundary layer effects, diffusion, and flow patterns will affect the microenvironment in the vicinity of a microbe. Most important of all, the microbes carry out biochemical reactions that alter the physical and chemical conditions in their immediate vicinity. Bulk measurements of substrates, metabolites, pH or oxygen in marine samples may therefore not reflect the heterogeneity of the natural situation.

Measurement of specific cell constituents may be used as biomarkers of microbial activity

Measurement of ATP (adenosine triphosphate) has been widely used as an estimate of total microbial community activity, for example in the vicinity of hydrothermal vents and in the analysis of microbial contamination of ballast water in ships. ATP occurs universally in all living organisms, and its presence is an indicator of active metabolism, since it is lost rapidly from dead cells. ATP is easily extracted from cells using boiling buffer or nitric acid, but to produce reliable results, care is needed in the processing of samples by filtration. A sensitive chemiluminescent assay for ATP depends on use of the enzyme luciferase, which catalyzes the reaction

$$\text{luciferin} + \text{ATP} \rightarrow \text{oxyluciferin} + \text{AMP} + \text{PP}_i + \text{CO}_2 + \text{light}$$

Luciferase was originally isolated from fireflies, but is now available as a recombinant product made in *Escherichia coli*. Light production is measured in sensitive photometers and these are available as handheld instruments for rapid monitoring of ATP levels.

Lipopolysaccharide (LPS), a unique component of the cell walls of Gram-negative bacteria (see *Figure 4.2*), can be detected by a sensitive assay using extracts from the horseshoe crab (*Limulus*), which causes an increase in optical density due to gel formation upon the addition of picogram amounts of LPS. The *Limulus* amoebocyte lysate assay has been widely used to estimate bacterial biomass in marine samples and to demonstrate the presence of Gram-negative bacteria in the tissues of the hosts of symbionts and pathogens. Great care is needed to prevent contamination during sample preparation. LPS and ATP measurements were commonly used in the 1980s, but have become less widely used since the introduction of nucleic-acid-based methods.

The diversity, distribution and concentration of photosynthetic pigments in the water column can be used to investigate the structure of phytoplankton. Chlorophylls, bacteriophylls, carotenoids and other pigments are analyzed by chromatography.

Measurement of some highly specific cell constituents can be useful. For example, analysis of unique archaeal lipids was a key factor in the discovery of the role of deep sea *Archaea* in ammonia oxidation (see *Research Focus Box 5.1*). The biomass of fungi in ecosystems such as saltmarshes and mangrove swamps can be determined by measuring ergosterol, which is a characteristic component of many fungal membranes (especially ascomycetes) and can be extracted by chromatography and measured by ultraviolet spectrophotometry.

Remote sensing and sampling permits analysis of microbial activities

At the opposite extreme from techniques to measure microenvironmental changes, the use of remote sensing has provided valuable information about the activities of planktonic organisms in the surface layers of the oceans. Whilst the ocean appears blue to the human eye, sensitive instruments detect subtle changes in ocean color due to the presence of plankton, suspended sediments, and dissolved organic chemicals. The first such satellite was the Coastal Zone Color Scanner (CZCS) launched in 1978, which collected light reflected from the sea surface at different wavelengths, including the absorption maximum for chlorophyll. This provided valuable information about the distribution of photosynthetic plankton and primary productivity (*Plate 8.1*). The CZCS ceased operations in 1986 and scanning is now carried out using the Sea-viewing Wide Field-of-View Sensor (SeaWiFS) system operated by the Orbital Sciences Corporation satellite, which makes data available to research agencies worldwide. The application of this technology in the monitoring of phytoplankton productivity is discussed in Chapter 9, and its use in monitoring algal blooms is considered in Chapter 11.

Microbiological sampling requires special techniques

The traditional method of collecting plankton is to tow a funnel-shaped net behind a boat. The smallest nylon mesh used is about 60 μm, which is clearly too large to capture most microbes. However, nets can be used to harvest larger phytoplankton and zooplankton to which microbes may be attached. Some aggregates of free-living microbes, such as clumps of filamentous cyanobacteria or diatoms, may also be large enough to collect in this way.

Obtaining samples of seawater and sediments is relatively straightforward in shallow coastal waters, with collection from a boat or by scuba divers. For most microbiological work, samples are collected in sterile plastic bags or bottles, taking great care to prevent contamination. Careful attention is needed in the choice of construction materials for sampling containers because many microbial processes can be affected by the presence of trace metals or rubber in hoses and stoppers.

Sampling in the open ocean requires the use of research vessels equipped with suitable sampling gear and onboard laboratory facilities. Research investigations are facilitated by the use of established long-term sampling programs combined with routine monitoring of oceanographic and atmospheric data. Well-known examples include the Pacific Ocean time-series site known as station ALOHA (A Long-term Oligotrophic Habitat Assessment) off Hawaii, the Bermuda Atlantic Time Series (BATS) site, and the Western Channel Observatory off Plymouth.

It is important to collect water samples in a form suitable for subsequent analysis together with appropriate environmental data about the sampling site. Because ocean water is highly heterogeneous, proper attention must be paid to the replication, frequency and location of sampling. For sampling within the water column, a device called a Niskin bottle is usually used. This is open at both ends and is lowered into the sea on a wire until it reaches the required depth, when it is closed by a weighted trigger that is sent down the cable from the surface. One of

? CAN BACTERIAL LIGHT BE SEEN FROM SPACE?

Miller et al. (2005) asked this question after looking at reports of the "milky sea" phenomenon observed by sailors in the Arabian Sea. The scientists obtained the location and time of a milky sea occurrence reported by a ship's captain some years earlier and noted this record in the ship's log: "The bioluminescence appeared to cover the entire sea area, from horizon to horizon ... and it appeared as though the ship was sailing over a field of snow or gliding over the clouds" (Nealson and Hastings, 2006). Miller et al. (2005) then checked the output from the meteorological satellites that monitor global cloudiness under daytime and nighttime conditions. After adjusting spectral frequencies of the output, they produced stunning images (*Plate 2.1*) of the bioluminescent bloom, showing it to cover over 15000 km^2 and estimated the total bioluminescent bacterial population of this milky sea to be about 4×10^{22} cells. Nealson and Hastings (2006) suggest that the bacteria responsible were probably *Vibrio harveyi* in association with massive blooms of *Phaeocystis* algae.

Figure 2.1 Procedures for sampling water and sediments. (A) A CTD (conductivity, temperature, depth) cast and rosette full of water sampling containers being deployed in Arctic waters. (B) The water filtration system on board *Sorcerer II*, the yacht used in the Global Ocean Survey. Samples of up to 200 liters are sequentially passed through filters of different pore sizes to trap different fractions of the microbial plankton. (C) Sampling of sediments by a remote-controlled underwater vessel at the Alfred Wegener Institute (AWI) "Hausgarten" experimental deep-sea observatory site in the Arctic Ocean. (D) A freefall deep-sea lander on board the AWI research vessel *Polarstern*, about to be deployed in the Arctic Ocean. Such automated landers can be equipped with sensors for profiling chemical and physical parameters, or with incubation chambers for *in situ* experiments. (Images courtesy of (A) The Hidden Ocean, Arctic 2005 Exploration, National Oceanic and Atmospheric Administration, USA; (B) Colin Munn; (C) and (D) Michael Klages/Alfred Wegener Institute.)

the most widely used systems is a circular rosette wheel equipped with sensors for conductivity, temperature and depth (CTD) so that the water properties can be monitored in real time. The wheel contains as many as 24 Niskin bottles that are closed selectively at specific points (*Figure 2.1A*). For some investigations, it may be necessary to collect several hundred liters of water, which can be concentrated by filtration or centrifugation (*Figure 2.1B*).

For collecting sediment samples, various types of grabs and corers are used. Corers bore a large tube into the sediment and are sealed with caps to keep the sample intact (*Figure 2.1C*). Small samples of coastal mud can be collected simply with a cutoff syringe, but special designs are needed for operation in deep waters and for collecting sands or loose sediments.

Exploration of the deep sea is extremely costly and is possible only with manned or remotely operated submersible vessels or landers (*Figure 2.1D*). The most famous of these manned submersible vehicles is *Alvin*, owned by the US Navy and operated by the Woods Hole Oceanographic Institute (*Plate 2.3*). *Alvin* has been rebuilt numerous times and made nearly 4500 deep-sea dives since its first launch in 1964. A typical dive can take two scientists and a pilot down to 4500 m, taking about two hours for *Alvin* to reach the seafloor and another two to return to the surface, with four hours of carefully planned recording and sampling work on the seafloor. Viewing ports and video cameras allow direct observation so that specific samples can be taken. A new manned vessel that will be able to reach over 99% of the ocean floor is planned. The Deepstar project in Japan uses a variety of submersible vessels and has made particular advances in the collection of deep-sea microbes under high pressure for biotechnological investigations. Other deep-sea submersibles include the *Nautile* operated by IFREMER in France, *MIR-1* and *MIR-2* operated by the Shirshov Institute of Oceanology in Russia, and the *Johnson Sea Link* vessel operated by the Harbor Branch Oceanographic

Institution, Florida. Special technology is needed to collect samples, especially from abyssal depths at great pressure and from high-temperature environments near hydrothermal vents. Other investigations are carried out using remotely operated vehicles (ROVs), which are usually equipped with camera and lights and linked by an "umbilical cable" to a surface vessel for control by a pilot. Improvements in battery technology and electronic systems is leading to the development of fully autonomous ROVs (AUVs or "rovers") that can travel for several months along preplanned routes either in the water column or on the seabed. ROVs can be equipped with a range of measuring instruments and sampling devices.

Mesocosm experiments attempt to simulate natural conditions

Carrying out controlled experiments is essential in order to understand the effects of factors such as nutrient additions, light, and temperature on community dynamics and microbial processes in the marine environment. Laboratory investigation of microbial processes in water uses "bottle experiments," in which volumes of water up to a few liters can be handled. Use of these techniques has led to great advances in our understanding of phototrophic and heterotrophic processes, for example by comparison of metabolic activities of samples incubated under light and dark conditions or the effects of temperature and nutrient additions. Such small-scale experiments are a vital first stage in the study of microbial processes in marine samples, but extending studies to the natural environment is difficult. Great care is needed to ensure that vessels used in these experiments are sterile and free of chemical contaminants. Adsorption effects on the walls of containers can result in the local concentration of nutrients and affect the composition and metabolic properties of the community. In general, the smaller the sample, the more likely it is that significant distortions due to these "bottle effects" will occur.

One way to overcome these problems is to use mesocosms—containers holding large volumes of seawater under controlled conditions in an attempt to simulate natural sea environments. Large tanks containing coastal seawater pumped in at controlled rates can be used in shore-based experimental stations. However, a more successful approach is the use of large enclosures constructed of polythene reinforced with vinyl and nylon—essentially, these are very big and very strong plastic bags immersed in the sea. The materials must be chosen carefully to simulate natural conditions, especially temperature and irradiances, as closely as possible. For example, the European Large Scale Facility for Marine Research, based at the University of Bergen, Norway, has operated a number of these mesocosm bags—each about 4 m deep and 2 m in diameter—suspended at various depths in the relatively calm, deep waters of a fjord (*Figure 2.2*). The bags are filled by pumping in unfiltered fjord water so that natural communities of phytoplankton, zooplankton, bacteria, and viruses are introduced. *Research Focus Box 7.1* describes examples of the experimental use of such mesocosms for the study of the effect of viruses on phytoplankton bloom dynamics.

Generally, it is impractical to conduct experiments in the open sea because of the continuous turbulence and dispersion of water and the organisms it contains. There are also considerable logistical problems in conducting research at sites sufficiently far from the continental shelf to avoid land effects. However, a notable exception is the series of experiments carried out to investigate the effect of iron additions on phytoplankton composition and productivity. The IRONEX experiments in the Pacific Ocean (1993, 1995) and the SOIREE (1999), SOFeX (2002), and LOHAFEX (2009) experiments conducted in the Southern Ocean involved the release of iron over areas up to 100 km^2. These experiments were possible because of the deployment of an inert tracer compound (sulfur hexafluoride, SF$_6$). The dispersion of SF$_6$ can be measured analytically, thus establishing the degree of dilution of iron in the water mass under study. The significance of

Figure 2.2 Mesocosm enclosures in a deep-water fjord at the Marine Biological Station, University of Bergen, Norway. The mesocosms are plastic containers, each with a volume of 4 m³ and equipped with mixers, pumps and sampling points. These have been used for many experiments on the effects of nutrients and CO_2 levels on phytoplankton, marine bacteria and viruses. (Image courtesy of Jean-Pierre Gattuso, CNRS, Laboratoire d'Océanographie de Villefranche, France.)

these experiments and plans for further large-scale investigations involving measurement of microbial parameters are discussed in *Research Focus Box 8.1*. There have been some attempts to develop mesocosm containers for use in open ocean conditions, although there are major technical difficulties in the construction of enclosures strong enough to withstand waves and currents.

Microelectrodes and biosensors are used to measure environmental changes

Special electrochemical microelectrodes constructed of glass and metal with tip diameters of 10 µm or less can be inserted using micromanipulators into certain habitats to measure changes in pH, oxygen, carbon dioxide, hydrogen, or hydrogen sulfide. Microsensors to detect small changes in light and temperature can also be constructed. These probes can be left in place for long periods, and when linked to recorders they provide continuous data about environmental changes and the rates of microbial and chemical processes. For example, the development and use of oxygen, sulfide, and pH microsensors made it possible to analyze the gradients of these factors in sediments and biofilms, which was used to show how sulfur bacteria (see p. 109) live in opposing gradients of sulfide and oxygen with only very small concentrations of both chemical species being present in the oxidation zone. Microsensor techniques are used to investigate oxygenic photosynthesis at a spatial resolution of less than 0.1 mm in biofilms and sediments. Recent innovations have included the development of microscale biosensors suitable for use in seawater and marine sediments. Here, biological components such as enzymes or antibodies are coupled to a signal converter so that the results of a reaction are amplified to produce an electronic reading. The development of such biosensors for various inorganic nitrogen species has been a major advance, permitting detailed study of the processes of nitrification and denitrification. Biosensors for methane and volatile fatty acids have recently been used to probe microbial transformations in anaerobic sediments and planktonic aggregates such as marine snow particles. Studies using oxygen microelectrodes have shown that even the tiniest aggregates of marine snow (see *Figure 1.9*) contain anaerobic niches, but it is not yet known if anaerobes such as methanogens grow in these ephemeral habitats.

Isotopes are used to study microbial transformations of compounds

One of the earliest applications of radioactive isotopes in marine microbial ecology was the introduction of ^{14}C-labeled bicarbonate into bottle experiments to measure rates of photosynthesis by picoplankton. Pioneers in the development of isotope incorporation methods in the late 1970s include David Karl and Jed Fuhrman, then working at the Scripps Institute of Oceanography. Fixation of the labeled precursor is detected by measuring incorporation of radioactivity into cells in a scintillation counter, or by autoradiography. Bacterial production in seawater samples or pure cultures is measured by the incorporation of [^{3}H]-adenine or [^{3}H]-thymidine into RNA or DNA, or by incorporation of labeled amino acids (e.g. [^{3}H]-leucine) into proteins. The flux of organic material in seawater can be measured using [^{14}C]-glucose. Many other microbial processes can be measured using appropriate isotopes. Appropriate correction factors for the discrimination of enzyme processes against the heavier radioisotopes must be included in calculations.

Stable isotopes are different forms of an element that are not radioactive but are acted on differentially during biological and chemical processes. For example, carbon in nature exists predominantly as ^{12}C, but a small amount exists as ^{13}C. Similarly, sulfur exists mainly as ^{32}S, with a small proportion as ^{34}S. Enzyme reactions tend to use the lighter isotope preferentially; so, for example, cells that have fixed CO_2 tend to have a higher ratio of ^{12}C:^{13}C than exists in nonbiological processes. Similarly, H_2S produced by sulfate-reducing bacteria has a higher proportion of the lighter isotope ^{32}S. Measuring the ratios of stable isotopes is thus a valuable way of distinguishing microbial (enzyme-catalyzed) processes from purely geochemical ones.

NanoSIMS (secondary ion mass spectrometry) is a recent exciting development that enables elemental transformations to be monitored at subcellular resolution. A beam of Cs^+ or O^- (the primary ion) is focused onto the surface of a sample with a resolution of 50 nm or less. The ions cause "sputtering" of a few atomic layers from the surface, and clusters of atoms are ejected—some of them become spontaneously ionized. The composition of these secondary ions is characteristic of the composition of the analyzed area. The secondary ion beam is mass filtered by a magnetic device, resulting in separation of up to seven different types of ion, each of which can be measured independently. NanoSIMS is proving especially useful in visualizing transformations such as nitrogen fixation or exchanges of carbon within cells or consortia of microbes (see *Plates 2.5* and *5.1*).

DIRECT OBSERVATION AND ENUMERATION OF MICROBES

Light and electron microscopy are used to study morphology and structure of microbes

The study of microbes began with the invention of the light microscope, and this remains an important method for the initial examination of plankton samples and cultured microbes. Eukaryotic organisms such as diatoms, dinoflagellates, and ciliates are large enough for microscopy to be useful in distinguishing morphological and structural features—in these groups, microscopic appearance is usually the main criterion for classification (for examples, see *Figures 6.2* and *6.4*). Direct light microscopy is also used for enumeration of eukaryotic plankton in seawater samples. Light microscopy reveals little more than the general shape and morphology of many smaller microbial cells, although the use of special dyes and illumination techniques can improve the amount of information revealed. However, the wavelength of visible light limits the effective magnification of the light microscope to about 1000–1500 times and it is not possible to resolve objects or structures smaller than about 200 nm. High-resolution light microscopy linked to digital recording equipment is important for the study of microbial movement and predation of bacteria by protists.

The development of the electron microscope enabled the study of the ultrastructure of cells and viruses. In the transmission electron microscope (TEM), a beam of electrons is focused onto an ultrathin section of the specimen in a vacuum, usually after staining with lead or uranium salts. Electrons are scattered as they pass through the specimen according to different densities of material in the cell. Because the wavelength of an electron beam is much smaller than that of light, the TEM has an effective magnification up to 1 million times and objects as small as 0.5 nm can be resolved (for example, see *Figure 4.6*). Various techniques such as shadowing, freeze-etching, and negative staining are used to visualize the membranes, internal structures, and surface appendages of cells. However, it must always be borne in mind that TEM images are only obtained after staining and fixing the cells and observing them in a vacuum. Therefore, the appearance of structures may not reflect their actual organization in the living cell. In the scanning electron microscope (SEM), a very fine electron beam scans the surface of the object, generating a three-dimensional image. SEM is, therefore, particularly useful for studying the structure of cell surfaces (for examples, see *Figures 1.3* and *6.5B*).

Exciting modern developments of electron microscopy include atomic force microscopy, in which a tiny probe is held in place very close to the surface of an object using weak atomic repulsion forces. Effective magnifications up to 100 million are possible and the atomic structure of molecules such as DNA or proteins can be visualized. A major advantage of this technique is that specimens do not need to be fixed or stained and can be examined in the living state. This technique has recently been used to reveal complex surface architectures at the nanometer to micrometer scale which connect planktonic microbes in microscale networks (*Plate 2.4*).

Epifluorescence light microscopy enables enumeration of marine microbes

Fluorescence occurs when material absorbs light at one wavelength (the excitation or absorption spectrum) and then re-emits it at a different wavelength (the emission spectrum). The specimen is illuminated with a tungsten-halogen or mercury vapor lamp. In direct fluorescence microscopy, a filter is placed between the light source and the specimen, allowing only light of the desired excitation wavelength to be transmitted, whilst a barrier filter placed between the specimen and the eyepiece transmits the emitted fluorescence and absorbs longer wavelengths. Epifluorescence microscopy depends on the use of dichroic mirrors as interference filters that transmit one set of wavelengths and reflect the others. Water samples are usually fixed with formalin or glutaraldehyde and filtered through membrane filters of a pore size appropriate to the group under study—0.22 µm filters are most commonly used for bacteria. For enumeration of viruses, larger particles are removed by prefiltering before trapping the viruses on 0.02 µm aluminum oxide filters. This method can therefore be used for observation and enumeration of all groups of marine microbes. The original stain (fluorochrome) used in plankton studies was acridine orange (AO, 3,6-bis[dimethylamino]acridinium chloride), which binds to DNA and RNA. Problems arise with AO owing to background fluorescence and difficulties in distinguishing microbes from inanimate particles, so the most widely used stain today is DAPI (4′,6′-diamido-2-phenylindole), which binds to DNA and largely overcomes these problems. DAPI is excited by ultraviolet light and emits bright blue light. The use of incident light also permits observation of microbes attached to suspended particles. Although viruses are below the limits of resolution of the light microscope, they may be visualized if stained with SYBR Green®, which emits a very bright fluorescence (*Plate 7.1*).

Some staining methods can distinguish between living and dead cells, based on differences in permeability. CTC (5-cyano-2,3-dilotyl tetrazolium chloride) is a fluorogenic redox dye that detects an active electron transport chain. Although there have been some criticisms of the technique, the CTC assay has been used to estimate the distribution of cell-specific metabolic activity in natural assemblages of marine bacteria. The Live-Dead® kit is a widely used test for bacterial viability; a mixture of SYTO 9® and propidium iodide stains bacteria fluorescent green if they have intact cell membrane, whereas bacteria with damaged membranes appear fluorescent red.

Confocal laser scanning microscopy enables recognition of living microbes within their habitat

Confocal microscopy uses a laser light source coupled to an optical microscope and computer-aided digital imaging system. The beam focuses on a narrow plane of the specimen. The advantage of confocal microscopy lies in its ability to be used on living specimens and to generate a three-dimensional image. Because the magnification is much less than that of electron microscopy, it is not as useful for revealing ultrastructural detail, but it is proving enormously useful in ecological studies of microbial communities, especially when combined with fluorescent *in situ* hybridization (FISH) techniques (see *Figures 2.8* and *Plate 5.1*), which enables specific microbes to be located on surfaces or within other organisms. In particular, understanding of the structure of biofilms and microbial mats has advanced considerably using this technique.

Flow cytometry measures the number and size of particles

Flow cytometry was originally developed for biomedical uses and was first applied to marine studies in the late 1970s. Since then, use of this technique has produced spectacular advances in enumeration and characterization of ocean microbes. A flow cytometer is an instrument that can separate and sort cells according to a specific fluorescence "signature," leading to a wide range of physiological and ecological information. Flow cytometry simultaneously measures and analyzes multiple physical characteristics of single particles, ranging in size from 0.2 to 150 μm, as they flow in a fluid stream through a beam of light. The coupling of the optical detection system to an electronic analyzer records how the particle scatters incident laser light and emits fluorescence. A flow cytometer is made up of three main components (*Figure 2.3A*).

THE "JELLYFISH GENE" IS A MAJOR TOOL IN MOLECULAR BIOLOGY

Green fluorescent protein (GFP) is produced in the jellyfish *Aequorea victoria* and has very wide applications in all areas of biology. Its discovery and application by Tsien, Chalfie, and Shimomura led to the award of Nobel Prize in 2008. The *gfp* gene can be easily inserted into the genome of many organisms and used as a marker of the expression of specific genes. Some examples of its use in marine microbiology include following the fate of *gfp*-marked bacteria after their ingestion by grazing protists or the process of infection of animal hosts by symbiotic or pathogenic bacteria. If specific genes are tagged, it is possible to visualize when they are "switched on" during the development of a microbial process.

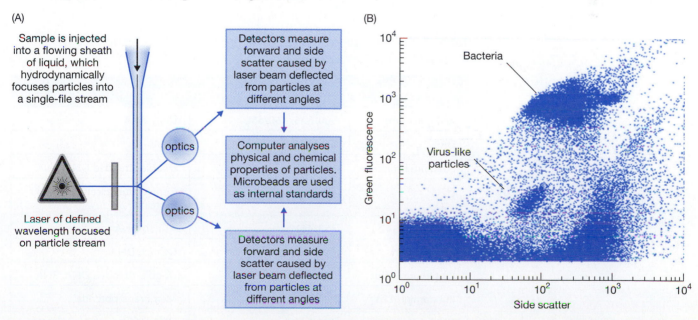

Figure 2.3 Flow cytometry. (A) Schematic diagram of the components of a flow cytometer. (B) A typical flow cytometry plot, in this case showing the separation of virus-like particles from bacteria in a culture of *Symbiodinum* after treatment with ultraviolet radiation. (Image courtesy of Jayme Lohr, PhD thesis, University of Plymouth, 2007.)

ⓘ DISCOVERY OF THE PLANET'S MOST ABUNDANT PHOTOSYNTHETIC ORGANISM

In 1988, Sallie Chisholm and Rob Olsen, then of Woods Hole Laboratory, were among the first scientists to use a portable flow cytometer on a research vessel. They were studying the cyanobacterium *Synechococcus* when they noticed unusual fluorescence signals from some very tiny cells present in huge numbers. These were subsequently identified as the cyanobacterium *Prochlorococcus*. This is now considered to be the commonest photosynthetic organism on Earth, responsible for production of at least 20% of atmospheric oxygen. It is remarkable that such an abundant organism has only been known for a little over 20 years.

The fluidics system injects particles into the instrument in a stream of liquid so that they pass in single file through a laser beam. The optics system consists of lasers—there may be four or more different wavelengths—to illuminate the particles in the sample stream and optical filters to direct the resulting light signals to the appropriate detectors. Each particle scatters light at different angles and may also emit fluorescence. The scattered and emitted light signals are converted to electronic pulses that can be processed by the electronics system. In a fluorescent-activated cell sorter (FACS), the electronics system can also transfer a charge and deflect particles with certain characteristics, so that different cell populations can be separated and collected. Flow cytometry is used for a range of specific measurements in marine microbiology, including quantifying different microbial groups and their diversity, estimation of cell dimensions and volume, determination of DNA content, and assessment of viability. Flow cytometry was first used for the analysis of phytoplankton cells, as they are naturally labeled with photosynthetic pigments such as chlorophyll and phycobilins, which are autofluorescent. However, all microbes can be detected in flow cytometry by using the same principles described for epifluorescence microscopy— "tagging" cells with fluorochromes that emit light of various wavelengths when illuminated by the appropriate laser. Flow cytometry has thus been particularly valuable in the analysis of bacteria and small eukaryotes in the picoplankton, especially following the development of portable instruments with stable optical and electronic systems that can be deployed on ships for direct examination of water samples. Recent improvements in the sensitivity of flow cytometry instruments and the introduction of new dyes mean that flow cytometry can now be used for detection and quantification of marine viruses. *Table 2.1* lists examples of fluorochromes used in flow cytometry and epifluorescence microscopy. Using an appropriate mixture of different fluorochromes and laser light of different wavelengths, different populations of microbes in aquatic ecosystems can be analyzed based on size, abundance, and specific properties. Exciting new advances in flow cytometry have been possible owing to a fusion with FISH technology (*Figure 2.8*), in which fluorescently labeled oligonucleotides can be used to discriminate specific taxa in heterogeneous natural marine microbial communities.

Table 2.1 Some representative fluorochromes used in epifluorescence microscopy and flow cytometry

Fluorochrome	Excitation/emission (nm)	Target
Acridine orange	500/526; 460/650	DNA, RNA
DAPI	358/461	DNA
Ethidium bromide	518/605	Double-stranded DNA, RNA
FITC	495/520	General fluorescent label, proteins
Hoechst 33342	350/461	AT-rich DNA
Mithramycin	425/550	GC-rich DNA
Propidium iodide	535/617	Double-stranded DNA, RNA
Rhodamine 123	480/540	Membrane potential
SYBR green®	494/521	DNA, can be used for viruses

AT, adenine and thymine; DAPI, 4',6'-diamido-2-phenylindole; FITC, fluorescein isothiocyanate; GC, guanine and cytosine.

CULTURE-BASED METHODS FOR ISOLATION AND IDENTIFICATION OF MICROBES

Different microbes require specific culture media and conditions for growth

There is a large discrepancy between the number of microbes that can be observed in direct counts of marine samples and those that can be cultured in the laboratory. Even those organisms that can be successfully cultured differ enormously in their physiological properties (see Chapter 3), so the composition of the culture medium and environmental conditions (temperature, pH, atmospheric conditions and pressure) must be chosen with great care.

Many autotrophic bacteria can be grown in a simple defined synthetic medium containing bicarbonate, nitrate or ammonium salts, sulfates, phosphates, and trace metals. Many heterotrophs can be grown on a similar defined medium with the addition of appropriate organic substrates such as sugars, organic acids, or amino acids. However, isolation and routine culture are often carried out using complex media made from semi-defined ingredients such as peptone (a proteolytic digest of meat or vegetable protein, which provides amino acids) and yeast extract (which provides amino acids, purines, pyrimidines, and vitamins). Many heterotrophs are more fastidious and require the addition of specific growth substances. An example of a widely used medium for routine culture of many marine bacteria is Zobell's 2216E medium, which contains low concentrations of peptone and yeast extract plus a mixture of various salts. It must be stressed that there is no single medium suitable for all types, and microbiologists employ a very wide range of recipes for different purposes. As noted in *Research Focus Box 2.1*, most marine bacteria have not yet been cultured.

Bacteria may be grown in liquid media (broth) or on media solidified with agar. Cultivation on agar plates (Petri dishes) permits samples to be streaked out to obtain single colonies, which can then be restreaked to form pure cultures. In the shake tube and roll tube techniques, samples are mixed with molten agar, which allows culture of some anaerobic bacteria. Agar is a polysaccharide obtained from particular types of seaweed and has certain advantages for microbiological work. It is colorless, transparent, and remains solid at normal incubation temperatures, but can be held in a molten state above 42°C. Its main advantage in microbiology is that it does not normally affect the nutritional status of the medium because most bacteria and fungi do not possess extracellular enzymes for its degradation. However, not surprisingly, many marine microorganisms can degrade this algal compound by production of agarase enzymes. When marine samples are plated onto agar, craters are frequently observed surrounding the colonies as a result of enzymic digestion of the agar.

The concentration of bacteria in a sample can be determined by plating appropriate tenfold dilutions on agar plates, either by mixing with molten agar (pour plate), spreading evenly over the surface of the agar (spread plate) or by pipetting small drops onto the surface (drop count or Miles–Misra technique). Colonies are counted after incubation at an appropriate temperature and the viable count is expressed as colony-forming units (CFU) per milliliter. When plating samples from natural sources, it is important to check plates at various intervals as slow-growing colonies may take days or weeks to develop and may be obscured by rapidly growing types. Selective media contain antibiotics or other chemicals that inhibit the growth of certain microorganisms, thus allowing only those of interest to multiply. They may also include dyes or chromogenic substances that give a differential reaction when metabolized by bacteria belonging to particular groups. These media are used as an enrichment method in liquid culture or to select colonies of the desired type on agar plates. In marine microbiology, selective media are used mainly for the isolation and growth of animal and human pathogens and nonindigenous indicators of pollution. These media frequently contain bile or synthetic detergents, which select for enteric bacteria and related groups. For example, the medium thiosulfate-citrate-bile-sucrose (TCBS) agar is widely used for isolation of vibrios, whilst lauryl sulfate broth is used in the enumeration of coliform bacteria in assessing fecal pollution of seawater (*Figure 13.4*). Selective antibiotics may be added to suppress the growth of bacteria on fungal isolation plates.

As well as plating, successive dilution in liquid media can be used to separate individual bacteria to obtain pure cultures. This method has been used successfully to grow previously nonculturable bacteria in seawater without the addition of nutrients (see *Research Focus Box 2.1*). The dilution-to-extinction method is also used in the most probable number (MPN) technique for quantification of microbial populations. One widely used application of this is in the determination of coliforms as indicators of fecal pollution in seawater or shellfish (*Figure 13.5*).

Pure culture methods for the cultivation of fungi and protists follow the same basic principles as those for bacteria. In general, photosynthetic microalgae are easier to maintain in culture, requiring mineral salts solution and appropriate conditions of temperature and light/dark cycles. However, the cultures can be very difficult to maintain, and there are numerous recipes for special media for growing and maintaining different species. Culture media can be made of defined chemicals or from natural seawater enriched with extracts from various sources. It is relatively easy to isolate single microalgae by streaking or making serial dilutions to obtain a clonal population. Because of their large size, individual cells may be isolated using a fine pipette. Phagotrophic dinoflagellates and other heterotrophic protists are more difficult to culture, and established cultures of only a few types are available. Because the nutritional requirements can vary so widely, considerable experimentation may be needed to optimize conditions for isolation and maintenance of different types. For many protists, especially diatoms and dinoflagellates, it is very difficult to obtain axenic cultures, that is cultures that are free of bacteria, fungi, or other protists. Axenic cultures can sometimes be obtained by extensive rinsing of the isolated protists, or treatment with antibiotics (see *Research Focus Box 12.3*).

A recent advance in pure culture methods allows the isolation of individual cells viewed under a microscope, which can be trapped using a laser beam and separated off for subsequent pure culture. Such "optical tweezers" provide a very powerful technique when used in conjunction with gene probes or antibody staining.

Enrichment culture selects for microbes with specific growth requirements

Enrichment culture depends on the provision of particular nutrients and incubation conditions to select for a certain group. For example, the addition of cellulose or chitin plus ammonium salts to a marine sample under aerobic conditions will enrich for *Cytophaga*-like bacteria, whilst sulfates with acetic or propionic acid under strictly anaerobic conditions will favor growth of *Desulfovibrio* and related types. The Winogradsky column is especially useful in investigating communities in marine sediments. A tall glass tube is filled with mud and seawater and various substrates added prior to incubation under appropriate conditions. A gradient of anaerobic to aerobic conditions will develop across the sediment and a succession of microbial types will be enriched. In the light, microalgae and cyanobacteria will grow in the upper parts of the column and will generate aerobic conditions by the production of oxygen. Depending on the substrate added, the activity of various groups will lead to the production of organic acids, alcohols, and hydrogen, which favor the growth of sulfate-reducing bacteria. In turn, the sulfide produced leads to the growth of anaerobic phototrophs that use hydrogen sulfide as an electron acceptor. A wide range of enrichment conditions may be adapted for particular groups, and the resulting communities may then

BOX 2.1   RESEARCH FOCUS

Culturing the uncultured

New methods allow culture of marine microbes known only by their genes

As discussed in Chapters 4 and 5, our knowledge of microbial diversity in the oceans has been revolutionized by the application of techniques, especially sequencing of 16S rRNA genes, which allow recognition and relatedness of groups of organisms based solely on their genetic sequence. We have little understanding of the properties of the great majority of microbes that inhabit our planet—many major divisions of the *Bacteria* and *Archaea* contain no known cultured species, so we often have no idea what physiological or biochemical properties these possess and the roles that they play in the ecosystem. As the number of previously unknown organisms identified by molecular analysis of the environment increased in the 1990s, many microbial ecologists assumed that these organisms are inherently unculturable and that it was futile to attempt to grow them. In the excitement generated by the rapidly accumulating results produced by new techniques, some molecular biologists were vociferous in their dismissal of the traditional methods applied to the culture of microbes in the laboratory, considering them "old-fashioned" and of little further use. As pointed out by Donachie et al. (2007), this attitude ignores important lessons from the history of microbiology; Beijerinck, Winogradsky and other pioneers of general microbiology used numerous approaches to overcome the difficult problems of cultivation of "unculturable" organisms more than 100 years ago. Provided sufficient effort and ingenuity are applied, there is every reason to be optimistic that we will be able to culture many more marine microbes. So, we should regard these organisms as "not yet cultured" rather than "unculturable."

The most likely reason that so many marine bacteria are so difficult to grow is that many are extreme oligotrophs, adapted to growing very slowly at low cell densities with very low nutrient concentrations. It is very difficult to reproduce these conditions in the laboratory, especially if we try to culture organisms quickly at thousands of times the density at which they normally exist. Common media used for the culture of marine heterotrophic bacteria typically contain nutrients at levels 10^6–10^9 times higher than are found in seawater. Button et al. (1993) developed techniques for culturing slow-growing marine bacteria by making successive dilutions of natural seawater in filtered, sterilized seawater, so that only one or two bacteria per tube are obtained. This leads to a successful enrichment of the dominant cell types, which grow to densities of about 10^5–10^6 ml^{-1}, similar to those occurring in seawater. Schut et al. (1997) used this technique to isolate an oligotrophic *Sphingomonas* sp. This approach was developed by Stephen Giovannoni and colleagues at Oregon State University in order to culture representatives of the SAR11 clade, which had been identified as the most abundant ocean bacteria (Morris et al., 2002). Rappé et al. (2002) developed a high-throughput method in microtiter trays containing sterile seawater supplemented with phosphate, ammonia, or low levels of organic compounds. The usual methods of determining cell growth (e.g. measuring turbidity) are useless at such low densities, so novel approaches such as microarray and FISH techniques were used to screen cultures. Bringing SAR11 into culture allowed the physiological properties of this ubiquitous organism to be investigated for the first time and enabled determination of the genome sequence (Giovannoni et al., 2005). The cultured isolate was named *Candidatus* Pelagibacter ubique. (See Glossary for definition of the *Candidatus* status.) Connon and Giovannoni (2002) then used this method to culture additional bacteria from groups previously regarded as unculturable. Cultivation of more strains of SAR11 proved difficult until the method was further modified by using ultraclean techniques and Teflon culture vessels (Giovannoni and Stingl, 2007). With these modifications, Stingl et al. (2007) isolated numerous additional strains of SAR11 and other *Alphaproteobacteria* from the Oregon coast and the Sargasso Sea.

Zengler et al. (2002) used a different approach, which involved encapsulation of single bacteria in microdroplets of gel to allow large-scale culture of individual cells. Concentrated seawater was emulsified in molten agarose to form very small droplets, and flow cytometry was used to sort droplets containing single bacteria. The droplets were transferred to growth columns (which filter out contaminating free bacteria) and various media and growth conditions were evaluated. Microbial diversity in the microcolonies within the droplets was examined using 16S rRNA gene typing. When organic media were used, the bacteria grown by this method resembled previously cultivated bacteria, but when unsupplemented filtered seawater was used as a growth substrate, a broad range of bacteria, including previously uncultured types, were isolated and grown. This technique permits physiological studies to be carried out on cultures with as few as 100 cells, and further development will have major implications for microbial ecology and biotechnology (Zengler, 2009).

Why, in the age of metagenomics, is it still so important to apply so much effort to cultivate microorganisms? Clearly, cultivation-independent and cultivation-dependent approaches to microbial community assessment are complementary. Metagenomic analysis of microbial communities relies heavily on data obtained from the sequencing of genomes of cultivated species. Study of the growth, physiology, and metabolism of live bacteria in culture allows the properties of the whole organism to be determined and related to field observations. This can also lead to many new opportunities for biotechnology (Keller and Zengler, 2004). Giovannoni et al. (2007) notes that "cultures often offer the best possibilities to test hypotheses that emerge from genome sequences," whilst Nichols (2007) argues that "novel cultivation-dependent approaches, if married to genomics, will likely provide the best ecological context with which to interpret the microbial world."

be subject to further enrichment or selective plating to obtain pure cultures. The results of enrichment cultures must be interpreted carefully because they often lead to overrepresentation of groups that may not be the dominant member of the natural community.

Phenotypic testing is used for identification and detailed characterization of many cultured bacteria

The initial characterization of cultures often begins with preparation of a Gram stain and observation of morphology under the microscope, but this does not provide sufficient information for identification and taxonomy. Bacteria that have been isolated in culture can be identified and characterized using a combination of growth characteristics and tests for the production of particular enzyme activities, particularly those involved in carbohydrate and amino acid metabolism. *Table 2.2* shows a summary of the methods used for bacterial identification and classification. Once it is certain that a pure culture has been obtained, it is usually easier to use 16S rRNA gene sequencing (see p. 40) to place a new isolate into a particular phylogenetic group, to define which phenotypic tests should be performed. Of course, phenotypic characterization based on biochemical test methods is suitable only for the small proportion of heterotrophic marine bacteria that can be easily cultured. Diagnostic tables and keys aid in the identification of unknown bacteria, although the inherent strain-to-strain variation in individual characteristics often causes problems for accurate identification and taxonomy. Commercial kits employing a battery of tests (e.g. API® or Micro-Bact®) offer advantages for standardization of methods and processing of results with the aid of databases developed using numerical taxonomic principles. However, most biochemical methods were developed for the identification of bacteria of medical importance and require adaptation for examination of marine environmental samples. These methods are most useful in areas such as identification of pathogens or fish spoilage organisms. For example, there are extensive databases and established diagnostic keys based on biochemical tests for the identification of *Vibrio* and related pathogens in aquaculture. Some tests are very characteristic of particular groups. For example, the detection of the enzymes β-galactosidase and β-glucuronidase using fluorogenic substrates is important in the identification of coliforms and *E. coli* in polluted waters (see *Figure 13.4*). The BIOLOG® system differentiates microbial taxa based on their nutritional requirements and incorporates up to 95 different substrates as a sole source of carbon or nitrogen, together with a tetrazolium salt. Different microbial groups use a range of substrates at characteristic levels and rates. Positive reactions are tested via a color reaction based on reduction of the tetrazolium salt. BIOLOG® systems have been developed specifically for Gram-positive bacteria, Gram-negative bacteria, and yeasts. Again, this method is not suitable for detection of nonculturable microorganisms, but it provides a less stressful environment than the surface of solid growth media and some investigators have used the BIOLOG® system for community characterization of metabolic activities in sediments or surface layers by direct inoculation without an intervening culture step.

Analysis of microbial components can be used for bacterial classification and identification

Analysis of the composition of bacterial membranes via gas chromatography of fatty acid methyl esters (FAME) is a powerful technique that can detect very small differences between species and strains. Individual taxa have a distinct fatty acid fingerprint. However, careful standardization of growth conditions is needed because membrane composition is strongly affected by environmental conditions and the nature of the culture medium. Certain taxonomic groups of bacteria and fungi can be identified by the presence of specific biomarker fatty acids (e.g. isoprenoid quinones) and other molecules, and this is used to monitor changes in microbial community composition as a result of environmental changes such as the introduction of pollutants into marine systems.

Table 2.2 Summary of techniques used for the identification and classification of cultured bacteria[a]

Technique	Usefulness at different taxonomic levels
Nucleic acid techniques	
DNA sequencing	All levels, for phylogenetic analysis
DNA–DNA hybridization	All levels above species, 70% hybridization level used for species definition
GC ratios	Comparison of species known to be closely related
PCR-based fingerprinting (e.g. RAPD, AFLP, rep-PCR)	Good discrimination at strain level
RFLP, ribotyping	Strain differentiation, but prone to variability
MLST	Reliable and robust; good discrimination of species and strains
16S rRNA gene sequencing	Higher taxonomic levels, for phylogenetic analysis
ITS region sequencing	Strain differentiation
Phenotypic characters	
Bacteriophage typing	Strain differentiation, highly specific
Biochemical tests (e.g. API® system, BIOLOG Phenotypic Microarrays™)	Species and strain differentiation, routine identification, community analysis
Morphology	Limited information except in some groups
Plasmid and protein profiles	Strain differentiation but highly variable
Serology	Strain (serotype) differentiation
Chemotaxonomic markers	
FAME	Rapid typing to genus level; further discrimination may be possible with careful standardization
PyMS	Rapid typing to genus level; further discrimination may be possible with careful standardization
Quinones and other biomarkers	Good differentiation above species level

[a]Many of these methods can also be used for fungi and protists. AFLP, amplified fragment length polymorphism; FAME, fatty acid methyl ester analysis; GC ratio, guanine–cytosine base pair ratio; ITS, internal transcribed spacer; MLST, multilocus sequence typing; PCR, polymerase chain reaction; PyMS, pyrolysis mass spectroscopy; RAPD, random amplification of polymorphic DNA; RFLP, restriction fragment length polymorphism.

Analysis of the protein profile of microbes can also be used for comparison of microbial species. Extracted proteins are dissociated with the detergent sodium dodecyl sulfate and separated by polyacrylamide gel electrophoresis (SDS-PAGE). However, gene expression and the resulting protein pattern produced are extremely dependent on environmental conditions and the technique needs to be used with caution in identification and taxonomy. Indeed, its main use is in assessing the effect of various factors on the synthesis of particular proteins, such as bacterial colonization or virulence factors during infection of a host.

Pyrolysis mass spectrometry (PyMS) generates a chemical fingerprint of the whole microbe by thermal degradation of complex organic material in a vacuum to generate a mixture of low-molecular-weight organic compounds that are separated by mass spectrometry. PyMS is easily automated and a high throughput of samples can be processed, but the equipment is expensive and is found only in a few specialized laboratories.

NUCLEIC-ACID-BASED METHODS

The use of nucleic-acid-based methods has had a major impact on the study of marine microbial diversity

Marine microbiologists use a wide range of techniques based on the isolation and analysis of nucleic acids. Many of these can be applied to the characterization of microbes in culture, for example for accurate identification and taxonomy or for diagnosis of disease, as shown in *Table 2.2*. The full genome sequence of many cultured microbes has also been determined, as shown in *Table 2.3*. However, the most dramatic advances in marine microbial ecology have occurred as a result of the recent development of "environmental genomics"—methods for the detection and identification of microbes based on direct extraction of genetic material from the environment, without the need for culture. First introduced in the mid-1980s, these culture-independent molecular techniques are very sensitive and can allow the recognition of very small numbers of specific organisms or viruses among many thousands of others. Direct analysis of the genes present in an environmental sample; or comparison of community profiles using "genetic fingerprinting" techniques allows us to make inferences about the diversity, abundance, and activity of microbes in the oceans. Since 2000, there have been rapid developments in techniques for isolation, amplification, and sequencing of DNA, together with advances in bioinformatics—the mathematical and computing approaches needed to understand the genomic data generated. It has become possible to assemble and analyze the genetic information obtained from microbial communities in the environment in a way comparable to the analysis of single genomes, leading to what we now term "metagenomics."

Sequencing of ribosomal RNA genes is the most widely used tool in studies of microbial diversity

Chapter 1 introduced the pioneering work of Woese and colleagues based on sequencing of ribosomal RNA, which led to revision of ideas about evolution of the major domains of life. The ribosome is composed of a large number of proteins and rRNA molecules, as shown in *Figure 2.4*. The rRNA molecules (16S for members of the *Bacteria* and *Archaea*; 18S for eukaryotes) in the small ribosomal subunit (SSU) quickly became the first choice for diversity studies for a number of reasons. First, the rRNAs are universally present in all organisms, because of the role of the ribosome in the essential function of protein synthesis. Second, the rRNA molecule has a complex secondary structure and mutations in parts of the gene that affect critical aspects of structure and function in the ribosome are often lethal; thus, changes in rRNA occur slowly over evolutionary time. Some parts of the rRNA molecules—and consequently the genes that encode them—are highly conserved, whilst others have a high degree of variability. Another advantage is that growing microbes often contain multiple copies of the rRNA genes, so use of the polymerase chain reaction (PCR; see below) allows them to be amplified from very small amounts of DNA.

Because PCR is carried out with primers directed against specific sequences in the SSU rRNA genes, gene fragments from different organisms can be selectively amplified from mixed DNA in a sample from the marine environment. Most environmental analyses use "universal" or broad-spectrum primers directed against short, highly conserved regions in the SSU rRNA genes and amplify the more

BREAKTHROUGH—DIRECT ISOLATION OF NUCLEIC ACIDS FROM THE SEA

Following Woese's use of rRNA for studies of relationships between organisms, parallel advances in nucleic acid sequencing technologies and computing led to the rapid generation of large databases of information linking sequences from known organisms to their taxonomic position. Because many marine microbes could not be cultured, microbial ecologists therefore asked whether it would be possible to obtain sequence information directly from environmental samples. The first application of direct sequencing of nucleic acids from the marine environment was in 1986 by Norman Pace and colleagues at the University of Colorado. Development of the polymerase chain reaction (PCR) in 1988 enabled rapid application in many different studies of diversity of ocean microbes. A number of research teams, most notably including those led by Stephen Giovannoni (Oregon State University), Ed DeLong (then at Monterey Bay Aquarium Research Institute), and Jed Fuhrman (University of Southern California), pioneered research in this field in the 1990s, providing the breakthrough that paved the way for the current importance of marine microbiology.

variable sequences between the primer annealing sites. The PCR amplification products can be cloned in *E. coli* to create a gene library. Commercial cloning vector kits are widely used; these ensure high efficiency of cloning and allow detection of transformant colonies containing DNA inserts. The most common detection system is "blue/white" screening, which is based on the detection of the enzyme β-galactosidase. This hydrolyzes a soluble, colorless substrate X-gal (5-bromo-4-chloro-3-indolyl-beta-d-galactoside) into an insoluble, blue product. If a DNA fragment is cloned into the *E. coli* vector gene encoding the β-galactosidase, production of the active enzyme is prevented, so that after transformation colonies appear white on selective agar containing X-gal. Alternatively, the PCR products can be separated by denaturing gradient gel electrophoresis (DGGE; see below). After determination of DNA sequences, data are compared to generate phylogenetic trees. A brief description of each of the main stages in this process is given in the following sections.

Although tremendous advances have been made through the use of SSU rRNA gene sequencing and this method dominates much of the discussion about diversity in this book, it should be remembered that is not the only approach and it has some major inherent limitations. The highly conserved nature of the genes and the small size of rRNA genes (1500–1800 bases) means that their useful information content is relatively low. SSU rRNA gene sequencing is best suited to comparison of higher-level microbial taxa, and there can be drawbacks when it is used to delineate genera and species. In theory, genes encoding proteins contain more information because the genetic code based on four nucleotides generates a protein sequence based on 20 amino acids. Increasingly, the sequences of genes encoding key structural proteins (e.g. protein synthesis elongation factors or enzymes) and enzymes (e.g. DNA polymerase or ATPase) are being determined for detailed phylogenetic comparisons. In addition, the use of probes directed against specific metabolic genes can provide information about the importance of particular microbial activities in the environment. For very fine discrimination between closely related lower taxonomic levels—such as between species or between strains within a species—it is often preferable to use noncoding regions between genes, such as the internal transcribed spacer (ITS) regions of nonfunctional RNA that occurs between the structural rRNA genes. These regions of DNA are not under the same constraints as functional genes and are often hypervariable. Multilocus sequence typing (MLST) is a technique for the typing of multiple

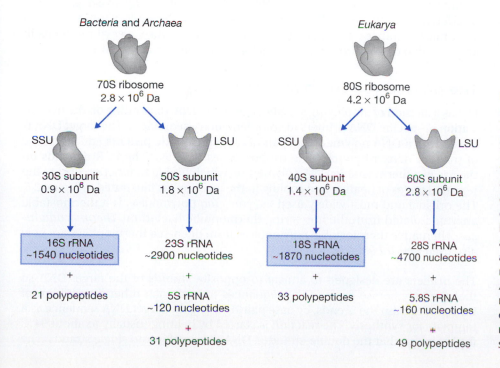

Figure 2.4 Structure of the ribosomes in the three domains. The subunits, proteins and rRNA molecules are classified by the Svedberg unit (S), a measure of the rate of sedimentation during centrifugation. Sedimentation rate depends on the mass and the shape of the particles; hence when the large (LSU) and small (SSU) subunits are measured separately, they do not add up to the S value of the intact ribosome. The 16S (*Bacteria* and *Archaea*) and 18S rRNA (*Eukarya*) molecules in the SSU are most commonly analyzed. Molecular masses of the ribosomes and their subunits are shown in daltons (Da).

? COULD GENE TRANSFER COMPLICATE OUR UNDERSTANDING OF PHYLOGENY?

A basic tenet in the use of SSU rRNA in determining phylogenetic relationships is the assumption that rRNA genes are not subject to horizontal gene transfer, because of the stringent constraints on the function of rRNA in the essential process of protein synthesis. However, it must be noted that there is a growing body of evidence that single organisms may contain heterogeneous SSU rRNA sequences. Several protist groups have been shown to contain two types of 18S rRNA genes with significant nucleotide differences (over 5%), and similar evidence of multiple 16S rRNA operons has been found in halophilic *Archaea* and in *Actinobacteria*. Since this phenomenon occurs in all three domains of life, it indicates that horizontal transfer of SSU rRNA genes may be more widespread than previously thought, and this may make the interpretation of rRNA-based phylogenetic data more complex in some groups.

loci, using 7–10 "housekeeping" genes for major metabolic processes. Internal 450–500-bp fragments are sequenced on both strands and separated by electrophoresis, providing allelic profiles that allow high discriminatory power. MLST has found particular applications for epidemiological studies of pathogenic bacteria, enabling characterization of isolates from different hosts and geographic regions.

The first step in all nucleic acid investigations involves the isolation of genomic DNA or RNA from the culture or community

Protocols for DNA or RNA extraction from cells in culture are usually straightforward and reproducible. By contrast, isolations from environmental samples are often problematic and require considerable optimization. Many methods are based on phenol-chloroform extraction or separation on solid-phase media. Commercially available kits provide advantages in speed and reproducibility, but are often not well suited to particular environmental applications. With a few PCR techniques (e.g. rep-PCR), preparation may involve simple procedures such as boiling a sample of culture, but for most methods, good-quality purified RNA or DNA is needed. Samples must usually be processed rapidly or preserved in such a way that the nucleic acids are protected from degradation (e.g. immediate freezing in liquid nitrogen or transfer to a –80°C freezer). RNA is particularly prone to degradation by nucleases, and special techniques for sample preservation and avoidance of contamination in the laboratory are needed. For extraction from planktonic microbes, it may be necessary to concentrate biomass by centrifugation or tangential flow filtration of large volumes of water. All nucleic-acid-based methods for analysis of diversity in the environment depend on the assumption that DNA is extracted in a more or less pure state equally from the different members of the community, but there are several reasons why this may not be the case. Microbes differ greatly in their sensitivity to the treatments used to break open the cells. In particular, the nature of the cell envelope varies enormously and cells may be protected by association with organic material or because they are inside the cells of other organisms as symbionts or pathogens, meaning that DNA may be extracted preferentially from some types according to the treatment used. Techniques such as freeze-thawing, bead beating (rapid shaking of the mixture with tiny glass beads), ultrasonic disintegration, or use of powerful lytic chemicals enhance the recovery of DNA, but can result in extensive degradation. Interference with the PCR often occurs because of the presence of inhibitors such as humic substances and metal ions, and these can cause special problems in amplification of DNA from samples such as sediments or animal tissue.

The polymerase chain reaction (PCR)

PCR is a method of amplifying specific regions of DNA, depending on the hybridization of specific DNA primers to complementary sequences. The target DNA is mixed with a DNA polymerase, a pair of oligonucleotide primers and a mixture of the four deoxyribonucleoside triphosphates (dNTPs). The PCR depends on the use of a thermostable DNA polymerase, which is able to function during the repeated cycles of heating and cooling in the reaction, as illustrated in *Figure 2.5*. The original and most widely used enzyme, *Taq* polymerase, is a thermostable enzyme isolated from the hot-spring thermophilic bacterium *Thermus aquaticus*. Alternative thermostable polymerases from deep-sea thermophiles are now also used (see *Table 14.3*).

The primers are designed to anneal to opposite strands of the target DNA so that the polymerase extends the sequences toward each other by addition of nucleotides from the 3′ ends, by base pairing using the target DNA sequence as a template for synthesis. The reaction is started by heating (usually to about 94°C for 1 min), so that the double-stranded DNA dissociates into single strands. The temperature is then lowered (usually to about 50–60°C for 0.5–1 min), so that the primers can attach to the complementary sequences on the DNA. The temperature is then raised to 72°C (the optimum for *Taq* polymerase), and primer extension begins. In the first stage of the reaction, the target DNA is copied for various distances until polymerization is terminated by raising the temperature again to 94°C. For the second cycle, the temperature is lowered and the other primer can now attach to the newly synthesized DNA strand. On raising the temperature again to 72°C, this strand is copied until the end of the first primer is reached. After three such cycles, the number of short duplexes containing the amplified sequence will increase exponentially; one target molecule will become 2^n molecules after n cycles. The heating and cooling cycles are carried out in specially designed electronically controlled thermal cyclers, in which the reaction tubes are placed in blocks, allowing very rapid and precise heating and cooling. Usually, 30–40 cycles take place, resulting in amplification of a single target sequence to over 250 million PCR products. The outcome of the reaction is usually checked by agarose gel electrophoresis. The DNA molecules are separated according to size and visualized under ultraviolet light after staining the gel with ethidium bromide or SYBR® green. Because of the high degree of conservation in the SSU gene, the PCR products will all have approximately the same size and should migrate as a single band.

UNDERSTANDING DIVERSITY DEPENDS ON COMBINING METHODS

The successful development of DNA-based culture-independent techniques led many molecular microbiologists to promulgate the common misconception that "cultivation methods can only describe 1% of microbial diversity." Donachie et al. (2007) provide strong evidence to show that such conclusions are misleading. They show that analysis of diversity based solely on 16S rRNA gene sequencing often overlooks a significant fraction of phylogenetic diversity that can be easily determined by cultivation methods and argue that, if we want to understand the full extent of microbial diversity, we need to include parallel culturing and molecular techniques. Whilst it is true that much 16S rRNA gene sequencing has led to the discovery of many new microbes that have not been cultured, some studies in which 16S rRNA methods were used alongside culture-based methods found that new bacterial taxa were represented in the culture libraries that were not present in the rRNA libraries. Differences in the efficiency of extraction of DNA and differential amplification of sequences in the PCR could account for these anomalies. Every method used to study microbial diversity has inherent biases and limitations.

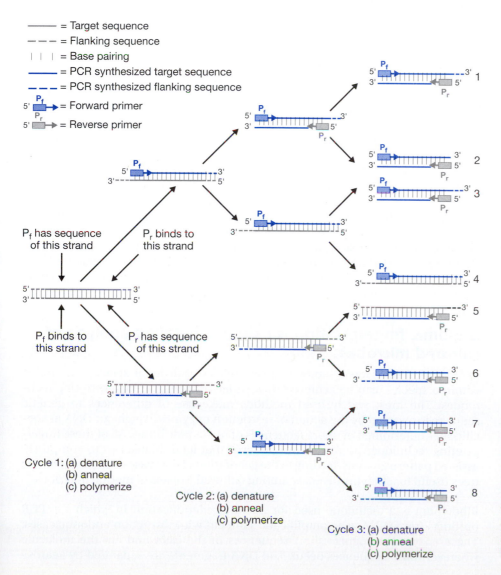

Figure 2.5 The polymerase chain reaction (first three cycles). Only after cycle 3 are there any duplex molecules of the exact length of the region to be amplified (molecules 2 and 7). After a few more cycles, these become the major product. (Reprinted from Turner et al. [2000] by permission of BIOS Scientific Publishers Ltd.)

MOLECULAR METHODS HAVE SOME INHERENT BIAS AND LIMITATIONS

A considerable amount of prior investigation is needed to ensure efficient extraction and amplification from samples. So-called "universal" primers targeted against specific groups can hybridize more favorably with some templates; for example, DNA templates from thermophilic organisms with a high content of G + C base pairs may denature less rapidly than those from mesophiles. An additional problem arises from the production of chimeric PCR products. These are artifacts caused by the joining together of the amplification products of two separate sequences and, if not detected, such sequences could suggest the presence of a novel nonexistent organism. This is a particular pitfall when multitemplate PCR is carried out using environmental samples of DNA. Various techniques are available to limit the formation of chimeras, and software to detect chimeric sequences should be used. Analysis of rRNA databases has revealed a number of chimeric sequences that have confused phylogenetic analyses.

Choosing a primer for use in the PCR obviously depends on knowing the sequence of the target DNA, but several factors need to be considered in order to ensure a successful PCR. For frequently studied genes, such as those for rRNA, there are "standard" primers and PCR conditions that are widely used. For studying novel genes, it is necessary to compare sequences in the databases to find a consensus sequence that recognizes conserved regions of a gene. An oligonucleotide, usually about 17–28 bases in length, is made using a DNA synthesizer—this is often carried out by commercial services. The ratio of G + C bases should usually be about 50–60%, depending on the gene, and primers should end in G or C to increase efficiency. Care needs to be taken to avoid complementary sequences within the primer, and between the pair of primers, as this can lead to preferential synthesis of primer dimers or secondary structures such as hairpins. Computer programs are used to design the appropriate sequence of bases to give good results and to calculate the temperatures to be used in the thermal cycler program.

Suitable positive and negative controls are always included in the reaction and each application of the PCR must be carefully optimized. This is especially important when trying to amplify particular sequences from a large mixed population of DNA, such as direct examination of environmental samples. Conditions such as the annealing temperature, amount of template DNA and concentration of Mg^{2+} are critical. For example, *Taq* polymerase is inactive in the absence of Mg^{2+}, but at excess concentration the fidelity of the polymerase is impaired and nonspecific amplification can result.

There are many variations of the basic PCR technique. In nested PCR, the level of specificity and efficiency of amplification is improved by carrying out a PCR for 15–30 cycles with one primer set, and then an additional 15–30 cycles with a second set of primers that anneal within the region of DNA amplified by the first primer set. Multiplex PCR involves the use of multiple sets of primers and results in the production of multiple products. This is often used as a quick screening method for the presence of certain organisms within water or infected tissue. The reverse transcriptase PCR (RT-PCR) is used to quantify messenger RNA (mRNA) and thus detects gene expression. This is a technique requiring great care in sample preparation and handling, since mRNA is very short-lived and the technique is prone to contamination from genomic DNA and nucleases. In quantitative real-time PCR (Q-PCR), the formation of a fluorescent reporter is measured during the reaction process. By recording the amount of fluorescence emission at each cycle, it is possible to monitor the PCR reaction during the exponential phase in which the first significant increase in the amount of PCR product correlates to the initial amount of target template. Q-PCR offers significant advantages for the diagnostic detection of microbes in the environment and in infected tissue samples, and handheld instruments for use in the field are now becoming available.

Genomic fingerprinting is used for detailed analysis of cultured microbes

Investigation of diversity among closely related individual species, or strains within a species, can be achieved by a variety of genomic fingerprinting techniques. The most widely used methods make use of differences in genetic sequences revealed by the action of restriction enzymes, which cut DNA at specific sites determined by short sequences of nucleotide bases. All of these fingerprinting techniques produce band patterns that lend themselves to computer-assisted pattern analysis, leading to the generation of databases and phylogenetic trees useful for studying diversity among cultured isolates of marine microbes.

Ribotyping is a technique used for bacterial identification in which the PCR products from rRNA gene amplification are cut with restriction endonucleases. These enzymes recognize short sequences in the DNA and cut the molecule wherever those sequences occur. The DNA fragments are separated by agarose electrophoresis and the resulting band pattern gives a rapid indication of similarities and differences between strains. Whist this method obviously lacks the resolution of sequence analysis, it is quick and relatively inexpensive.

Restriction fragment length polymorphism (RFLP) analysis involves cutting DNA with a combination of restriction endonucleases and separating the fragments according to size by gel electrophoresis. Fragments containing a specific base sequence can be identified by Southern blotting, by transfer to a sheet of nitrocellulose and addition of a radioactively labeled complementary DNA probe. Hybridization of the probe is detected using autoradiography or with an enzyme-linked color reaction. This enables identification of the restriction fragment with a sequence complementary to that of the DNA probe. Fragments of different length caused by mutations in duplicate copies of the gene can be detected and used for strain differentiation.

The RFLP method can be combined with PCR amplification of specific genes, such as SSU rRNA genes, followed by restriction endonuclease digestion and electrophoresis. This avoids the need for blotting and radiography but is not always successful in providing useful markers for strain differentiation. The RAPD (random amplified polymorphic DNA) technique is a more commonly used PCR marker technique for strain differentiation. It uses a single short primer in a PCR reaction, leading to a fingerprint of multiple bands generated by single nucleotide differences between individuals that prevent or allow primer binding. The method is quick and easy to perform but does not always give reliable results for detecting strain differences. AFLP (amplified fragment length polymorphism) is now a widely used technique with better resolution than either RFLP or RAPD analysis. Genomic DNA is digested using a pair of restriction endonucleases, one of which cuts at common sites and the other at rare sites. Adapter sequences are ligated to the resulting fragments and a PCR performed with primers homologous to the adapters plus selected additional bases. A subset of the restriction fragments is amplified, resulting in a large but distinct set of bands on a polyacrylamide gel. Band patterns are compared using imaging systems, and similarities and differences between strains can be computed to generate phylogenetic trees. Despite its advantage, AFLP analysis is a technique requiring very highly purified DNA and its use is generally restricted to laboratories specializing in taxonomy.

Microsatellite markers are repeat regions of short nucleotide sequences that can be analyzed using PCR amplification of unique flanking sequences and separation of the resulting bands by gel electrophoresis. Microsatellite markers are used extensively in the population analysis of eukaryotic organisms such as microalgae.

The genomic fingerprinting method for bacteria known as rep-PCR is a simple and reproducible method for distinguishing closely related strains and deducing phylogenetic relationships. The method is based on the fact that bacterial DNA naturally contains interspersed repetitive elements, such as the REP, ERIC and BOX sequences, for which standard PCR primers are available. The method requires very simple preparation of DNA samples without the need for extensive purification and is especially useful for the initial screening of cultures collected from marine samples.

Pulsed-field gel electrophoresis (PFGE) is a method in which the orientation and duration of an electric field is periodically changed, allowing large-molecular-weight fragments of DNA (up to 1000 kb) to migrate through the gel. Bacteria can be embedded in the gel and lysed *in situ* into large fragments, using a restriction enzyme that recognizes rare restriction sites. PFGE typing detects small changes in the genome resulting from the insertion or deletion of gene sequences or mutations that alter the restriction enzyme sites. PFGE is also very useful for analysis of virus diversity.

Determination of GC ratios and DNA–DNA hybridization is used in bacterial taxonomy

Determination of the ratio of nucleotide base pairs is often used in bacterial taxonomy. The principle of this method is that two organisms with very similar DNA sequences are likely to have the same ratio of guanine (G) + cytosine (C) to the total bases. DNA is extracted from the bacterium and the concentration of the bases is determined by high-performance liquid chromatography (HPLC) or by determining the "melting point" (T_m) of DNA. This is the temperature at which double-stranded DNA dissociates; because the hydrogen bonds connecting G–C pairs are stronger than those connecting adenine (A) and thymidine (T), the T_m increases with higher ratios of G + C. Closely related organisms have very similar GC ratios, and this can be used to compare species within a genus. However, bacteria possess a wide range of GC ratios and completely unrelated organisms can share a similar GC ratio.

A more useful measure of relatedness for the definition of species is DNA–DNA hybridization. Purified DNA from two organisms to be compared is denatured by melting and mixing their DNA. When cooled, DNA with a large number of homologous sequences will reanneal to form duplexes. The amount of hybridization can be measured if the DNA from one of the organisms is labeled by prior incorporation of bases containing ^{14}C or ^{32}P. The amount of radioactivity in the reannealed DNA collected on a membrane filter is measured and the percentage hybridization calculated by comparison with suitable controls. Taxonomists often use the benchmark of 70% or greater hybridization (under carefully standardized conditions) as the definition of a bacterial species. The difficulties associated with the concept of a bacterial species are discussed in Chapter 4.

DNA sequencing is a major tool in marine microbiology

The most commonly used method of obtaining the sequence of DNA is based on the well-established "dideoxy" technique of chain termination during synthesis of a copy of the DNA, developed by Frederick Sanger. A sequencing primer is added to a single-stranded DNA molecule in four separate reaction mixtures and chain extension begins using the normal nucleotides (dNTPs). The different reactions contain a small amount of dideoxy nucleotides (ddNTPs), which are synthetic analogs of the nucleotides lacking a –OH group on the sugar. Because of this, the DNA chain is extended only up to the point at which one of the ddNTPs is incorporated. The products from the four reactions are compared by electrophoretic separation, and the sequence of the DNA is determined by reading the sequence of nucleotides at which the chain extension is terminated. The original technique was modified to allow direct sequencing of PCR products as double-stranded DNA using dNTPs labeled with different fluorescent dyes, which can be read by lasers. Automated DNA sequencers allow robotic handling and high throughput of multiple samples, generating color-coded printouts of the sequence and data suitable for direct analysis by computer. For community analysis, it is often necessary to obtain only a partial sequence. A recent improvement of the Sanger method involves the application of microfluidic separation devices so that all steps in the process are fabricated onto an electronic wafer to produce a "lab on a chip."

"Next-generation" technologies allow inexpensive high-throughput sequencing

Whilst automated sequencing based on the Sanger method is still widely used, there has been intensive research to develop new methods that permit rapid and inexpensive DNA sequencing. The development of new methods by biotechnology companies is occurring at a very rapid rate and only a brief summary of some of the main methods is given here. Images and animations of the methods are

---

ℹ HIGH-THROUGHPUT SEQUENCING IS REVOLUTIONIZING BIOLOGY

Much of the commercial impetus for the development of new DNA sequencing technologies has been driven by biomedical companies seeking to develop personalized medicine based on individual human genome sequencing. Whilst sequencing the first human genome involved multinational consortia and cost several billion dollars, it is likely that it will soon be possible to have one's own genome sequenced for $1000 or less. The spin-off for marine microbiologists is that it is becoming possible for individual laboratories to conduct metagenomic analyses or to sequence multiple strains of cultured microbes for a few dollars within a very short period. As noted by Mardis (2008), "this paradigm shift promises to radically alter the path of biological inquiry." Making sense of the vast number of data acquired by different researchers has required new approaches. A major advance has been the development of the open-source RAST service developed by Rob Edwards and coworkers—this is a high-throughput pipeline available to all researchers interested in using metagenomics (Meyer et al., 2008). The pipeline compares protein and nucleotide databases to produce automated functional assignments of metagenomic sequences.

available on the company websites. The method of pyrosequencing was developed in 2005 by the company 454 Life Sciences and quickly became the main method for metagenomic studies in marine microbiology. The process depends on PCR-amplification in minute individual reactions within beads of an emulsion. The genomic DNA is sheared and oligonucleotide adaptors are attached, enabling each fragment to be attached to tiny primer-coated beads (~28 μm diameter) in droplets of an oil–water emulsion containing the polymerase enzyme and NTPs for PCR, generating multiple copies of the same DNA sequence on each bead. The beads are distributed in picoliter-sized wells in a fabricated substrate, and pyrosequencing is performed in parallel on each DNA fragment. The four different nucleotides are added stepwise to the immobilized primed DNA template; each time a nucleotide (A, C, G, or T) is incorporated into the growing chain, inorganic pyrophosphate (PP_i) is released. ATP sulfurylase first converts the PP_i to ATP, which is then converted to ADP (adenosine diphosphate) by luciferase, with the emission of light. After each nucleotide addition, a washing step is done to allow iterative addition of the four bases. When first introduced, the "read length" of sequences was very short, but improvements in the technology mean that fragments up to 1000 base pairs can be read.

Illumina™ sequencing technology involves the attachment of genomic DNA fragments to a transparent surface of a flow cell, where they are amplified to create millions of clusters, each containing ~1000 copies of the same template. These templates are sequenced using DNA sequencing by synthesis with DNA polymerase, employing reversible terminators with fluorescent dyes that can be removed after each round of synthesis. The instrument scans the incorporation of each nucleotide using a laser and an algorithm assigns the base sequence, using a reference sequence for internal quality control.

The Applied Biosystems SOLiD™ sequencer is based on sequencing catalyzed by DNA ligase. DNA fragments are linked by adaptor sequences to complementary oligonucleotides on 1-mm magnetic beads. After amplification by emulsion PCR, the beads are attached to the surface of a specially treated slide inside a fluidics cassette. A universal sequencing primer that is complementary to the adapters on the DNA fragments is added, with sequential addition of a set of four fluorescently labeled oligonucleotides that hybridize to the DNA fragment sequence adjacent to the universal primer, so that DNA ligase can join the nucleotides. Multiple cycles of ligation, detection, and cleavage are performed, with the number of cycles determining the eventual read length.

Each method has advantages for different applications. The long and accurate read lengths obtained with Sanger methods mean that they continue to be used for sequencing new genomes or sequences with repetitive regions, but progress in this field is so rapid that the new technologies are certain to dominate marine microbiology in the next few years.

Sequence data are used for phylogenetic analysis

Sequence data are used to determine the degree of phylogenetic relatedness between organisms. Data are derived from both microbes cultured in the laboratory and sequences obtained by direct community analysis of the environment. Data processing depends on the availability of publicly accessible databases (e.g. GENBANK, Ribosome Database Project, and EMBL) and specialized computer software. A newly generated sequence is first compared and aligned to previously described sequences using programs such as SEQUENCE_MATCH, BLAST, CLUSTAL, and ARB, available via the internet. Various mathematical methods and computer algorithms are used for the construction of phylogenetic trees, but details of the theory of each technique and its merits are beyond the scope of this book. In one common method (neighbor-joining tree) the different sequences are compared in order to determine the evolutionary distance between all the permutated pairs of sequences in the dataset by calculating the percentage of nonidentical sequences. Statistical corrections to allow for the possibility of multiple mutations at a given site are included in the calculation. A distance matrix is then constructed and a phylogenetic tree is drawn by grouping the most similar sequences and then adding less similar groups of sequences. The lengths of the lines in the tree are in proportion to their evolutionary distances apart (e.g. see *Figure 4.1*). In another approach, called parsimony, the tree is constructed after calculating the minimum number of mutational events needed for each pair to transform one sequence into the other. The tree is then drawn using the minimum number of such parsimonious steps. This method relies on the inherent assumption that evolution proceeds in one direction by the fastest route; this is undoubtedly an oversimplification. The validity of trees can be tested by a computational technique known as "bootstrapping," in which multiple iterations are made of trees based on subsamples of the sites in an alignment.

Sequence analysis of rRNA molecules shows that certain nucleotide sequences are highly conserved in particular regions. The nucleotide bases are numbered in a standard notation and "signature sequences" derived that are characteristic of groups of organisms at different taxonomic levels. For example, the sequence CACYYG (where Y is any pyrimidine) occurs at approximate position 315 in the 16S rRNA molecule in almost all *Bacteria,* but does not occur in *Archaea* or *Eukarya*. Sequence CACACACCG at position 1400 is distinctive of *Archaea*, but does not occur in *Bacteria* or *Eukarya*. Signature sequences can therefore be used to design domain-specific oligonucleotide probes for identification of any member of the respective group in a marine sample. Other sequences may be diagnostic for phyla, families, or even genera. This is a very powerful technique for culture-independent community analysis, especially in combination with fluorescent *in situ* hybridization (FISH; see below).

Denaturing gradient gel electrophoresis (DGGE) and terminal restriction fragment length polymorphism (TRFLP) are widely used to assess composition of microbial communities

In DGGE, PCR products are run in a polyacrylamide gel under a linear gradient of denaturing conditions—typically, 7 M urea and 40% formamide at a temperature of 50–65°C. Small sequence variations result in different migration of DNA fragments. Double-stranded DNA migrates through the gel until it reaches denaturing conditions sufficient to cause separation of the double helix ("melting"). DGGE primers must contain a guanine- and cytosine-rich sequence (GC clamp), which is resistant to complete denaturation under the conditions used, so that the band is "locked" in place according to the T_m. In analysis of community DNA, a set of bands will be generated representing variations in the sequence of the target gene (*Figure 2.6*). A typical analysis of an environmental sample will generate more than 50 bands, and many different samples can be compared on a single gel. This provides a rapid and efficient method of "community fingerprinting," (i.e. monitoring qualitative changes in community composition at different locations or under different conditions). Sometimes, the results of DGGE community analysis are interpreted in a semiquantitative fashion, using the assumption that the intensity of bands indicates the relative abundance of the species producing them. However, great caution is needed in drawing such conclusions because the techniques used for DNA extraction, the presence of inhibitors, the nature of the primers, and the PCR reaction conditions can all have significant effects on the results obtained. Temperature gradient gel electrophoresis (TGGE) works using similar principles, but the concentration of denaturing chemicals remains the same whilst the temperature of the gel is increased gradually and uniformly.

In terminal restriction fragment length polymorphism (TRFLP) analysis, PCR is performed on DNA extracted from the mixed community with one of the primers labeled with a fluorescent probe. After PCR, the products are cut using specific

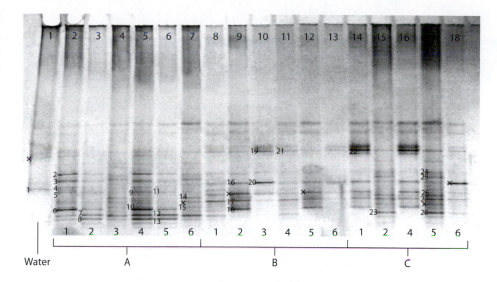

Figure 2.6 Example of denaturing gradient gel electrophoresis (DGGE) gels for microbial community analysis. The gels show 16S rRNA gene fragments from water and replicate tissue slurry samples from three colonies of the coral *Pocillopora damicornis*. (Reproduced from Bourne and Munn [2005] by permission of John Wiley & Sons, Inc.).

restriction enzymes. Each PCR product produces a fluorescent molecule with the probe at the end. The size of the fragments is determined by differences in the presence or absence of restriction sites in a particular region of the molecule or by deletions and insertions in the region. Like DGGE, TRFLP can provide a quick semiquantitative picture of community diversity, but results must, again, be interpreted carefully.

Elucidating the full genome sequence of microbes has provided major insights into their functional roles

The whole genome shotgun sequencing method pioneered in the 1990s by J. Craig Venter and colleagues (see *Research Focus Box 4.1*) has been applied to an increasing number of microbes. Automated sequencing methods and robotic handling of samples mean that it is now possible to determine the complete sequence of microbial genomes within a very short period. Nevertheless, alignment of contiguous nucleotide sequences, genome mapping and annotation of the sequences—the prediction of gene sequences and attachment of biological information—often relies on collaborative efforts between different laboratories and is most useful when different strains of the microbe are sequenced. The genomes of numerous marine microbes have now been mapped and analyzed. Initial attention was focused on culturable hyperthermophilic and psychrophilic *Bacteria* and *Archaea* because of their potential uses in biotechnology (see Chapter 14), but many other organisms important in ocean processes, disease, and symbiotic interactions have now been sequenced. Some important examples illustrating how genomics has provided insights into the physiology, adaptations, and ecology of marine microbes are shown in *Table 2.3*. Most *Bacteria* and *Archaea* possess single circular chromosomes, but there are many exceptions to this rule which can lead to complications in the interpretation of genomic data. Advances in DNA isolation methods and sequence analysis have enabled the full or almost complete genome sequence of some microbes to be determined without the need for isolation and culture. In particular, this has led to major advances in our understanding of the function of symbiotic bacteria in marine animals (see *Research Focus Box 10.1*).

Table 2.3 Examples of *Bacteria* and *Archaea* for which full genome sequences have been published

Microbe[a]	Genome size (Mbp)[b]	Genes[c]	Notable features
Nanoarchaeum equitans (A)	0.49	552	Parasitic on another archaeon Smallest nonsymbiotic cellular genome
Cand. Ruthia magnifica (B) (M)	1.16	1076	First intracellular symbiont to be sequenced Chemolithoautotroph in clams Genome reduction showing loss of genes for free-living lifestyle
Igniococcus hospitalis (A)	1.30	1442	Anaerobic, hyperthermophilic, chemoautolithotroph Host organism for *Nanoarchaeum equitans*
Cand. Pelagibacter ubique (B)	1.31	1354	Cultured representative of a group constituting 25% of ocean bacterioplankton. Smallest genome of any free-living bacterium. Many adaptations for oligotrophy
Aquifex aeolicus (B)	1.6	1522	Hyperthermophile One of the earliest diverging members of the *Bacteria*
Methanocaldococcus jannaschii (A)	1.66	1738	Methanogenic, hyperthermophilic chemolithotroph using H_2 and CO_2
Thermotoga maritima (B)	1.86	1868	Deeply branching thermophilic bacterium Crystal structure of many proteins determined
Pyrococcus furiosus (A)	1.9	2065	Adaptations for extreme hyperthermophily Strictly anaerobic chemoorganotroph
Prochlorococcus marinus (B)	2.41	2272	Major contributor to ocean photosynthesis Genomes of different ecotypes are adapted to different light and nutrient conditions
Synechococcus sp. (B)	2.43	2526	Widely distributed cyanobacterium Evidence of extensive gene transfer and adaptation to different light niches
Cand. Endoriftia persephone (B) (M)	~3.2	?	Intracellular symbiont of hydrothermal vent tubeworm Genome reveals capacity for chemolithotrophic metabolism in the host and heterotrophy in the environment
Nitrosococcus oceani (B)	3.52	3132	Chemolithoautotrophic, ammonia-oxidizing bacterium with major role in nitrogen cycling in upper ocean
Synechocystis sp. (B)	3.57	3168	Mixotrophic cyanobacterium Genes controlling circadian rhythms
Desulfovibrio desulfuricans (B)	3.73	3784	Sulfate-reducing bacterium abundant in sediments
Pseudoalteromonas haloplanktis (B)	3.85 (C × 2)	3487	Cold-adapted bacterium from Antarctic Protective mechanisms against reactive oxygen species
Caulobacter crescentus (B)	4.01	3763	Stalked bacterium colonizing surfaces Genome analysis reveals control of life cycle and differentiation of cell types
Vibrio cholerae (B)	4.03 (C × 2)	3887	Complex adaptations for life in the human gut and the aquatic environment Extensive mechanisms for gene acquisition
Silicibacter pomeroyi (B)	4.15	3869	Major role in sulfur cycle via DMSP metabolism Many amino and carboxylic acid transporters
Vibrio fischeri[d] (B)	4.28 (C × 2)	3802	Bioluminescence and specific association with squid host Symbiont with many genes resembling those in pathogenic vibrios
Roseobacter denitrificans (B)	4.33	4149	Representative of group carrying out aerobic anoxygenic photosynthesis

Table 2.3 cont.

Microbe[a]	Genome size (Mbp)[b]	Genes[c]	Notable features
Vibrio splendidus (B)	4.96 (C × 2)	4498	Dominant *Vibrio* sp. in seawater High genetic diversity Invertebrate pathogen
Aeromonas salmonicida subsp. *salmonicida* (B)	5.04	4578	Fish pathogen Genome shows adaptations to host specificity
Vibrio vulnificus (B)	5.12 (C × 2)	4537	Human pathogen associated with oysters and coastal sediments Numerous virulence factors
Photobacterium profundum (B)	6.4 (C × 2)	5480	Deep sea bacterium Adaptations to pressure through differential gene expression and membrane composition
Pirellula baltica (B)	7.14	7325	First sequence from *Planctomycetes* Absence of peptidoglycan. Many genes similar to those of eukaryotes
Trichodesmium erythraeum (B)	7.75	5076	Major diazotrophic cyanobacterium Segregates photosynthesis and nitrogen fixation without heterocysts

[a]Values are given for one representative strain; note that where multiple strains have been sequenced, there can be very significant differences in the composition of genomes. [b]Total genome size in Mbp. (C × 2) indicates the presence of two chromosomes. [c]Value is based on the open reading frames identified in primary annotation of the genome. [d]Recently reclassified as *Aliivibrio fischeri*. (A), *Archaea*; (B), *Bacteria*; *Cand.*, *Candidatus* (interim taxonomic status, as full phenotypic description is not available); (M), based on metagenome of uncultured bacteria; DMSP, dimethylsulfoniopropionate.

Once sequences are complete and gaps are closed, autoannotation of the genome sequence is carried out by computer software that predicts the open reading frames (ORFs), which can be recognized by searching for characteristic conserved sequences at the promoter region (where DNA polymerase binds), ribosomal binding sites, and start and stop codons on the transcribed mRNA. DNA sequences—and the predicted amino acid sequences of the proteins that they would encode—are compared with those in DNA and protein databases. More detailed examination using other evidence is used to refine the annotation of the genome. Often, many of the predicted ORFs cannot be reliably linked to a function using current knowledge of bioinformatics. Therefore, the next stages in molecular analysis are functional genomics and proteomics. This involves inactivation of specific regions of the genome and high-resolution separation and identification of proteins. Such post-genomic analyses of marine microbes are revealing many previously unknown properties and activities. It is important to bear in mind that each genome sequence is only a "snapshot" of one or two individual strains of microbe. As reduced sequencing costs enable analysis of genomes from multiple isolates of the same species, very large variations in genome properties can be revealed. Indeed, such comparative genomics can yield important information about adaptations to specific environments (e.g. light regimes in *Prochlorococcus*, see p. 194; host specificity in *Vibrio fischeri*, see *Research Focus Box 10.2*).

Metagenomics is revolutionizing our understanding of marine microbial ecology

As described previously, assessment of microbial diversity using PCR amplification of rRNA gene sequences obtained straight from the environment without the need for culture has been a major tool of marine microbiologists since the late 1980s. Despite its unquestioned importance, the method has a number of

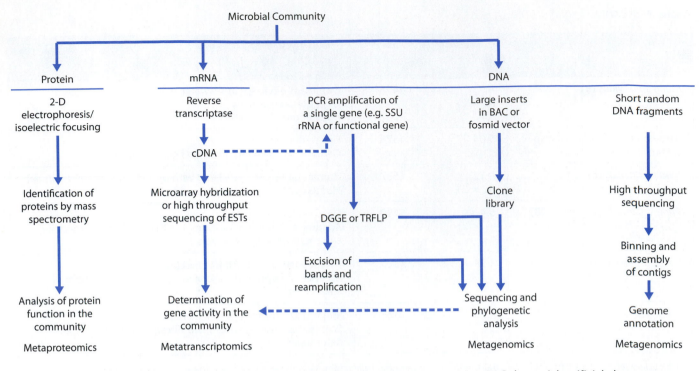

Figure 2.7 Strategies for molecular analysis of microbial communities. BAC, bacterial artificial chromosome; DGGE, denaturing gradient gel electrophoresis; EST, expressed sequence tag; SSU, small subunit; TRFLP, terminal restriction fragment length polymorphism.

inherent limitations, owing to the selective targeting of a single gene. The concept of metagenomics developed in the late 1990s as a method of analyzing collections of genes sequenced from microbial communities in the environment. In the first studies, methods for handling large DNA fragments up to 300 kb were developed, so that pieces of DNA containing genes localized to the SSU genes could be isolated and recognized by detection of their signature sequence of SSU rRNA (or other marker gene) using Southern blotting (*Figure 2.7*). The large fragments can be cloned directly into the large F plasmids of *E. coli* to create bacterial artificial chromosomes (BACs), which have the additional advantage that expression of some of the inserted genes may occur in the *E. coli* host harboring the vector.

The development of metagenomics has been hastened dramatically by the application of shotgun sequencing, a method originally developed by J. Craig Venter and colleagues to accelerate progress in the Human Genome Project. DNA is extracted from environmental samples and broken into small DNA fragments of 2–50 kb, which are cloned to generate plasmid libraries. The clones are then sequenced from both ends of the insert and bioinformatics software is used to assemble the contiguous sequences (contigs) from the overlapping sequences based on sequence similarity. Contigs are ordered and oriented into larger (but not necessarily continuous) pieces of a genome, called scaffolds. The resulting assemblies can link genes to their regulatory elements, guide investigations of biological pathways, and connect unknown sequences with taxonomic markers to suggest evolutionary relationships. One of the first successes of this approach was the identification of proteorhodopsin genes in diverse marine bacteria (*Research Focus Box 3.1*).

The next step in the development of metagenomics involved the sequencing of large numbers of random fragments of DNA, made possible by the deployment of multiple DNA sequencing machines and the development of new bioinformatics software to reconstruct the contigs arising from many small fragments.

The frequency of specific DNA sequences gives an indication of abundance of a particular organism, and even very rare organisms should be represented if enough sequences are analyzed. However, reconstruction of a whole genome becomes more likely when that genome is among the most abundant in the sample. The metagenomic analysis of multiple samples collected during circumnavigation of the globe (the Global Ocean Survey) is discussed in *Research Focus Box 4.1*. Full or even partial genome sequences allow an assessment of the phylogenetic diversity of the community and the ecological functions of dominant organisms. Metagenomics is a very rapidly developing field owing to the development of new high-throughput sequencing technologies, discussed previously, as well as the application of new bioinformatics tools. It is even possible to amplify DNA isolated from a single microbial cell and to determine its genome sequence.

Fluorescent hybridization (FISH) allows visualization and quantification of specific microbes

The acronym FISH is no accident. This technique permits an investigator to "fish" for a specific nucleic acid sequence in a "pool" of unrelated sequences. Originally pioneered by Rudolf Amann of the Max Planck Institute for Marine Microbiology, it has proved to be one of the most useful culture-independent techniques for identification of particular organisms or groups of organisms in the marine environment. Another major advantage is the ability to quantify different members of a microbial community and their dynamics. As noted above, oligonucleotide probes can be designed to recognize and hybridize to complementary sequences that are unique to specific microbial groups. In FISH, the oligonucleotide probe is covalently linked to a fluorescent dye such as fluorescein, Texas Red, or indocarbocyanines (*Figure 2.8A*). Hybridization with the target within a

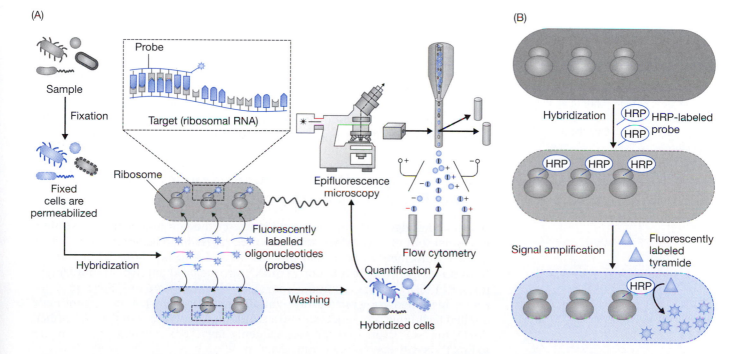

Figure 2.8 Fluorescent *in situ* hybridization (FISH). (A) Principles of the basic method. The sample is fixed to stabilize cells and make the membrane permeable to the oligonucleotide probe, which attaches to its intracellular targets. After washing, single cells can be identified by epifluorescence microscopy or flow cytometry. (B) Principles of catalyzed reporter deposition (CARD-FISH). Hybridization involves a single oligonucleotide that is covalently linked to horseradish peroxidase (HRP). The signal is amplified because multiple tyramide molecules are radicalized by a single HRP enzyme. (Reprinted by permission of Macmillan Publishers Ltd, © Nature, Amann and Fuchs [2008]).

morphologically intact microbial cell can therefore be visualized by fluorescence or confocal microscopy. The probe signal that is produced after hybridization and washing away excess unbound probe is proportional to the amount of target nucleic acid present in the sample. This allows an investigator to visualize and quantify the microbes in marine samples and determine cell morphology and spatial distributions *in situ*. FISH has been particularly useful for examining microbes attached to particles and spatial localization of microbes in biofilms, in symbiotic associations, and in infected tissues. The widespread use of SSU rRNA genes for diversity studies means that there are extensive databases of sequence information that can be used to design probes as phylogenetic stains. Cells usually contain thousands of copies of rRNA molecules and they are relatively stable. However, many marine microbes have very small cells and may be slow growing or dormant and can give variable results because they may have low numbers of ribosomes, or have membrane modifications that lower the accessibility of the probe. This places some limits on the uses of FISH in investigation of oligotrophic ocean environments, but recent modifications have improved the efficiency of hybridization; multiple probes, different sequences, and different fluorochromes can be applied to single samples to improve sensitivity and to assess genetic diversity. Enzyme-linked methods such as catalyzed reporter deposition (CARD) FISH (*Figure 2.8B*) can be much more sensitive. Here, hybridization involves an oligonucleotide that has been covalently crosslinked to the enzyme horseradish peroxidase (HRP). Because the HRP molecule is much larger than fluorochromes, permeabilization requires harsher enzymatic or chemical treatments, conducted in agarose gels to avoid loss of cell shape. Fluorescently labeled tyramide (a derivative of the amino acid tyrosine) is then added, and multiple molecules are activated by a single HRP enzyme and bind permanently to the cells, resulting in amplification to give a strong fluorescent signal. Besides detection of rRNA, the FISH technique can be extended to search for a wide range of specific gene sequences characteristic of particular structures or metabolic functions, assuming that sufficient copies of the nucleic acid are present in the sample.

The combination of microautoradiography (MAR) with FISH is an extremely sensitive technique for following the metabolism of specific compounds by individual cells and is not subject to biases that can result from DNA extraction or PCR methods. After incubation of the samples and negative controls with radioactively labeled substrate, the cells are fixed to a membrane filter or slide. After washing and permeabilizing, FISH probes are added and the sample is visualized by overlaying with photographic emulsion. Where isotope has been incorporated, the film will be exposed and the pattern can be analyzed.

Metatranscriptomics and metaproteomics reveal metabolic activities in the environment

Whilst knowledge of the metagenome indicates the potential activities within a microbial community, it does not reveal the expression of genes and hence their actual involvement in particular processes. As shown in *Figure 2.7*, metatranscriptomics involves the isolation of mRNA and the production of a complementary DNA (cDNA) copy using the enzyme reverse transcriptase. Specific genes can be identified using microarray hybridization (see below). A more commonly applied technique is to sequence short segments (500–800 bp) of the transcribed cDNA; these are known as expressed sequence tags (ESTs). Databases are built up to link ESTs with specific gene functions.

Even knowing which genes are expressed is one step short of knowing which proteins are present in a community, because structural proteins and enzymes may all be modified after translation owing to complex regulatory processes. Proteomics is usually carried out by two-dimensional separation of the proteins present in a mixed sample using a combination of SDS-PAGE and isoelectric focusing. The protein spots are digested in the electrophoreis gel and analyzed by mass spectrometry. Peptide patterns can be compared with those in databases, or full protein sequences can be obtained and homology with well-characterized proteins can be determined.

Microarrays enable assessment of gene activity in the environment

One of the major methods used in functional genomics is DNA microarray technology. DNA sequences identified from microbial genomes are attached (or synthesized directly) on the surface of a silicon chip in a highly ordered array to produce a device that can act as a probe for thousands of genes. For some bacteria, chips with probes for every expressed gene in the genome have been designed. The target nucleic acids (mRNA or cDNA) are labeled with fluorescent reporter groups and lasers detect hybridization of complementary sequences to the oligonucleotide on the microarray chip. Microarrays are particularly useful for determining the effects of changing conditions on the patterns of gene expression, for example following nutrient changes in mesocosm experiments or during different phases of interactions with a host (see *Research Focus Box 10.2*). The results of microarray experiments are often presented as color-coded patterns indicating which genes are upregulated, downregulated, or unaltered during the transition.

Conclusions

This chapter has shown that marine microbiologists use a very wide range of methods to study the diversity and activity of microbes, and examples of their application in research will be found throughout the book. The most significant advances in marine microbiology have occurred as a result of the development of new molecular methods, but longer-established techniques such as culture and microscopy remain as important as ever. The next few years will undoubtedly be dominated by increasing application of metagenomics and the other "omics" technologies, with an exponential growth in information fuelled by the continued fall in sequencing costs. The challenge for microbiologists will be to interpret the huge datasets that are now accumulating and to translate genomic information into real understanding of the ecology of marine microbes and their role in diverse activities.

? CAN WE DEVELOP INSTRUMENTS FOR *IN SITU* MONITORING AT SEA?

The use of mobile platforms and ROVs or AUVs is widespread in oceanographic studies, especially for physical and chemical measurements. Researchers at the Monterey Bay Research Institute have developed a range of automated sensors for measurements of chemical and physical factors linked to biological processes, with a network of sensors linked to a network (www.mbari.org/rd/). Development of solid-state lasers means that new types of automated flow cytometers might be developed or deployed for such applications (Paul et al., 2007). The greatest advances will come from *in situ* application of molecular biological methods, and several instruments have already proved useful. For example, robotic devices can filter and concentrate water samples before the microbial cells are disrupted and applied to a microarray. Hybridization of an oligonucleotide probe to DNA sequences of interest can be detected using a luminescence assay and positive results visualized by a charge-coupled device, with results signaled to a receiving station. Some success has been achieved in measuring gene expression in harmful algal blooms (see p. 279) using an autonomous sensor to detect and characterize mRNA sequences.

References

Amann, R. and Fuchs, B. M. (2008) Single-cell identification in microbial communities by improved fluorescence *in situ* hybridization techniques. *Nat. Rev. Microbiol.* 6: 339–348.

Bourne, D. G. and Munn, C. B. (2005) Diversity of bacteria associated with the coral *Pocillopora damicornis* from the Great Barrier Reef. *Environ. Microbiol.* 7: 1162–1174.

Button, D. K., Schut, F., Quang, P., Martin, R. and Robertson, B. R. (1993) Viability and isolation of marine bacteria by dilution culture: theory, procedures and initial results. *Appl. Environ. Microbiol.* 59: 881–891.

Connon, S. A. and Giovannoni, S. J. (2002) High-throughput methods for culturing microorganisms in very low-nutrient media yield diverse new marine isolates. *Appl. Environ. Microbiol.* 68: 3878–3885.

Donachie, S. P., Foster, J. S. and Brown, M. V. (2007) Culture clash: challenging the dogma of microbial diversity—Commentaries. *ISME J.* 1: 97–99.

Gilbert, J. A., Field, D., Huang, Y., Edwards, R., Li, W., Gilna, P. and Joint, I. (2008) Detection of large numbers of novel sequences in the metatranscriptomes of complex marine microbial communities. *PLoS ONE* 3: e3042, doi:10.1371/journal.pone.0003042

Giovannoni, S. and Stingl, U. (2007) The importance of culturing bacterioplankton in the "omics" age. *Nat. Rev. Microbiol.* 5: 820–825.

Giovannoni, S. J., Foster, R. A., Rappé, M. S. and Epstein, S. (2007) New cultivation strategies bring more microbial plankton species into the laboratory. *Oceanography* 20: 62–69.

Giovannoni, S. J., Tripp, H. J., Givan, S., Podar, M., Vergin, K. L., et al. (2005) Genome streamlining in a cosmopolitan oceanic bacterium. *Science* 309: 1242–1425.

Keller, M. and Zengler, K. (2004) Tapping into microbial diversity. *Nat. Rev. Microbiol.* 3: 141–150.

Mardis, E. R. (2008) Next-generation DNA sequencing methods. *Annu. Rev. Genomics Hum. Genet.* 9: 387–402.

Meyer, F., Paarmann, D., D'Souza, M., et al. (2008) The metagenomics RAST server—a public resource for the automatic phylogenetic and functional analysis of metagenomes. *BMC Bioinformat.* 9: 386, doi:10.1186/1471-2105-9-386.

Miller, S. D., Haddock, S. H. D., Elvidge, C. D. and Lee, T. F. (2005) Detection of a bioluminescent milky sea from space. *Proc. Natl Acad. Sci. USA* 102: 14181–14184.

Morris, R. M., Rappé, M. S., Connon, S. A., Vergin, K. L., Siebold, W. A., Carlson, C. A. and Giovannoni, S. J. (2002) SAR11 clade dominates ocean surface bacterioplankton communities. *Nature* 420: 806–810.

Nealson, K. H. and Hastings, J. W. (2006) Quorum sensing on a global scale: massive numbers of bioluminescent bacteria make milky seas. *Appl. Environ. Microbiol.* 72: 2295–2297.

Nichols, D. (2007) Cultivation gives context to the microbial ecologist. *FEMS Microbiol. Ecol.* 60: 351–357

Paul, J., Scholin, C., van den Engh, G. and Perry, M. J. (2007) *In situ* instrumentation. *Oceanography* 20: 70–78.

Rappé, M. S., Connon, S. A., Vergin, K. L. and Giovannoni, S. J. (2002) Cultivation of the ubiquitous SAR11 marine bacterioplankton clade. *Nature* 418: 630–632.

Schut, F., Gottschal, J. C. and Prins, R. A. (1997) Isolation and characterisation of the marine ultramicrobacterium *Sphingomonas* sp. strain RB2256. *FEMS Microbiol. Rev.* 20: 363–369.

Stingl, U., Tripp, H. J. and Giovannoni, S. J. (2007) Improvements of high-throughput culturing yielded novel SAR11 strains and other abundant marine bacteria from the Oregon coast and the Bermuda Atlantic Time Series study site. *ISME J.* 1: 361–371.

Turner, P. C., White, M. R. H., McLennan, A. G. and Bates, A. D. (2000) *Instant Notes in Molecular Biology.* BIOS Scientific Publishers Ltd, Oxford.

Zengler, K. (2009) Central role of the cell in microbial ecology. *Microbiol. Molec. Biol. Rev.* 73: 712–729

Zengler, K., Toledo, G., Rappé, M. S., Elkins, J., Mathur, E. J., Short, J. M. and Keller, M. (2002) Cultivating the uncultured. *Proc. Natl Acad. Sci. USA* 99: 15681–15686.

Further reading

Books, general reviews, and websites

Cold Spring Harbor Laboratory. Dolan DNA Learning Centre. Animations of molecular biology methods including PCR, sequencing and microarrays. http://www.dnalc.org/resources/animations/index.html [Accessed 13 December 2010]

Leadbetter, J. (ed.) (2005) *Environmental Microbiology (Methods in Enzymology).* Elsevier Academic Press, San Diego.

Osborn, M. and Smith, C. (eds) (2005) *Molecular Microbial Ecology (Advanced Methods).* Taylor & Francis, Abingdon.

Paul, J. (ed.) (2001) *Marine Microbiology. Methods in Microbiology,* Vol. 30. Academic Press, San Diego.

Robinson, I. S. (2003) *Measuring the Oceans from Space: The Principles and Methods of Satellite Oceanography.* Springer–Praxis Books

Spencer, J. F. T. and Ragout de Spencer, A. L. (eds) (2004) *Environmental Microbiology: Methods and Protocols (Methods in Biotechnology).* Humana Press, New Jersey.

Microbial imaging, FISH, and flow cytometry

Amann, R. I., Ludwig, W. and Schleifer, K. H. (1995) Phylogenetic identification and *in-situ* detection of individual microbial-cells without cultivation. *Microbiol. Rev.* 59: 143–169.

Gerdts, G. and Luedke, G. (2006) FISH and chips: marine bacterial communities analyzed by flow cytometry based on microfluidics. *J. Microbiol. Methods* 64: 232–240.

Invitrogen (2010) *Molecular Probes. The Handbook—A Guide to Fluorescent Probes and Labeling Technologies.* http://www.probes.invitrogen.com

Kuypers, M. M. M. and Jørgensen, B. B. (2007) The future of single-cell environmental microbiology. *Environ. Microbiol.* 9: 6–7.

Malfatti, F. and Azam, F. (2009) Atomic force microscopy reveals microscale networks and possible symbioses among pelagic marine bacteria. *Aquat. Microb. Ecol.* 58: 1–14.

Ormerod, M. G. (2008) *Flow Cytometry—A Basic Introduction.* http://flowbook.denovosoftware.com/

Savidge, T. and Charalabos, P. (eds) (2004) *Microbial Imaging. Methods in Microbiology,* Vol. 24. Academic Press, San Diego.

Tujula, N. A., Holmstrom, C., Mussmann, M., Amann, R., Kjelleberg, S. and Crocetti, G. R. (2006) A CARD-FISH protocol for the identification and enumeration of epiphytic bacteria on marine algae. *J. Microbiol. Methods* 65: 604–607.

DNA extraction, PCR, and ribosomal RNA

Amann, R. and Ludwig, W. (2000) Ribosomal RNA-targeted nucleic acid probes for studies in microbial ecology. *FEMS Microbiol. Rev.* 24: 555–565.

Baker, G. C., Smith, J. J. and Cowan, D. A. (2003) Review and re-analysis of domain-specific 16S primers. *J. Microbiol. Methods* 55: 541–555.

Luna, G. M., Dell'Anno, A. and Danovaro, R. (2006) DNA extraction procedure: a critical issue for bacterial diversity assessment in marine sediments. *Environ. Microbiol.* 8: 308–320.

Ribosome Database Project. Centre for Microbial Ecology, Michigan State University. http://rdp.cme.msu.edu [Accessed 13 December 2010]

Electrophoresis and fingerprinting methods

Brown, M. V., Schwalbach, M. S., Hewson, I. and Fuhrman, J. A. (2005) Coupling 16S-ITS rDNA clone libraries and automated ribosomal intergenic spacer analysis to show marine microbial diversity: development and application to a time series. *Environ. Microbiol.* 7: 1466–1479.

Danovaro, R., Luna, G. M., Dell'Anno, A. and Pietrangeli, B. (2006) Comparison of two fingerprinting techniques, terminal restriction fragment length polymorphism and automated ribosomal intergenic spacer analysis, for determination of bacterial diversity in aquatic environments. *Appl. Environ. Microbiol.* 72: 5982–5989.

Eiler, A. and Bertilsson, S. (2006) Detection and quantification of *Vibrio* populations using denaturant gradient gel electrophoresis. *J. Microbiol. Methods* 67: 339–348.

Moeseneder, M. M., Winter, C., Arrieta, J. M. and Herndl, G. J. (2001) Terminal-restriction fragment length polymorphism (T-RFLP) screening of a marine archaeal clone library to determine the different phylotypes. *J. Microbiol. Methods* 44: 159–172.

Muyzer, G. and Smalla, C. (1998) Application of denaturing gradient gel electrophoresis (DGGE) and temperature gradient gel electrophoresis (TGGE) in microbial ecology. *Antonie Leeuwenhoek* 73: 127–141.

Popa, R., Popa, R., Mashall, M. J., Nguyen, H., Tebo, B. M. and Brauer, S. (2009) Limitations and benefits of ARISA intra-genomic diversity fingerprinting. *J. Microbiol. Methods* 78: 111–118.

Satoshi, I. and Sadowsky, M. J. (2009) Applications of the rep-PCR DNA fingerprinting technique to study microbial diversity, ecology and evolution. *Environ. Microbiol.* 11: 733–740.

DNA sequencing

Goldberg, S. M. D., Johnson, J., Busam, D., et al. (2006) A Sanger/pyrosequencing hybrid approach for the generation of high-quality draft assemblies of marine microbial genomes. *Proc. Natl Acad. Sci. USA* 103: 11240–11245.

Looi, M.-K. (2009) Wellcome Trust Feature: Genomics—the next generation. [Includes animations of sequencing technologies.] http://www.wellcome.ac.uk/News/2009/Features/WTX056032.htm

Rogers, Y. H. and Venter, J. C. (2005) Genomics: massively parallel sequencing. *Nature* 437: 326-327.

Shendure, J. and Ji, H. (2008) Next-generation DNA sequencing methods. *Nat. Biotechnol.* 26: 1135–1145.

Genomics and metagenomics

DeLong, E. F. (2005) Microbial community genomics in the ocean. *Nat. Rev. Microbiol.* 3: 459–469.

Edwards, R. A. and Dinsdale, E. A. (2007) Marine environmental genomics: unlocking the ocean's secrets. *Oceanography* 20: 56–61.

Edwards, R. A. and Rohwer, F. (2005) Viral metagenomics. *Nat. Rev. Microbiol.* 3: 504–510.

Gaasterland, T. (1999) Archaeal genomics. *Curr. Opin. Microbiol.* 2: 542–547.

Koonin, E.V. and Galperin, M. Y. (1997) Prokaryotic genomes: the emerging paradigm of genome-based microbiology. *Curr. Opin. Genet. Dev.* 7: 757–763.

Moran, M. A. (2008) Genomics and metagenomics of marine prokaryotes. In: *Microbial Ecology of the Oceans*, 2nd Edn (ed. D. L. Kirchman). John Wiley & Sons, Inc: Hoboken, NJ. pp. 91–129.

National Research Council (2005) *Committee on Metagenomics: Challenges and Functional Applications. The New Science of Metagenomics: Revealing the Secrets of Our Microbial Planet.* National Academies Press, Washington.

Microarrays

Gentry, T. J., Wickham, G. S., Schadt, C. W., He, Z. and Zhou, J. (2006) Microarray applications in microbial ecology research. *Microb. Ecol.* 52: 159–175.

Gingeras, T. R. and Rosenow, C. (2000) Studying microbial genomes with high-density oligonucleotide arrays. *ASM News* 66: 463–469.

Chapter 3

Metabolic Diversity and Ecophysiology

Microbes have been evolving for between three and four billion years, resulting in a huge diversity of metabolic types. Organisms have evolved to obtain energy from light or various inorganic substances or by breaking down different organic compounds to their basic constituents. Some species obtain key elements for building cellular material directly from inorganic minerals, whilst others require complex "ready-made" organic compounds. The selective advantage of the ability to utilize particular nutrients under a particular set of physical conditions has led to the evolution of the enormous metabolic diversity that we see today. Such diversity results in the occurrence of microbes in every conceivable habitat for life. Microbial metabolism fuels biogeochemical cycles, transformation of nutrients, and other processes that maintain the web of life in the ocean and underpin marine ecology. The first part of this chapter reviews the major metabolic pathways by which microbes obtain and transform energy and carbon for cellular growth. This is followed by consideration of the major physiological factors affecting microbial growth in the plankton and on surfaces.

Key Concepts

- All life depends on the activity of autotrophic organisms using light or chemical energy to fix CO_2 into cellular material.
- Microbes harness energy from light by a range of mechanisms to support autotrophic and heterotrophic metabolism.
- Chemolithotrophic activity is widespread in the water column, as well as in sediments, vents, and seeps.
- Diverse mechanisms are responsible for microbial transformations of carbon, sulfur, and nitrogen.
- Most planktonic microbes are genetically adapted to slow growth with minimum nutrients, whilst others can exploit patches of high nutrient concentration.
- Intercellular communication and antagonism is a major feature of biofilm communities.
- Most marine microbes are adapted to growth under conditions of low temperature and high pressure.

All cells need to obtain energy and conserve it in the compound ATP

It is convenient to classify microbes in terms of their energy source as either chemotrophs (which produce ATP by the oxidation of inorganic or organic compounds) or phototrophs (which use light as a source of energy for the production of ATP). Chemotrophs may be further divided into those that can obtain energy solely from inorganic compounds (chemolithotrophs) and those that use organic compounds (chemoorganotrophs). Chemolithotrophy is unique to members of the *Bacteria* and *Archaea*, whilst chemoorganotrophy is found in all groups of cellular microbes.

All cells need carbon as the major component of cellular material

Those organisms that can use carbon dioxide as the sole source of carbon are termed "autotrophs," whilst those that require organic compounds as a source of carbon are termed "heterotrophs." Autotrophs are primary producers that fix carbon from CO_2 into cellular organic compounds, which are then used by heterotrophs. Most chemolithotrophs and phototrophs are also autotrophic, but some (e.g. purple non-sulfur photosynthetic bacteria and some sulfur-oxidizing chemolithotrophs) can switch between heterotrophic and autotrophic metabolism. The term "mixotroph" is used to describe organisms with a mixed mode of nutrition. For example, some sulfur-oxidizing bacteria are not fully lithotrophic as they require certain organic compounds because they lack key biosynthetic enzymes. Many protists, including numerous types of flagellates and dinoflagellates, are termed mixotrophic because they derive nutrition both from photosynthesis and from the engulfment of prey such as bacteria. Indeed, among the protists there is a spectrum from absolute heterotrophy to absolute autotrophy. *Table 3.1* shows a summary of these nutritional categories.

Phototrophy involves conversion of light energy to chemical energy

Virtually all life on Earth depends directly or indirectly on solar energy, and it is likely that simple mechanisms for harvesting light energy developed very early in the evolution of life. Phototrophs contain light-sensitive pigments, which use energy from sunlight to oxidize electron donors. Phototrophic protists and cyanobacteria use H_2O as electron donor resulting in the production of O_2 (oxygenic photosynthesis), whilst other phototrophic bacteria use H_2S, S^0, $H_2SO_2^{2-}$ or H_2, meaning that no O_2 is generated (anoxygenic photosynthesis). In the process of photophosphorylation, light energy creates a charge separation across membranes and "traps" energy from the excited electrons into a chemically stable molecule (ATP) and also generates NADPH (nicotinamide adenine dinucleotide phosphate [reduced]; *Figure 3.1*). Many phototrophs are also autotrophic and use ATP and NADPH for the fixation of CO_2 into cellular material, in which case the process meets the strict definition of photosynthesis.

Although many microbes possess photochemical mechanisms, true photosynthesis occurs only in cells containing the magnesium-containing pigments; either chlorophylls (found in oxygenic cyanobacteria and chloroplasts) or bacteriochlorophylls (found in anoxygenic phototrophic bacteria). These molecules are closely related, apart from a small region of the porphyrin molecule. There are also small differences in other parts of the molecule, resulting in structures that absorb different wavelengths of light. A variety of accessory pigments are also present, notably carotenoids, which function primarily to protect cells from harmful photooxidation reactions that can occur in bright light, such as the generation of toxic forms of oxygen. Carotenoids such as fucoxanthin may be important in light harvesting for photosynthesis in eukaryotic algae. Antenna pigments

Table 3.1 Nutritional categories of microorganisms

Energy source	Carbon source	Hydrogen or electron source	Representative examples
Photolithoautotrophy			
Light	CO_2	Inorganic	Cyanobacteria, purple sulfur bacteria, phototrophic protists
Photoorganoheterotrophy			
Light	Organic compounds	Organic compounds or H_2	Purple non-sulfur bacteria, aerobic, anoxygenic bacteria, proteorhodopsin-containing bacteria, and archaea[a]
Chemolithoautotrophy			
Inorganic	CO_2	Inorganic	Sulfur-oxidizing bacteria, hydrogen bacteria, nitrifying bacteria, and archaea
Chemoorganoheterotrophy			
Organic compounds	Organic compounds	Organic compounds	Wide range of bacteria and archaea, all fungi, phagotrophic protists
Mixotrophy (combination of lithoautotrophy and organoheterotrophy)			
Inorganic	Organic compounds	Inorganic	Sulfur-oxidizing bacteria, e.g. *Beggiatoa*
Mixotrophy (combination of photoautotrophy and organoheterotrophy)			
Light + organic compounds	CO_2 + organic compounds	Inorganic or organic	Phagotrophic, photosynthetic protists (some flagellates and dinoflagellates)

[a]The role of proteorhodopsin in nutrition is uncertain (see *Research Focus Box 3.1*).

surround complexes of 50–300 molecules of chlorophyll or bacteriochlorophyll combined with proteins (reaction centers). These arrays occupy a large surface area to trap the maximum amount of light of different wavelengths and transfer its energy to the reaction center. In cyanobacteria, the most important antenna pigments are the phycobilins phycoerythrin and phycocyanin. Thus, phototrophs contain a diversity of different pigments with different light-absorbing properties. This has great ecological significance, because it determines the optimum niche within the water columns and allows organisms with different combinations of pigments to coexist in illuminated habitats by absorbing a particular fraction of the light energy that other members of the community are not absorbing.

The photosynthetic pigments are contained within special membrane systems that create the structure necessary for generation of a proton-motive force. In photosynthetic bacteria, there are usually extensive invaginations of the cytoplasmic membrane; these can be complex and multilayered in cyanobacteria. In chloroplasts, the photosynthetic pigments are attached to stacks of sheet-like membranes called thylakoids.

Oxygenic photosynthesis involves two distinct but coupled photosystems

Oxygenic photosynthesis (in which H_2O is the electron donor leading to production of O_2) is the most important contributor to primary productivity and the major roles played by cyanobacteria, diatoms, and dinoflagellates in ocean

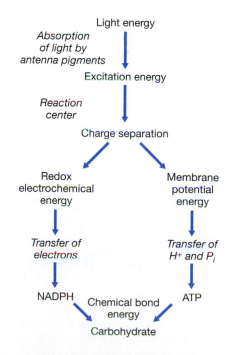

Figure 3.1 Transformation of light energy to chemical energy in photosynthesis.

processes. It is characterized by the presence of two coupled photosystems operating in series. As shown in *Figure 3.2*, photosystem I (PS I) absorbs light at long wavelengths (>680 nm) and transfers the energy to a specialized reaction center chlorophyll (P700). Photosystem II (PS II) traps light at shorter wavelengths and transfers it to chlorophyll P680. Absorption of light energy by PS I leads to a very excited state, which then donates a high-energy electron via chlorophyll *a* and iron-sulfur proteins to ferredoxin. The electron can then pass through an electron transport chain, returning to P700 and leading to ATP synthesis via generation of a proton-motive force (cyclic photophosphorylation). ATP formation and reduction of $NADP^+$ to NADPH also occurs via a noncyclic route involving both photosystems. These processes are known as the light reaction of photosynthesis. In the light-independent (dark) reaction, ATP and NADPH molecules are used to reduce CO_2, leading to its fixation into carbohydrate. The reaction may be represented as follows, where (CH_2O) represents the basic unit of carbohydrates

$$CO_2 + 3ATP + 2NADPH + 2H^+ + H_2O \rightarrow (CH_2O) + 3ADP + 3P_i + 2NADP^+$$

There are some important differences between the photosynthetic machinery of *Prochlorococcus* and other cyanobacteria, which are discussed in Chapter 4.

Anaerobic anoxygenic photosynthesis uses only one type of reaction center

Purple phototrophic bacteria contain bacteriochlorophyll *a*, but lack PS II and therefore cannot use H_2O as an electron donor in noncyclic electron transport. The light reaction generates insufficient reducing power to produce NADPH, and these bacteria therefore use H_2S, H_2 or S^0 because they are better electron donors. For this reason, they do not generate O_2 during photosynthesis and most are strict anaerobes found in shallow sediments and microbial mats. This mode of photosynthesis was almost certainly the first to evolve, before the development of PS II by cyanobacteria, enabling water to be used as electron donor. Interestingly, some cyanobacteria growing in sulfide-rich habitats can carry out anoxygenic photosynthesis because they rely only on PS I.

Aerobic anoxygenic phototrophy is widespread in planktonic bacteria

Another form of phototrophy that does not result in generation of O_2 has recently been discovered to be widespread and important. Up to 20% of the bacteria in the photic zone may be aerobic anoxygenic phototrophs (AAnP). This metabolism is particularly important in the alphaproteobacterial *Roseobacter* clade, which is very abundant in surface waters. Like the purple bacteria, these bacteria do

Figure 3.2 Electron flow in oxygenic photosynthesis. The light-driven electron flow generates an electrochemical gradient across the membrane, leading to ATP synthesis. Electrons flow through two photosystems, PS1 and PS2. Electrons are provided by water and the oxygen-evolving complex (OEC) generates oxygen. A, chlorophyll *a*; Cyt, cytochrome; FAD, flavin adenine dinucleotide; FeS, non-heme iron-sulfur protein; Fd, ferredoxin; PC, plastocyanin; Pheo, pheophytin; P680 and P700, the reaction centers of PSI and PSII, respectively, Q, quinone.

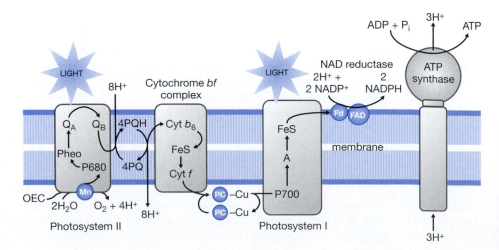

not generate O_2 during photosynthesis but, unlike them, they do grow in aerobic conditions. Sufficient transfer of electrons to drive ATP production by cyclic photophosphorylation only occurs in the presence of O_2. These organisms are photoheterotrophs and lack the enzymes for fixation of CO_2 into cell material. They use O_2 for respiratory metabolism of organic carbon and for the synthesis of bacteriochlorophyll *a*. However, AAnP bacteria metabolize more efficiently when light is available, suggesting that in oligotrophic waters they can use energy from both light and scarce nutrients simultaneously. In these bacteria, light inhibits bacteriochlorophyll *a* synthesis, so its activity diminishes during the day.

Some phototrophs use rhodopsins as light-harvesting pigments

True photosynthesis is unknown in the *Archaea*, although some extremely halophilic types contain a specialized protein called bacteriorhodopsin, which is conjugated to the carotenoid pigment retinal. This appears as purple patches in the membrane of cells grown under low oxygen concentration and high light intensity. Light energy is used to form a proton pump, which generates ATP to provide sufficient energy for slow growth when other energy-yielding reactions are not possible. Bacteriorhodopsin also functions to maintain the Na^+/K^+ balance in the cell in highly saline conditions. A similar light-mediated ATP synthesis system based on rhodopsin-like molecules has recently been shown to be very widespread amongst marine *Bacteria* and *Archaea* in the photic zone. Proteorhodopsin is thought to provide energy to support heterotrophic metabolism when nutrients are scarce. The discovery of proteorhodopsin is discussed in *Research Focus Box 3.1* and its ecological significance of is considered in Chapter 8.

Chemolithotrophs use inorganic electron donors as a source of energy and reducing power

Chemolithotrophs derive energy for the synthesis of ATP using an electron transport chain for the oxidation of inorganic molecules, and these bacteria play a major role in biogeochemical cycling in marine habitats. Until quite recently, it was thought that chemolithotrophy was only of major significance in sediments and around hydrothermal vents, but genomic data have shown that some of these processes are also highly important in the water column. The energetics of chemolithotrophy depend on the free-energy yield of coupled reactions between the electron donor and electron acceptor, but a general feature is that the yield of ATP from oxidation of inorganic substances is very low. Hence, the ecological and biogeochemical importance of chemolithotrophs is magnified because they need to oxidize large quantities of material in order to grow. Many chemolithotrophs are obligate aerobes, because oxygen is the most energetically favorable electron acceptor. However, anaerobic processes using sulfate, nitrate, and nitrite as electron acceptor are very important in anoxic sediments and deep ocean waters.

Thiotrophic bacteria use sulfur compounds as electron donor

Many types of bacteria are able to use elemental or reduced forms of sulfur (S^0, S^{2-}, $S_2O_3^{2-}$) as electron donor, with O_2 as electron acceptor. Bacteria such as *Thiobacillus*, *Thiothrix*, and *Thiovulum* are strict aerobes found in the top few millimeters of marine sediments, which are rich in sulfur. They frequently show chemotaxis to seek out the desired gradient of O_2 and sulfur compounds. They are also very prominent at hydrothermal vents and cold seeps, both as free-living forms and as symbionts of animals (see Chapter 10), where they form the base of the food chain in the absence of input from photosynthesis. *Beggiatoa*, which is very common in microbial mats, obtains energy from the oxidation of

---

ⓘ AAnP MAY CONTRIBUTE SIGNIFICANTLY TO OCEAN ENERGY

The process of AAnP had been known since the 1970s but was thought to be unimportant until Kolber et al. (2000) provided biophysical evidence demonstrating that the process is widespread in surface waters of the Pacific and Atlantic oceans. They measured the vertical distribution of bacteriochlorophyll *a* and fluorescence signals and found that photosynthetically competent AAnP bacteria comprise at least 11% of the total microbial community in the photic zone. At first, it was assumed that this type of metabolism was restricted to a very small number of types within the *Alphaproteobacteria*. However, Beja et al. (2002) prepared large insert genomic libraries and screened them for *puf* genes, which code for the subunits of the light-harvesting and reaction-center complexes. To identify groups containing the photosynthetic genes in natural populations, they used reverse transcriptase PCR of mRNA. Recently, metagenomic analyses have confirmed that genes involved in AAnP are present in many diverse groups, including *Alpha-*, *Beta-* and *Gammaproteobacteria*. Marine AAnP populations are complex and dynamic, forming a significant component of bacterioplankton in certain oceanic areas (Yutin et al., 2007). The discovery that these bacteria may be contributing significantly to photosynthesis in the oceans necessitates a radical rethink of ocean energy budgets and carbon cycling.

BOX 3.1   RESEARCH FOCUS

Microbes can top up energy with solar power
Genomic techniques reveal previously unknown types of phototrophy

One of the first demonstrations of the power of metage-nomics in revealing previously unknown types of metabolism in ocean processes was the discovery of proteorhodopsin in marine bacteria. Rhodopsin is a light-absorbing pigment found in the eye of animals. It has been known for some time that membranes of the archaean genus *Halobacterium* contain analogs of this pigment, termed "bacteriorhodopsins," which use light energy to synthesize ATP. This was thought to be a process unique to these extremely halophilic *Archaea*. However, whilst carrying out sequence analysis of DNA from an uncultivated group of the gammaproteobacteria known as SAR86, Beja et al. (2000) found evidence of a genetic sequence with strong sequence homology to the archaeal rhodopsins. Using bioinformatic tools, Beja and colleagues made a structural model of the protein believed to be encoded by the gene and identified the transmembrane domains that are a critical feature associated with the rhodopsin function. Phylogenetic analysis showed that the gene for the proteobacterial protein (which they called proteorhodopsin, PR) is phylogenetically distinct from the archaeal protein bacteriorhodopsin. The gene for the PR protein was then cloned in *Escherichia coli*, in which it was expressed as an active protein and acted as a proton pump.

To determine whether this laboratory phenomenon was relevant in the natural environment, Beja et al. (2001) used a technique known as laser-flash photolysis (which measures conformational changes in the protein as it absorbs light) to measure PR-like properties of membranes isolated from marine bacterioplankton concentrated from seawater by filtration. Their results showed that marine bacteria do indeed possess a functional reaction cycle similar to that seen in *E. coli* expressing the recombinant SAR86 gene. Subsequently, genes encoding PR have been shown to be present in a very diverse range of marine microbes. A PR gene was found in the genome sequence of the first cultured strain (*Cand.* Pelagibacter ubique) of the highly abundant SAR11 clade (see *Research Focus Box 2.1*), and Giovannoni et al. (2005) showed that the gene was expressed in culture. Sabehi et al. (2005) inserted large fragments of DNA from the Mediterranean Sea into special cloning vectors (bacterial artificial chromosomes) and calculated that up to 13% of bacteria in the photic zone contained a PR gene. As more surveys were conducted, it became clear that proteorhodopsin is not restricted to the *Proteobacteria*, or even to the *Bacteria*, because Frigaard et al. (2006) demonstrated the bacteria-like PR in large insert DNA clones of *Archaea* extracted from various depths in the Pacific Ocean. The GOS metagenomic datasets (see *Research Focus Box 4.1*) provided further evidence of the wide distribution of PRs in diverse microbes and phylogenetic analysis of the genes required for synthesis of the PR protein, and the associated retinal pigment provided strong evidence of lateral gene transfer (LGT) between organisms. McCarren and DeLong (2006) showed that acquisition of a cassette of just six genes confers the ability to harvest biochemical energy from light on a microbe and suggested that this could happen easily in a single genetic transfer.

Further studies showed that there is sequence variation in PR from bacteria isolated from different regions and depths, leading to altered biophysical properties. Beja et al. (2001) found that PR isolated from SAR86 bacteria in surface waters absorbs light maximally at 527 nm (green light), whereas at 75 m the absorption maximum is 490 nm (blue light), indicating an ecophysiological adaptation to ensure maximum absorption of blue light, which penetrates better in deeper waters. This spectral tuning is due to small changes in amino acid composition at a critical region of the molecule. A similar pattern was observed in the Mediterranean Sea, with seasonal variations in the distribution of different spectral types (Sabehi et al., 2007), but, surprisingly, samples from the Sargasso Sea showed only blue-adapted ecotypes. The reason for this is unclear.

Given that PRs are so widespread in diverse microbes and that the protein constitutes a considerable fraction of the cell membrane (Beja et al., 2000), it seems logical to assume that PR must provide a significant ecological advantage to planktonic bacteria and archaea that possess it. However, the physiological benefits that PR provides are still uncertain. Giovannoni et al. (2005) found that *Cand.* P. ubique (SAR11) grew equally well in seawater cultures in dark and light conditions, whereas Gómez-Consarnau et al. (2007) showed that *Dokdonia* sp. (a PR-containing member of the *Flavobacteria*) showed about a sixfold increase in growth yield when grown in the light of the appropriate wavelength. The growth-enhancing effect of light was most effective at low concentrations of organic matter. Although the proton pump generated by PR drives ATP synthesis, it is not coupled to the production of reducing power to fix carbon dioxide. Walter et al. (2007) hypothesized that PR-based phototrophy might become important to a cell as an alternative energy source if normal respiratory metabolism was unavailable. They showed that when respiration was inhibited (with sodium azide) in genetically modified *E. coli* expressing PR, and cells were illuminated with light of the correct wavelength, the proton pump could power the bacterial flagellum. Interestingly, analysis of genome sequences reveals that PR genes are present in copiotrophic bacteria like some *Vibrio* spp., and phylogenetic analysis indicates that it has been acquired by LGT. Gómez-Consarnau et al. (2010) showed increased long-term survival of a *Vibrio* strain when starved in seawater exposed to light rather than held in darkness. However, no differences in survival were observed in a PR-deficient mutant strain. PR may have a range of physiological roles (Fuhrman et al., 2008) and it remains to be seen how much contribution PR makes to the nutrition of planktonic microbes in the ocean. Use of a "solar power top-up" certainly seems to be an effective strategy of bacteria to supplement energy when organic nutrients are scarce, conferring a fitness advantage for bacteria to survive periods of resource deprivation.

inorganic sulfur compounds, and different strains are very diverse in their metabolism, varying from chemorganotrophy to chemolithoautotrophy. An outline of the biochemical process in sulfur oxidation is shown in *Figure 3.3*. *Thioploca* and *Thiomargarita* obtain energy by using sulfide as an electron donor, coupled with NO_3^- as electron acceptor and can grow autotrophically or mixotrophically using organic molecules as carbon source. They grow in nitrate-rich waters above anoxic bottom waters with high levels of organic matter. Each cell contains a very thin layer of cytoplasm around the periphery and a liquid vacuole that constitutes 80% of the cell volume and stores very high concentrations of NO_3^-.

Many chemolithotrophs use hydrogen as an electron donor

Hydrogen is a common product of the breakdown of organic compounds and many bacteria are capable of using it as the electron donor. In aerobic hydrogen bacteria, electrons from H_2 are transferred to quinone and proceed via a cytochrome chain, ending with the reduction of O_2 to H_2O. The reaction has a high yield of energy, supporting the synthesis of ATP. Examples found in marine habitats include *Alcaligenes*, *Pseudomonas*, and *Ralstonia*, most of which can fix CO_2 to grow autotrophically, although they are usually capable of growing with reduced organic compounds. Hydrogen-oxidizing bacteria are typically associated with sediments and suspended particles where an oxic–anoxic interface provides the optimum conditions for growth. Some thermophilic chemolithotrophs found at hydrothermal vents use NO_3^- to oxidize H_2, S^0, or $S_2O_3^{2-}$.

Nitrification by *Bacteria* and *Archaea* is a major process in the marine nitrogen cycle

The formation of nitrate via the aerobic oxidation of ammonia or nitrite is known as nitrification; nitrifying bacteria are present in suspended particles and in the upper layers of sediments. The ammonia oxidizers *Nitrosomonas* and *Nitrosococcus* are obligate chemolithoautotrophs, whilst the nitrite oxidizers *Nitrobacter* and *Nitrococcus* can also be mixotrophic, using simple organic compounds. These reactions yield little energy, and oxidation of 35 ammonia molecules or 15 nitrite molecules is required to produce fixation of one molecule of CO_2. Because of these activities, nitrifying bacteria play a major role in nitrogen cycling in the oceans, especially in shallow coastal sediments and beneath upwelling areas such as the Peruvian coast and the Arabian Sea. Like the phototrophs, nitrifying bacteria have extensive internal structures to increase the surface area of the membrane. It is difficult to obtain estimates of the abundance and community structure of nitrifying bacteria. Although most can be cultivated in the laboratory, the energetics of this mode of chemolithotrophy mean that the bacteria grow slowly and are difficult to work with. Immunofluorescence and genomic

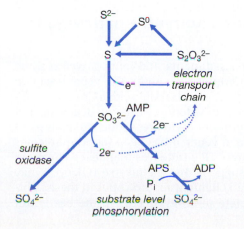

Figure 3.3 Oxidation of reduced sulfur compounds by chemolithotrophs. Sulfite is most commonly oxidized with the enzyme sulfite oxidase (left side of diagram) resulting in the direct generation of ATP via the electron transport chain and a proton motive force generated across the membrane that leads to ATP synthesis by ATPase. Some sulfur-oxidizing bacteria use the enzyme adenosine phosphosulfate (APS) reductase (right side).

methods reveal that *Nitrosococcus oceani* and similar strains are widespread in many marine environments, with worldwide distribution, at concentrations between 10^3 and 10^4 cells ml^{-1}. This organism is thought to be responsible for significant oxidation of ammonia in the open ocean. *Nitrospira* also seems to be distributed worldwide. Study of their activities and contribution to nitrogen cycling is usually carried out using isotopic methods with $^{15}NO_2^-$ or $^{15}NH_4^+$ or by using various inhibitors of nitrification enzymes (e.g. nitrapyrin inhibits ammonia monooxygenase). Nitrification is a strictly aerobic process and sufficient O_2 usually only penetrates a few millimeters into sediments. Activity of burrowing worms can increase O_2 availability to deeper levels of sediments.

Until recently, oxidation of ammonia (the first step in nitrification) was believed to be carried out exclusively by the specialized aerobic *Bacteria* named above. The discovery that nitrification is also carried out by the archaeal division *Crenarchaeota* is explored in *Research Focus Box 5.1*. This is of particular importance because this group is so abundant, especially in the deep ocean.

The Calvin–Benson cycle is the main method of carbon dioxide fixation in autotrophs

The great majority of photoautotrophs and chemolithoautotrophs incorporate CO_2 into cellular material via the cycle of reactions first discovered in plants by Calvin, Benson, and Bassham in the 1950s. The key enzyme in this pathway is ribulose bisphosphate carboxylase/oxygenase (RuBisCO), which is unique to autotrophs. In cyanobacteria and some chemolithoautotrophs, large amounts of RuBisCO are contained in crystal-like inclusion bodies (carboxysomes) surrounded by a thin envelope. These structures increase the efficiency of CO_2 fixation. If genetic mutations are generated, resulting in structural changes or loss of the carboxysomes, cells require much higher concentrations of CO_2 than normal for growth.

Each turn of the Calvin cycle incorporates one molecule of CO_2 by carboxylation of ribulose bisphosphate (RUBP). The intermediate molecule is unstable and immediately dissociates to two molecules of 3-phosphoglycerate. This is then phosphorylated to form glyceraldehyde-3-phosphate, which is a key intermediate of the glycolytic pathway. Therefore, monomers of hexose sugars are synthesized by a reversal of glycolysis. As shown in *Figure 3.4A*, the overall reaction mechanism leads to the incorporation of six molecules of CO_2 to generate one molecule of hexose sugar. It requires 12 molecules of NADPH and 18 molecules of ATP. Hexose monomers are converted by various reactions to other metabolites and the building blocks of macromolecules. If energy and reducing power are in excess supply, hexoses are converted to storage polymers such as starch, glycogen, or polyhydroxybutyrate, and deposited as cellular inclusions. The final step in the Calvin cycle is the regeneration of one molecule of RUBP by another important enzyme, phosphoribulokinase.

Some *Archaea* and *Bacteria* use alternative pathways to fix CO_2

Some *Archaea* and *Bacteria* in the deep sea, sediments, and vents, as well as some symbionts of marine invertebrates (see Chapter 10), can fix CO_2 via at least two other possible routes. The reverse (reductive) tricarboxylic acid (TCA) cycle (*Figure 3.4B*) uses ferredoxin-linked enzymes leading to the formation of acetate and was first described in *Aquifex* and *Chlorobium*. By using sequence analysis of microbial communities from hydrothermal vents, it has been found that genes encoding the key enzyme of this pathway (ATP citrate lyase) appear to be more common than genes for RuBisCO. This suggests that the reverse TCA cycle may be the main mechanism of CO_2 fixation in chemolithoautotrophs in these habitats. Another route, the hydroxypropionate pathway, occurs in the bacterium

(A)

(B)

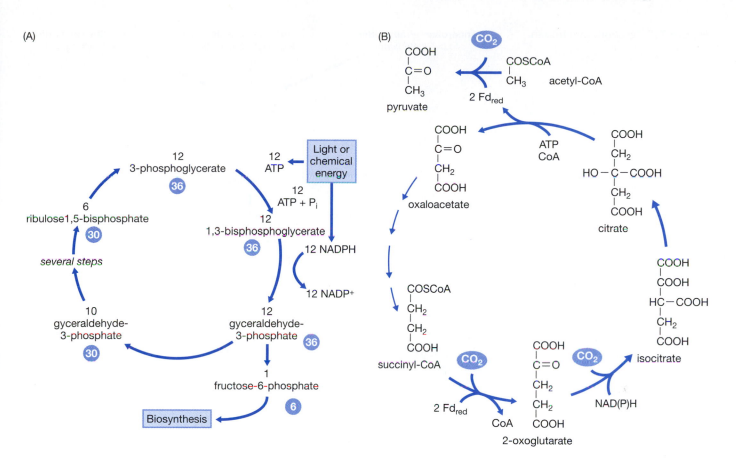

Chloroflexus and in some *Archaea* from terrestrial hot springs and volcanic habitats, but it is not known if this pathway also occurs in marine organisms. Anammox bacteria (*Planctomycetes*) use the acetyl-coA pathway, in which acetate is formed by combining the methyl group formed from reduction of one molecule of CO_2 with the carbonyl group from reduction of another CO_2 molecule with nitrite as the electron acceptor. The acetyl coA pathway for reduction of CO_2 is also used by methanogens and by acetate-producing chemoorganotrophs growing in anaerobic sediments. In this case, H_2 serves as the electron donor.

Fixation of nitrogen makes this essential element available for building cellular material in all life

The direct incorporation of atmospheric nitrogen into cellular material—known as diazotrophy—is very restricted in its distribution. It occurs only in certain species of *Bacteria* and *Archaea*, including photolithotrophs, chemolithotrophs, and chemoorganotrophs. It depends on the nitrogenase complex, consisting of the enzymes dinitrogenase (which contains Fe) and dinitrogenase reductase (which contains Fe and Mo). Nitrogen fixation is tightly regulated by different processes. Transcription of the *nif* genes encoding these enzymes is regulated by levels of O_2 and NH_3. Dinitrogenase is also very labile in the presence of O_2. This poses no problem under anaerobic conditions, but aerobic nitrogen fixers such as photosynthetic cyanobacteria must employ special strategies for protecting the enzyme from O_2. Furthermore, many bacteria limit the flow of electrons to the nitrogenase complex in the presence of excess ammonium in the environment, thus shutting down the process. The reduction of N_2 to NH_4^+ is extremely energy demanding, because N_2 is a very unreactive molecule due to the stability of the dinitrogen triple bond. *Figure 3.5* shows the reactions in nitrogen fixation, and the overall equation is

Figure 3.4 Alternative pathways of carbon fixation in autotrophs. (A) The Calvin–Benson cycle. One molecule of hexose sugar is produced from each six molecules of CO_2 incorporated; figures in circles show the total number of carbon atoms involved. (B) The reverse (reductive) tricarboxylic acid cycle used in some chemolithotrophic autotrophs. Each cycle results in the fixation of one molecule of CO_2 and production of one pyruvate molecule, which is subsequently phosphorylated using ATP, leading to the biosynthesis of cellular material. Intermediate steps are omitted for clarity.

$$N_2 + 6\,H^+ + 6\,e^- \rightarrow 2\,NH_3$$

Figure 3.5 Reactions in nitrogen fixation.

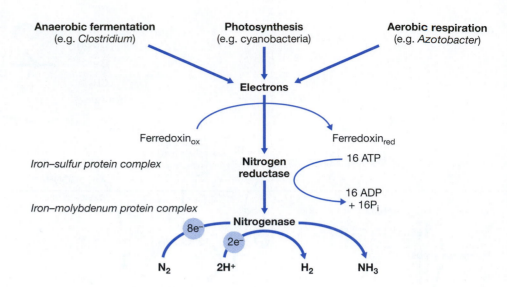

Life on Earth depends mostly on the activity of these organisms incorporating atmospheric N_2 into amino acids and the formation of other cellular material, which is then recycled by other organisms (about 5–8% of natural N_2 fixation occurs as a result of lightning activity). The aerobic cyanobacteria and the anaerobic archaeal methanogens play a particularly significant role in nitrogen cycling in the oceans, as discussed in Chapter 9.

Many marine microbes obtain energy by the fermentation of organic compounds

All energy-yielding metabolic processes depend on coupled reduction–oxidation (redox) reactions, with conservation of the energy in the molecule ATP using the reaction

$$ADP + P_i + energy \rightleftharpoons ATP + H_2O$$

Fermentation occurs only under anaerobic conditions and is the process by which substrates are transformed by sequential redox reactions without the involvement of an external electron acceptor. Fermentative microbes can be either strict or facultative anaerobes; the latter can often grow using respiration when O_2 is present. Fermentation yields much less ATP than aerobic respiration for each molecule of substrate and growth of facultative anaerobes is better in the presence of O_2. A very wide range of fermentation pathways exist in marine *Bacteria* and *Archaea*, especially those associated with degradation of organic material in sediments, animal guts, and microbial mats. Carbohydrates, amino acids, purines, and pyrimidines can all serve as substrates. Some individual species are adapted to use only a limited range of substrates, whilst others can carry out fermentations with various starting materials.

Aerobic and anaerobic respiration use external electron acceptors

In respiration, electrons are transferred via a sequence of redox reactions through an electron transport chain located in the cell membrane, which leads to the transfer of protons to the exterior of the membrane and generates a proton-motive force. Electron transport chains in most organisms contain flavoproteins, iron-sulfur proteins and quinones. In obligate and facultatively aerobic bacteria, the iron–containing cytochrome proteins are key components, but they are not

---

? DO MARINE PLANTS AND ANIMALS DEPEND ON SYMBIOTIC NITROGEN FIXATION?

In terrestrial biology, the symbiotic partnership of leguminous plants such as peas, beans, and alfalfa with symbiotic bacteria belonging to the genus *Rhizobium* has been extensively investigated, because of its importance in soil fertility. Does something similar happen in the marine environment? It appears that the high productivity of seagrass beds and mangroves is linked to close coupling with nitrogen-fixing bacteria in a biofilm of bacteria growing on the roots (the rhizosphere; review, Welsh, 2000). They make fixed nitrogen available to the plants. In recent years, additional examples of nitrogen-fixing symbionts have been discovered in diatoms (Bergman et al., 2007), corals (Lesser et al., 2004), and sponges (Mohamed et al., 2008). Anaerobic bacteria inhabiting the gut of invertebrates such as planktonic crustacean larvae may also be a significant source of fixed nitrogen (Braun et al., 1999).

present in all aerobic *Archaea*. The terminal electron acceptor in aerobic respiration is O_2. In chemoorganotrophs, respiration occurs via oxidation of organic compounds as substrates. Again, the great diversity of microbes that employ respiratory metabolism is based on specialization in the use of particular substrates. Aerobic chemolithotrophs oxidize inorganic compounds such as NH_4^+, H_2S and Fe^{2+}. The nitrifying bacteria and sulfide-oxidizing bacteria, considered above, are especially important marine examples.

Anaerobic respiration is a process in which terminal electron acceptors other than O_2 are used. The most important types of anaerobic respiration in marine systems are denitrification, sulfate reduction, methanogenesis, and methane oxidation, considered below.

Reduction of nitrate and denitrification result in release of nitrogen and other gases

Many organisms use nitrate (NO_3^-) as an electron acceptor in anaerobic respiration. The complete process of reduction is known as dissimilative reduction of nitrate and results in the formation of the gases nitric oxide, nitrous oxide, and nitrogen.

$$NO_3^- \rightarrow NO_2^- \rightarrow NO \rightarrow N_2O \rightarrow N_2$$

The synthesis of the enzymes involved in denitrification is repressed by molecular O_2 and induced by high levels of NO_3^-. The gases are lost from the ocean to the atmosphere, so these processes are critical in marine nitrogen cycling, especially in oxygen minimum zones (see Chapter 9). A major recent discovery is that a group of autotrophic bacteria belonging to the *Planctomycetes* can provide an alternative route for the return of fixed nitrogen to the atmosphere. These bacteria carry out the process of anaerobic ammonia oxidation (anammox) using nitrite as the electron acceptor.

$$NH_4^+ + NO_2^- \rightleftharpoons N_2 + 2H_2O$$

The reactions occur inside a specialized structure known as the anammoxosome, which is surrounded by a membrane containing unique lipids that aggregate to form a dense, almost impermeable structure. This membrane protects the cell from hydrazine, the highly reductive intermediate formed during the reactions.

Sulfate reduction is a major process in marine sediments

Sulfate is abundant in seawater and sulfate-reducing bacteria are very widely distributed in anoxic marine sediments, as members of microbial mat communities, and at hydrothermal vents. Although most marine *Bacteria* and *Archaea* can assimilate sulfate for biosynthesis of cellular compounds, those known as sulfate-reducing bacteria (SRB) have the ability to link the oxidation of substrates to ATP generation, using sulfate as an electron acceptor, leading to the production of hydrogen sulfide (dissimilative sulfate reduction). Although SRB are found in several phyla of the *Bacteria*, the most important types are members of the *Deltaproteobacteria*. A wide range of compounds (including H_2, organic acids, and alcohols produced by fermentation) are used by different SRB as electron donors. Thus, organic compounds generated by many different members of the microbial community can be used by SRB. The first step is activation of sulfate by ATP to adenosine-5'-phosphosulfate (APS). In dissimilative reduction, APS is reduced directly to sulfite (SO_3^{2-}) and H_2S. In dissimilative reduction, an electron transport chain leads to the production of ATP. In the presence of H_2, some SRB grow autotrophically by incorporation of CO_2 using the acetyl coA pathway. In assimilative sulfate reduction, another P is added to APS from ATP to form phosphoadenosine-5'phosphosulfate. This is reduced to SO_3^{2-} and then to

HS⁻ for the synthesis of the amino acids cysteine and methionine, and coenzyme A plus other cell factors that contain sulfur. Some organisms reduce elemental sulfur to H_2S.

Methanogenesis is a special type of metabolism carried out only by a group of *Archaea*

Many members of the *Euryarchaeota* produce methane as the final step in the anaerobic biodegradation of organic material. Most methanogens use H_2, with CO_2 serving as both the oxidant for energy generation and for incorporation into cellular material. The energy-yielding reaction used to generate ATP (via generation of a proton-motive force) is

$$CO_2 + 4H_2 \rightarrow CH_4 + 2H_2O$$

This reaction occurs in a number of steps, depending on unique coenzymes that act as carriers for the C_1 unit from the substrate (CO_2) to the product (CH_4). Other coenzymes transfer the electrons from hydrogen or other donors, if used. These compounds are unique to this group of *Archaea*. The final step of reduction leads to the formation of a single molecule of ATP. Some organisms also form methane using methyl compounds (such as methanol, methylamine, and dimethyl sulfide), formate, acetate, pyruvate, or carbon monoxide. Molecules such as sugars and fatty acids are not directly used for methane generation, but because methanogens exist in syntrophic communities with fermentative bacteria, virtually any organic compound can eventually be converted to methane. Because the energetics of sulfate reduction are better than those of methanogenesis, SRB compete with methanogens for these substrates, so methanogens are usually found in deeper sediments where sulfate levels are lower. Bacterial fermentation produces H_2, CO_2 and acetate as end products, which are then used by methanogens. The physiology and ecology of the methanogenic *Euryarcheaota* is considered further in Chapter 5.

Aerobic catabolism of methane and other C_1 compounds is widespread in coastal and oceanic habitats

Methylotrophs are able to use various one-carbon (C_1) compounds both as a source of carbon and as an electron donor. Of these, dimethylsulfoniopropionate (DMSP) is the most significant because of its importance in nutrient cycling in the oceans (see Chapter 9). A very wide range of bacteria in different phylogenetic groups can carry out this process. However, some bacteria in the *Alphaproteobacteria* and *Gammaproteobacteria* are obligate methylotrophs and use only C_1 compounds in their metabolism. A subset known as the methanotrophs can grow only on methane and a few other simple C_1 compounds. The methanotrophs possess a unique copper-complexed enzyme, methane monooxygenase, which leads to the formation of methanol and its subsequent conversion to formaldehyde by methanol dehydrogenase in the reactions

$$CH_4 + O_2 + XH_2 \rightarrow CH_3OH + H_2O + X \text{ (where X is a reduced cytochrome)}$$

$$CH_3OH \rightarrow HCHO + 2e^- + 2H^+$$

Some methanotrophs use the unique ribulose monophosphate pathway for assimilating C_1 (in the form of formaldehyde) into cellular material, whereas other types use the less efficient serine pathway for carbon assimilation. Apart from the importance of free-living forms in methane oxidation, methanotrophs also occur as symbionts of mussels found near "cold seeps" of methane-rich material on the ocean floor, providing the animals with a direct source of nutrition (see Chapter 10). Methanotrophs are also important in bioremediation of low-molecular-weight halogenated compounds, an important process in contaminated marine sediments.

Use of complex macromolecules requires the synthesis of extracellular enzymes

As we have seen, the carbon source for many marine microbes can be simple molecules such as CO_2 or CH_4. Heterotrophs must obtain their carbon from preformed organic compounds, and these may include carbohydrates, amino acids, peptides, and organic acids. The origin of these compounds in the marine environment as a component of dissolved organic matter (DOM) is discussed in Chapter 8. Generally, complex macromolecules cannot be assimilated by bacterial or archaeal cells, so many produce extracellular enzymes including chitinases, amylases, and proteases for degradation of polymers into monomers. Chitinase is especially significant in marine bacteria, since chitin (a polymer of *N*-acetyl glucosamine, NAG) is such an abundant compound in the sea. Chitin is a major structural component of the exoskeleton of many invertebrates (especially crustaceans), and the cell walls of bacteria and fungi. Many marine bacteria possess surface-associated chitinases and the released NAG is an excellent source of both carbon and nitrogen. As in all cases in which bacteria degrade macromolecules in their environment, mechanisms must exist for transporting the monomers produced into the cell. In a natural bacterial assemblage, degradation of complex macromolecules by one type of microbe may make nutrients available to many others. Such processes are particularly important in particles of marine snow and may lead to a plume of DOM as the particle falls through the water column, because colonizing bacteria produce DOM more quickly than they can use it (see *Figure 1.9*). Bacteria inhabiting this plume probably grow at much faster rates than those in bulk seawater.

Acquisition of iron is a major challenge for marine microbes

Iron is an essential constituent of cytochromes and iron–sulfur proteins, which have crucial roles in electron transport processes, especially photosynthesis and nitrogen fixation. Although iron is one of the most common elements in the Earth's crust, the amount of free iron in the open ocean can be virtually undetectable (picograms per liter) without ultraclean assay techniques. In oxygenated seawater, iron occurs in the ferric Fe(III) state and forms highly insoluble compounds, so it is a major challenge for marine microbes to obtain sufficient iron for growth. Many bacteria, both autotrophs and heterotrophs, secrete chelating agents known as siderophores that bind to iron, preventing its oxidation and allowing transport into the cell via specific receptors on the cell surface. In culture, these compounds are produced under conditions of iron restriction and some structures have been characterized. These include phenolics and derivatives of amino acids or hydroxamic acid. The first siderophores to be characterized from oceanic bacteria are structurally quite different from those previously described and behave as self-assembling amphiphilic molecules. Aquachelins from *Halomonas aquamarina* and marinobactins from *Marinobacter* both have a water-insoluble fatty-acid region and a water-soluble region of amino acids. In laboratory studies in the absence of iron, these compounds form clusters of molecules attached together by their fatty-acid tails. Upon binding to Fe(III), the micelles come together to form vesicles. It is not yet known whether this phenomenon of aggregation occurs in the natural environment, but it might be important for bacteria associated with particles containing local high concentrations of organic matter. The mechanism by which the cell takes up these siderophores is also unknown. As discussed in Chapter 9, iron cycling has a major role in the primary productivity of oceans and "iron fertilization" of the oceans is considered a potential method to mitigate the effects of elevated levels of atmospheric CO_2.

Iron deprivation is also a problem for bacteria growing in the tissues of vertebrate animal hosts, which produce iron-binding proteins as a defense mechanism. For

---

> ### (i) MOST MARINE BACTERIA HAVE A REQUIREMENT FOR SODIUM
>
> A particular feature of the majority of marine bacteria is their absolute requirement for sodium (usually in the range 0.5–5.0%) and they fail to grow when K^+ is substituted in the culture medium. Sodium is required by transport proteins (permeases), which bring substrates into the cell via Na^+-substrate symport across the membrane and also helps to stabilize the outer membrane found in Gram-negative bacteria. This distinguishes them from closely related terrestrial and freshwater species. As might be expected, most grow optimally at a concentration of NaCl similar to that found in seawater (about 3.0–3.5% NaCl), although they differ greatly in the lower and upper limits for growth. Gram-positive bacteria (which are much less common in seawater) do not seem to have this requirement.

example, *Vibrio vulnificus* and *Vibrio anguillarum* produce siderophores, known as vulnibactin and anguibactin respectively, and these are important factors in virulence (see Figure 11.7).

? WHY IS IRON SO SCARCE, YET SO IMPORTANT?

Most parts of the open ocean contain very little iron to support marine microbial life, yet it is an essential element required for a number of key biological processes, including photosynthesis and nitrogen fixation. Why does life depend on such a scarce nutrient? The iron-containing proteins that are critical to these reactions are highly conserved across diverse phylogenetic lineages, suggesting that they evolved very early in the history of Earth. The early evolution of life occurred under anoxic conditions, but the development of oxygenic photosynthesis about 2.4 billion years ago resulted in a sudden transition of the atmosphere and oceans to an oxidizing state. Iron precipitated in large quantities and the amount of soluble iron in seawater today is extremely low. Aquatic microbes, including the cyanobacterial ancestors that caused the change in oxygen content of the atmosphere, faced an iron-shortage crisis and this necessitated the development of efficient mechanisms for scavenging the increasingly low levels in the environment. Other transitional metals are occasionally found to replace iron, but they are equally scarce in the oceans, and iron is uniquely suited to its function.

The growth of bacterial cells depends on availability of nutrients and environmental factors

During growth, a bacterial cell must synthesize additional cell membranes, cell wall components, hundreds of different RNA molecules, thousands of proteins, and a complete copy of its genome. The coordination of these activities is subject to complex regulation of gene expression. With a few exceptions (such as the budding and stalked bacteria), almost all bacteria reproduce by binary fission into two equal-sized cells. When growing under optimal or near-optimal conditions, the cell doubles in mass before division; regulation of DNA replication and cell septation are largely controlled by the rate of increase in cell mass. The growth cycle of culturable heterotrophs in the laboratory is well known. Upon introduction to the growth medium, unbalanced growth (without cell division) occurs during a lag phase. During this period, the cell synthesizes metabolites, enzymes, coenzymes, ribosomes, and transport proteins needed for growth under the prevailing conditions. Under ideal conditions, bacteria then enter a logarithmic phase and grow exponentially. Different species vary enormously in the time taken for the population to double (generation time). Environmental factors such as temperature, pH and O_2 availability (for aerobes) have a major influence on the rate of growth. Growth rate increases with temperature; most marine bacteria isolated from water at 10°C can grow in the laboratory with generation times of about 7–9 h, whilst those isolated at 25°C typically double every 0.7–1.5 h. All organisms have a minimum, optimum, and maximum temperature for growth and this reflects their evolution in a particular habitat. Above the minimum temperature, metabolic reactions increase in rate as temperature increases, and so the generation time reduces. The maximum temperature is usually only a few degrees above the optimum; this is determined by the thermostability of enzymes and membrane systems. Some marine bacteria have very low generation times. For example, *Vibrio parahaemolyticus* can divide every 10–12 min at 37°C (although this temperature is not encountered in its normal estuarine habitat) and the hyperthermophile *Pyrococcus furiosus* doubles every 37 min at 100°C. At the other extreme, many oligotrophs and psychrophiles growing just above 0°C may have generation times of several days.

Bacteria adapt to starvation by a series of coordinated changes to cell metabolism

In culture, bacteria continue exponential growth until they enter the stationary phase because (i) a limiting nutrient is exhausted, (ii) toxic by-products of growth accumulate to become self-limiting, or (iii) some density-dependent signal limits growth. Cells entering the stationary phase do not simply "shut down" but undergo a number of genetically programmed changes that involve the synthesis of many new starvation-specific proteins induced via global regulatory systems. When cells of laboratory-cultured marine bacteria are starved, they become smaller and spherical. In the 1980s, Morita coined the term "ultramicrobacteria" to describe the very small cells (0.3 μm diameter or less) that develop in this way. To achieve reduction in cell size, some proteins and ribosomes are degraded in a controlled fashion, but 30–50 new proteins are induced in response to starvation. These include proteins that are essential for survival with limited sources of energy and nutrients as well as proteins that enable cells to acquire scarce substrates at very low concentrations. Cell surfaces become more hydrophobic, which increases the adhesion of bacteria to surfaces, and surface association in biofilms appears to be a major adaptation to nutrient limitation. Changes in the fatty acid content of membranes promote better nutrient uptake and provide increased stability, whilst cell walls become more resistant to autolysis. Starved cells have a much-reduced ATP content , but maintain a proton-motive force across the membrane. Starved cells are more resistant to a variety of stress conditions, and mutations in any of the genes in the survival induction pathway may lead to death in the stationary phase. As noted in Chapter 1, most pelagic bacteria are also very small and have similar morphology to the ultramicrobacteria formed by starvation of cultured marine isolates in the laboratory. For some time, the prevailing paradigm was that marine bacteria are "normal" cells that have undergone reduction in size and are in a state of perpetual near-starvation. This led to the view that bacterioplankton are virtually inactive in the natural environment. In fact, we now know that the naturally occurring small cells found in the sea are *more* active than larger bacteria on a per volume basis and a large fraction of supposedly inactive cells can be induced to high metabolic activity by addition of substrates. However, determination of *in situ* growth rates of marine heterotrophic microbes remains one of the most difficult aspects. Recent techniques have relied on the use of different fluorochromes in flow cytometry (see *Table 2.1*) to determine cell biomass. Some protocols attempt to distinguish dead and live bacteria by employing fluorochromes directed against nucleic acids, whilst others use fluorochromes that detect respiratory activity. One method combines nucleic acid staining with a permeable dye and an impermeable dye both attached to gene probes to measure membrane integrity and DNA content. Cell viability can also be assessed by measuring the uptake of radiolabeled compounds (such as amino acids) by microautoradiography (see p. 54) or the use of fluorescent electron acceptors to monitor the presence of an active electron transport chain. Contrary to earlier ideas, planktonic microbes in the open sea are metabolically active and their small size is an evolutionary (genotypic) adaptation to ensure efficient use of scarce nutrients in their oligotrophic environment.

Most marine microbes are adapted to an oligotrophic lifestyle and grow very slowly

Estimating the growth rate of bacteria in natural seawater is difficult because of the complicating effects of viral lysis and protistan grazing, but there is no doubt that growth rates are very low and nothing like the rapid doubling times observed with those bacteria that can be easily grown in laboratory culture. Many planktonic bacteria in the oceanic gyres probably divide once a day or less, whilst those associated with marine snow particles or patches of dissolved nutrients will divide much more rapidly. Many of the particle-associated microbes are relatively easy to culture, but only a small fraction of the oligotrophs have so far been cultured, using the special methods discussed in *Research Focus Box 2.1*. Most planktonic bacteria are metabolically active oligotrophs, but their small size and slow growth rates are an adaptation to the chronic low levels of nutrients. By contrast, the particle-associated bacteria may be considered as copiotrophs with a "feast or famine" lifestyle. Rapid growth rates and increased cell size can be induced in response to increased levels of nutrients. Conversely, as noted in the preceding section, they show size reduction and other adaptations in response to nutrient limitation.

Some bacteria enter a "viable but nonculturable" state in the environment

The terms "viable but nonculturable" (VBNC) and "somnicell" ("sleeping" cells) were originally introduced in the 1980s by Rita Colwell and coworkers, who observed that *Vibrio cholerae* does not die off when introduced into the environment. Since then, the VBNC state has been studied extensively and shown to occur in a wide range of Gram-negative (and a few species of Gram-positive) bacteria, including pathogens of humans, fish, and invertebrates. The phenomenon

---

ⓘ USING METAGENOMICS TO PREDICT THE NUTRITIONAL TYPES OF BACTERIA

Studying the factors that determine uptake and utilization of nutrients by an organism is obviously very difficult if it cannot be cultured. To get around this problem, Lauro et al. (2009) compared the full genome sequences of *Sphingopyxis alaskensis* (an oligotroph) and *Photobacterium angustum* (a copiotroph) and identified clusters of orthologous groups of genes that characterize each type. They extended the analysis to include another 32 genome sequences and developed a bioinformatics model based on 43 genomic markers to predict the trophic type accurately. They found that the copiotrophs have larger diversity of outer membrane proteins and secreted proteins (including extracellular enzymes) than oligotrophs. Oligotrophs have a smaller diversity of nutrient transporter proteins than copiotrophs, but they have high affinity for a broad range of substrates. Copiotrophs have more genes associated with motility, defense mechanisms, and signal transduction, which explains their ability to regulate their metabolism in response to nutrients and other extracellular stimuli. This approach can be used to screen the large numbers of metagenomic data now being generated in order to assess rapidly changes in the major characteristics of indigenous populations.

VBNC BACTERIA— JUST RESTING, OR SUICIDAL?

One of the most controversial aspects of the VBNC state is whether cells can be resuscitated to a form that grows on agar plates in the laboratory. Many studies have found that addition of nutrients or temperature changes can cause VBNC bacteria to revert, provided that the cells have not been in the VBNC state for too long. Others refute these findings and state that the effect is due to the regrowth of a few remaining culturable cells in the population (review, Oliver, 2005). Bloomfield et al. (1998) proposed that bacteria possess a kind of "self-destruct" mechanism, occurring when exponentially growing bacteria face a sudden insult such as the deprivation of nutrients, resulting in the decoupling of growth from metabolism. Consequently, cells may suffer a burst of oxidative metabolism, resulting in lethal free radicals. Bacteria can avoid this if they can induce changes to protect DNA, proteins, and cell membranes. The shock of sudden transfer of cells into a rich medium when they are still in the process of adaptation to life in the aquatic environment could result in rapid death. The inability of cells in the VBNC state to detoxify peroxides and other free radicals commonly present in culture media or induced within the cells themselves during exposure to rich media seems to be a key explanation for the phenomenon of nonculturability. In *Vibrio vulnificus*, Smith and Oliver (2006) showed that this is due to repression of the gene encoding periplasmic catalase, and studies in the author's laboratory have shown that *Vibrio shiloi* can be resuscitated on media containing catalase, which breaks down toxic peroxide (Vattakaven et al., 2006).

is thus especially important in studies of the ecology of indigenous pathogenic marine bacteria and the survival characteristics of sewage-associated pathogens and indicator organisms. This term has been used by some microbiologists to explain the discrepancy between the number of bacterial cells that can be visualized by direct observation (e.g. epifluorescence or immunological detection) and those that can be enumerated using viable plate counts. The VBNC state could constitute a dangerous reservoir of pathogens, which cannot easily be detected, and it is also relevant to evaluation of the survival of introduced genetically modified organisms into the environment. A variety of factors, especially nutrient deprivation and changes in pH, temperature, pressure, or salinity can initiate the cascade of cellular events leading to the VBNC state. The VBNC state remains a matter of considerable controversy among microbiologists; whilst many microbiologists believe these events to be an inducible response to promote survival under adverse conditions, many remain skeptical about the validity of the VBNC state as an adaptive stage.

Several methods can be used to detect VBNC cells. For example, radiolabeling techniques can be used to show that VBNC cells are metabolically active (i.e. they incorporate a suitable substrate); and demonstration of the reduction of reduced tetrazolium salts can be used to indicate an active electron transport chain. One method, which has been widely used to distinguish metabolically active from inactive bacteria, is the "direct viable count," in which cell growth is stimulated by the addition of yeast extract whilst cell division is inhibited by the antibiotic nalidixic acid. This results in greatly enlarged cells that can be easily seen in the microscope. Another method that is widely used is the use of nucleic-acid-binding fluorochromes in conjunction with epifluorescence microscopy or flow cytometry. Some dyes have differential ability to penetrate dead and "alive" (i.e. metabolically active) cells.

It is of fundamental importance to distinguish the transition shown by bacteria that can be grown in laboratory culture to a starvation or VBNC state from the "unculturability" of most marine bacteria. As discussed in *Research Focus Box 2.1*, these are better described as "not yet cultured," since, although they cannot be cultured using conventional methods, there has been considerable progress using dilution methods.

Nutrients are acquired via specialized transport mechanisms

In laboratory cultures, most nutrients are usually provided at concentrations well in excess of those required for balanced growth. Therefore, the rate of growth is independent of the nutrient concentration, although the final growth yield (number of cells or cell mass) does increase with higher concentrations of nutrient. Bacterial cells possess very efficient mechanisms for transporting nutrients into the cell. However, the growth rate will reduce at very low threshold concentrations of the limiting nutrient.

Bacteria synthesize special proteins that span the membrane to form a channel through which nutrients (and certain other substances) pass into the cell. Often, these transport proteins are specific to particular substrates. The proton-motive force drives simple transport across the membrane. Facilitated diffusion can occur when the external concentration of nutrients is higher than in the cell and active transport enables the cell to acquire nutrients even when the external concentration is very low, as will occur in most aquatic environments. As the substrate is transported across the membrane, the conformation of the transport protein changes so that the substrate is released on the inner side of the membrane. This process requires energy in the form of ATP hydrolysis or the proton-motive force. In the latter case, H$^+$ must be transported into the cell at the same time (symport). In another method of transport, the substrate is chemically modified during entry to the cell. Many bacteria accomplish this using the phosphotransferase system, a group of enzymes that phosphorylate molecules such as sugars. In many Gram-negative bacteria, amino acids, peptides, and sugars may cross the outer membrane through porins into the periplasmic space between the outer and cytoplasmic membranes (see *Figure 4.2*), where they bind to specific binding proteins. These proteins have extremely high affinities for their substrate, which means that the cell can acquire nutrients when present at micromolar or nanomolar concentrations in the external environment. The periplasmic binding proteins then transfer the nutrient to a membrane-spanning protein for active transport across the membrane. Such systems are known as ABC transporters, derived from the term ATP-binding cassette. Hundreds of different ABC transporters exist for the uptake of different groups of organic compounds such as sugars and amino acids and for inorganic molecules such as sulfate and phosphate. Oligotrophic bacteria must possess very efficient transport systems in order to ensure that nutrients can enter the cell against a very steep concentration gradient. In the case of *Cand*. Pelagibacter ubique, genomic analysis has shown a very high proportion of genes encoding membrane transporters and proteomic analysis reveals very high levels of periplasmic substrate-binding proteins. In conjunction with maintaining a very small cell size and large periplasmic space, this maximizes nutrient uptake (see *Research Focus Box 2.1*).

Growth efficiency of many marine bacteria is probably low

Growth efficiency can be defined as the ratio of yield of biomass produced (grams) to amount of substrate utilized (moles). Efficiency can also be expressed in terms of the amount of energy in the form of ATP (y_{ATP}) produced per mole of substrate. In studies of bacterial culture under constant conditions (chemostat culture), the value of y_{ATP} is always much lower (usually <50%) than that predicted by theoretical calculations based on the energetics of biochemical pathways. This is because organisms require a certain amount of nutrient for cell maintenance such as transport, maintenance of internal pH and a proton gradient across the membrane, DNA repair, and other essential cell functions. These functions take priority over biosynthesis for growth. At low growth rates due to nutrient limitation, the cell will use proportionately more of the substrate for maintenance energy. Below a certain threshold, cells will divert all the substrate to just "staying alive" rather than growing. Uncoupling of catabolic and anabolic processes often occurs at low growth rates when growth is limited by a substrate other than the energy source. Maintaining the maximum possible energy state of the cell enables it to resume growth rapidly when conditions change. These effects of nutrient levels and growth efficiency are important concepts in assessing the role of microbes in carbon flux and productivity in marine systems. Because nutrient concentrations in the sea are generally low, growth rates are necessarily slow and a large proportion of energy intake is used for maintenance energy, especially active transport and synthesis of extracellular enzymes for acquiring nutrients. Therefore, net growth will be low, despite considerable turnover of organic material. However, as noted in Chapter 1, there is an increasing recognition that the ocean environment is very heterogeneous with respect to nutrients and that many marine microbes are associated with local concentrations of organic matter.

Microbes use a variety of mechanisms to regulate cellular activities

Even the simplest microbial cells contain hundreds of genes, encoding enzymes and other macromolecules required for the hundreds of biochemical reactions in metabolism. To ensure efficient use of resources and to respond to changes in environmental conditions, cells need to regulate both activity of enzymes and the level of macromolecular synthesis. Regulation of the activity of enzymes is a very rapid process, occurring by various mechanisms such as feedback allosteric inhibition, in which an excess of the end product of a biosynthetic pathway often inhibits the first enzyme of the pathway. Some other enzymes are modified by covalent addition of groups, which inhibits catalytic activity. Although some proteins are required under all conditions and are said to be constitutive, there are many mechanisms by which cells regulate the production of enzymes or other proteins that are only required under certain conditions, by controlling the level of gene expression. Because mRNA is very short-lived, switching transcription on or off provides a very rapid response to changing circumstances. Also, bacterial and archaeal genes are often grouped together in operons, so that expression of related genes (e.g. several enzymes in a biosynthetic pathway) is regulated together under the control of a single promoter and operator region, upstream of the structural genes. As noted above, copiotrophic planktonic microbes possess many regulatory systems, and the same applies to microbes that grow on surfaces or colonize animals or plants. The expression of a gene involves the binding of RNA polymerase to a specific region of the DNA (the promoter). Many genes are subject to negative control, when regulatory compounds bind to specific DNA-binding proteins. This causes an altered conformation of the protein, allowing it to bind to the DNA at the operator region (between the promoter and the gene) leading to the blocking of transcription. Genes can also be regulated positively—in this case, an activator protein acts to help the RNA polymerase bind efficiently to the promoter region of the DNA. Often, several operons will be regulated by the same repressor or activator, so they are expressed coordinately. Sometimes, regulation of a master gene results in a cascade of subsequent regulatory steps, so that expression of hundreds of genes is up- or downregulated in response to a single stimulus.

One of the most common methods used by cells to respond to external stimuli is via two-component regulatory systems. In this case, the signal does not act directly to control transcription, but is detected by a sensor kinase protein in the cell membrane, which then transduces the signal to a response regulator protein in the cytoplasm. The kinase enzyme is composed of two domains, one of which is exposed to the exterior and detects an environmental or chemical signal. When this occurs, the cytoplasmic domain of the protein becomes phosphorylated. The phosphoryl group is then transferred to a second specific protein called the response regulator. When phosphorylated, the response regulator interacts with specific regions of the DNA and can either activate or repress transcription, of specific genes, depending on the system. In Chapters 11 and 12 we shall see important examples of such signaling and global control networks when considering infection of animals by symbiotic or pathogenic bacteria from the aquatic environment.

Some bacteria use motility in the quest for nutrients and optimal conditions

The ability to swim toward high concentrations of favorable nutrients (chemotaxis) is an important property of many marine bacteria, which swim due to the presence of one or more flagella. These are long helical filaments attached to the cell by means of a hook-like structure and basal body embedded in the membrane. In marine bacteria, the most common arrangement is attachment of a single flagellum to one end of the cell (polar flagellation). Some groups have peritrichous flagella, in which the flagella are arranged all over the cell surface. The flagellum in *Bacteria* is about 20 nm in diameter and is made of many subunits of a single protein (flagellin) synthesized in the cytoplasm and passed up the central channel of the filament to be added at the tip. Although the basic structure of flagella in *Archaea* is similar to that of *Bacteria*, the flagellum is made of several different proteins and is much thinner (about 13 nm in diameter), possibly too thin for the subunits to pass up the central channel. Genomic analysis shows little homology between bacterial and archaeal flagellum proteins. The basal body of the flagellum consists of a central rod and a series of rings embedded in the cell wall and membrane, composed of about 20 different proteins. Movement is caused by the rotation of the basal body acting as a tiny motor, which couples the flow of ions to generation of force. The flagellum is rigid and behaves like a propeller. The remarkable mechanics of the motor and the process of self-assembly of the fine flagella tubules have attracted the attention of engineers for possible exploitation in nanotechnology. The mechanism has been extensively studied in *Escherichia coli*, in which rotation is caused by the movement of H^+ ions from the outside to the inside of the cell through protein subunits (Mot) surrounding the ring in the membrane. However, most motile marine bacteria that have been investigated seem able to use either H^+ or Na^+ ions, which is of advantage in alkaline (pH 8) seawater. Sodium-driven flagellar motors are exceptionally fast—vibrios can rotate their flagella at 1700 revolutions per second and swim at up to 400 µm per second. When *Vibrio parahemolyticus* or *Vibrio alginolyticus* are grown on solid surfaces or in viscous environments, they undergo a remarkable change in morphology. The cells stop dividing normally and elongate up to about 30 µm. At the same time, they start synthesizing large numbers of lateral or peritrichous flagella as well as the normal polar flagellum (*Figure 3.6*). The polar flagellum is effective when the cell is in the free-living, planktonic state, whilst the lateral flagella enable cells to move better through viscous environments. This can be observed in the laboratory by the phenomenon of swarming on the surface of agar plates. In the natural environment, differentiation to the swarmer form enables vibrios to adopt a sessile lifestyle, which is significant in the formation of biofilms on surfaces or the colonization of host tissues. The polar flagellum acts as a sensor and increased viscosity slows down its rotation, leading to induction of the lateral phenotype. Experimentally induced interference of the Na^+ ion flux with chemical agents also induces differentiation to the swarmer state. Genetic analysis of the Na^+ motor is leading to an understanding of this mechanism. However, little is known at present about how the signal from the flagellum induces changes in the expression of genes for altered cell division and synthesis of different flagella. In vibrios, the polar (Na^+-driven) flagellum is enclosed in a sheath formed from the outer membrane. It is not known whether the flagellum rotates within the sheath or whether the filament and sheath rotate as a single unit. The lateral (H^+) flagella do not appear to possess a sheath.

The swimming behavior of many marine bacteria is very different to the "classical" system, which has been studied extensively in *E. coli*. When motile bacteria like *E. coli* are observed in the microscope, they move in an apparently random fashion. They swim on a relatively straight path ("run") before executing a brief "tumble" and swimming off in another direction. The tumble is caused by a change in direction of rotation of the flagellum and reorientation of the cell

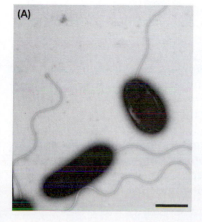

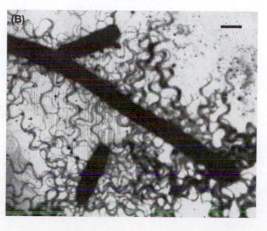

Figure 3.6 Electron micrographs of *Vibrio parahaemolyticus*. (A) Grown in liquid showing the swimmer cells with single, sheathed polar flagellum. (B) Grown on a surface showing the elongated swarmer cell with numerous lateral flagella. Bars represent ~1 µm (see McCarter, 1999). (Image courtesy of Linda McCarter, University of Iowa.)

due to bombardment by water molecules (Brownian motion). The presence of attractants or repellents modifies the frequency of tumbling. Cells tumble less frequently and therefore spend longer swimming up a gradient of increasing concentration of an attractant and show a net movement toward the source of the substance. The bacterium senses minute changes in concentration with time, through a series of chemoreceptors on the cell surface and a chemical signaling system (involving changes in the methylation state of proteins) that relays information to a tumble generator at the base of the flagellum. This complex process involves many different proteins encoded by numerous genes. The movement of bacteria occurs over tiny, micrometer distances, whereas energy flow into the ocean through convection, wind, and planetary rotation creates turbulence on a scale many orders of magnitude greater. However, the effects of turbulence translate down to powerful micrometer-scale shear forces. Furthermore, because most free-living marine bacteria are very small (as little as 0.2 μm diameter), the effects of Brownian motion (random bombardment of the bacterium by molecules in the suspending fluid) are very great. Therefore, marine bacteria need to swim very fast and show very rapid responses to altered conditions if motility and chemotaxis are to be effective. Between 40 and 70% of marine bacteria may be motile, although the observation of motility shows diurnal and seasonal variations and depends on the water composition. Marine bacteria can indeed swim in short bursts at staggering speeds: up to 400 μm (equivalent to several hundred body lengths) per second. Studies of marine vibrios show that the Na^+-driven motor enables rotation speeds hundreds of times faster than the H^+-driven system. Analysis of the polar flagellar motor shows that the Na^+-driven motor has additional components that are responsible for detecting the speed of rotation or the flux of Na^+ ions. Very high speeds enable small marine bacteria to swim in a relatively straight line before stopping and reversing direction. The reversal of flagellar rotation is almost instantaneous, so that the bacterium is not "knocked off course" by shear or Brownian motion. Most high-speed bacterial community members use both H^+ and Na^+ ion motors simultaneously. Obviously, chemotaxis is only relevant for copiotrophic bacteria that have the ability to respond to localized concentrations of organic carbon. Clusters of bacteria can be observed around microscopic nutrient sources such as algal cells or particles of organic matter. The "run and reverse" motility means that, if a bacterium finds itself moving away from a nutrient "hot spot," it can rapidly move back to the favorable area.

Movement toward light (phototaxis) is common in phototrophic marine bacteria, which detect changes in the concentration of visible light of different wavelengths and swim toward higher illumination. The extremely halophilic archaeon *Halobacterium salinarum* possesses rhodopsin molecules, which act as light sensors influencing rotation of the flagella. Other tactic movements include those toward or away from high concentrations of oxygen (aerotaxis) and ionic substances (osmotaxis). Magnetotaxis is a specialized response seen in certain species of bacteria found in marine and freshwater muds which contain intracellular inclusion bodies called magnetosomes. These bacteria orient themselves in the Earth's magnetic field, although their movements are primarily in response to gradients of chemicals and oxygen.

Gliding motility does not involve flagella and only occurs when bacteria are in contact with a solid surface. It is especially important in the responses of cyanobacteria and members of the *Cytophaga* and *Flavobacteria* genera to gradients of oxygen and nutrients in microbial mats. The mechanism of gliding motility is not fully understood. The most common mechanism appears to be a kind of jet propulsion through the exclusion of slime via minute pores in the cell surface. However, in *Flavobacterium* it is thought that rotatory motors, similar to flagellar motors, transmit force to special outer membrane proteins, which results in a ratchet-like movement of the cell envelope, analogous to the movement of a tank on caterpillar tracks.

? HOW CAN BACTERIAL SWIMMING AFFECT GLOBAL PROCESSES?

In a generally low-nutrient environment such as the open ocean, it is energetically impossible for bacteria to seek out nutrients by biased random movement. Instead, they must rely on turbulent flow to bring them within a patch of increased nutrient concentration and then use the "run and reverse" movement strategy to maximize the likelihood that they will remain there (Mitchell and Kogure, 2006). Is this chemotactic behavior efficient enough to provide nutritional advantage to the bacteria before the patch is dissipated by physical effects? Stocker et al. (2008) used a microfluidic device to measure the response of *Pseudoalteromonas haloplanktis* to nutrient patches mimicking those produced by a point source such as a lysed algal cell or to plumes generated by a falling marine snow particle (see *Figure 1.9*). The bacteria concentrated in a nutrient "hot spot" within a few seconds and the fastest 20% of the population gained a tenfold higher nutrient concentration than nonmotile cells. In response to a nutrient plume, motile bacteria gained a fourfold increase in nutrient exposure. Bacterial motility in the ocean has great ecological and geochemical significance because of the heterogenous distribution of nutrients. It is intriguing that study of such microscale processes may improve our understanding of global-scale processes like CO_2 flux and climate change.

Formation of biofilms is an important step in microbial colonization of surfaces

The physiological properties of bacteria in the sea are greatly affected by population density and interactions with other microbes. There are significant differences between free planktonic cells and those that are associated as biofilms on surfaces and with particles such as marine snow. Although biofilm formation has been recognized for more than a century, it is only in recent years that significant advances in the study of biofilm physiology have been achieved, largely because of the application of confocal microscopy and FISH techniques (see *Figure 2.8*). Bacteria and diatoms both initiate biofilm formation as a result of environmental cues, particularly nutrient availability. Bacteria may undergo considerable developmental changes during transition from the suspended, planktonic form to the sessile, attached form. The events occurring in biofilm formation by single species of bacteria have been studied extensively in laboratory investigations, and the process can be regarded as a form of cellular development toward a multicellular lifestyle (some researchers liken it to a tissue). Organic molecules (largely polysaccharides and proteins) will coat any surface with a conditioning film within minutes of placing it into seawater. The initial stage in colonization by bacteria usually involves motility toward the surface, and changes in microviscosity as the cell approaches the surface may cause motility to slow. This causes a transient, reversible adsorption to the surface, mediated by electrostatic attraction and van der Waals forces. Genetic studies involving the creation of biofilm-deficient mutants have shown the importance of motility due to flagella and type IV pili in the movement of bacteria across the surface in order to contact other cells to form microcolonies.

During the process of biofilm formation, expression of specific genes is induced—this depends very much on the nature of the substrate. In experimental studies, bacteria will quickly colonize inert surfaces such as pieces of plastic, glass, or stainless steel immersed in seawater, but the genes expressed are quite different from those expressed when they colonize organic substrates such as chitin. Some marine heterotrophic bacteria express chitin binding and chitinase genes selectively when they encounter chitin, so that they can begin to utilize it as a nutrient source. This is especially important in the colonization of the surface of crustacean animals (see *Research Focus Box 12.1*).

The most important developmental change following attachment is the expression of copious quantities of exopolymeric substances (EPS). These provide a strong and sticky framework that cements the cells together. The chemical and physical properties of EPS are very variable and the nature and amount produced depends on the bacterial species, concentration of specific substrates, and environmental conditions. The majority are polyanionic because of the presence of acids (e.g. D-glucuronic, D-galacturonic, or D-mannuronic acids), ketal-linked pyruvate, or the presence of phosphate or sulfate residues. The EPS often form a complex network of long, interlinked strands surrounding the cell (glycocalyx). Both rigid and flexible properties can be conferred on the biofilm, depending on the secondary structure of the EPS, and interactions with other molecules such as proteins and lipids may produce a gel-like structure. The mature biofilm often has a complex architecture composed of pillars and channels. Stalked bacteria and diatoms may increase the length of their stalk when growing in dense biofilms. Once established, the dense packing of bacteria and the diversion of metabolism to the production of EPS may mean that the cells are metabolically active but divide at very slow rates.

In nature, biofilms are rarely composed of single species. Bacteria, algae, flagellates, ciliates, other protists, and viruses will all interact in the mature biofilm, and there is growing evidence that horizontal gene transfer between different microbial species is greatly enhanced within biofilms. This is of great significance in the evolution of organisms with altered characteristics.

(i) BACTERIA STICK TOGETHER TO WARD OFF PREDATORS

One of the advantages of the biofilm mode of life for bacteria seems to be protection against predation (grazing) by phagotrophic protists. Matz et al. (2008) compared the grazing efficiency of two flagellate protists common in coastal waters, the surface feeding *Rhynchomonas nasuta* and the suspension feeding *Cafeteria roenbergensis*, against a large number of bacteria isolated from the surface of a seaweed. *C. roenbergensis* increased in density in response to planktonic cells of the bacteria and predated the bacteria with high efficiency. By contrast, when grown as biofilms, most of the bacteria were resistant or toxic to the surface-feeding flagellate *R. nasuta*. One of the most effective antigrazing compounds produced by biofilm cells was identified as the alkaloid violacein, which inhibits protozoan feeding at nanomolar concentrations by inducing a conserved eukaryotic cell death program. The authors concluded that biofilm-specific resistance against predation contributes to the successful persistence of biofilm bacteria in various environments. Unraveling the interactions between bacterial biofilms and eukaryotic cells may reveal important clues to the evolution of compounds that affect cell function.

Pili are important for bacterial attachment to surfaces and exchange of genetic information

Pili (also known as fimbriae) are fine hair-like protein filaments on the surface of many bacteria. Pili are typically about 3–5 nm in diameter and 1 μm long. They are composed of a single protein, although a cell may have different types and the amino acid sequence of each type can show significant strain-to-strain variation. The most common pili are those involved in the attachment of bacteria to surfaces; some microbiologists restrict the use of the term fimbriae to these adhesive structures. Their function has been particularly well studied in the interaction of pathogenic bacteria with mucous membranes of animals. There is often a specific receptor recognition site on the pili, which can explain why certain strains of bacteria attach specifically to particular hosts or tissues. Host immune responses to bacterial attachment are often important in resisting infection. Adhesive pili are often encoded by bacteriophages (viruses) or plasmids (small elements of DNA that replicate independently of the main chromosome) and may be exchanged by transduction or conjugation, respectively. The human pathogen *V. cholerae* provides a good example: here, pili play a crucial role in infection, and the acquisition and expression of genes are important factors in explaining the transition of this bacterium from its estuarine or marine habitat to the human gut. Pili are also involved in colonization of the squid light organ by *Vibrio fischeri* (see *Research Focus Box 10.2*). Another example is the attachment of *Roseobacter* spp. to invertebrate larvae, where pili develop into a form of holdfast. More generally, pili play a key role in the attachment of bacteria to inanimate surfaces as the first stage in biofilm formation, discussed below.

As well as the more common adhesive pili, some cells possess structures known as type IV pili. These are responsible for a type of cell movement on surfaces known as twitching motility. This involves reversible extension and retraction of the pili and can result in a sudden jerky movement of a few micrometers. Twitching motility is thought to be important in the formation of microcolonies during the initial establishment of biofilms.

Sex pili are quite different structures, often several micrometers long, formed only by donor ("male") bacteria involved in the process of conjugation. Sex pili and the machinery needed for replication and transfer of DNA from donor to recipient cells are encoded by conjugative plasmids. Sex pili are important in the transfer of genes, such as those encoding antibiotic resistance or degradative ability, between cells of bacteria in the marine environment. In addition, sex pili are the receptors for certain bacteriophages, and an application of this is the use of F+ RNA bacteriophages as an indicator of fecal pollution in seawater.

Antagonistic interactions between microbes occur on particles or surfaces

As noted above, the activity of extracellular enzymes leads to the dissolution of organic material from marine snow particles as they sink through the water column, and much of this DOM becomes available to other members of the plankton. Extracellular enzyme activity is likely to provide little "return" for free-living bacteria at low density. Composition of the bacterial community inhabiting the particle and the plume of DOM that follows it could therefore have significant effects on the release and subsequent utilization of nutrients. We know that there are extensive antagonistic interactions between microbes in dense communities inhabiting the soil or the gut of animals, owing to the production of antibiotics that inhibit growth of organisms unrelated to the producer strain, but there have been few such investigations of antagonistic interactions between marine bacteria. However, recent studies indicate that a large proportion of marine-particle-associated bacterial isolates possess antibiotic activity against other pelagic bacteria. Such antibiotic interactions are likely to be widespread in sediments, microbial mats, and biofilms on the surface of plants and animals.

? ARE INTERACTIONS BETWEEN BACTERIA IMPORTANT IN CORAL HEALTH?

As discussed in Chapters 10 and 11, corals contain complex communities of microbes and disturbance of the resident community is often associated with disease. Interactions between coral-associated bacteria have been investigated in the author's laboratory. Nissimov et al. (2009) showed that several bacteria isolated from mucus of the Mediterranean coral *Oculina patagonica* inhibited the growth of *Vibrio shiloi*, *V. coralliilyticus* and other pathogens. *In vitro* experiments showed that biofilms of *Roseobacter* sp. and *Pseudoalteromonas* sp. inhibited growth of the pathogens, supporting the concept of a probiotic effect on microbial communities associated with corals. In another study, Tait et al. (2010) showed that vibrios isolated from diverse corals differ greatly in the production of quorum sensing signal molecules. The high diversity of vibrios and the differential effects of temperature on signal production and inhibition may partly explain the complexity of community changes in responses of corals to environmental factors, although analysis of signal production and disruption *in situ* is needed to understand fully the role of quorum sensing in coral health.

Quorum sensing is an intercellular communication system for regulation of gene expression

Quorum sensing (QS) is a mechanism of intercellular communication—the controlled production, release, and detection of threshold concentrations of signal molecules.

QS is widely used by a variety of bacteria to coordinate communal behavior. Usually, this involves the coordinated expression of certain genes in response to population density. Study of this phenomenon is leading to new insights into the physiology and ecology of attached marine microbial communities. The signal molecules used in QS are chemically diverse, but the commonest and best studied are N-acyl homoserine lactones (AHLs; *Figure 3.7*), which are widespread in diverse bacteria. AHLs have different chain lengths, and presence or absence of oxo- and hydroxyl groups. In a mixed bacterial population, there may be many different structures present.

QS was first studied in detail in the bacterium *V. fischeri*, in which it is used to control bioluminescence (see p. 104). When grown in laboratory broth cultures, the bacteria emit no light until the bacteria enter the late logarithmic or stationary phase and the population reaches a certain critical density (typically about 10^7 cells ml^{-1}). This is because the bacteria synthesize a freely diffusible autoinducer molecule and release it into the medium. Low-density cultures can be induced to show bioluminescence by the addition of supernatants from high-density cultures. The autoinducer is an AHL synthesized from S-adenosylmethionine by the protein LuxI. As illustrated in *Figure 3.8*, AHL is produced in the cytoplasm and passively diffuses through the bacterial membrane. AHL diffuses back into the cell, binding to the protein LuxR, a polypeptide of about 250 amino acids comprising two domains; the N-terminal domain binds the AHL and the C-terminal domain binds to a palindromic sequence (*lux* box) upstream of the *lux* operon promoter. Thus, when a certain threshold concentration is reached, the bioluminescence genes encoding the various proteins required for the bioluminescence system are expressed by activation of transcription of the operon. The LuxI/LuxR system is very widespread and used by many Gram-negative bacteria to regulate the activity of many genes, including colonization and virulence factors. This description of the LuxI/LuxR QS mechanism is somewhat simplified, as bacteria differ considerably in the details of the regulatory circuits and other factors are often involved.

Elucidation of the regulation of bioluminescence in another marine vibrio, *V. harveyi*, has revealed two separate systems. Like *V. fischeri*, *V. harveyi* synthesizes and responds to an AHL molecule (in this case termed AI-1, synthesized by the *luxL* and *luxM* genes). However, it also possesses a separate autoinducer (AI-2, a boron-containing furanosyl diester; *Figure 3.7B*) synthesized by the *luxS* gene. As shown in *Figure 3.9*, the two autoinducers are recognized by sensor kinase proteins LuxN and LuxQ, which have a histidine kinase domain and a response regulator domain. The "signal" is thus transduced by a phosphorylation mechanism to protein LuxU and onto LuxO, which is a negative regulator of the *luxCDABEGH* operon. Phosphorylation-dephosphorylation of LuxO determines the expression of the structural genes for bioluminescence. Highly conserved *luxS* homologs occur in many Gram-negative and Gram-positive bacteria. AI-2 synthase genes have been found in about half of all sequenced bacterial genomes and AI-2 is often considered a universal signaling molecule, which bacteria use for communication both within and between species.

Gram-positive bacteria do not use AHLs, but rely on peptide signals via two-component response regulator proteins resembling those of *V. harveyi*. In this case, the autoinducers are usually short peptides and are highly species specific. They are actively exported from the cells by ATP-binding transporters.

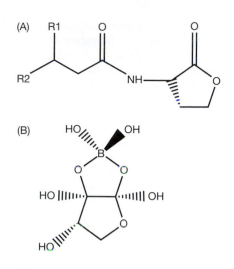

Figure 3.7 Structures of some quorum-sensing signal molecules in marine bacteria. (A) Core molecule of acyl homoserine lactone (AHL)-like inducers, where R1 can be H, O or OH and R2 is a side chain unit (CH_2) repeated 0, 1, 2, 3, 4 or more times. For example, in *Vibrio fischeri* 3-oxo-C6-homoserine lactone, R1 = $CH_3(CH_2)_2$, R2 is absent. In *Vibrio harveyi* 3-hydro-C4-homoserine lactone, R1 = CH_3, R2 is absent. (B) Autoinducer-2 (AI-2) a furanosyl borate diester produced by a wide range of Gram-negative and Gram-positive bacteria, including *V. harveyi*. (Note that *V. fischeri* has been reclassified as *Aliivibrio fischeri* (Urbanczyk et al., 2007) and the strain of *V. harveyi* used for elucidation of quorum sensing has been reclassified as *Vibrio campbellii* (Lin et al., 2010); the original names are used for ease of reference to published work). (Reprinted from Dobretsov et al. [2009] by permission of Taylor & Francis Ltd.)

Figure 3.8 Quorum sensing in *Vibrio fischeri*. Bioluminescence is controlled by two regulatory genes *luxI* and *luxR* in two different operons. The *luxI* gene encodes the enzyme responsible for synthesis of the vibrio autoinducer (VAI; 3-oxo-C6-homoserine lactone). The upper diagram (A) represents a cell at low cell density, in which there is a low constitutive transcription of *luxI* and *luxR* (*blue arrows*). As cell density increases (B), VAI accumulates in the area surrounding the cells. LuxR is a repressor of the *lux* operon promoter (–). When LuxR binds the autoinducer VAI, it binds strongly to the *lux* box upstream of the promoter and transcription of the right operon is enhanced (+). Production of VAI (*gray arrows*) increases exponentially, giving an autocatalytic feedback loop as well as initiation of bioluminescence. Genes *luxA* and *luxB* encode the α and β subunits of the enzyme luciferase. Genes *luxC*, *luxD* and *luxE* encode enzymes for the synthesis of the aldehyde substrates from fatty acids. The LuxR–VAI complex also binds at the *luxR* promoter, but in this case it represses transcription, resulting in a compensatory negative feedback.

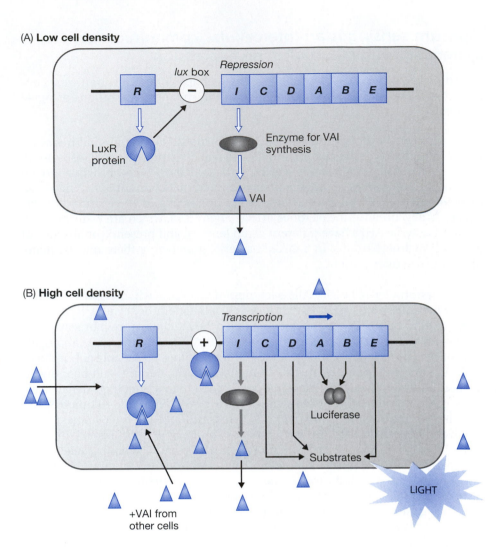

Following its discovery and detailed investigation in marine vibrios, QS has emerged as one of the most important mechanisms of gene regulation in bacteria, and application of this knowledge extends across the whole field of microbiology. Experimental studies of single-species biofilms have shown that the production of AHLs is important in determining the three-dimensional structure of mature biofilms. AHL mutants produce densely packed biofilms that are more easily dislodged from the surface. In multispecies marine biofilms bacteria may utilize signaling molecules produced by other species or actively inhibit or degrade them by production of specific enzymes. Marine-snow-associated bacteria have recently been shown to produce AHLs, suggesting that QS is important in the production of extracellular enzymes and antagonistic interactions in these densely populated habitats.

Most marine microbes grow at low temperatures

Over 90% of the ocean has a temperature of 5°C or colder. The temperature of seawater in the deep sea and in polar regions ranges from –1 to 4°C, whilst internal fluids in sea ice can be as low as –35°C in winter. With the exception of sea ice, these temperatures are very stable and little affected by seasonal changes. The overwhelming majority of marine microorganisms are adapted to this cold environment and conditions can only be considered "extreme" from a human perspective. Psychrophilic (cold-loving) microbes are defined as those with an optimum growth temperature of less than 15°C, a maximum growth temperature of 20°C, and a minimum growth temperature of 0°C or less. In fact, many deep sea and polar bacteria have quite a narrow temperature growth range and

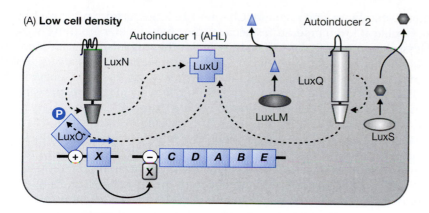

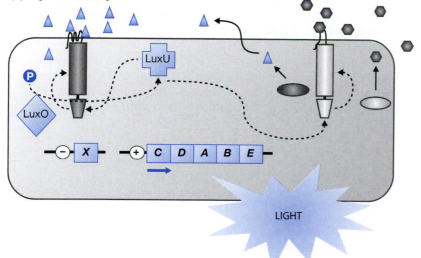

Figure 3.9 Quorum sensing in *Vibrio harveyi*. There are two separate quorum-sensing systems linked by a phosphorelay system (*dotted arrows*). Signaling occurs via the two-component proteins LuxN and LuxQ, which contain a sensor kinase domain (*rectangle*) and response regulator domain (*trapezoid*). Signals from both sensors are passed to the shared integrator LuxU protein (*cross*) to the LuxO protein (*diamond*). (A) represents a cell in a low-density population. When phosphorylated (denoted by P), protein LuxO activates an unidentified repressor (shown as ☒) of the *luxCDABE* operon so that the luciferase enzyme subunits and reaction substrates are not expressed. (B) Illustrates a cell in a high-density population and shows the accumulation of the autoinducers (AI-1, AI-2) synthesized by proteins LuxLM and LuxS. These bind to the sensor proteins (in the case of AI-2 via a periplasmic binding protein) and switch the activities of the sensor kinases into phosphatases. LuxO is dephosphorylated so that repression of the *lucCDABE* operon is relieved.

may lose viability after brief exposure to typical laboratory temperatures. Therefore, special precautions are needed in their collection, transport, and culture. Psychrotolerant bacteria are those that can grow at temperatures as low as 0°C, but have optima of 20–35°C; many organisms from shallow seawater or coastal temperate regions fall into this category. Apart from understanding their considerable importance in nutrient cycling and ocean processes, there has also been recent interest by astrobiologists in psychrophiles because of the planned future exploration and search for possible life on Europa, the ice-covered moon of Jupiter. Biotechnologists also study psychrophiles because of the industrial potential of their fatty acids and extracellular polymer-degrading enzymes such as chitinase, chitobiase, and xylanase (Chapter 14). Proteins from psychrophiles are more flexible at low temperatures because they have greater amounts of α-helix and lesser amounts of β-sheet than those from other organisms. In addition, specific amino acids are often located at particular regions of the active site of the enzymes, ensuring better access of the substrate. Study of several enzymes has been facilitated by using recombinant DNA technology for their expression and manipulation in *E. coli*. *Colwellia psychroerythrea* (domain *Bacteria*), an obligate psychrophile that is widely distributed in Antarctic and Arctic regions, was one of the first cold-adapted organisms for which the genome sequence was elucidated. Comparison of the genome with that of mesophiles confirms the importance of amino acid composition and the prevalence of certain residues for protein function. Since the discovery of the abundance of the *Archaea* (especially *Crenarchaeota*) in deep ocean waters, there have been many studies of proteins from these organisms. For example, the elongation factor 2 of psychrophilic methanogens is overproduced and has higher affinity for GTP (guanosine 5′-triphosphate) compared with that of mesophilic relatives, permitting efficient protein synthesis at low temperatures. Active transport of substances across membranes must also be efficient at low temperatures. This is achieved by the incorporation of large amounts of unsaturated fatty acids into the membrane, which helps to maintain membrane fluidity and transport processes. Omega-3-polyunsaturated fatty acids, once thought to be nonexistent in bacteria, have been found in Antarctic and deep-sea isolates and have considerable biotechnological potential (see Chapter 14).

Microbes growing in hydrothermal systems are adapted to very high temperatures

Only some *Bacteria* and *Archaea* can grow at temperatures above 60°C. Such thermophilic organisms are found in the marine environment in areas of geothermal activity including shallow submarine hydrothermal systems, abyssal hot vents (black smokers) and active volcanic seamounts (see *Figure 1.10*). In deep-sea vent systems, the temperature of seawater can exceed 350°C. As this superheated water mixes with cold seawater, a temperature gradient is established and diverse communities of thermophilic organisms with different temperature optima occur. Those organisms that can grow above temperatures of 80°C are termed "hyperthermophiles." *Table 3.2* shows the temperature growth ranges of representative marine species. The majority of hyperthermophiles described to date belong to two major groups of *Archaea*. About 70 species of hyperthermophilic *Archaea* have been described and the full genome sequences of several have now been elucidated. Only two major genera of *Bacteria* (*Aquifex* and *Thermotoga*) are hyperthermophilic. In both domains, hyperthermophiles occupy very deep branches of the phylogenetic tree (see Figures 4.1 and 5.1).

As shown in *Table 3.2*, hyperthermophiles show a range of physiological types and can be aerobic or anaerobic, chemoorganotrophic, or chemolithotrophic. Their enzymes and structural proteins are adapted to show optimum activity and stability at high temperatures. The overall structure of proteins from hyperthermophiles often shows relatively little difference from that of homologous proteins in mesophiles. However, variation in a small number of amino acids at critical locations in the protein seems to affect the three-dimensional conformation, permitting greater stability and function of the active site of enzymes. Intracellular proteins from hyperthermophiles also contain a high proportion of hydrophobic regions and disulfide bonds, which improve thermostability. There is great interest in the use of enzymes from hyperthermophiles in high-temperature industrial processes (see Chapter 14). Adaptations of the cell membrane also occur to ensure stability and effective nutrient transport at high temperatures. The membranes of *Archaea* contain ether-linked isoprene units and hyperthermophiles usually possess monolayer membranes, which appear to be more stable at high temperatures.

Table 3.2 Growth conditions of hyperthermophilic marine *Archaea* and *Bacteria*

Species	Growth temperature (°C)			Nutrition	Aerobic (Ae)/ Anaerobic (An)
	Min.	Optimum	Max.		
Archaea					
Archeoglobus fulgidus	60	83	95	CL	An
Ferroglobus placidus	65	85	95	CL	An
Igniococcus sp.	65	90	103	CL	An
Methanocaldococcus jannaschii	46	86	91	CL	An
Methanococcus igneus	45	88	91	CL	An
Methanopyrus kandleri	84	98	110	CL	An
Pyrobaculum aerophilum	75	100	104	CL	Ae/An
Pyrococcus furiosus	70	100	105	CO	An
Pyrodictium occultum	82	105	110	CL	An
Pyrolobus fumarii	90	106	113	CL	An
Staphylothermus marinus	65	92	98	CO	An
Thermococcus celer	75	87	93	CO	An
"Strain 121"[a]	85	106	121	CL	An
Bacteria					
Aquifex pyrophilus	67	85	95	CL	Ae
Thermotoga maritima	55	80	90	CO	An

[a]Strain 121 is related to *Pyrodictium*. CL, chemolithotrophic; CO, chemoorganotrophic. Data from Huber and Stetter (1998), and Kashefi and Lovley (2003).

Microbes that inhabit the deep ocean must withstand a very high hydrostatic pressure

Pressure increases by one atmosphere (atm = 0.103 megapascals) for every 10 m water depth and over 75% of the ocean's volume is more than 1000 m deep. As we now know that *Bacteria* and *Archaea* are distributed in great numbers throughout the water column, as well as in marine sediments many hundreds of meters deep, growth under conditions of very high pressure is the normal state of affairs for the vast majority of marine microbes. Zobell and Morita pioneered the study of deep-sea bacteria in the 1940s and 1950s. Recent advances in sampling and cultivation methods together with the application of molecular techniques to the study of diversity and physiology are leading to some significant new insights. Those *Bacteria* and *Archaea* that have been isolated and cultured in the laboratory are usually found to be barotolerant (i.e. they can grow over a wide range of pressures from 1 to 400 atm). For such organisms, high pressure usually results in lower growth rates and metabolic activity. Many organisms isolated from coastal environments can only tolerate pressures up to about 200 atm. By contrast, some organisms can only grow at pressures greater than 400 atm; these are known as obligate or extreme barophiles (the alternative term "piezophiles" is also used). Many of these organisms die if brought to the surface, and exposure to light exacerbates this lethal effect. By the use of special isolation techniques involving collection in pressurized chambers and cultivation in solid silica gel media, an increasing number of species of obligate barophiles have been cultured in recent years, including some from the deepest habitats such as the Marianas Trench in the Pacific Ocean (10500 m). Genetic studies of extreme barophiles indicate that many have a close resemblance to common barotolerant or barosensitive species (e.g. the γ-proteobacterial genera *Shewanella, Photobacterium, Colwellia*, and *Moritella*), although some unique taxa have also been discovered. At abyssal depths (>4000 m), barophiles appear to be ecologically dominant over bacteria from shallow waters carried there by sinking organic matter. The most abundant sources of obligate barophiles are nutrient-rich niches such as decaying animal carcasses or the gut of deep-sea animals. However, oligotrophic barophiles adapted to very low nutrient concentrations also occur in seawater. Some obligately barophilic chemolithotrophic *Archaea* have been found near hydrothermal vents, but little is known about their physiology.

A range of adaptations seems to be present in deep-sea organisms. It is important to note that very low temperatures (apart from hydrothermal vents) and very low nutrient conditions (apart from the localized occurrence of concentrations of organic matter) characterize the deep sea, as well as high pressure. We must therefore consider adaptation of deep-sea microorganisms as a response to the combined effects of these factors.

Protection of enzymes from the effects of pressure seems to be due mainly to changes in the conformation of proteins. Proteins in barophiles seem less flexible and less subject to compression under pressure, owing to a decreased content of the amino acids proline and glycine. Cells grown under high pressure also contain high levels of osmotically active substances, which are thought to protect proteins from hydration effects of high pressure. The most well-studied adaptation to high pressure and low temperature is a change in membrane composition. Membranes of barophiles contain a higher proportion of polyunsaturated fatty acids and have a more tightly packed distribution of fatty-acyl chains. Pressure also affects DNA secondary structure.

Application of molecular genetics to the study of two barophilic bacteria, *Photobacterium profundum* and *Shewanella* sp., has revealed some insight into the mechanisms of regulation of the pressure response. When *P. profundum* is shifted from atmospheric to high pressure, the relative abundance of two outer membrane proteins (OmpH and OmpL) is altered. These proteins act as porins for the transport of substances across the outer membrane. An increased production of OmpH probably provides a larger channel, suggesting that the pressure response enables the bacteria to take up scarce nutrients more easily (as would occur in the deep-sea environment). A pair of cytoplasmic membrane proteins regulate transcription of the *ompL* and *ompH* genes. Interestingly, these have a high-sequence homology to the genes encoding ToxR and ToxS proteins, which were first discovered in *V. cholerae*, where they detect changes in temperature, pH, and salinity during the transition from the aquatic environment to the host, leading to the expression of virulence factors (see *Figure 12.1*). The homology indicates some common ancestry of this environmental sensing system that has evolved to perform different functions according to habitat. Some other genes have also been shown to be important in the response to pressure in a deep-sea *Shewanella* sp. and these may be grouped into pressure-regulated operons. Use of gene probes directed against these genes could aid in the identification of new species of barophiles that we cannot currently culture.

Microbes vary in their requirements for oxygen or tolerance of its presence

Obligate aerobes always require the presence of oxygen and use it as the terminal electron acceptor in aerobic respiration. Facultative aerobes can carry out anaerobic respiration or fermentation in the absence of oxygen or aerobic respiration in its presence. Even though oxygen is not required, the growth of facultative organisms is better in its presence on account of the greater yield of ATP from aerobic respiration. Microaerophiles carry out aerobic respiration but require an O_2 level lower than that found in the atmosphere. Obligate anaerobes carry out fermentation or anaerobic respiration and many are killed by exposure to oxygen, although some are aerotolerant and survive (but do not grow) in its presence. Examples of all categories occur in marine *Bacteria* and *Archaea*.

Oxygen can exist in various forms that are highly reactive and toxic to all cells unless they possess mechanisms to destroy them. Singlet oxygen ($^1O_2$) is a high-energy state that causes spontaneous oxidation of cellular materials. In particular, singlet oxygen forms during photochemical reactions, and phototrophs usually contain carotenoid pigments, which convert singlet oxygen to harmless forms (quenching). For this reason, nonphototrophic marine organisms exposed to bright light (such as those inhabiting clear surface waters) are also often pigmented. Various other toxic oxygen species form during respiration (*Figure 3.10*). Superoxide and hydroxyl radicals are particularly destructive and react rapidly with cellular compounds. The evolution of mechanisms for the removal of toxic oxygen species was a major step in the transition of the biosphere from anaerobic to aerobic following the development of oxygen-evolving photosynthesis. Organisms capable of aerobic growth usually contain the enzymes catalase ($2 H_2O_2 \rightarrow H_2O + O_2$), superoxide dismutase ($2O_2^- + 2H^+ \rightarrow H_2O_2 + O_2$), and peroxidase ($H_2O_2 + NADH + H^+ \rightarrow 2H_2O + NAD^+$). Superoxide reductase is an enzyme originally found in *Pyrococcus furiosus* and thought to be unique to the *Archaea*, but genome sequence analysis has shown that it may be widely distributed in obligate anaerobes in place of superoxide dismutase. This enzyme reduces superoxide to H_2O_2 without the formation of O_2 ($O_2^- + 2H^+ + cyt\ c_{reduced} \rightarrow H_2O_2 + cyt\ c_{oxidized}$).

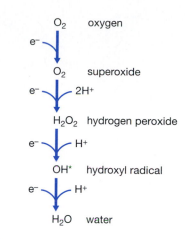

Figure 3.10 Formation of toxic intermediates during reduction of oxygen to water.

Ultraviolet irradiation has lethal and mutagenic effects

Research into the effects of ultraviolet (UV) radiation on marine microbes is needed because of growing evidence that UV radiation is increasing at certain locations on Earth, as a result particularly of ozone depletion in the upper atmosphere and the formation of an ozone "hole" over Antarctica and the Southern Ocean. The lethal and mutagenic effects of UV radiation result from damage to DNA. UV-B causes direct damage to DNA through the formation of pyrimidine dimers, whilst the main effects of UV-A are due to formation of toxic oxygen and hydroxyl radicals. Various mechanisms for the repair of UV-induced damage exist, including nucleotide excision repair and light-activated enzyme repair (photoreactivation). Studies of DNA damage in bacteria in surface waters show that there is a pronounced effect over the course of the day, with maximal damage evident in the late afternoon and repair occurring during the night. We do not yet fully understand the ecological significance of these processes. Bacteria produce a range of UV-screening products such as mycosporine-like amino acids and scytonemin, a complex aromatic compound formed in the sheath of some cyanobacteria. Some bacteria isolated from corals in very clear surface waters show extreme resistance to the effects of UV radiation by enhancing the activity of NAD(P)H quinine oxidoreductase, a powerful antioxidative enzyme. These mechanisms could have significant biotechnological potential in human health, as products for skin-protection treatments and for overcoming the effects of oxidative stress during aging.

Microbes are protected from osmotic damage by various mechanisms

Several genera of *Archaea* are extreme halophiles that grow at very high NaCl concentrations (15–35%) found in salterns, brine pockets within sea ice, and submarine brine pools (see *Research Focus Box 1.2*). Extreme halophilicity is rare in the *Bacteria*, but *Salinibacter ruber* is an exception. To protect themselves from dehydration due to loss of water from the cell to the external environment, marine microbes must maintain the concentration of intracellular solutes at a high level. One way of achieving this is by accumulating noninhibitory substances known as compatible solutes or osmoprotectants. Usually, these types of sugars, alcohols, or amino acids are extremely soluble in water. For example, many Gram-negative bacteria synthesize compounds such as glycine-betaine, ectoine, glutamate or α-glucosylglycerol. These substances are released when cells lyse and some bacteria accumulate glycine-betaine from the environment rather than synthesizing it themselves. Most Gram-positive bacteria accumulate the amino acid proline as an osmoprotectant. In algae, DMSP is the main osmoprotectant and plays a major role in the ocean sulfur cycle (Chapter 9).

The extremely halophilic *Archaea* use a different method to prevent water loss. They have an active mechanism for pumping K^+ ions into the cell until the internal concentration balances the high concentration of Na^+ outside. In some species, a large proportion of the proton-motive force for the ion pump derives from light-mediated generation of ATP via the pigment-containing protein bacteriorhodopsin (see *Research Focus Box 3.1*). Extreme halophiles also have other adaptations for growth at high NaCl concentrations. Their enzymes and structural proteins have a high proportion of acidic amino acids, which protects the conformation from disruption by high salt concentrations. Internal cellular components, such as the ribosomes and DNA-replication enzymes, require high K^+ concentrations for their integrity and activity.

Conclusions

This chapter has explored the very wide range of metabolic activities in marine microbes. We have seen a large diversity of mechanisms by which they obtain energy from light or chemical sources and various ways in which they use this to fuel their biosynthetic processes. Some types of metabolism, with major significance in global ecology, have been discovered only in the last few years and there are undoubtedly many more surprises to be discovered. Investigation of phenomena such as antagonism and intercellular communication is revealing information about the ways in which different microbial species interact and the interdependence of their metabolic activities. In subsequent chapters, numerous examples of the activities of such communities in marine habitats will become apparent.

References

Beja, O., Aravind, L., Koonin, E. V., et al. (2000) Bacterial rhodopsin: evidence for a new type of phototrophy in the sea. *Science* 289: 1902–1906.

Beja, O., Spudich, E. N., Spudich, J. L., Leclerc, M. and DeLong, E. F. (2001) Proteorhodopsin phototrophy in the ocean. *Nature* 411: 786–789.

Beja, O., Suzuki, M. T., Heidelberg, J. F., et al. (2002) Unsuspected diversity among marine aerobic anoxygenic phototrophs. *Nature* 415: 630–633.

Bergman, B., Rai, A. N. and Rasmussen, U. (2007) Cyanobacterial associations. In: *Associative and Endophytic Nitrogen-Fixing Bacteria and Cyanobacterial Associations* (eds C. Elmerich and W. E. Newton). Springer, Netherlands, pp. 257–301.

Bloomfield, S. F., Stewart, G. S. A. B., Dodd, C. E. R., Booth, I. R. and Power, E. G. M. (1998) The viable but non-culturable phenomenon explained? *Microbiology* 144: 1–3.

Braun, S. T., Proctor, L. M., Zani, S., Mellon, M. T. and Zehr, J. P. (1999) Molecular evidence for zooplankton-associated nitrogen-fixing anaerobes based on amplification of the *nifH* gene. *FEMS Microbiol. Ecol.* 28: 273–279.

Dalsgaard, T., Thamdrup, B. and Canfield, D. E. (2005) Anaerobic ammonium oxidation (anammox) in the marine environment. *Res. Microbiol.* 156: 457–464.

Dobretsov, S., Teplitski, M. and Paul, V. (2009) Mini-review: quorum sensing in the marine environment and its relationship to biofouling. *Biofouling* 25: 413–427.

Fenchel, T. and Glud, R. N. (1998) Veil architecture in a sulphide-oxidizing bacterium enhances countercurrent flux. *Nature* 394: 367–369.

Frigaard, N. U., Martinez, A., Mincer, T. J. and DeLong, E. F. (2006) Proteorhodopsin lateral gene transfer between marine planktonic *Bacteria* and *Archaea*. *Nature* 439: 847–850.

Fuhrman, J. A., Schwalbach, M. S. and Stingl, U. (2008) Opinion—proteorhodopsins: an array of physiological roles? *Nat. Rev. Microbiol.* 6: 488–494.

Giaquinto, L., Curmi, P. M. G., Siddiqui, K. S., Poljak, A., DeLong, E., DasSarma, S. and Cavicchioli, R. (2007) Structure and function of cold shock proteins in *Archaea*. *J. Bacteriol.* 189: 5738–5748.

Giovannoni, S. J., Bibbs, L., Cho, J. C., et al. (2005) Proteorhodopsin in the ubiquitous marine bacterium SAR11. *Nature* 438: 82–85.

Gómez-Consarnau, L., Gonzalez, J. M., Coll-Lladó, M., et al. (2007) Light stimulates growth of proteorhodopsin-containing marine *Flavobacteria*. *Nature* 445: 210–213.

Gómez-Consarnau, L., Akram, N., Lindell, K., et al. (2010) Proteorhodopsin phototrophy promotes survival of marine bacteria during starvation. *PLoS Biol.* 8: e1000358, doi:10.1371/journal.pbio.1000358.

Henke, J. M. and Bassler, B. L. (2004) Three parallel quorum-sensing systems regulate gene expression in *Vibrio harveyi*. *J. Bacteriol.* 186: 6902–6914.

Huber, H. and Stetter, K. O. (1998) Hyperthermophiles and their possible potential in biotechnology. *J. Biotechnol.* 64: 39–52.

Kashefi, K. and Lovley, D. R. (2003) Extending the upper temperature limit for life. *Science* 301: 934.

Kolber, Z. S., Van Dover, C. L., Niederman, R. A. and Falkowski, P. G. (2000) Bacterial photosynthesis in surface waters of the open ocean. *Nature* 407: 177–179.

Kuenen, J. G. (2008) Anammox bacteria: from discovery to application. *Nat. Rev. Microbiol.* 6: 320–326.

Lauro, F. M., McDougald, D., Thomas, T., et al. (2009) The genomic basis of trophic strategy in marine bacteria. *Proc. Natl Acad. Sci. USA* 106: 15527–15533.

Lesser, M. P., Mazel, C. H., Gorbunov, M. Y. and Falkowski, P. G. (2004) Discovery of symbiotic nitrogen-fixing cyanobacteria in corals. *Science* 305: 997–1000.

Lin, B., Wang, Z., Malanoski, A. P., et al. (2010) Comparative genomic analyses identify the *Vibrio harveyi* genome sequenced strains BAA-1116 and HY01 as *Vibrio campbellii*. *Environ. Microbiol. Rep.* 2: 81–89.

Matz, C., Webb, J. S., Schupp, P. J., et al. (2008) Marine biofilm bacteria evade eukaryotic predation by targeted chemical defense. *PLoS ONE* 3, doi:10.1371/journal.pone.0002744.

McCarren, J. and DeLong, E. F. (2006). Proteorhodopsin photosystem gene clusters exhibit co-evolutionary trends and shared ancestry among diverse marine microbial phyla. *Environ. Microbiol.* 9: 846–848.

McCarter, L. (1999) The multiple identities of *Vibrio parahaemolyticus*. *J. Mol. Microbiol. Biotechnol.* 1: 51–57.

Mitchell, J. G. and Kogure, K. (2006) Bacterial motility: links to the environment and a driving force for microbial physics. *FEMS Microbiol. Ecol.* 55: 3–16.

Mohamed, N. M., Colman, A. S., Tal, Y. and Hill, R. T. (2008) Diversity and expression of nitrogen fixation genes in bacterial symbionts of marine sponges. *Environ. Microbiol.* 10: 2910–2921.

Muyzer, G., Yildirim, E., van Dongen, U., Kuhl, M. and Thar, R. (2005) Identification of "*Candidatus* Thioturbo danicus," a microaerophilic bacterium that builds conspicuous veils on sulfidic sediments. *Appl. Environ. Microbiol.* 71: 8929–8933.

Nissimov, J., Rosenberg, E. and Munn, C. B. (2009) Antimicrobial properties of resident coral mucus bacteria of *Oculina patagonica*. *FEMS Microbiol. Lett.* 292: 210–215.

Oliver, J. D. (2005) The viable but nonculturable state in bacteria. *J. Microbiol.* 43: 93–100.

Sabehi, G., Loy, A., Jung, K. H., et al. (2005) New insights into metabolic properties of marine bacteria encoding proteorhodopsins. *PLoS Biol.* 3: e273, doi:10.1371/journal.pbio.0030273.

Sabehi, G., Kirkup, B. C., Rozenberg, M., Stambler, N., Polz, M. F. and Beja, O. (2007) Adaptation and spectral tuning in divergent marine proteorhodopsins from the eastern Mediterranean and the Sargasso Seas. *ISME J.* 1: 48–55.

Smith, B. and Oliver, J. D. (2006) *In situ* and *in vitro* gene expression by *Vibrio vulnificus* during entry into, persistence within, and resuscitation from the viable but nonculturable state. *Appl. Environ. Microbiol.* 72: 1445–1451.

Stocker, R., Seymour, J. R., Samadani, A., Hunt, D. E. and Polz, M. F. (2008) Rapid chemotactic response enables marine bacteria to exploit ephemeral microscale nutrient patches. *Proc. Natl Acad. Sci. USA* 105: 4209–4214.

Tait, K., Joint, I., Daykin, M., Milton, D. L., Williams, P. and Camara, M. (2005) Disruption of quorum sensing in seawater abolishes attraction of zoospores of the green alga *Ulva* to bacterial biofilms. *Environ. Microbiol.* 7: 229–240.

Tait, K., Hutchison, Z., Thompson, F. L. and Munn, C. B. (2010) Quorum sensing signal production and inhibition by coral-associated vibrios. *Environ. Microbiol. Rep.* 2: 81–89.

Thar, R. and Kuhil, M. (2002) Conspicuous veils formed by vibrioid bacteria on sulfidic marine sediment. *Appl. Environ. Microbiol.* 68: 6310–6320.

Urbanczyk, H., Ast, J. C., Higgins, M. J., Carson, J. and Dunlap, P. V. (2007) Reclassification of *Vibrio fischeri*, *Vibrio logei*, *Vibrio salmonicida* and *Vibrio wodanis* as *Aliivibrio fischeri* gen. nov., comb. nov., *Aliivibrio logei* comb. nov., *Aliivibrio salmonicida* comb. nov. and *Aliivibrio wodanis* comb. nov. *Int. J. Syst. Evol. Microbiol.* 57: 2823–2829.

Vattakaven, T., Bond, P., Bradley, G. and Munn, C. B. (2006) Differential effects of temperature and starvation on induction of the viable-but-nonculturable state in the coral pathogens *Vibrio shiloi* and *Vibrio tasmaniensis*. *Appl. Environ. Microbiol.* 72: 6508–6513.

Walter, J. M., Greenfield, D., Bustamante, C. and Liphardt, J. (2007) Light-powering *E. coli* with proteorhodopsin. *Proc. Natl Acad. Sci. USA* 104: 2408–2412.

Welsh, D. T. (2000) Nitrogen fixation in seagrass meadows: regulation, plant–bacteria interactions and significance to primary productivity. *Ecol. Lett.* 3: 58–71.

Yutin, N., Suzuki, M. T., Teeling, H., Weber, M., Venter, J. C., Rusch, D. B. and Beja, O. (2007) Assessing diversity and biogeography of aerobic anoxygenic phototrophic bacteria in surface waters of the Atlantic and Pacific Oceans using the Global Ocean Sampling expedition metagenomes. *Environ. Microbiol.* 9: 1464–1475.

Further reading

Autotrophy

Nakagawa, S. and Takai, K. (2008) Deep-sea vent chemoautotrophs: diversity, biochemistry and ecological significance. *FEMS Microbiol. Ecol.* 65: 1–14.

Shively, J. M., van Keulen, G. and Meijer, W. G. (1998) Something from almost nothing: carbon dioxide fixation in chemoautotrophs. *Ann. Rev. Microbiol.* 52: 191–230.

Wood, A. P., Aurikko, J. P. and Kelly, D. P. (2004) A challenge for 21st century molecular biology and biochemistry: what are the causes of obligate autotrophy and methanotrophy? *FEMS Microbiol. Rev.* 28: 335–352.

Phototrophy

Bryant, D. A. and Frigaard, N. U. (2006) Prokaryotic photosynthesis and phototrophy illuminated. *Trends Microbiol.* 14: 488–496.

Moran, M. A. and Miller, W. L. (2007) Resourceful heterotrophs make the most of light in the coastal ocean. *Nat. Rev. Microbiol.* 5: 792–800.

Zubkov, M. V. (2009) Photoheterotrophy in marine prokaryotes. *J. Plankton Res.* 31: 933–938.

Nitrogen, sulfur, and iron cycles

Butler, A. and Martin, J. D. (2005) The marine biogeochemistry of iron. *Met. Ions Biol. Syst.* 44: 21–46.

Cabello, P., Roldan, M. D. and Moreno-Vivian, C. (2004) Nitrate reduction and the nitrogen cycle in archaea. *Microbiology* 150: 3527–3546.

Campbell, B. J., Engel, A. S., Porter, M. L. and Takai, K. (2006) The versatile ε-proteobacteria: key players in sulphidic habitats. *Nat. Rev. Microbiol.* 4: 458–468.

Friedrich, C. G., Bardischewsky, F., Rother, D., Quentmeier, A. and Fischer, J. R. (2005) Prokaryotic sulfur oxidation. *Curr. Opin. Microbiol.* 8: 253–259.

Karavaiko, G. I., Dubinina, G. A. and Kondrat'eva, T. F. (2006) Lithotrophic microorganisms of the oxidative cycles of sulfur and iron. *Microbiology* 75: 512–545.

Muyzer, G. and Stams, A. J. M. (2008) The ecology and biotechnology of sulphate-reducing bacteria. *Nat. Rev. Microbiol.* 6: 441–454.

Tortell, P. D., Maldonado, M. T., Granger, J. and Price, N. M. (1999) Marine bacteria and biogeochemical cycling of iron in the oceans. *FEMS Microbiol. Ecol.* 29: 1–11.

Vraspir, J. M. and Butler, A. (2009) Chemistry of marine ligands and siderophores. *Annu. Rev. Mar. Sci.* 1: 43–63.

Weber, K. A., Achenbach, L. A. and Coates, J. D. (2006) Microorganisms pumping iron: anaerobic microbial iron oxidation and reduction. *Nat. Rev. Microbiol.* 4: 752–764.

Quorum sensing

Decho, A. W., Norman, R. S. and Visscher, P. T. (2010) Quorum sensing in natural environments: emerging views from microbial mats. *Trends Microbiol.* 18: 73–80.

Gram, L., Grossart, H.-P., Schlingloff, A. and Kiørboe, T. (2002) Possible quorum sensing in marine snow bacteria: production of acylated homoserine lactones by *Roseobacter* strains isolated from marine snow. *Appl. Environ. Microbiol.* 68: 4111–4116.

Biofilms and extracellular products

Stoodley, P., Sauer, K., Davies, D. G. and Costerton, J. W. (2002) Biofilms as complex differentitated communities. *Ann. Rev. Microbiol.* 56: 187–209.

Sutherland, I. (2001) Biofilm exopolysaccharides: a strong and sticky framework. *Microbiology* 147: 3–9.

Chapter 4

Marine *Bacteria*

Many members of the domain *Bacteria* can be grown and studied in the laboratory, but many more organisms are known only by their genetic "signatures" in environmental samples. Thus, there is a large discrepancy in the extent of our knowledge of the properties of various *Bacteria*. For those that cannot yet be cultured, we can infer some of their likely properties by considering their relationship to well-studied species, their habitats, and geochemical evidence relevant to their activities. Advances in genome sequencing mean that we are beginning to be able to predict the metabolic nature of uncultured types through analysis of genes encoding key enzymes. In addition, new techniques are enabling the culture of marine *Bacteria* previously regarded as unculturable. The first section of this chapter contains an overview of some of the main features of *Bacteria*, followed by a synopsis of diversity and phylogenetic studies. In the subsequent sections, the major types of *Bacteria* are grouped according to their phenotypic properties. Important groups and genera are discussed with reference to activities of particular ecological or applied importance.

Key Concepts

- Most *Bacteria* have a simple cell structure, with distinctive membranes and cell wall structure, but there are many variations in cellular organization and components.

- A relatively small number of major clades dominate most samples of ocean plankton, but extensive microdiversity enables colonization of specific niches.

- *Bacteria* that are phylogenetically closely related can differ greatly in their metabolic and physiological characteristics.

- There is no universally accepted concept of bacterial species, which makes it difficult to appreciate the true level of diversity in the *Bacteria*.

- Genomic analysis of cultured *Bacteria* and metagenomic analysis of environmental samples is leading to new insights into bacterial diversity and ecology.

OVERVIEW OF DIVERSITY OF THE *BACTERIA*

The domain *Bacteria* contains about 80 phyla, many of which have no cultivated members

The most important breakthroughs in the understanding of marine microbial diversity have occurred as a result of the application of molecular techniques to the direct analysis of environmental samples, without the need to isolate and culture microorganisms. The cloning of 16S rRNA gene sequences and the more recent use of metagenomics have led to a complete reevaluation of the importance of marine *Bacteria* and *Archaea*, which are both more abundant and more diverse than we could possibly have imagined before the advent of these techniques. Microbiologists have long realized that there is a large discrepancy between the numbers of organisms counted using direct microscopic observation and those recovered on culture media. Analysis of 16S rRNA gene sequences direct from the environment shows that culture methods reveal only a small fraction of the total diversity of marine *Bacteria* and that some groups contain no currently cultivable relatives. However, it is important to recognize that use of 16S rRNA methods *alone* also provides only a partial picture of the true diversity. As indicated in Chapter 1, grouping organisms on the basis of core genes like those for 16S rRNA is usually a good indicator of evolutionary relatedness over long periods, but it is does not take account of significant differences in modern organisms that have arisen as a result of specialization for a particular ecological niche. Many examples given in this chapter show how the recent use of metagenomic approaches and the genome sequencing of multiple strains reveal a large amount of microdiversity within taxonomic groups as currently defined.

The *Bacteria* can be grouped into approximately 80 major divisions (phyla) based on the phylogeny of 16S rRNA genes, of which about 20 contain culturable representatives. Since the first studies in the early 1990s, a number of research groups have conducted studies in diverse marine habitats throughout the world and established databases containing thousands of sequences from marine samples. Different investigators, using variations of the basic methods and sampling different geographic regions and depths have repeatedly found similar phylogenetic groups, which can be clustered into these major clades. *Figure 4.1* shows the best-characterized major phyla—representatives of most of these groups are found in marine habitats. The view presented here is considerably simplified because the phylogenetic classification of *Bacteria* is at present in a state of great flux. It is very important to emphasize that molecular methods often show that *Bacteria* grouped together because of shared major physiological properties (e.g. nitrification or phototrophy) may be quite distantly related.

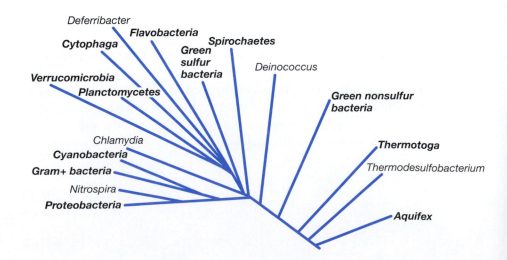

Figure 4.1 Simplified phylogenetic tree of the domain *Bacteria* showing relationships revealed by 16S rRNA sequencing. Representatives (*bold type*) of almost all the major divisions are found in marine habitats.

There is no generally accepted concept for the definition of bacterial species

The concept of species in bacteriology has always been a difficult one. Microbiologists have adopted the binomial system developed by Linnaeus for classification of plant and animal species. In plants and animals, the presence of distinct morphological differences, sexual reproduction, and geographic separation can all be used to explain the concept of species as a group of individuals that can produce fertile offspring and are reproductively isolated from other species. This definition is meaningless for bacteria, which are haploid and do not have sexual reproduction. Microbiologists use a polyphasic approach to naming and classifying bacteria, based on observing a combination of numerous phenotypic, genotypic, and phylogenetic properties in strains of interest. A variety of characteristics must be considered before a new species can be defined, and strict rules are applied by an International Code of Nomenclature. The three most important criteria for determining that two bacteria belong to the same species are to show that (i) they have more than 70% DNA–DNA cross-hybridization (see p. 45); (ii) the 16S rRNA sequences are more than 97% similar; and (iii) they share a high degree of phenotypic similarity, with characteristics that distinguish them from other species. All of these criteria have inherent flaws. Uncultured bacteria cannot be assigned to a species because we do not know their phenotype. However, in some cases, it is possible to assign a provisional species name with *Candidatus* designation if genetic evidence points to significant differences from other recognized species.

Although the above methods provide a pragmatic solution to identifying and naming bacteria, they do not address the fundamental question of what constitutes a species, and there have been many attempts to develop a meaningful theoretical concept for bacteria that is underpinned by evolutionary and ecological processes.

Bacteria show a variety of cell forms and structural features

As introduced in Chapter 1, life may be organized into three phylogenetic domains, the *Bacteria*, *Archaea*, and *Eukarya*. The *Bacteria* are considered to be unicellular organisms with a simple cell structure. Under the microscope, *Bacteria* show various cell shapes and examples of all these types occur in marine habitats, as shown in *Figure 1.3*. The basic cell forms are spherical or ovoid cells (cocci), rods (bacilli), and spiral cells (spirilla). Some groups of marine *Bacteria* have distinctive morphologies such as tightly coiled spirals (spirochaetes), cells with buds, stalks, or hyphae (e.g. *Planctomycetes* and *Caulobacter*), or filamentous forms (*Actinomycetes*). As evident from *Table 1.2*, marine *Bacteria* show enormous diversity in cell size, ranging from about 0.1 μm to 750 μm in diameter. All cells possess a 6–8 nm-wide cytoplasmic membrane composed of a phospholipid bilayer containing hydrophobic fatty acids linked to glycerol moieties via ester linkages, plus a wide range of proteins. The membrane acts as a selectively permeable barrier controlling movement of ions and molecules in and out of the cell. Transport mechanisms are needed for nutrient uptake, excretion, and secretion of extracellular products. As well as controlling the entry/exit of substances to/from the cell by its selective permeability and its specific transport proteins, the cytoplasmic membrane is the most important site of energy generation and conservation. During metabolism, electron carriers associated with the membrane bring about a charge separation across the membrane. H^+ ions (protons) become concentrated on the exterior surface of the membrane and OH^- ions concentrate on the inner surface. This charge separation creates an electrochemical potential that can be converted to chemical energy. As explained in Chapter 3, *Bacteria* vary enormously in the oxidation–reduction reactions used to produce this gradient, but the common "currency" of energy exchange is always the molecule ATP, used for synthesis and mechanical work by the cell. When protons cross the membrane through a specific port, the contained energy is captured in the conversion of ADP to ATP.

The membrane is stabilized by Mg^{2+} and Ca^{2+} and by sterol-like compounds known as hopanoids, derived from hopane. Hopanoids are quite resistant to breakdown and large quantities have accumulated in the environment over the many millennia of bacterial evolution. The total global mass of carbon deposited in hopanoids is estimated at about 10^{12} tons, approximately the same as the mass of organic carbon in all organisms living today. A very large fraction of fossil fuels such as petroleum and coal are composed of hopanoids, much of which has accumulated from the settlement of bacteria in the oceans. Because different bacteria produce characteristic hopanoid structures, these compounds are frequently used as markers for the analysis of microbial communities in marine sediments.

One of the original defining features of the *Bacteria* as "prokaryotic" cells (see p. 4) was that they have a simple cellular organization with no membrane-bound nucleus or organelles such as mitochondria or chloroplasts. However, this distinction is no longer valid, as many bacterial cells do show extensive internal compartmentalization. Members of the *Planctomyctes* clearly belong to the *Bacteria*, but have a membrane-bound nucleus and a membrane-bound organelle (the anammoxosome). Invaginations or aggregates of membrane vesicles are especially important in phototrophs and chemolithotrophs with very high respiratory activity. Phototrophs grown in low light intensity will produce more internal membranes and pigments than those grown under high illumination, in order to harvest available light with the greatest efficiency. Gas vesicles are also important in the ecology of marine phototrophs such as *Cyanobacteria*. Gas vesicles have a 2-nm-thick, single-layered, gas-permeable wall composed of two hydrophobic proteins, surrounding a central space. These form a very strong structure into which gas diffuses from the cytoplasm, resulting in increased buoyancy of the cell. This enables phototrophic bacteria to maintain themselves in the desired zone of light intensity in the water column. Organisms from deep habitats have narrow vesicles as they are more resistant to hydrostatic pressure. Many types of autotrophic *Bacteria* contain crystal-like regular hexagons known as carboxysomes. These are about 120 nm diameter and surrounded by a thin envelope. They act as stores of the enzyme RuBP carboxylase, the key enzyme used by many autotrophs for the fixation of carbon dioxide (see p. 66). Packaging of the carboxylase enzyme enhances its efficiency, and genetic mutations causing the loss or structural changes of the carboxysomes result in cells requiring much higher concentrations of carbon dioxide than normal. Many bacteria also contain granules of organic or inorganic material used as a store of energy or structural components. Sometimes they occur as simple concentrations of material free within the cytoplasm, but they are often enclosed in a thin, single-layered lipid membrane. For example, a wide range of marine *Bacteria* produce polyphosphate granules and carbon and energy storage polymers such as polyhydroxyalkanoates. Large inclusions of elemental liquid sulfur up to 1.0 μm in diameter occur in sulfur-oxidizing bacteria and can constitute up to 30% of the cell weight (see *Figure 10.4B*). *Cyanobacteria* also contain granules composed of large amounts of α-1,4-linked glucan, which is the polymerized carbohydrate product of photosynthesis. They also contain granules composed of polypeptides rich in the amino acids arginine and aspartic acid, which appear to be produced as a reserve of nitrogen.

The cell wall is an important feature of bacterial cells

Almost all marine *Bacteria* possess some type of cell wall external to the cytoplasmic membrane. The cell wall is responsible for the shape of the cell and provides protection in an unfavorable osmotic environment. The key component of the cell wall in *Bacteria* is peptidoglycan, a mixed polymer composed

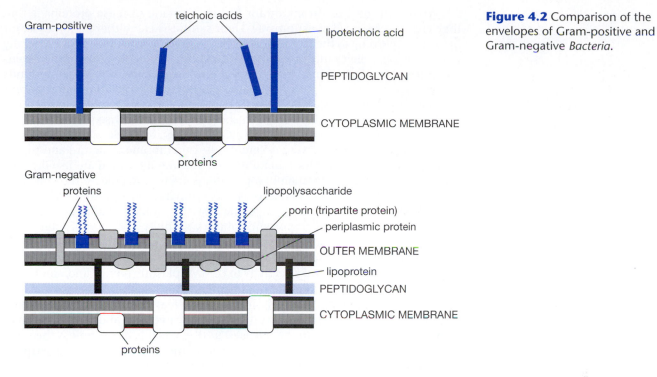

Figure 4.2 Comparison of the envelopes of Gram-positive and Gram-negative *Bacteria*.

of alternating residues of two amino sugars (*N*-acetyl glucosamine and *N*-acetyl muramic acid) with a tetrapeptide sidechain composed of a small range of amino acids. The great mechanical strength of peptidoglycan lies in the formation of cross-links between the peptide sidechains, forming a mesh-like molecule. Although variation in the types of amino acids and the nature of cross-linking occurs, peptidoglycan with essentially the same structure occurs in all *Bacteria* with the exception of the *Planctomycetes*. Many *Bacteria* possess an additional surface layer (S-layer) external to the peptidoglycan-containing cell wall, composed of arrays of a single protein or glycoprotein arranged in a tetragonal or hexagonal lattice-like crystalline structure (see *Figure 11.8*). Interestingly, S-layers are found in almost all phylogenetic groups of both *Bacteria* and *Archaea* and seem to represent one of the simplest biological membranes, which probably developed very early in evolution. Again, the *Planctomycetes* present an important exception as members of this group lack peptidoglycan and possess only an S-layer.

For more than a century, it has been traditional to use the Gram stain to differentiate Gram-positive and Gram-negative types. This basic distinction remains a useful first stage in the examination of unknown bacteria in culture and underpins some important physiological differences, but it has no phylogenetic significance. As shown in *Figure 4.2*, Gram-positive bacteria have a relatively thick, simple wall composed of peptidoglycan, together with ribitol or phosphate polymers known as teichoic acids. Gram-negative bacteria have a thinner but more complex wall with a very thin peptidoglycan layer that lacks teichoic or teichuronic acids, together with an additional outer membrane. The outer membrane of cells of Gram-negative *Bacteria* is a lipid–protein bilayer complex, but is very different from the cytoplasmic membrane. The outer membrane is anchored tightly to the peptidoglycan layer via a lipoprotein. It also contains lipopolysaccharide (LPS), a compound unique to Gram-negative *Bacteria*, which possesses some very important properties.

The proteins of the outer membrane have very distinctive properties. The major proteins, called "porins," are trimeric structures that act as channels for the diffusion of low-molecular-weight solutes across the outer membrane. Other proteins act as specific receptors for iron and other key nutrients and their

---

(i) LPS HAS SPECIAL IMPORTANCE IN PATHOGENIC BACTERIA

LPS is composed of lipid A covalently bound to a core polysaccharide and strain-specific polysaccharide sidechains. These are immunogenic and are known as the O-antigen. Serological typing based on differences in the sequence of sugars in the O-antigen is important in the identification of closely related bacteria, and strain-to-strain variation in immunological specificity is important in the epidemiology of many bacterial diseases. LPS stimulates the activation of complement and related host responses, which is very important in the interactions of pathogenic or symbiotic bacteria with their animal hosts. LPS is also known as endotoxin because the lipid A molecule is responsible for hemorrhage, fever, shock, and other symptoms of diseases caused by Gram-negative pathogens of vertebrate animals.

transport into the cell. An important feature of many of these proteins is that they are not produced constitutively. Their synthesis can either be repressed or induced, according to the concentration of the specific substances. An especially important example of this in marine bacteria is in the acquisition of the essential nutrient iron, which may be imported into the cell via secreted compounds known as siderophores (see p. 71). Expression of genes encoding siderophores and outer membrane receptors for their entry into the cell (as well as many other genes) may only occur when iron is in short supply. As a consequence, bacteria grown in laboratory culture, which provides excess iron unless special precautions are taken, may have very different outer-membrane structure and physiological properties from those in natural environments, which are usually very iron-depleted. Other outer membrane proteins are responsible for the detection of environmental changes such as pressure, temperature, and pH, and communication of this information via signaling systems. In later chapters, numerous examples are given of the importance of the variability of the outer membrane and its role in monitoring the physical and chemical status of the cell's immediate environment. Because the outer membrane is an additional exterior permeability barrier with different selectivity to the cytoplasmic membrane, some substances cross one membrane, but not the other. This means that the periplasmic space, between the membranes, is an important feature of Gram-negative bacteria. Many binding and transport proteins are located here and proteins such as enzymes and toxins may not be released until the cell is lysed, unless there are specific excretion mechanisms. Vesicles of outer membrane containing periplasmic contents may also be released from the cell surface.

Many *Bacteria* produce a glycocalyx or capsule

In addition to the cell wall, many marine *Bacteria* secrete a slimy or sticky extracellular matrix. It is usually composed of polysaccharide, in which case it may form a network extending into the environment and is termed the "glycocalyx." In other cases, there may be a more distinct rigid layer (capsule), which may contain protein. The main importance of the glycocalyx in marine systems is in the attachment of bacteria to plant, animal, and inanimate surfaces, leading to the formation of biofilms. The bacterial glycocalyx is critical for the settlement of algal spores and invertebrate larvae; the significance of these processes in biofouling of surfaces in the marine environment is considered in Chapter 13. The release of these sticky exopolymers is also very important in the aggregation of bacteria and organic detritus in the formation of "marine snow" particles (p. 14). Furthermore, a large proportion of the dissolved organic carbon in ocean waters probably derives from the glycocalyx of bacteria. Slime layers can act as a protective layer, preventing the attachment of bacteriophages and penetration of some toxic chemicals. In pathogens, the presence of a capsule can inhibit engulfment by host phagocytes, and in free-living forms it is commonly assumed to inhibit ingestion by protists. However, many grazing flagellates seem to feed voraciously on capsulated planktonic bacteria, probably because the capsule contains additional nutrients.

Phylogenetic studies of planktonic *Bacteria* reveal a small number of major clades

The most abundant clones of 16S rRNA gene sequences from *Bacteria* found in ocean plankton do not correspond to cultured species at all. When phylogenetic trees are constructed based on genetic similarity, more than three quarters of the marine *Bacteria* revealed by this approach belong to about 20 phylogenetic groups with worldwide distribution. These form clusters of related types, rather than single lineages, and genetic variability within these populations is very high. In some clusters, differences in 16S rRNA sequences may be due to genetic variability associated with adaptation to different depths of water. *Figure 4.3* shows the frequency of the most common gene clusters based on PCR based analysis of 16S rRNA sequences from plankton. Subsequent analyses using metagenomic

THE *PROTEOBACTERIA* —A DOMINANT MARINE GROUP

The group is further divided into five classes distinguished by Greek letters (α, β, γ, δ, ε); namely the *Alphaproteobacteria, Betaproteobacteria, Gammaproteobacteria, Deltaproteobacteria* and *Epsilonproteobacteria*. There are marine representatives of all of these groups, with the α and γ types being especially important in the bacterioplankton. Culture-independent molecular methods reveal an enormous diversity not evident from culture-based studies and show that representatives of the *Alphaproteobacteria* are among the most dominant bacteria in marine waters. Conversely, members of the *Gammaproteobacteria* are most common when culture methods are used, but are less well represented in databases constructed using molecular methods. Most of the easily cultured marine bacteria fall within this group. The *Deltaproteobacteria* and *Epsilonproteobacteria* are especially important in sediments, but some sequences are recovered from plankton samples.

(A)

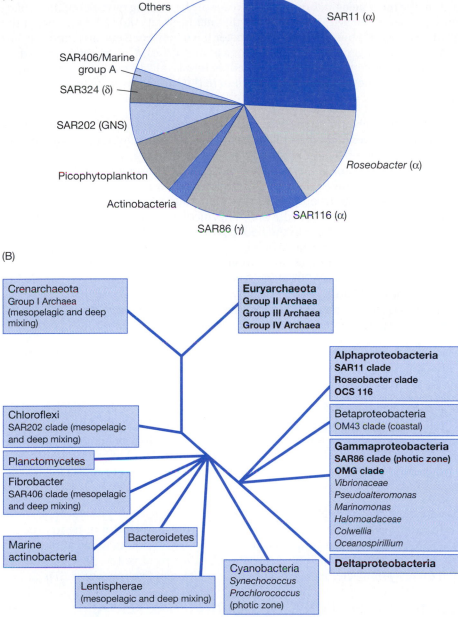

Figure 4.3 Diversity of *Bacteria* and *Archaea* in marine plankton. (A) Frequency of the most common 16S rRNA gene clusters of *Bacteria*, based on PCR amplification of 16S rRNA genes from 578 clones. (Redrawn from data in Giovannoni and Rappé [2000]). The "SAR" prefix indicates the original identification of sequences from Sargasso Sea samples. Symbols α, γ and δ indicate groups of the *Proteobacteria*; GNS, green nonsulfur bacteria. (B) Schematic illustration of the phylogeny of the major clades of *Bacteria* and *Archaea*. Groups in bold type appear to be ubiquitous in seawater; for other groups, labels indicate those found predominantly in the mesopelagic zone or in surface waters exposed to deep mixing during polar winters, coastal ecosystems or the photic zone. (Redrawn from Giovannoni and Stingl [2005].)

approaches have confirmed the dominance of *Alphaproteobacteria*. Bacterial diversity in suspended particles (marine snow) and free picoplankton is very different and varies with depth and environmental conditions. Diversity in marine sediments is very high, but has not been investigated as thoroughly as that of plankton.

The *Roseobacter* clade (belonging to the family *Rhodobacteriaceae*) are especially abundant and widely distributed in plankton, as well as occurring as in sediments, microbial mats, sea ice, and in association with algae or invertebrate animals. Up to 25% of marine bacteria may belong to this clade. Nearly 40 genera containing culturable representatives have been identified, including *Roseobacter* and *Silicibacter*. Members of this group are very important in the carbon and sulfur cycles (see Chapters 8 and 9).

The SAR 11 cluster of 16S rRNA sequences has been found in almost every pelagic environment ranging from shallow coastal waters to depths of over 3000 m. It is probably the most abundant group of microorganisms in the sea and on average

The International Census of Marine Microbes (ICOMM) is coordinating efforts to increase our knowledge of microbial diversity in the oceans. One of the most significant and surprising early discoveries of ICOMM has been the realization that much of the diversity in the deep sea is accounted for by rare microorganisms that are present in low abundance. Sogin et al. (2006) used 454-tag sequencing (see p. 46) to examine water from depths of 550–4100 m at various sites in the Atlantic Ocean and diffuse hydrothermal vent sites in the Pacific Ocean. This method enabled very high-level analysis of thousands of short sequences of 16S rRNA genes; overcoming the bias inherent in previous studies of molecular diversity, in which dominant populations mask the detection of low-abundance types. Sogin and colleagues concluded that the number of different kinds (operational taxonomic units, OTUs) of bacteria in the oceans could exceed five to ten million and that the "rare biosphere is very ancient and may represent a nearly inexhaustible source of genomic innovation." However, a later critique of the method of pyrosequencing by Quince et al. (2009) found that it can be prone to multiple errors in a few reads, which can lead to apparently unique sequences. Kunin et al. (2010) have cautioned that careful analysis and interpretation of pyrosequencing data is needed to prevent overestimation of species richness.

accounts for nearly a third of the cells present in surface waters and a fifth of the cells in the mesopelagic zone. SAR11 is a deeply branching member of the *Alphaproteobacteria* and is phylogenetically distinct from all cultured members of the group. Some members of the SAR11 cluster have now been cultured and this has allowed study of genomic and physiological properties, as discussed in *Research Focus Box 2.1*. Members of the α-proteobacterial SAR116 clade are also ubiquitous, although they may be more prevalent in the surface layers of the oceans and shallower coastal waters.

As noted previously, many members of the *Gammaproteobacteria* can be easily cultivated and were regarded historically as the most dominant marine bacteria. Although members of this group are the most easily isolated from waters all over the world because many will form colonies on commonly used agar culture media, 16S rRNA studies show that their dominance as free-living heterotrophs has been overestimated compared with other groups. However, some of the most well-known genera fall within the *Gammaproteobacteria* and are of special significance for discussions later in the book, because of their interactions with other marine organisms, their established role in ecological processes, their pathogenicity for humans, and their biotechnological potential. Members of the abundant γ-proteobacterial SAR86 clade appear to be photoheterotrophs adapted to oligotrophic conditions, and have not yet been cultured. This seems to be a very broad group with several subsets adapted for specific niches and is phylogenetically very distant from the culturable members of the *Gammaproteobacteria*.

The *Actinobacteria* clade forms a smaller proportion of sequences and is a branch of the high G + C Gram-positive bacteria of gene clones, only distantly related to culturable representatives. They occur primarily in the upper, photic layers of the ocean.

Three major groups appear to be primarily associated with the aphotic zone of the deep ocean. These are the SAR202 clade (a deeply branching group of the green nonsulfur bacteria), the marine group A clade (probably a unique, previously undiscovered branch of the *Bacteria*) and the marine group B clade (*Deltaproteobacteria*).

The *Cyanobacteria*, notably the genera *Prochlorococcus* and *Synechococcus*, have been extensively studied in view of their importance in contributing to productivity through photosynthesis. They are very widely distributed and abundant in the photic zone, especially in tropical and subtropical waters. *Trichodesmium* and other cyanobacteria play a major role in nitrogen fixation (see p. 67).

SOME MAJOR TYPES OF MARINE *BACTERIA*, GROUPED BY PHENOTYPE

Several groups of bacteria carry out anoxygenic photosynthesis

The phototrophs known as "purple bacteria" have variable morphology (rods, ovoid cocci, and spirals)—representatives of this group occur in the *Alpha-*, *Beta-* and *Gamma-* divisions of the *Proteobacteria*. Unlike cyanobacteria, algae and plants, these bacteria do not evolve oxygen during photosynthesis, and the discovery of this group was of major importance in the development of a unifying theory for the mechanism of photosynthesis. Purple phototrophs contain bacteriochlorophylls as the photosynthetic pigment, which together with additional carotenoid pigments give the bacteria their distinctive color. The pigments are located inside multifolded invaginations of the cytoplasmic membrane, which allow the bacteria to make efficient use of available light. Various types of bacteriochlorophyll occur and these absorb light of different wavelengths. This, together with the light-absorbing properties of associated proteins and the source of electrons used to reduce carbon dioxide during photosynthesis, determines the habitat and ecological role of the various species.

The purple sulfur bacteria utilize hydrogen sulfide or other reduced sulfur compounds as the source of electrons. The overall reaction for photosynthesis can be represented as

$$CO_2 + H_2S + H_2O \rightarrow (CH_2O) + S + H_2O$$

(where CH_2O represents reduced carbon compounds that are the initial products of photosynthesis). Granules of sulfur are deposited inside the cells. Purple sulfur bacteria are most commonly found in anaerobic sediments of shallow lakes and sulfur springs, but several types (e.g. *Thiocapsa* and *Ectothiorhodospira*) also occur in shallow marine sediments where anaerobic conditions and high levels of hydrogen sulfide occur at depths to which sufficient light of the appropriate wavelength can penetrate. Another group, known as the purple nonsulfur bacteria can grow aerobically in the dark using organic compounds or molecular hydrogen as electron donor. Many can grow either photoautotrophically using carbon dioxide and hydrogen, or photoheterotrophically using a range of organic compounds as carbon source. Despite their name, some can use low levels of H_2S as a source of electrons. Marine genera include *Rhodospirillum* and *Rhodomicrobium*.

Until recently, it was assumed that anoxygenic bacterial photosynthesis is a strictly anaerobic process. However, the discovery of representatives of a new bacterial group that contain bacteriochlorophyll, yet grow aerobically, has forced a revision of this view. As noted above, the *Roseobacter* clade is very abundant in plankton. Although they grow aerobically, these types do not produce oxygen from photosynthesis. Like the purple bacteria, the aerobic anoxygenic phototrophic (AAnP) bacteria contain a range of carotenoid pigments. However, the photosynthetic apparatus is not as well structured and the complex membrane invaginations typical of the anaerobic phototrophs are not seen in the aerobic types. This probably explains why the AAnP bacteria cannot use light as a sole source of energy and rely on various organic compounds as a source of carbon and energy. It is now known that the *Roseobacter* clade contains both culturable and nonculturable types, as well as phototrophic and nonphototrophic representatives. These are widely distributed in coastal and oceanic plankton and they occur in a wide range of associations with other marine organisms. In molecular studies, the *Roseobacter* clade emerges as the second most abundant 16S rRNA gene clone type (>30% of clones). Their association with blooms of algae and dinoflagellates has been particularly studied, and they may play a role in the formation of dinoflagellate toxins (see *Research Focus Box 12.2*). Despite their obvious ubiquitous presence in ocean water, it is at the moment difficult to be sure of the role of this group in ecological processes. The diverse metabolic properties of the group undoubtedly play a large part in nutrient cycling. For example, *Roseobacter* and *Silicibacter* have a major role in the breakdown of dimethylsulfoniopropionate (DMSP), which has great significance in global climate processes (see *Research Focus Box 9.2*).

The group known as green sulfur bacteria forms a separate lineage distinct from the *Proteobacteria*, but resembles the purple sulfur bacteria in metabolism. However, sulfur is produced outside of the cell rather than as intracellular granules. Also, in addition to bacteriochlorophyll *a*, green sulfur bacteria contain bacteriochlorophyll *c*, *d* or *e*. These pigments are contained in a membrane-bound structure known as the chlorosome. Members of the family *Chromatiaceae* are common on intertidal mudflats and as members of consortia in microbial mats and sediments with high levels of sulfide.

Nitrifying bacteria grow chemolithotrophically using reduced inorganic nitrogen compounds as electron donors

Marine examples, which are present in suspended particles and in the upper layers of sediments, include *Nitrosomonas* and *Nitrosococcus* (which oxidize

BOX 4.1   RESEARCH FOCUS

The Global Ocean Survey
How an entrepreneurial scientist with a passion for sailing revolutionized
understanding of marine microbial diversity

From sequencing his own genome to sequencing ocean metagenomes

J. Craig Venter is a biochemist and molecular biologist whose initiatives in the development of genomics have revolutionized biology and medicine. He developed the use of expressed sequence tags (ESTs; see p. 54) and whole genome shotgun sequencing (WGS; see p. 49). In 1992, he founded The Institute for Genomic Research (TIGR), and his team published the first full genome sequence of a free-living organism, the bacterium *Haemophilus influenzae*, followed by numerous other genomes of important microbes and animals. In 1998, frustrated by the slow progress of the Human Genome Project, he advocated the use of WGS as the fastest way to complete the task. Despite opposition from many geneticists, who believed that WGS would not be accurate enough, Venter founded the company Celera and invested heavily in mass automated DNA sequencing facilities, enabling completion of the draft sequence of the human genome derived from five humans in 2000. In 2007, the first full diploid sequence of *Homo sapiens*—Venter himself—was published (Levy et al., 2007). As he notes in his autobiography: "As the sequencers in my new facility [The J. Craig Venter Institute, JCVI] were churning out more reads of my genome, I turned my attention to a project that combined two great loves of my life; science and sailing" (Venter, 2007).

Sorcerer II sets sail for a pilot study in the Sargasso Sea

Venter bought a 95-foot sailing yacht *Sorcerer II* to embark on a sampling expedition. The boat was equipped with pumps to take samples of the surface water and pass them through filters of different sizes (see *Figure 2.2B*) to trap the microbes. The filters were frozen until they could be transported back to the JCVI laboratories for extraction of the total DNA and sequencing. Again, many scientists were skeptical that WGS would work, but Venter proved them wrong when the results of a pilot study conducted in the Sargasso Sea near Bermuda were published (Venter et al., 2004). Venter had chosen this region for a trial run because it is known to be nutrient poor and was therefore expected to have relatively low diversity. However, more than 1.2 million genes and an estimated 1800 microbial "species" were obtained. Of particular interest was the discovery of more than 800 variations of the proteorhodopsin gene, encoding a light-driven proton pump for phototrophic energy production by heterotrophic bacteria (see *Research Focus Box 3.1*). The Sargasso Sea metagenome showed that the process is widespread and could help to explain the unexpectedly high diversity of species in the nutrient-poor Sargasso Sea.

Like much of his previous work, Venter's spectacular results attracted a lot of criticism from fellow scientists. Some felt that the technology-driven approach lacked oceanographic context and did not fully consider factors like temperature, ocean currents, and seasonality. The sheer scale of the data seemed overwhelming, with more than one billion bases of genetic information deposited in the GenBank database—Tress et al. (2006) argued that the large number of hypothetical sequence fragments would lead to problems with identification of structure and function in other studies. Critics also highlighted the surprising finding that nearly complete genome scaffolds were assembled from *Burkholderia cepacia*, an organism more usually regarded as terrestrial organisms from high-nutrient environments. Whilst Venter and colleagues recognized the unusual nature of their findings, they highlighted the rigorous sampling procedure and suggested that a high-nutrient source in marine snow or marine mammal vectors might explain the finding. However, DeLong (2005) argued that the most likely explanation was shipboard contamination, because the sequences (and those of another genus, *Shewanella*) were essentially clonal and found in only one sample. DeLong (2005) emphasized the need for caution when interpreting metagenomic datasets, and subsequent analyses have omitted these data (e.g. Stingl and Giovannoni, 2005). Falkowski and de Vargas (2004) gave a positive, but guarded, commentary on the paper by Venter et al. (2004), arguing that the diversity may have been biased by sampling "hot spots" of microbial growth associated with zooplankton feces and the like, and commenting that, "despite their huge sequencing effort, Venter and collaborators were able to reconstruct only two, almost-complete genomes [*Burkholderia* and *Shewanella*], and this was with the help of fully sequenced templates." Nevertheless, Falkowski and de Vargas (2004) acknowledged the importance of the discoveries in increasing the awareness of the vast genetic diversity and complexity present in the oceans.

The Global Ocean Survey (GOS)

Sorcerer II then embarked on a global circumnavigation expedition that is reminiscent of the first great oceanographic voyage by *HMS Challenger* in 1872. The expedition started in Canada in 2003 and proceeded along the US east coast, the Gulf of Mexico, and through the Panama Canal to the Galapagos Islands, where extensive sampling was conducted. Venter's team then sampled across the central and south Pacific, around Australia, the Indian Ocean to South Africa, across the Atlantic and back to the United States, ending in Florida in 2006. Aspects of this sampling cruise also proved controversial, with some countries

BOX 4.1   RESEARCH FOCUS

attempting to prevent sampling in their waters or the publication of data obtained, accusing Venter of biopiracy by seeking to privatize biodiversity found in national waters. In fact, all the results of the GOS were published as a special collection of articles and videos in the open-access journal *PLoS Biology* (Ocean Metagenomics Collection, 2007). Data, comprising 7.7 million sequencing reads representing more than six billion base pairs, were obtained from 41 sampling sites taken about every 330 km on an 8000-km transect. The analysis was conducted on near-surface marine microbial plankton in the 0.2–0.8-μm size range, and Nealson and Venter (2007) point out that larger organisms and microbes attached to larger particles were not analyzed and that this is just one niche of hundreds in the oceans.

Making sense of a vast number of data

Development of new bioinformatics approaches to analyze the GOS data is an ongoing requirement. Rusch et al. (2007) found that most sequence reads could not be assembled, indicating that the samples contained great microbial diversity. They then used the 584 microbial genomes available in databases as points of reference. By relaxing the stringency of the search parameters, most of the GOS sequences could be loosely aligned to the reference genomes, but it was difficult to draw meaningful conclusions about phylogeny. When the search criteria were made more rigorous, by requiring higher similarity of the full length of sequence reads to a reference genome, only 30% of the GOS data was recruited to known genomes. Most of these aligned to three of the most abundant marine bacteria—*Candidatus* Pelagibacter, *Synechococcus*, and *Prochlorococcus*. Further

analysis revealed that a very high level of sequence diversity occurred in the subtypes of related DNA sequences, reflecting adaptations to local environments.

Yooseph et al. (2007) compared amino-acid sequences predicted from the GOS data sequences from known proteins already deposited in the databases. By the use of novel statistical matching techniques, they identified nearly six million proteins (1.8 times the number already in public databases). About 2000 clusters of protein sequences appeared unique to the GOS dataset and many of these were thought to be involved with photosynthesis or electron transport. The GOS data set revealed a very high number of viral-like sequences within the bacterial samples (Rusch et al., 2007; Williamson et al., 2008) and there were a large number of proteins of probable viral origin in the protein data (Yooseph et al., 2007). This reinforces the growing realization of the importance of viruses as agents of genetic diversity through lateral gene transfer and makes the assembly of complete microbial genomes from environmental samples very difficult (Breitbart et al., 2007).

As more genome sequences of marine microbes become available, and new bioinformatics techniques are developed, retrospective analysis of the GOS dataset is possible. In 2009, *Sorcerer II* embarked on another major sampling expedition across the North Atlantic, obtaining samples from the North, Baltic, and Mediterranean Seas, and we will soon have a further major release of metagenomic data as a resource. These data will continue to be an invaluable resource, to be "mined" for information about marine microbes many years to come.

---

ammonia to nitrite) and *Nitrosobacter, Nitrobacter* and *Nitrococcus* (which oxidize nitrite to nitrate). No organisms that can carry out both reactions are known. The ammonia-oxidizers are obligate chemolithoautotrophs and fix carbon via the Calvin cycle. The nitrite-oxidizers are usually chemolithoautotrophic, but can be mixotrophic using simple organic compounds heterotrophically. Because of these activities, nitrifying bacteria play a major role in nitrogen cycling in the oceans, especially in shallow coastal sediments and beneath upwelling areas such as the Peruvian coast and the Arabian Sea. Previously, nitrifying bacteria were classified mainly on morphological characteristics. However, 16S rRNA gene analysis shows that they occur in several branches of the *Proteobacteria*, and one type, *Nitrospira*, forms a distinct bacterial phylum. Only members of the *Gammaproteobacteria* appear important in marine environments. Like the phototrophs, nitrifying bacteria have extensive internal structures in order to increase the surface area of the membrane.

A wide range of *Proteobacteria* can grow chemolithotrophically using reduced sulfur compounds

The rod-shaped *Thiobacillus* is the best-known genus, using hydrogen sulfide, elemental sulfur, or thiosulfate as electron donors. Filamentous bacteria such as *Beggiatoa, Thiothrix,* and *Thiovulum* are also well represented in the marine environment. These bacteria are usually strict aerobes found in the top few millimeters of marine sediments, which are rich in sulfur. They frequently show

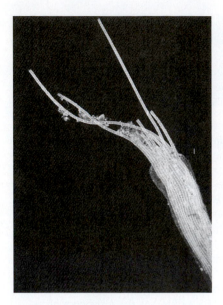

Figure 4.4 *Thioploca araucae* showing filamentous cells surrounded by a sheath. (Image courtesy of Victor Gallardo, Universidad de Concepcíon, Chile.)

chemotaxis to seek out the desired gradient of oxygen and sulfur compounds and are also very prominent at hydrothermal vents and cold seeps, both as free-living forms and as symbionts of animals (see Chapter 10), where they form the base of the food chain. *Beggiatoa* and other filamentous forms commonly show gliding motility and become intertwined to form dense microbial mats, often with a complex community structure containing sulfate-reducers and phototrophs. Although *Beggiatoa* obtains energy from the oxidation of inorganic sulfur compounds, it does not possess the enzymes needed for autotrophic fixation of carbon dioxide and therefore uses a wide range of organic compounds as a carbon source.

Thioploca, Thiothrix, and *Thiomargarita* are filamentous sulfur-oxidizing chemolithotrophs whose importance in the oxidation of sulfide in anaerobic sediments has only recently been discovered. *Thioploca* spp. are multicellular filamentous bacteria that occur in bundles surrounded by a common sheath and contain granules of elemental sulfur (*Figure 4.4*). Thioploca is one of the largest bacteria known, with cell diameters from 15 to 40 µm and filaments many centimeters long, containing thousands of cells. Several species, including *T. chileae*, *T. araucae*, and *T. marina* have been described. In the late 1990s huge communities of *Thioploca* spp. were discovered along the Pacific coast of South America, where upwelling creates areas of nitrate-rich water, with bottom waters becoming anoxic. Blooms of *Thioploca* can be very dense, up to 1 kg wet weight per square meter. Anoxic reduction of hydrogen sulfide is coupled to the reduction of nitrate. Each cell contains a very thin layer of cytoplasm around the periphery and a liquid vacuole that constitutes 80% of the cell volume. The vacuole stores very high concentrations of nitrate, which is used as an electron acceptor for sulfide oxidation. The bacteria can grow autotrophically or mixotrophically using organic molecules as a carbon source. The filaments stretch up into the overlying seawater, from which they take up nitrate, and then glide down 5–15 cm deep into the sediment through their sheaths to oxidize sulfide formed by intensive sulfate reduction.

Thiomargarita namibiensis was discovered in 1999 and holds the current record as the largest known bacterium. The spherical cells are normally 100–300 µm wide, but some reach diameters of 750 µm (*Figure 1.4C*). They occur in large numbers in coastal sediments off Namibia, and occur in filaments with a common mucus sheath. Microscopic granules of sulfur reflect incident light, and the name derives from their resemblance to a string of pearls. The hydrographical conditions off this coast bring large quantities of nutrients to the surface and massive phytoplankton growth results in settlement of organic material to the seabed, where it is degraded by bacteria, forming large amounts of hydrogen sulfide. *Thiomargarita* oxidizes sulfide using nitrate and, like *Thioploca*, the interior of the cell is filled with a large vacuole. The nitrate stored in the vacuole and the sulfur stored in the peripheral cytoplasm act as nutrient reserves that allow these bacteria to grow for several months in the absence of external nutrients.

Aerobic methanotrophs and methylotrophs are widespread in coastal and oceanic habitats

These bacteria occur especially in the top layers of marine sediments, where they utilize methane produced by anaerobic methanogenic *Archaea* (see p. 123). A very wide range of bacteria in different phylogenetic groups can use methane as a source of carbon and energy, using the reactions described on p. 70. The membranes of methanotrophs contain sterols, a feature that is very unusual and shown in only one other bacterial group, the mycoplasmas. Methanotrophs such as *Methylomonas, Methylobacter*, and *Methylococcus* (in the *Gammaproteobacteria*) contain intracytoplasmic vesicles and use the unique ribulose monophosphate pathway for carbon fixation, whereas genera such as *Methylosinus* and *Methylocystis* (in the *Gammaproteobacteria*) contain membranes running around the periphery of the cell and use the serine pathway for carbon assimilation. Apart from the importance of free-living forms in methane oxidation, methanotrophs also occur as symbionts of mussels found near "cold seeps" of methane-rich material on the ocean floor, providing the animals with a direct source of nutrition (see p. 209). Methanotrophs are also important in bioremediation of low-molecular-weight halogenated compounds, an important process in contaminated marine sediments.

The pseudomonads are a heterogeneous group of chemoorganotrophic, aerobic, rod-shaped *Proteobacteria*

This group, which comprises representatives in both the α and β divisions of the *Proteobacteria* includes the well-known genera *Pseudomonas* and *Burkholderia*. These genera are usually found in soils and plant material, and some are human pathogens. Whilst they can be isolated from coastal waters, they are probably not indigenous marine organisms. However, many salt-requiring organisms with some similar properties and relatedness to *Pseudomonas* can be isolated from coastal and ocean seawater, and in association with marine plants and animals. Of these, *Alteromonas* and *Shewanella* are probably the best known. As with other groups, 16S rRNA studies are leading to large-scale reclassification of these genera, and they are probably members of the large clade that includes the vibrios and enteric bacteria. *Alteromonas* and *Pseudoalteromonas* spp. are frequently isolated on marine agar and are often distinguished by brightly colored colonies due to the production of various pigments. Because of their dominance in culture-based surveys, it is assumed that they play a major role in heterotrophic nutrient cycling. However, these genera are not well represented in molecular-based surveys, so it is difficult to determine their ecological importance. *Shewanella* spp. are frequently isolated from the surfaces of marine algae, shellfish, fish, and marine sediments. Some are extreme barophiles. *Shewanella* spp. show great metabolic versatility and can use a wide range of compounds as electron acceptors. Some are important in the spoilage of fish.

Free-living aerobic nitrogen-fixing bacteria are important in sediments

Nitrogen fixation by *Azotobacter* is especially important in estuarine, intertidal, seagrass, and salt-marsh sediments. Apart from some members of the *Cyanobacteria*, *Azotobacter* is one of the few marine aerobic nitrogen fixers. The oxygen-sensitive nitrogenase is protected because of the very high respiratory rate of *Azotobacter* and the presence of a slimy capsule. Nitrogenase is also complexed with a protective protein. *Azotobacter* is heterotrophic and can use a wide range of carbohydrates, alcohols, and organic acids as a growth substrate. In soil, *Azotobacter* forms cysts, a resting stage with reduced metabolic activity and resistance to adverse environmental factors, but it is not clear if cysts are formed in marine habitats.

The *Enterobacteriaceae* is a large and well-defined family of *Gammaproteobacteria*

The *Enterobacteriaceae* are so called because they are best known as commensals and pathogens in the gut of warm-blooded animals, and include genera such as *Escherichia*, *Salmonella*, *Serratia*, and *Enterobacter*. They are fermentative, facultatively anaerobic bacteria and are oxidase-negative and Gram-negative rods, usually motile by means of peritrichous flagella. These properties help to distinguish them from other Gram-negative rods such as *Vibrio*, *Pseudomonas*, or *Alteromonas*. Enterobacteria can be isolated from coastal waters containing sewage or runoff from terrestrial sources (see Chapter 12). They are also found in the gut of fish and marine mammals, where some (e.g. *Edwardsiella tarda*) may be pathogenic. *Serratia marcescens* has been reported as the causative agent of a disease of corals, white pox (see p. 227).

Table 4.1 Some principal species of *Aliivibrio*, *Vibrio*, and *Photobacterium*

Light organ symbionts (chapter 10)	Human pathogens (chapter 12)	Coral pathogens (chapter 11)	Fish pathogens (chapter 11)	Molluskan pathogens (chapter 11)	Crustacean pathogens (chapter 11)
A. fischeri (V. fischeri)[a] P. leiognathi P. phosphoreum	V. alginolyticus V. cholerae V. cincinnatiensis V. damsela V. fluvialis V. furnissii V. hollisae V. metschnikovii V. mimicus V. parahaemolyticus V. vulnificus	V. shiloi (V. shiloni, V. mediterranei)[c] V. coralliilyticus V. harveyi V. splendidus	A. salmonicida (V. salmonicida)[a,b] P. damselae P. damselae subsp. piscicida V. anguillarum V. ichthyoenteri V. vulnificus (biotype 2) V. wodanis	V. pectenicida V. splendidus V. tapetis V. tubiashii	V. campbellii V. harveyi V. penaeicida V. proteolyticus V. parahaemolyticus

[a]The genus affiliation has changed very recently, so the original designations as members of the *Vibrio* genus continue to be widely used in the literature. [b]Avoid confusion with *Aeromonas salmonicida*. [c]The original name continues to be used in the literature, despite taxonomic correction (F. L. Thompson et al., 2001).

Vibrio and related genera have worldwide distribution in coastal and ocean water and sediments

Members of the family *Vibrionaceae* (*Gammaproteobacteria*) are oxidase-positive and facultatively anaerobic bacteria, which typically form curved rods with sheathed polar flagella. Collectively known as the vibrios, the family comprises the genera *Vibrio*, *Aliivibrio*, *Enterovibrio*, *Grimontia*, *Photobacterium*, and *Salinivibrio*. They are commonly associated with the surfaces of many marine animals and plants, and suspended organic matter. They play a major role in initial colonization of surfaces and biofilm formation and as symbionts and pathogens, as summarized in *Table 4.1*. Because of their importance, especially as pathogens, the vibrios have been investigated extensively and taxonomy of the group has been frequently revised. Ribosomal RNA typing does not provide a sufficiently accurate picture, and application of high-resolution molecular typing and genomic methods have led to recognition of a number of major clades, containing more than 130 species. Recent use of genomic analysis, multilocus sequence analysis, and supertree analyses has resulted in the distinction of four major groups (*Vibrio* core group, *V. cholera–V. mimicus*, *Aliivibrio* spp., and *Photobacterium* spp.). The advent of inexpensive, high-throughput DNA sequencing will allow rapid identification using this multiparameter genomic approach. The related family *Aeromonadaceae* includes the genus *Aeromonas*; this is usually regarded as a freshwater organism. However, *Aeromonas salmonicida* is a major pathogen of farmed salmon and trout in both freshwater and marine systems (see p. 241).

Some members of the Vibrionaceae are bioluminescent

Luminous bacteria are very common in the marine environment and occur as free-living forms in seawater, on organic debris, as commensals in the gut of many marine animals, and as light-organ symbionts (see *Research Focus Box 4.2*). Some light-organ symbionts have never been cultured, but most of the marine bioluminescent bacteria that can be isolated and cultured are members of the family *Vibrionaceae*—the commonest types being *Photobacterium phosphoreum*, *Photobacterium leiognathi*, *Vibrio fischeri* (recently reclassified as *Aliivibrio fischeri*), and *Vibrio harveyii*. The reaction mechanism for bacterial bioluminescence is shown in *Figure 4.5*. The enzyme responsible, luciferase, is a mixed-function oxidase that simultaneously catalyzes the oxidation of reduced flavin mononucleotide ($FMNH_2$) and a long-chain aliphatic aldehyde (RCHO)

BOX 4.2   **RESEARCH FOCUS**

Let there be light … but why?

Study of mutants leads to new ideas about why some bacteria are bioluminescent

The evolution of bioluminescence in bacteria is something of an enigma, as it is difficult to explain its biological role. How did this process, which can consume up to 20% of the cell's energy, evolve? What benefits do bacteria derive from emitting light? The discovery of sophisticated mechanisms of regulation of bioluminescence by density-dependent quorum sensing adds a further complication to any attempt to explain evolution of the process. It is tempting to say that quorum sensing "makes sense" from the bacterium's point of view, as there is clearly no advantage in single or well-isolated bacterial cells initiating the energetically expensive process of light emission because the amount of light will be too small. In nature, bioluminescence will only occur when the bacteria are numerous enough to be seen, such as in a symbiotic light organ or on a particle. It could be that luminescent bacteria are not truly free-living, as they usually seem to be associated with particulate aggregates or as commensals or symbionts of animals. Bacteria will be shed from animals in feces or exudates, and these aggregate to form marine snow particles. Bacteria growing in these particles could cause them to glow and make them more attractive as food items for animals, thus aiding transmission of the bacteria between hosts.

The ecological benefits are obvious for an animal host harboring bioluminescent symbionts, such as the association between vibrios and fish or squid (see *Research Focus Box 10.2*), and selection pressure over many millennia must also have been a major evolutionary force in the development of such a complex process. O'Grady and Wimpee (2008) screened hundreds of *Vibrio* species in coastal water and found that large numbers contained *lux* genes, but were not visibly luminescent in culture. They concluded that planktonic bioluminescent strains are very common by sheer number and diversity, yet the selective pressure that maintains this process is not obvious. Nackerdien et al. (2008) examined the interactive effects of mutations in *lux* genes and genes for different components of the quorum sensing system of *Vibrio harveyi* and showed that quorum sensing influences growth rate in a manner that cannot be explained fully by drain of energy from light emission.

Some studies have indicated that bioluminescence may promote DNA repair. Ultraviolet (UV) light causes the formation of pyrimidine dimers that prevent replication of the DNA. In the presence of blue light, a process called photoreactivation occurs; the enzyme DNA photolyase binds to the dimers and excises them and other enzymes restore the damaged segment of DNA. Induction of the DNA repair process is initiated by a complex regulatory system called SOS. Czyz et al. (2000) found that random mutagenesis of *V. harveyi* led to many UV-sensitive mutants that were also unable to emit light. When UV-irradiated *V. harveyi* cells were incubated in the light, more cells survived than when they were grown in the dark. When *luxA* or *luxB* mutants were cultivated in the dark after UV-irradiation, their survival was greatly reduced. Transfer of the *lux* operon from *V. harveyi* to *Escherichia coli* by cloning gave some protection to *E. coli* from the lethal effects of UV when cells were subsequently incubated in the dark. When wild-type (bioluminescent) bacteria and *lux⁻* mutants were grown in mixed cultures, the *lux⁻* mutants dominated the culture after a few days. However, when cultures were subjected to low UV-irradiation, the wild-type bacteria predominated. In the absence of UV, cells possessing the *lux* genes reproduced a little more slowly (perhaps because of the additional genetic burden). However, under the selection pressure of UV-irradiation, possession of the *lux* genes is advantageous. Czyz et al. (2000) suggest that the cell's own light emission stimulates DNA repair.

Other research indicates that bioluminescence may protect bacteria against the toxic effects of oxygen. As well as bacteria, bioluminescence is present in some fungi and diverse groups of animals. It appears to have multiple evolutionary origins because there is a very wide diversity of enzymes and substrates for the bioluminescent reaction mechanism. Apart from light emission, the only common feature is the requirement for oxygen. Rees et al. (1998) argue that bioluminescent reactions evolved primarily as a mechanism for detoxification of the highly toxic derivatives of molecular oxygen (see *Figure 3.10*), with the initial evolutionary drive being the nature of the substrates rather than the luciferase enzymes. Czyz and Wegrzyn (2001) found that *luxA* and *luxB* mutants (which do not make functional luciferase) were more sensitive to hydrogen peroxide than the wild type, whereas *luxD* mutants (which make luciferase but not the acyltransferase enzyme needed for fatty-acid substrate synthesis) were not. This suggests that the luciferase enzyme may play a role in detoxification of hydrogen peroxide. In contrast to this hypothesis, Ruby and McFall-Ngai (1999) proposed that bacterial luciferase may have evolved from a simpler reaction, which does not produce light but generates superoxide. These authors argue that the principal function of luciferases is the generation of superoxide, leading to damage of host tissues and nutrient release, and that this is a common feature of many symbiotic and pathogenic bacteria.

THE PANGENOME OF VIBRIOS

Classification and identification of the *Vibrionaceae* was problematic until the development of new molecular methods such as amplified fragment length polymorphism (AFLP) and multilocus sequence analysis (MLSA) pioneered by Fabiano Thompson and colleagues (see F. L. Thompson et al., 2005, 2006). The genomes of different strains of the same species of bacteria can differ by as much as 20–30%. C.C. Thompson et al. (2009) compared 32 different *Vibrio* genomes and found that the overall pangenome (the full complement of genes found across all strains) contained 26504 genes. This represents huge genetic diversity; by comparison, the human genome contains about 25000 genes. However, the core *Vibrio* genome consists of only 488 genes that are present in all strains of all species. As the number of genomes sequenced increases, the size of the pangenome increases, but the size of the core genome decreases. Thus, the pangenome represents the full genetic potential of the community of related organisms, but individual strains might have only a small fraction of these genes that equip it to occupy a particular niche. Martin Polz and colleagues found that *Vibrio splendidus* isolates that had over 99% similarity in 16S rRNA sequences from a single coastal site differed by as much as a megabase in genome size (J.R. Thompson et al., 2005) and comprised at least a thousand distinct genotypes, each occurring at extremely low environmental concentrations. Subsequently, Hunt et al. (2008) mapped the likely habitats of *V. splendidus* strains onto a phylogenetic tree and used this to develop a model showing the evolution over millions of years of numerous ecologically distinct populations from a common ancestor.

such as tetradecanal. Blue-green light with a wavelength of about 490 nm is emitted because of the generation of an intermediate molecule in an electronically excited state. Luciferases from all bioluminescent bacteria are dimers of α (~40 kDa) and β (~35 kDa) subunits, encoded by the *luxA* and *luxB* genes that occur adjacently in the *lux* operon, along with other gene-encoding enzymes, leading to the synthesis of the aldehyde substrate via fatty-acid precursors. The overall process is dependent on ATP and NADPH. The *lux* operon genes from several vibrios have been cloned and seem to have very similar structures. The α- and β-subunits of strains show about 30% identity in amino acid sequences and the β-subunit probably arose by gene duplication. Recombinant *lux* gene technology is widely used as a reporter system for monitoring gene expression, with important biotechnological applications. Inhibition of bioluminescence in *V. fischeri* is also used in a proprietary test for measuring environmental pollution (see p. 303). Bioluminescence is regulated by quorum sensing (see *Figures 3.8* and *3.9*).

The *Oceanospirillales* are characterized by their ability to break down complex organic compounds

As the name suggests, this order was originally characterized by their spiral cell shape, but they are very diverse in their physiology and ecology. This group apparently consists of uniquely marine types, as reflected in the names given to the various genera shown in *Table 4.2*, with many of the species named in honor of famous microbial ecologists. New members of this group have been discovered in symbiotic association with *Osedax* worms and in sediments in the vicinity of whale falls (see *Plate 10.4*). *Oceanospirillum* spp. are aerobic and motile and undoubtedly play a major role in the heterotrophic cycling of nutrients in seawater. Some other members in this group are important in the sulfur cycle, especially through degradation of DMSP. Some are active in the biodegradation of hydrocarbons and may have applications in bioremediation. There is wide variation in physiological properties such as optimum growth temperature, halophilicity, and utilization of substrates. The taxonomy of these genera is in a state of considerable flux, and 16S rRNA methods indicate that, although related, they represent a number of deeper phylogenetic groups. As shown in *Table 4.2*, even the "spirilla" designation is not a reliable distinguishing feature, as several genera are rod-shaped rather than helical.

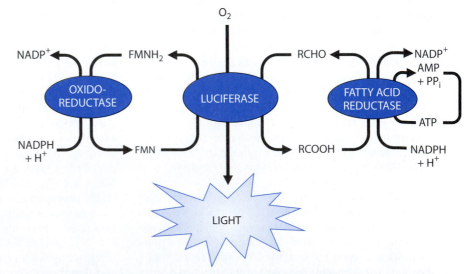

Figure 4.5 Reactions in bacterial bioluminescence. The *luxAB* genes encode the α and β subunits of luciferase. The *luxCDE* genes code for the transferase, synthetase and reductase enzymes used for the formation of the long-chain aldehyde substrate (e.g. tetradecanal) from fatty acids.

Table 4.2 Properties of some representative *Oceanospirillum, Marinospirillum, Alcanivorax* and *Neptunomonas* species

Species	Characteristic properties	Habitat/important properties
O. beijerincka O. maris O. linum O. multiglobuliferum	Helical; bipolar flagellar tufts; OGT 25–32°C; ONC 0.5–8%; %GC 45–50	Found in gut of shellfish, coastal seawater, seaweed
M. minutulum M. megaterium	Helical; polar or bipolar flagellar tufts; OGT 15–25°C; ONC 2–3%; %GC 42–45	Found in nutrient-rich environments such as fish and shellfish guts
A. borkumenis A. jadensis	Rods; nonmotile; OGT 20–30°C; ONC 3–10%; %GC 53–64	Common in seawater or sediments contaminated with oil. Degrade n-alkanes
M. hydrocarbono-clasticus M. aquaeoli	Rods; single polar flagellum (or nonmotile); OGT 30–32°C; ONC 3–6%; %GC 56–58	Isolated from oil-contaminated environments, hydrothermal vents, sulfide-rich sediments. Degrade n-alkanes
N. napthovorans	Rods; %GC 46	Isolated from oil-contaminated sediments. Degrade PAHs
M. communis M. mediterranea	Rods; single or bipolar flagella; OGT 20–25°C; ONC 0.7–3.5%; %GC 46–49	Isolated in open ocean and coastal seawater. Produce polyphenol oxidase (melanin biosynthesis)
M. georgiense M. stanieri M. jannaschii	Rods; single flagellum; OGT 37°C; ONC 0.6–2.9%; %GC 55	Isolated from nutrient rich coastal seawater. Degrade DMSP and other sulfur compounds

DMSP, dimethylsulfoniopropionate; OGT, optimum growth temperature; ONC, optimum NaCl concentration for growth; %GC, DNA mol % guanine + cytosine ratio; PAHs, polyaromatic hydrocarbons.

Magnetotactic bacteria orient themselves in the Earth's magnetic field

The rod-shaped or spiral cells of magnetotactic bacteria contain chains of magnetic particles comprised of magnetite (Fe_3O4) and greigite (Fe_3S_4) anchored within the cell (*Figure 4.6*). A magnetic field imparts torque on the chain of magnetosomes, so that the cell becomes aligned with magnetic field lines. The bacteria swim in this direction using polar flagella. The bacteria use this behavior in conjunction with aerotactic responses to locate a favorable zone in sediments of optimal O_2 and sulfide concentrations, which they require for growth. Although the best-studied examples are found in freshwater mud, magnetotactic bacteria

Figure 4.6 Transmission electron microscopy images of *Magnetospirillum* sp. AMB-1. (A) Grown under iron-rich conditions, showing a chain of magnetite crystals. (B) Grown in the absence of iron, showing empty magnetosome vesicles. (Reprinted from Komeili et al. [2004] by permission of the National Academy of Sciences.)

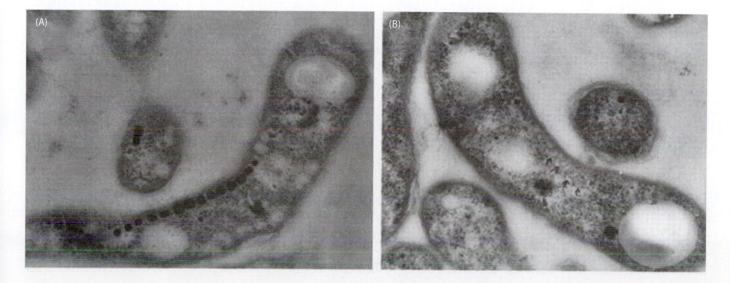

(A)   (B)

are widespread in salt marshes and other marine sediments. Recent molecular studies indicate that magnetotaxis is not restricted to a small, specialized lineage as previously thought. Magnetotactic bacteria can be isolated quite easily by applying a magnetic field to mixed environmental samples, although they are difficult to cultivate. Analysis of the genome of these bacteria reveals that a number of genes are required for the acquisition of iron via siderophores (see p. 71) and energy-dependent transport into the cell, followed by its conversion to magnetite or greigite within the magnetosomes. Microfossils of magnetite crystals found in deep-sea marine sediments are thought to originate from magnetotactic bacteria at least 50 million years ago. The discovery of very similar chains of magnetite in a meteorite in 2001 has been interpreted by some experts as evidence of ancient life on Mars.

A unique magnetotactic bacterium named *Candidatus* Magnetoglobus multicellularis (a member of the *Deltaproteobacteria*) forms a compact spherical aggregate of highly organized flagellated bacterial cells that swim in either helical or straight trajectories (*Figure 4.7*). The bacteria behave as a truly multicellular organism, as the aggregate grows by enlarging the size of its cells and doubling the volume of the whole structure. Cells divide synchronously and separate into two identical spherical aggregates.

Bdellovibrio is a predator of other bacteria

Bdellovibrio is genus of small spiral cells in the *Deltaproteobacteria*, which has the unusual property of preying on other Gram-negative bacteria (*Figure 4.8*). Free-living *Bdellovibrio* swims at high speed and uses a chemosensory system to detect high concentrations of prey. Different strains are highly specific in targeting particular hosts. It attaches to its prey and retractable fibers bring it into close contact with the bacterial cell. A cocktail of enzymes degrades proteins, lipids, and carbohydrates to create an opening in the cell wall. *Bdellovibrio* then enters the periplasmic space between the outer and inner membranes. It replicates in the periplasmic space, producing extracellular enzymes to liberate nutrients for growth. The cytoplasm of the prey is consumed and the *Bdellovibrio* cell elongates into a large filamentous cell. When all the nutrients are exhausted, the filament differentiates into about 15 motile cells, which then seek out new hosts. *Bdellovibrio* spp. are very widespread in the marine environment and may control populations of other bacteria, although their full ecological role is unknown.

Budding and stalked *Proteobacteria* show asymmetric cell division

This group of bacteria is distinguished by the presence of extrusions or appendages of the cytoplasm, called prosthecae. Unlike almost all other bacteria, members of this group show a life cycle with unequal cell division, in which the "mother cell" retains its shape and morphological features whilst budding off a smaller "daughter" cell. This occurs owing to polarization of the cell, that is, new cell wall material is grown from a single point rather than by intercalation as occurs in other bacteria. Some, such as *Hyphomicrobium* and *Rhodomicrobium*, bud off progeny cells from hypha-like extensions, whilst others, such as *Caulobacter*, have distinctive stalks. This is a particular advantage in aquatic environments, because it means that these bacteria often attach firmly to algae, stones, or other surfaces. The progeny cells are motile and swim away to colonize fresh surfaces (*Figure 4.9*). *Caulobacter crescentus* has been intensively studied as a model of cellular differentiation. The cell cycle is controlled by a set of three master regulator proteins that determine the coupling of gene expression to cell-cycle events with extraordinary precision. The increased surface area:volume ratio of stalked bacteria probably enables them to thrive in nutrient-poor waters and the prosthecae also enable these aerobic bacteria to remain in well-oxygenated environments by avoiding sinking into sediments. Phylogenetically, most representatives

---

? HOW DO BACTERIA MAKE NANO-SIZED MAGNETS?

The individual nano-sized magnetosomes in magnetotactic bacteria are too small to behave as magnets on their own and must be aligned into a linear chain so that they function like a compass needle. The physical tendency of magnetic dipoles would be to agglomerate in order to lower the magnetostatic energy, so the magnetosomes must be stabilized with an organic structure to anchor the chain in place. How is this achieved? Scheffel et al. (2006) identified a key gene in *Magnetospirillum gryphiswaldense* that encodes an acidic protein MamJ, which forms a novel filamentous structure. A mutant strain in which the gene *mamJ* was deleted produced normal magnetite crystals, revealed by electron microscopy, but these assembled into compact clusters instead of the chains found in the wild-type strain. The mutants showed reduced magnetic orientation. Production of a GFP-fusion protein (see p. 33) and three-dimensional imaging showed that the MamJ protein is laid down as a cytoskeleton-like structure from pole to pole of the cell, and this guides the formation of the magnetosomes and magnetite crystals like beads on a string. In a parallel study in *Magnetospirillum magneticum*, Komeili et al. (2004) concluded that magnetosome vesicles are positioned in the filaments by another protein, MamK (see *Figure 4.6B*).

Figure 4.7 Electron microscopy images of "*Candidatus* Magnetoglobus multicellularis." (A) Transmission electron micrograph showing the multicellular consortium consisting of about 40 bacterial cells containing lipid inclusions (*asterisks*) and magnetosomes surrounding an acellular space (Ac). The cells and the magnetosomes are specially arranged so that the total magnetic moment aligns the whole structure so it can swim along field lines. Bar = 1 μm. (Reprinted from Martins et al. [2007] by permission of Wiley–Blackwell.) (B) Scanning electron micrograph showing the spherical morphology and the tightly bound cells. Bar = 2 μm. (Reprinted from Abreu et al. [2007] by permission of the International Union of Microbiological Societies.)

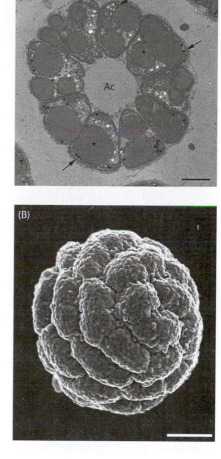

belong to the *Alphaproteobacteria*. *Caulobacter* and *Hyphomicrobium* are chemoorganotrophic, whilst *Rhodomicrobium* and *Rhodopseudomonas* are phototrophic. These bacteria are often the first to colonize bare surfaces, and they are especially important in the formation of biofilms, with significant consequences for larval settlement and biofouling (Chapter 13).

Sulfur- and sulfate-reducing bacteria have a major role in the sulfur cycle

Most members of the sulfur- and sulfate-reducing bacteria (SRB) belong to the *Deltaproteobacteria*, and their activities are highly significant in the sulfur cycle in anoxic marine environments. SRB acquire energy for metabolism and growth by utilizing organic compounds or hydrogen as electron donors coupled to dissimilative sulfate reduction (see p. 69). SRB produce H_2S, resulting in the characteristic stench and blackening (due to deposits of FeS) of decomposition in anoxic mud, sediments, and decaying seaweed. H_2S is highly toxic and adversely affects many forms of marine life, but it can be used by a wide range of chemotrophic and phototrophic bacteria, as described in earlier sections, completing the cycling of sulfur. Numerous types of SRB have been isolated from marine sediments and these are classified on the basis of morphological and physiological properties as well as 16S rRNA typing. More than 25 species have been described (they also occur in soils, animal intestines, and freshwater habitats), and representative examples and their properties are shown in *Table 4.3*, illustrating the diversity in physiological and morphological properties. The main division of the SRB is between: (i) those that couple the oxidation of substrates such as acetate or ethanol to reduction of elemental sulfur (but not SO_4^{2-}) to H_2S; and (ii) those that can reduce SO_4^{2-}, using a range of organic compounds or H_2. It should be noted that some Gram-positive bacteria and *Archaea* (p. 126) also carry out sulfate reduction.

Figure 4.8 Composite image of scanning electron micrographs showing the life cycle of *Bdellovibrio bacterivorans* (*light-colored cell*) infecting a Gram-negative host bacterium (*dark-colored cell*). Infection begins by attachment of a flagellated *Bdellovibrio* cell (1, 2). During active penetration and entry to the host cell, the flagellum is shed (3–5). *Bdellovibrio* replicates in the periplasmic space, producing extracellular enzymes to liberate nutrients for growth and elongating into a large filamentous cell (6–7). When nutrients are exhausted, the *Bdellovibrio* differentiates into several motile cells (8–9), which then swim to seek out new hosts (10). (Image courtesy of Snejzan Rendulic, Juergen Berger, and Stephan Schuster, Max Planck Institute for Developmental Biology.)

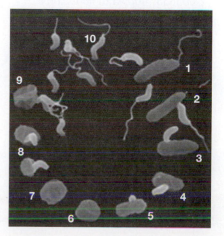

Figure 4.9 Composite image of scanning electron micrographs showing the life cycle of *Caulobacter crescentus*. The cycle begins with settlement of a motile swarmer cell on a surface (1), which loses its flagellum (2) and synthesizes a stalk, with a highly adhesive holdfast (3). An asymmetric cell division (4) results in formation of a stalked "mother" cell and a flagellated swarmer cell, which separate (5). The attached stalked cell undergoes further division cycles. (Image courtesy of Yves Brun, Indiana University.)

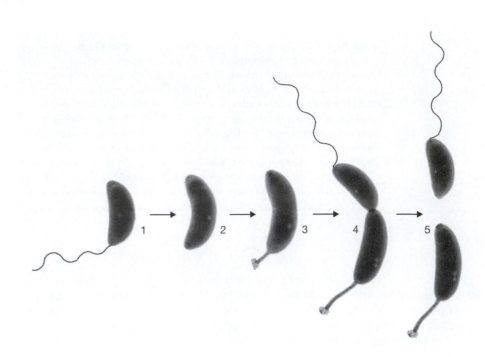

? COULD SRB BE USED AS BATTERIES ON THE SEA BED?

Marine sediments contain large amounts of organic material—could these provide a source of energy? If a graphite rod electrode (anode) is placed in anoxic marine sediments and connected to a cathode in the O_2-containing seawater above, a significant electrical current is generated. By comparing the communities that developed in the presence or absence of the electrical connection, Bond et al. (2002) showed that the sediment becomes enriched with SRB. The most prevalent organism enriched during the electrical process is *Desulfuromonas acetoxidans*, which grows anaerobically by coupling the oxidation of acetate to reduction of sulfur. They also studied the generation of electricity with pure cultures and showed that the bacteria grow as a result of acetate oxidation coupled to reduction of the electrode. Marine sediments are unlikely to make a significant contribution to the world's energy problems, but the electricity produced in the initial studies has been sufficient to power small fuel cells on the seabed (Tender et al., 2002). An obvious and immediate application is for powering oceanographic monitoring equipment. The mechanism could almost certainly be harnessed in conjunction with bioremediation of sediments contaminated with organic compounds.

Recently, SRB have been shown to form syntrophic consortia with sulfide-oxidizing bacteria in an animal symbiosis (see *Figure 10.4*) and with *Archaea* in the anaerobic oxidation of methane (See *Plate 5.1*). Despite being found most commonly in anoxic sediments, there is increasing evidence that many SRB are tolerant of oxygen and some can coexist with oxygenic *Cyanobacteria* in microbial mats.

The *Cyanobacteria* carry out oxygenic photosynthesis

The *Cyanobacteria* is a large and diverse group, and members are characterized by their ability to carry out photosynthesis in which oxygen is evolved, although some anoxygenic *Cyanobacteria* have recently been described. *Cyanobacteria* contain chlorophyll *a*, together with accessory photosynthetic pigments called phycobilins. This group was formerly known as the "blue-green algae" (because of the presence of the blue pigment phycocyanin together with the green chlorophyll) and is still treated as a division of the algae by many marine biologists and phycologists, although the *Cyanobacteria* clearly form one of the major phyla of the domain *Bacteria*. Furthermore, many marine genera contain phycoerythrins, which give the cells a red-orange, rather than blue-green, color. Fossil evidence, in the form of morphological structures and distinctive biomarkers typical of the group (hopanoids) indicates that organisms resembling *Cyanobacteria* may have evolved about three billion years ago and the evolution of oxygen from the photosynthetic activities of *Cyanobacteria* (or their ancestors) was probably responsible for changes in the Earth's early atmosphere. *Cyanobacteria* occupy very diverse habitats in terrestrial and aquatic environments, including extreme temperatures and hypersaline conditions. In the marine environment, habitats include the plankton, sea ice, and shallow sediments, as well as microbial mats on the surface of inanimate objects, algae, or animal tissue.

The *Cyanobacteria* are morphologically very diverse, ranging from small undifferentiated rods to large branching filaments showing cellular differentiation. Unicellular *Cyanobacteria* divide by binary fission, whilst some filamentous forms multiply by fragmentation or release of chains of cells. Many types are surrounded by mucilaginous sheaths that bind cells together. The chlorophyll *a* is contained within lamellae called phycobilosomes, which are often complex and multilayered. *Cyanobacteria* show remarkable ability to adapt the arrangement of their photosynthetic membranes and the proportion of phycobilin proteins to

Table 4.3 Selected genera of marine sulfur- and sulfate-reducing bacteria

Genus	Morphology	Optimum temp. (°C)	DV	DNA (mol% G + C)
Sulfate reducers—do not utilize acetate				
Desulfovibrio	Curved rods, motile	30–38[a]	+	46–61
Desulfomicrobium	Motile rods	28–37	–	52–57
Desulfobacula	Oval to coccoid cells	28	ND	42
Sulfate reducers—oxidize acetate				
Desulfobacter	Oval or curved rods, may be motile	28–32	–	45–46
Desulfobacterium	Oval, may have gas vesicles	20–35	–	41–59
Dissimilatory sulfur reducers—do not reduce sulfate				
Desulfuromonas	Rods, motile	30	–	50–63
Desulfurella	Short rods	52–57	–	31

[a]One species is thermophilic. DV, desulfovoridin, a pigment used as a chemotaxonomic marker.

maximize their ability to utilize light of different wavelengths. Many *Cyanobacteria* form intracellular gas, which enable cells to maintain themselves in the photic zone. Gas vesicles, as well as mucilaginous sheaths and pigments, also protect cells from extreme effects of solar radiation. Gliding motility is very important in *Cyanobacteria* that colonize surfaces. Gliding movement, up to 10 μm per second, occurs parallel to the cell's long axis and involves the production of mucilaginous polysaccharide slime. There are two possible mechanisms by which gliding occurs. One is the propagation of waves moving from one end of the filament to the other, created by the contraction of protein fibrils in the cell wall. The other mechanism is secretion of mucus by a row of pores around the septum of the cell. Some types, such as *Nostoc*, are only motile during certain stages of their life cycle, when they produce a gliding dispersal stage known as hormogonia. *Synechococcus* also seems able to swim in liquid media without using flagella.

Until the use of 16S rRNA analysis established *Cyanobacteria* as a group within the *Bacteria*, they were classified by botanists into about 150 genera and 1000 species based on morphological features (*Table 4.4*). Subsequent phylogenetic analysis shows that these groupings are very unreliable and many genera are polyphyletic. Since bacteriologists can now grow many of the *Cyanobacteria* in pure culture, analysis of biochemical characteristics and molecular features are taking over as a basis of classification, and a major revision of the group is in progress. Pure culture studies show that the physiological properties of *Cyanobacteria* are more variable than previously thought. Many are capable of anaerobic growth, some can use hydrogen, hydrogen sulfide, or reduced organic compounds as electron donors and some can grow heterotrophically in culture; but the significance of these modes of nutrition in natural marine environments is unknown.

Many marine *Cyanobacteria* carry out nitrogen fixation

As discussed in later chapters, nitrogen fixation is of fundamental significance in primary production in the oceans. The bond in molecular nitrogen is very stable, and its reduction to ammonia is an extremely energy-demanding process. Most

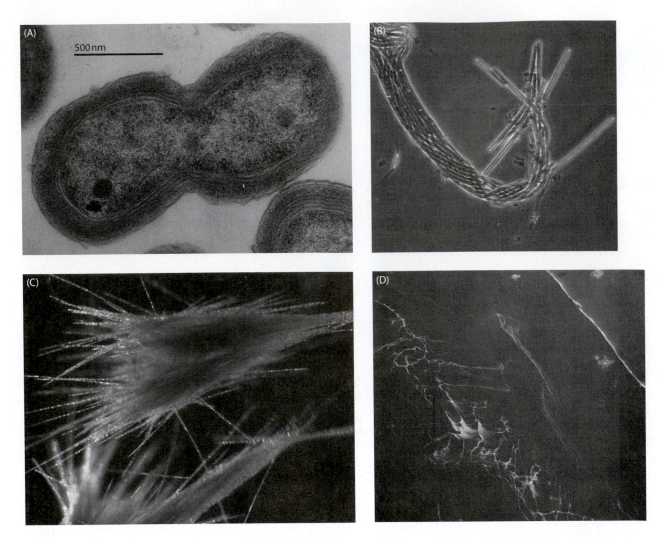

Figure 4.10 Images of representative cyanobacteria. (A) Transmission electron micrograph of ultrathin section of dividing cell of *Prochlorococcus marinus*. (B) Light micrograph of strings of cyanobacteria from a saltmarsh microbial mat. (C) Light micrograph of *Trichodesmium erythraeum*. (D) Photograph of *Trichodesmium* plankton bloom observed off northwest Australia from the International Space Station. (Images courtesy of (A) Chisholm Lab, Massachusets Institute of Technology; (B) Rolf Schauder (University of Frankfurt), Mark Schneegurt (Wichita State University), and Cyanosite (www.cyanosite.bio.purdue.edu); (C) S. Kranz, Alfred Wegener Institute; (D) NASA Earth Observatory.)

nitrogen-fixers are anaerobic, but the *Cyanobacteria* are aerobic, generating oxygen by photosynthesis. This poses something of a problem, because the key enzyme nitrogenase is strongly inhibited by oxygen. In some types, nitrogen fixation is restricted to the night when no oxygen is generated. Many of the more efficient nitrogen-fixers contain differentiated cells known as heterocysts within the filament. Because the heterocysts contain no photosytem II, they provide an oxygen-free environment, which protects the enzyme nitrogenase. However, one of the most prolific marine nitrogen-fixers is *Trichodesmium*, which does not contain heterocysts but is able to switch the two processes of the oxygen-producing photosynthetic system and the oxygen-sensitive nitrogen-fixing system on and off over timescales of a few minutes. There also appears to be a spatial separation of oxygen evolution and nitrogen fixation, because nitrogen fixation is limited to certain parts of the cell. *Trichodesmium* forms dense filamentous masses that are responsible for large blooms, especially in tropical seas (*Figure 4.10C, D*), but the amount of nitrogen it fixes does not seem sufficient to match global estimates. Recently, small cyanobacteria (3–10 μm), including *Crocosphaera*, have been shown to carry out a substantial fraction of oceanic nitrogen fixation.

The genera *Prochlorococcus* and *Synechococcus* dominate the picoplankton in large areas of the Earth's oceans

Synechococcus and *Prochlorococcus* are major contributors to the carbon cycle through photosynthetic CO_2 fixation, accounting for between 15 and 40% of carbon input to ocean food webs. Marine *Synechococcus* species are abundant in

Table 4.4 Examples of marine *Cyanobacteria*

Order	Features	Major marine genera
Chroococcales	Unicellular or aggregates of single cells; may be motile	*Prochlorococcus Synechococcus Synechocystis*
Pleurocapsales	Aggregates of single cells; reproduce by small spherical gliding cells (baeocytes) formed by multiple fission	*Cyanocystis Pleurocapsa*
Oscillatorales	Filamentous cells (trichome), often sheathed. Intercalary binary division at right angles to long axis; motile	*Trichodesmium Lyngbya*
Nostocales	Filamentous trichome with heterocysts	*Nostoc*
Stigonematales	Branching clusters, filamentous with heterocysts	Mainly freshwater or terrestrial

the upper 25 m of the ocean, with different physiological types present in coastal waters and the open ocean.

Prochlorococcus is a very small cyanobacterium about 0.6 μm diameter (*Figure 4.10A*), which was not discovered until 1988, despite the fact that it inhabits large parts of the oceans between 25 and 200 m at a density between 10^5 and 10^6 ml^{-1}, making it the most abundant photosynthetic organism on Earth. *Prochlorococcus* is most abundant in the region from 40°S to 40°N, temperature range 10–33°C. *Prochlorococcus* contains modified forms of chlorophyll (divinyl chlorophylls *a* and *b*), but lacks phycobilins. The photosynthetic apparatus seems to be adapted to allow *Prochlorococcus* to grow at considerable depths, where the amount of light is very low (<1% of that at the surface). The small cell size gives a large surface area:volume ratio, which helps *Prochlorococcus* obtain scarce nutrients in oligotrophic ocean waters. Other organisms with similar properties include *Prochloron*, which is an intracellular symbiont of certain marine invertebrates. These organisms were originally placed in a group called the prochlorophytes, but phylogenetic analysis shows that this is not a distinct lineage within the *Cyanobacteria*. Prochlorophytes appear to have evolved divinyl chlorophylls *a* and *b*, which allows them to harvest longer wavelengths of blue light which penetrate deeper waters. Indeed, studies on *Prochlorococcus* cultures have shown that there are genetically distinct populations of ecotypes, which occupy distinct light niches, and the basis of this has been confirmed by analysis of the genome sequences.

If *Prochlorococcus* is isolated from different depths in the water column, it is clear that there are distinctly different types adapted for growth in different ecological niches. Those in the upper part of the water column (~25–100 m depth) are adapted to high light intensities, but this region contains relatively low levels of nutrients. Those strains found toward the bottom of the photic zone (~80–200 m depth) are growing in very low light levels, but nutrients are more abundant. The two ecotypes have different ratios of chlorophyll a_2 to b_2 and differ in their optimal irradiances for photosynthesis. These significant differences in ecotype are determined by genetic differences of only about 2% in 16S rRNA sequences, yet analysis of the full genome sequences shows that they differ markedly in the number of genes they possess.

The discovery of *Prochlorococcus* is also highly significant for evolutionary theory. It has been thought for many years that the chloroplasts present today in algae and plants evolved from *Cyanobacteria* in accordance with the endosymbiosis theory. Phylogenetic analyses indicate that prochlorophytes, despite resembling chloroplasts without phycobilins, are not the immediate ancestral origin of the chloroplast. It is possible that the prochlorophytes and the rest of the *Cyanobacteria* may have evolved from ancestors that contained phycobilins and more than one type of chlorophyll. Prochlorophytes may have lost their phycobilins and the other *Cyanobacteria* may have lost their chlorophyll *b* during evolution, whilst the eukaryotic chloroplast evolved from the hypothetical ancestor of both groups. We do not know when this divergence occurred.

Cyanobacteria are important in the formation of microbial mats in shallow water

Complex stratified communities of microorganisms develop at interfaces between sediments and the overlying water. Filamentous *Cyanobacteria* such as *Phormidium*, *Oscillatoria*, and *Lyngbya* are often dominant members of the biofilm in association with unicellular types such as *Synechococcus* and *Synechocytsis* (*Figure 4.10B*). Steep concentration gradients of light, oxygen, sulfide, and other chemicals develop across the biofilm. The mat becomes anoxic at night and H_2S concentrations rise. *Cyanobacteria* (and other motile bacteria in the biofilm) can migrate through the mat to find optimal conditions. Anoxygenic phototrophs as well as aerobic and anaerobic chemoheterotrophs are also present.

Stromatolites are fossilized microbial mats of filamentous bacteria and trapped sediment. These ancient structures were widespread in shallow marine seas more than three billion years ago. Ancient stromatolites were probably formed by anoxygenic phototrophs, but modern stromatolites are dominated by a mixed community of *Cyanobacteria* and heterotrophic bacteria. Growth of modern marine stromatolites represents a dynamic balance between sedimentation and intermittent lithification of cyanobacterial mats (*Figure 4.11*). Rapid sediment accretion occurs when the stromatolite surfaces are dominated by pioneer communities of gliding filamentous *Cyanobacteria*. During intermittent periods, surface films of exopolymer are decomposed by heterotrophic bacteria, forming thin crusts of microcrystalline calcium carbonate. Other types of *Cyanobacteria* modify the sediment, forming thicker stony plates.

The *Firmicutes* are a major branch of Gram-positive *Bacteria*

The *Firmicutes* are unicellular *Bacteria* distinguished by a thick cell wall composed mainly of peptidoglycan and genomes with a low G + C ratio. The genera *Bacillus* and *Clostridium* contain a large number of diverse species and are best known as soil saprophytes, but they are also a major component of marine sediments. Although relatively little information exists on their abundance and distribution, some species have been named because of their initial isolation from marine sediments (e.g. *Bacillus marinus*) and it is likely that great diversity in marine representatives remains to be discovered. Their most distinctive feature is the production of extremely resistant endospores, which allow them to resist high temperatures, irradiation, and desiccation, and endospores may persist for thousands of years. Some *Bacillus* spp. bind and oxidize metals such as manganese and copper and the spores may have applications in remediation of polluted marine sediments. Some members of the related genus *Paenibacillus* produce a powerful chitinolytic activity, which may be useful for the breakdown of shrimp and crab shell waste. Both genera are also used as probiotics in aquaculture (see p. 326). *Bacillus* spp. are usually aerobic whilst *Clostridium* spp. are strict anaerobes. *Clostridium* has a wide range of fermentation pathways leading to the formation of organic acids, alcohols, and hydrogen. Some types are also efficient in the fixation of nitrogen. Clostridia play a major role in decomposition and nitrogen cycling in anoxic marine sediments. One species, *C. botulinum*, is important as a source of toxins associated with fish products (see p. 268).

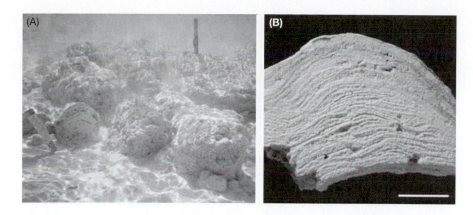

Figure 4.11 Stromatolites. (A) Columnar build-ups in shallow water, Highborne Cay, Bahamas. (B) Vertical section showing lamination; scale bar = 2 cm. (From Reid et al. [2000] reprinted by permission of Macmillan Publishers Ltd.)

Other genera of *Firmicutes* include *Staphylococcus*, *Lactobacillus*, and *Listeria*, which are aerobic, catalase-positive cocci and rods with typical respiratory metabolism. They are occasionally isolated in marine samples, but are probably very minor members of the marine bacterial community. They can be important as agents of fish spoilage and food-borne intoxication following processing (p. 286). *Streptococcus iniae* and some other species are important pathogens in warm-water fish and dolphins. *Renibacterium salmoninarum* is an obligate pathogen of salmonid fish (p. 244).

Epulopiscium fishelsoni and related species are giant bacteria with a unique "viviparous" lifestyle

Epulopiscium spp. are often found in large numbers in the intestinal tracts of various species of herbivorous surgeonfish on the Great Barrier Reef and Red Sea. They seem to have a limited host range and may be mutualistic symbionts, which aid in the breakdown of food ingested by the fish. *Epulopiscium* is one of the largest bacteria known. Based on 16S rRNA gene sequencing, these are a family of morphologically diverse bacteria and cell size is variable; in populations of large cells, cells up to 400 × 80 μm are common and some rare cells can reach 600 μm. In terms of cell volume, these giant cells are millions of times bigger than most marine bacteria (*Table 1.2*). When originally discovered, it was thought to be a eukaryotic protist because of its large size and complex intracellular structure (*Figure 4.12*). *Epulopiscium* spp. have a unique reproductive process; they are "viviparous," meaning that new cells are formed inside the parent cell, which undergoes a localized cell lysis to release the active progeny. One of the most remarkable features of the life cycle is that it follows a circadian rhythm. Samples of surgeonfish gut contents taken early in the morning show small internal offspring (between 2 and 7, depending on the species) appearing near the tips of the mother cell. The offspring cells grow during the day and are released at night through partial lysis of the mother cell, which then dies. This process is

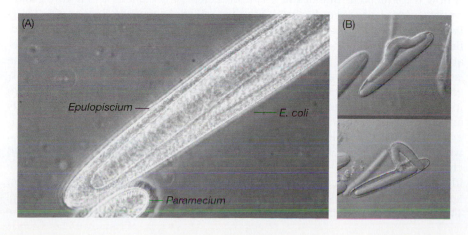

Figure 4.12 *Epulopiscium fishelsoni*— one of the largest known bacteria. (A) Light micrograph showing comparative sizes of *Epulopiscium*, *Paramecium* (a protist) and *Escherichia coli* (a "typical" bacterium). (B) "Viviparous" release of daughter cell during reproduction. (Image courtesy of Esther R. Angert, Cornell University, USA.)

reminiscent of the process of endospore formation, which involves the partition of the spore from the parent cell. Although it has not been cultured, 16S rRNA sequencing indicates that one of the closest-known relatives of *Epulopiscium* spp. is *Metabacterium polyspora*, which typically produces multiple endospores as a means of propagation in its animal host, the guinea pig. Based on the phylogenetic affiliation and morphological observations, it has been hypothesized that daughter-cell formation in *Epulopiscium* evolved from the process of endospore formation.

The *Actinobacteria* is a large phylum including the mycobacteria and actinomycetes

A second major group of the high G + C-containing Gram-positive bacteria is the phylum *Actinobacteria*, containing more than 30 families. The phylum is well represented in clone libraries, especially those from coastal and estuarine waters. *Mycobacteria* are slow-growing, aerobic, rod-shaped organisms distinguished by unusual cell wall components, which render them acid-fast in a staining procedure. They are widely distributed as saprophytes on surfaces such as sediments, corals, fish, and algae. Some species (e.g. *Mycobacterium marinum*) have been identified as pathogens of fish and marine mammals which can be transmitted to humans (p. 269).

The *Actinomycetes* is a large and diverse group of bacteria including *Actinomyces* and related genera. They have various cell morphologies, ranging from coryneform (club-shaped, e.g. *Corynebacterium*, *Arthrobacter*) to branching filaments with reproductive conidiospores (e.g. *Streptomyces* and *Micromonospora*). Actinomycetes are widely distributed in marine sediments. Because of their high abundance in soil, it is possible that actinomycetes in coastal sediments are derived from terrestrial runoff, but they are also found in deep-sea samples. Only one exclusively marine species, *Rhodococcus marinonascens*, has been cultured and studied in detail, but 16S rRNA studies suggest that there are many diverse marine actinomyctes. Their main ecological importance is in decomposition and heterotrophic nutrient cycling, owing to their production of extracellular enzymes, which break down polysaccharides, proteins, and fats. The actinomycetes are also an exceptionally rich source of secondary metabolites. Many of the most widely used antibiotics (e.g. aminoglycosides and tetracyclines) come from actinomycetes found in soil. Extensive surveys of the diversity of marine actinomycetes have been undertaken in the search for unique compounds with biotechnological potential as antimicrobial or anticancer drugs (Chapter 14).

The *Cytophaga–Flavobacterium–Bacteroides* group is morphologically and metabolically diverse

This collection of diverse, aerobic, facultative, or anaerobic chemoheterotrophs is now recognized as one of the major branches of the *Bacteria* (Phylum *Bacteroidetes*). Many of the key genera (*Cytophaga*, *Flavobacterium*, *Bacteroides*, *Flexibacter*, and *Cellulophaga*) are polyphyletic and their taxonomy is very confused. Based on analysis of the *gyrB* gene, the new genus *Tenacibaculum* has been created to accommodate marine types formerly known as *Flexibacter*. Many marine isolates of *Cytophaga* and *Flavobacterium* have unusual flexirubin and carotenoid pigments and can be easily isolated as colored colonies on agar medium inoculated from sediments, marine snow, and the surfaces of animals and plants and incubated aerobically at ambient temperatures. Their most distinctive properties are gliding motility and the production of various extracellular enzymes, which are responsible for degradation of polymers such as agar, cellulose, and chitin. Agar is normally resistant to bacterial degradation, which is why it is an ideal gelling agent for culture plates, but softening or the formation of craters on agar plates is often observed with marine isolates belonging to the CFB group. The production of hydrolytic enzymes is of major ecological significance in the

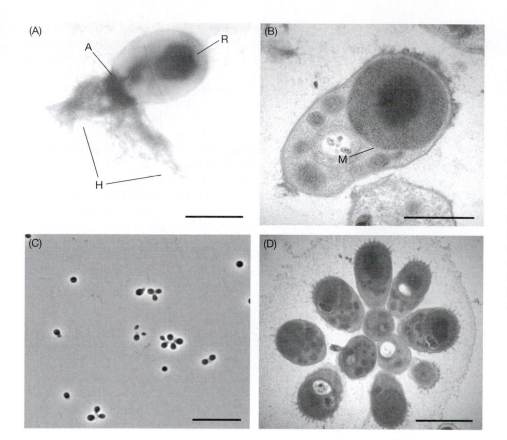

Figure 4.13 Microscopic images of *Rhodopirellula baltica*. (A) Electron micrograph image of a single cell displaying the polar organization of *R. baltica* cells. A, attachment pole; H, holdfast substance; R, reproduction pole. Bar = 0.5 μm. (B) Electron micrograph image displaying the intracellular compartmentalization. M, membrane engulfing the pirellulosome-like structures. Bar = 0.5 μm. (C) Light microscopic image of rosette-like aggregates. Bar = 10 μm. (D) Electron micrograph image of a rosette showing the attachment of cells via the attachment poles. Bar = 1 μm. (Reproduced from Schlesner et al. [2004] by permission of the International Union of Microbiological Societies.)

degradation of complex organic materials such as the cell walls of phytoplankton and exoskeletons of crustaceans. Some species are pathogenic for fish and invertebrates. Many are psychrophilic, being commonly isolated from cold-water marine habitats and sea ice. The normal habitat of the genus *Bacteroides* is the gut of mammals and they may be present in sewage-polluted waters and persist for quite long periods in the sea, prompting investigation of their use as indicators of water quality (p. 293).

The *Planctomycetes* are a group with cells that show some similarities to eukaryotes

Like the stalked *Proteobacteria*, members of the phylum *Planctomycetes* have a life cycle comprising a stalked sessile stage and a motile stage, but here the stalk is a separate protein appendage rather than an extension of the cell. Unusually among the *Bacteria*, they also lack peptidoglycan, with the cell wall being an S-layer composed of cysteine- and proline-rich protein. The phylum currently contains a single family, *Planctomycetaceae*. An increasing number of marine planctomycetes are recognized from 16S rRNA analysis and several types can be cultured using simple organic media, including the genera *Planctomyces*, *Pirulella*, *Rhodopirellula* (Figure 4.13), *Blastopirellula*, *Isosphaera*, and four *Candidatus* genera, (Kuenenia, Brocadia, Scalindua, and Anammoxoglobus). The species that have been cultured are aerobic heterotrophs that grow very slowly, so the group is probably underrepresented in culture-based investigations. Although they are not one of the major groups of planktonic bacteria in pelagic waters, because of their attachment mechanism planctomycetes are especially associated with marine snow and they are often present in large numbers in association with blooms of diatoms or other algae. Recent genome analysis indicates that they possess numerous genes encoding sulfatases, enzymes that could be responsible for the breakdown of sulfated heteropolysaccharides, which are produced in large quantities in marine environments (e.g. by fish and algae). Several enzymes are necessary for the breakdown of these complex compounds. The planctomycetes may also be important in the metabolism of C_1 compounds.

Despite obvious membership of the domain *Bacteria*, the group *Planctomycetes* shows some features that are reminiscent of those in eukaryotes; namely, they have membrane-bound compartments in the cell, in which metabolic and genetic components are separated. Some members of the group contain a specialized structure in which the anaerobic oxidation of ammonia reaction anammox is carried out (see p. 69); this is of major importance in the oxygen-minimum zones (see *Research Focus Box 9.1*). Of particular note is the fact that there is a nucleus containing a very large genome bound by a unit membrane; also, some features of genetic organization and transcription are similar to those of eukaryotes. *Bacteria* and *Archaea* have traditionally been described as prokaryotic, with the absence of true nucleus being the defining feature underlying use of this term. The observation of the nucleus and organelle-like structures in the *Planctomycetes*—and some other related groups such as *Poribacteria* associated with sponges—gives support to the view that the prokaryote–eukaryote distinction is no longer valid (see p. 4).

Verrucomicrobia is a poorly characterized phylum of *Bacteria*

Members of this group have diverse habitats, as shown by 16S rRNA gene sequence analysis of environmental samples. However, only a very few strains from soil have been isolated in culture. Despite the occurrence of sequences from marine sediments and aggregates, nothing is known of the physiological properties and importance of this group in marine habitats. This group is linked with the *Planctomycetes* and *Chlamydia* by some taxonomists into a superphylum.

The spirochetes are Gram-negative, tightly coiled, flexuous bacteria distinguished by very active motility

Motility occurs through the possession of internal flagella, located in the periplasmic space between the cell membrane and cell wall. The internal flagella rotate as a rigid helix, like other bacterial flagella, causing the protoplasmic cylinder to rotate in the opposite direction, leading to flexing and jerky movement. Genera such as *Cristispira* and *Spirochaeta* are widespread in marine habitats, but little is known about their ecological role. Many are strictly anaerobic and found in sediments, and it is likely that they are also important components of the gut microbiota in marine animals. *Cristispira* occurs in the digestive tract of certain mollusks, but has not been cultured.

Aquifex and *Thermotoga* are hyperthermophiles

As can be seen in *Figure 4.1*, *Aquifex* and its relatives form a deeply branching root of the phylogenetic tree of *Bacteria*. Analysis of sequences of 16S rRNA and several other genes confirms that this group is closest to the hypothetical ancestor of all *Bacteria*. Because *Aquifex* spp. such as *A. pyrophilus* and *A. aeolicus* are extremely thermophilic (maximum growth temperature can be as high as 95°C) and chemolithotrophic, they have a major role in primary production in marine hydrothermal vents. They grow using hydrogen, thiosulfate, or sulfur as the electron donor and oxygen or nitrate as the electron acceptor, fixing carbon via the reductive citric acid cycle (*Figure 3.4B*). The extremely thermophilic properties of these bacteria are obviously of great interest from a biotechnological perspective. *A. aeolicus* was one of the first marine bacteria for which a full genome sequence was obtained; it was found to have a very small genome, only about one third that of *E. coli*. A related species, *Hydrogenothermus marinus* has also been isolated from deep-sea and coastal hydrothermal systems.

Thermotoga also forms a deeply branching, phylogenetically distant group of the *Bacteria*. As well as evidence from gene sequences, the function of the ribosome is very different from other *Bacteria*, and is not affected by rifampicin and other antibiotics that affect protein synthesis. The name of the genus derives from a unique outer membrane ("toga"), which balloons out from the rod-shaped cells. The cells are Gram-negative, but the amino acid composition of the peptidoglycan is unlike that of other *Bacteria*, and there are unusual long-chain fatty acids in the lipids. *Thermotoga* is widespread in geothermal areas and occurs in shallow and deep-sea hydrothermal vents. Different species vary in their temperature optima, with a range from 55°C up to 80–95°C for the hyperthermophilic species *T. maritima* and *T. neapolitana*. These are fermentative, anaerobic chemoorganotrophs and use a wide range of carbohydrates. They also fix N_2 and reduce sulfur to H_2S. Like *Aquifex*, these organisms have considerable biotechnological potential, and the *T. maritima* genome has been sequenced. A large number of the genes are involved in the transport and utilization of nutrients, in keeping with its ability to use a wide range of substances for growth. The genomes of both *Aquifex* and *Thermotoga* contain a high proportion of genes with homology to archaeal genes, indicating that large-scale lateral gene transfer has occurred in the evolution of these bacteria.

Conclusions

This Chapter has illustrated the great diversity of morphology, metabolism, physiology, and genetic makeup of members of the domain *Bacteria*. This diversity has enabled *Bacteria* to colonize every conceivable habitat in the marine realm, and their activities affect all aspects of marine ecology and biogeochemistry. The evolutionary history and classification of the *Bacteria* remains problematic. Only a small fraction of the organisms known from molecular-based studies have been cultured and studied in detail. Despite more than a century of microbial ecology, completely new types of bacteria are still being discovered at ever-increasing rates and the oceans undoubtedly hold many more surprises.

References

Abreu, F., Martins, J. L., Silveira, T. S., Keim, C. N., de Barros, H. G. P., Frederico Filho, J. G. and Lins, U. (2007) "Candidatus Magnetoglobus multicellularis", a multicellular, magnetotactic prokaryote from a hypersaline environment. *Int. J. Syst. Evol. Microbiol.* 57: 1318–1322.

Angert, E. R. and Clements, K. D. (2004) Initiation of intracellular offspring in *Epulopiscium*. *Molec. Microbiol.* 51: 827–835.

Bond, D. R., Holmes, D. E., Tender, L. M. and Lovley, D. R. (2002) Electrode-reducing microorganisms that harvest energy from marine sediments. *Science* 295: 483–485.

Breitbart, M., Thompson, L. R., Suttle, C. A. and Sullivan, M. B. (2007) Exploring the vast diversity of marine viruses. *Oceanography* 20: 135–139.

Bryant, D. A. (2003) The beauty in small things revealed. *Proc. Natl Acad. Sci. USA* 100: 9647–9649.

Coleman, M. L., Sullivan, M. B., Martiny, A. C., Steglich, C., Barry, K., DeLong, E. F. and Chisholm, S. W. (2006) Genomic islands and the ecology and evolution of *Prochlorococcus*. *Science* 311: 1768–1770.

Czyz, A. and Wegrzyn, G. (2001) On the function and evolution of bacterial luminescence. In: *Bioluminescence and Chemiluminescence* (ed. J. F. Case, P. J. Herring, B. H. Robinson, S. D. H. Haddock, L. J. Kricka and P. E. Stanley). World Science Publishing Company, Singapore, pp. 31–34.

Czyz, A., Wrobel, B. and Wegrzyn, G. (2000) *Vibrio harveyi*

bioluminescence plays a role in stimulation of DNA repair. *Microbiology* 146: 283–288.

DeLong, E. F. (2005) Microbial community genomics in the ocean. *Nat. Rev. Microbiol.* 3: 459–69.

Dunny, G. M., Brickman, T. J. and Dworkin, M. (2008) Multicellular behavior in bacteria: communication, cooperation, competition and cheating. *BioEssays* 30: 296–298.

Falkowski, P. G. and de Vargas, C. (2004) Shotgun sequencing in the sea: a blast from the past. *Science* 304: 58–60.

Flint, J. F., Drzymalski, D., Montgomery, W. L., Southam, G. and Angert, E. R. (2005) Nocturnal production of endospores in natural populations of *Epulopiscium*-like surgeonfish symbionts. *J. Bacteriol.* 187: 7460–7470.

Fuchsman, C. A. and Rocap, G. (2006) Whole-genome reciprocal BLAST analysis reveals that *Planctomycetes* do not share an unusually large number of genes with *Eukarya* and *Archaea*. *Appl. Environ. Microbiol.* 72: 6841–6844.

Giovannoni, S. and Rappé, M. (2000) Evolution, diversity and molecular ecology of marine prokaryotes. In: *Microbial Ecology of the Oceans* (ed. D. L. Kirchman) Wiley–Liss Inc., pp. 47–84.

Giovannoni, S. J. and Stingl, U. (2005) Molecular diversity and ecology of microbial plankton. *Nature* 437: 345–348.

Hunt, D. E., David, L. A., Gevers, D., Preheim, S. P., Alm, E. J. and Polz, M. F. (2008) Resource partitioning and sympatric differentiation among closely related bacterioplankton. *Science* 320: 1081–1085.

Komeili, A., Vali, H. and Newman, D. K. (2004) Magnetosome vesicles are present before magnetite formation, and MamA is required for their activation. *Proc. Natl Acad. Sci. USA* 101: 3839–3844.

Kunin, V., Engelbrektson, A., Ochman, H. and Hugenholtz, P. (2010) Wrinkles in the rare biosphere: pyrosequencing errors can lead to artificial inflation of diversity estimates. *Environ. Microbiol.* 12: 118–123.

Levy, S., Sutton, G., Ng, P. C., et al. (2007) The diploid genome sequence of an individual human. *PLoS Biol.* 5: 2113–2144.

Martins, J. L., Silveira, T. S., Abreu, F., Silva, K. T., da Silva-Neto, I. D. and Lins, U. (2007) Grazing protozoa and magnetosome dissolution in magnetotactic bacteria. *Environ. Microbiol.* 9: 2775–2781.

Martiny, A. C., Kathuria, S. K. and Berube, P. (2009) Widespread metabolic potential for nitrite and nitrate assimilation among *Prochlorococcus* ecotypes. *Proc. Natl Acad. Sci. USA* 106: 10787–10792.

Nackerdien, Z. E., Keynan, A., Bassler, B. L., Lederberg, J. and Thaler, D. S. (2008) Quorum sensing influences *Vibrio harveyi* growth rates in a manner not fully accounted for by the marker effect of bioluminescence. *PLoS ONE* 3: e1671, doi:10.1371/journal.pone.0001671.

Nealson, K. H. and Venter, J. C. (2007) Metagenomics and the global ocean survey: what's in it for us, and why should we care? *ISME J.* 1: 185–187.

Ocean Metagenomics Collection (2007) *PLoS Biol.*, accessed September 3, 2010, http://www.ploscollections.org/home.action.

O'Grady, E. A. and Wimpee, C. F. (2008) Mutations in the lux operon of natural dark mutants in the genus *Vibrio. Appl. Environ. Microbiol.* 74: 61–66.

Quince, C., Lanzén, A., Curtis, T. P., Davenport, R. J., Hall, N., Head, I. M., Read, L. F. and Sloan, W. T. (2009) Accurate determination of microbial diversity from 454 pyrosequencing data. *Nat. Methods* 6: 639–641.

Rees, J. F., De Wergifosse, B., Noiset, O., Dubuisson, M., Janssens, B. and Thompson, E. M. (1998) The origins of marine bioluminescence: turning oxygen defence mechanisms into deep-sea communication tools. *J. Exp. Biol.* 201: 1211–1221.

Reid, R. P., Visscher, P. T., Decho, A. W., et al. (2000) The role of microbes in accretion, lamination and early lithification of modern marine stromatolites. *Nature* 406: 989–992.

Ruby, E. G. and McFall-Ngai, M. J. (1999) Oxygen-utilizing reactions and symbiotic colonization of the squid light organ by *Vibrio fischeri. Trends Microbiol.* 7: 414–420.

Rusch, D. B., Halpern, A. L., Sutton, G., et al. (2007) The Sorcerer II Global Ocean Sampling expedition: northwest Atlantic through eastern tropical Pacific. *PLoS Biol.* 5: e77, doi:10.1371/journal.pbio.0050077.

Scheffel, A., Gruska, M., Faivre, D., Linaroudis, A., Plitzko, J. M. and Schussler, D. (2006) An acidic protein aligns magnetosomes along a filamentous structure in magnetotactic bacteria. *Nature* 440: 110–114.

Schlesner, H., Rensmann, C., Tindall, B. J., Gade, D., Rabus, R., Pfeiffer, S. and Hirsch, P. (2004) Taxonomic heterogeneity within the *Planctomycetales* as derived by DNA–DNA hybridization, description of *Rhodopirellula baltica* gen. nov., sp. nov., transfer of *Pirellula marina* to the genus *Blastopirellula* gen. nov. as *Blastopirellula marina* comb. nov. and emended description of the genus *Pirellula. Int. J. Syst. Evol. Microbiol.* 54: 1567–1580.

Sogin, M. L., Morrison, H. G., Huber, J. A., et al. (2006) Microbial diversity in the deep sea and the underexplored "rare biosphere". *Proc. Natl Acad. Sci. USA* 103: 12115–12120.

Stingl, U. and Giovannoni, S. J. (2005) Molecular diversity and ecology of microbial plankton. *Nature* 437: 343–348.

Tender, L. M., Reimers, C. E., Stecher, H. A., et al. (2002) Harnessing microbially generated power on the seafloor. *Nat. Biotechnol.* 20: 821–825.

Thompson, C. C., Vicente, A. C. P., Souza, R. C., et al. (2009) Genomic taxonomy of vibrios. *BMC Evol. Biol.* 9: 258, doi:10.1186/1471-2148-9-258.

Thompson, F. L., Hoste, B., Thompson, C. C., Huys, G. and Swings, J. (2001). The coral bleaching *Vibrio shiloi* is a later synonym of *Vibrio mediterranei* Pujalte and Garay 1986. *Syst. Appl. Microbiol.* 24: 516–519.

Thompson, F. L., Gevers, D., Thompson, C. C., Dawyndt, P., Naser, S., Hoste, B., Munn, C. B. and Swings, J. (2005) Phylogeny and molecular identification of vibrios on the basis of multilocus sequence analysis. *Appl. Environ. Microbiol.* 71: 5107–5115.

Thompson, F. L., Austin, B. and Swings, J. (2006) *The Biology of Vibrios.* ASM Press, Washington.

Thompson, J. R., Pacocha, S., Pharino, C., Klepac-Ceraj, V., Hunt, D. E., Benoit, J., Sarma-Rupavtarm, R., Distel, D. L. and Polz, M. F. (2005) Genotypic diversity within a natural coastal bacterioplankton population. *Science* 307: 1311–1313.

Tress, M. L., Cozzetto, D., Tramontano, A. and Valencia, A. (2006) An analysis of the Sargasso Sea resource and the consequences for database composition. *BMC Bioinformatics* 7: 213, doi:10.1186/1471-2105-7-213.

Venter, J. C. (2007) *A Life Decoded.* Viking Penguin Books, New York.

Venter, J. C., Remington, K., Hoffman, J., et al. (2004) Environmental genome shotgun sequencing of the Sargasso Sea. *Science* 304: 66–74.

Williamson, S. J., Rusch, D. B., Yooseph, S., et al. (2008) The Sorcerer II Global Ocean Sampling expedition: metagenomic characterization of viruses within aquatic microbial samples. *PLoS ONE* 3: doi:10.1371/journal.pone.0001456.

Woebken, D., Teeling, H., Wecker, P., et al. (2007) Fosmids of novel marine *Planctomycetes* from the Namibian and Oregon coast upwelling systems and their cross-comparison with planctomycete genomes. *ISME J.* 1: 419–435.

Yooseph, S., Sutton, G., Rusch, D. B., et al. (2007) The Sorcerer II Global Ocean Sampling expedition: expanding the universe of protein families. *PLoS Biol.* 5: e16, doi:10.1371/journal.pbio.0050016.

Further reading

Achtman, M. and Wagner, M. (2008) Microbial diversity and the genetic nature of microbial species. *Nat. Rev. Microbiol.* 6: 431–440.

Bergey's Manual Trust (2010) http://www.bergeys.org/ [contains links to the most recent volumes of *Bergey's Manual of Systematic Bacteriology* and other taxonomic resources].

Dworkin, M., Falkow, S., Rosenberg, E., Schleifer, H.-F. and Stackebrandt, E. (2006) *The Prokaryotes*, 3rd Edn, Vols 1–7. Springer-Verlag, Berlin.

Gevers, D., Cohan, F. M., Lawrence, J. G., et al. (2005) Re-evaluating prokaryotic species. *Nat. Rev. Microbiol.* 3: 733–739.

Martiny, J. B. H., Bohannan, B. J. M., Brown, J. H., et al. (2006) Microbial biogeography: putting microorganisms on the map. *Nat. Rev. Microbiol.* 4: 102–112.

Pedros-Alio, C. (2006) Marine microbial diversity: can it be determined? *Trends Microbiol.* 14: 257–263.

Staley, J. T. (2006) The bacterial species dilemma and the genomic-phylogenetic species concept. *Philos. Trans. R. Soc. Lond. B Biol. Sci.* 361: 1899–1909.

Chapter 5

Marine *Archaea*

As discussed in Chapter 1, the application of 16S rRNA gene sequencing methods led to the establishment of the *Archaea* as a completely separate domain of life. Other information supports the concept that the *Archaea* are a monophyletic group, especially the nature of their ribosomes and mechanisms of transcription and translation. We now know that the previous view of "archaebacteria" as a rather uncommon, specialized subset of bacteria occupying extreme habitats is completely wrong. Although many of the most thermophilic and halophilic microbes are indeed members of the *Archaea*, species belonging to this domain are very abundant and diverse, and can be found everywhere in the marine environment. This is especially true in the deep ocean, where chemolithotrophic metabolism linked to oxidation of ammonia plays a major role in the carbon and nitrogen cycles. This chapter explores the diversity of the *Archaea* and introduces their role in biogeochemical processes.

Key Concepts

- The basic metabolism and organization of cells of *Archaea* resembles those of *Bacteria*, but there are important differences in membrane structure, organization and replication of DNA, and protein synthesis.

- Production of methane by members of the *Euryarchaeota* is the final step in the anaerobic biodegradation of organic material and leads to formation of massive reserves of methane in deep-sea sediments.

- *Archaea* can oxidize methane in anoxic sediments by reverse methanogenesis, in association with sulfate-reducing bacteria.

- Numerous types of chemolithotrophic and organotrophic *Euryarchaeota* and *Crenarchaeota* active at hydrothermal vents show adaptations for growth at very high temperatures.

- *Crenarchaeota* are abundant in the deep sea, where one of their main functions is oxidation of ammonia coupled to chemolithotrophy.

Several aspects of cell structure and function distinguish the *Archaea* from the *Bacteria*

Whilst the basic organization of the cells of *Bacteria* and *Archaea* is similar, there are some important differences, which have significant effects on the physiology of *Archaea*. Many features of cell structure and function of *Archaea* show much closer similarities to those of *Eukarya* than they do to *Bacteria*. There is a major difference between the cell membrane of *Bacteria* and *Archaea*. Specifically, the fatty acids of the cell membrane of *Bacteria* are linked to the glycerol moieties via ester linkages, whereas most *Archaea* contain hydrocarbons with repeating 5-carbon units (isoprenoids) rather than fatty acids, and these are linked to glycerol via ether linkages. Many *Archaea*, especially those from high-temperature environments such as hydrothermal vents, have monolayer membranes composed of a single layer of diglyceride tetraethers.

Although *Archaea* possess small compact genomes like those of *Bacteria*, organization of the nuclear material is more complex and variable in *Archaea*. In some species, packaging and stabilization of the DNA structure is achieved either by supercoiling (using the enzyme DNA gyrase), as in *Bacteria*. However, other species possess positively charged proteins known as histones, which have amino-acid sequences homologous to those found in eukaryotes. Short sequences of DNA are wound around clusters of histones to form tetrameric clusters called nucleosomes. Some of the extremely thermophilic *Archaea* have an additional type of DNA gyrase that induces positive supercoiling in the DNA to protect it against denaturation. The process of DNA replication is also different from *Bacteria*. Although most *Archaea* possess single circular chromosomes, there are often multiple origins of replication, and the structure of the replication complex and polymerase enzymes again resembles that found in eukaryotes. Conversely, regulation of gene expression in *Archaea* is very similar to that in *Bacteria*, involving the use of repressor or activator proteins to control the level of transcription. Translation of mRNA and protein synthesis in *Archaea* shares many features with *Bacteria*, but many of the component proteins of *Archaea* and *Eukarya* show closer homology than they do with those from *Bacteria*.

The *Euryarchaeota* and *Crenarchaeota* form the major branches of the *Archaea*

Application of 16S rRNA gene sequencing led to the early recognition of two major branches (phyla) of the phylogenetic tree, namely the *Euryarchaeota* and *Crenarchaeota*, as shown in *Figure 5.1*. Many of the marine *Archaea* are known only by gene sequences isolated from environmental samples and have not been cultured. As noted in previous sections, despite the undoubted usefulness of 16S rRNA gene sequencing, reliance on a single gene to infer phylogenetic relationships can lead to problems. In recent years, phylogenetic analysis has been extended by the study of other conserved key genes, such as those encoding proteins involved with transcription and translation, as well as comparison of whole genome sequences of cultured representatives. These approaches have confirmed the *Euryarchaeota* and *Crenarchaeota* as valid major phyla. The major physiological groups within these phyla are considered in the subsequent sections.

Many members of the *Euryarchaeota* produce methane

Production of methane (methanogenesis) is the final step in the anaerobic bio-degradation of organic material. Methanogens show high physiological diversity in morphology, cell-wall constituents, and physiology, and are grouped into a number of orders and genera using these criteria (*Table 5.1*). Methanogens are mesophilic or thermophilic obligate anaerobes and can be isolated using strict

---

(i) ARCHAEA, ARCHAEA, EVERYWHERE

The discovery that *Archaea* are a major component of ocean microbiota ranks as one of the most significant surprises to emerge from the application of methods to directly sequence 16S rRNA isolated from planktonic biomass (see p. 41). The first clues came in 1992, when Jed Fuhrman (University of Southern California) and Ed De Long (then at Woods Hole Oceanographic Institution) independently discovered archaeal sequences in seawater samples. Fuhrman et al. (1992) found sequences in pelagic water samples from the Pacific Ocean (100 and 500 m depth) that were only distantly related to those of any organisms previously characterized. The closest match of some of these sequences was to extreme thermophiles, which had been assigned to the new domain *Archaea* (at that time, often still referred to as archaebacteria). DeLong (1992) also discovered widespread archaeal rRNA sequences in coastal surface waters off the east and west coasts of North America. Before these discoveries, it was assumed that *Archaea* are only found in extreme environments, inhospitable to other life forms. Over the next few years, numerous studies showed that these as-of-yet uncultivated *Archaea* are highly abundant in surface and deep surface waters in all the major ocean basins (e.g., see Fuhrman et al. 1994; DeLong et al., 1994).

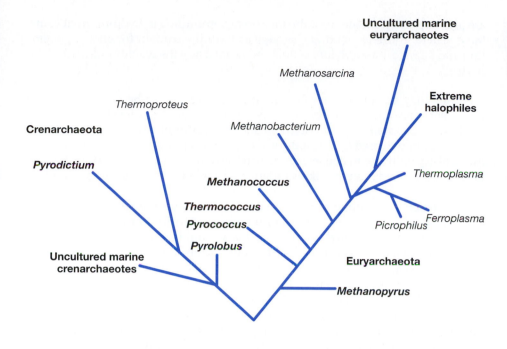

anaerobic techniques from a wide range of habitats including the gut of animals, anoxic sediments, and decomposing material. They also occur as endosymbionts of anaerobic protozoa found especially in the hindgut of termites. It is likely that shipworms and other marine invertebrates that digest wood and cellulose also harbor ciliates with archaeal endosymbionts. Thermophilic methanogens

Table 5.1 Properties of some representative methanogens found in marine sediments and hydrothermal vents

Order and genus	Morphology	Major substrates	Optimum temp. (°C)
Methanococcales			
Methanococcus	Irregular cocci	H_2, formate	35–40
Methanothermococcus	Cocci	H_2, formate	60–65
Methanocaldococcus	Cocci	H_2	80–85
Methanotorris	Cocci	H_2	88
Methanomicrobiales			
Methanoculleus	Irregular cocci	H_2, formate	20–55
Methanogenium	Irregular cocci	H_2, formate	15–57
Methanolacinia	Irregular rods	H_2	40
Methanospirillum	Spirilla	H_2, formate	30–37
Methanosarcinales			
Methanosarcina	Irregular cocci in groups	H_2, methylamines, acetate	35–60
Methanococcoides	Irregular cocci	methylamines	23–35
Methanopyrales			
Methanopyrus	Rods in chains	H_2	98

? HOW IMPORTANT ARE MARINE SOURCES OF METHANE?

Methane production by methanogens results in the release of 500–600 million tonnes of gas into the atmosphere each year. The biggest source (about 20%) is methane emitted in the "burps" of ruminant animals, which matches that produced from coal mining and extraction of petroleum and "natural gas" for fuel. Natural wetlands, peat bogs, and paddy fields contribute about half of all methane released. Methane emissions from ocean *Archaea* constitute only 1–2% of the total because most methane in sediments is consumed by methanotrophs or anaerobic consortia. However, methane production over millennia has resulted in the production of vast quantities of frozen methane hydrate (clathrate), which exists as solid deposits under pressure in the deep sediments. There may be about 10 times as much methane in clathrate as current natural gas reserves, making it an attractive energy source, but the logistics of extraction from deep marine sediments are formidable. Methane is about 20 times as powerful as CO_2 in its "greenhouse gas" effects. There is strong geological evidence that the mass extinction events of 250 million and 56 million years ago were due to sudden release of methane into the atmosphere, leading to sudden global warming. Natural seeps of methane occur from several parts of the ocean floor. Increased sea temperatures could increase the release rate of this methane. However, the release of vast quantities of methane from melting of the Arctic permafrost is probably of much greater immediate concern, with the potential of a catastrophic positive feedback.

are also important members of the microbial community at hydrothermal vents. Most genera utilize H_2, with CO_2 serving as both the oxidant for energy generation and for incorporation into cellular material using the reactions described on p. 70.

Although they are strict anaerobes, methanogens can also be found in surface microbial mats and ocean waters, which can contain high levels of dissolved methane. Presumably, they exist in the anoxic interior of particles in which oxygen has been depleted by respiratory activity of other organisms. Methanogenesis is also significant in deeper waters with upwelling of nutrients. Here, intense heterotrophic oxidation of sinking organic matter leads to oxygen depletion.

The extremely thermophilic archaeon *Methanococcus jannaschii* has been recognized as one of the most important members of hydrothermal vent communities and has been studied extensively. It is a primary consumer of H_2 and CO_2 produced by geochemical activity at the vents. Analysis of the complete genome sequence reveals a 1.7-Mb circular chromosome containing about 1700 genes. Sequence analysis of the genes encoding key metabolic pathways and cellular processes shows that these are similar to those found in *Bacteria*, whereas those encoding protein synthesis and DNA replication show more similarity to eukaryotic genes. However, a large number of genes in *M. jannaschii* have little homology to bacterial or eukaryotic genes, suggesting that many new cellular processes remain to be discovered. About 20 other genomes of methanogens have been sequenced; most are about the same size as *M. jannaschii*, but the genomes of *Methanosarcina* spp. are about four times as big, presumably reflecting more metabolic versatility. The function of many of the putative genes identified in these species is not known.

Methanopyrus is one of the most thermophilic organisms known, with rapid growth (generation time of 1 h) at a maximum temperature of 110°C. It is found in the walls of black smoker chimneys in hydrothermal vents. Although it is a methanogen (utilizing H_2 + CO_2 only), it is phylogenetically distant from the rest of this group and has very unusual membrane lipids and thermostabilizing compounds in the cytoplasm.

Archaea in deep sediments can carry out anaerobic oxidation of methane coupled to sulfate reduction

For many years, the fate of the large amounts of methane produced in marine sediments has been something of a mystery. As described above, production of methane is an anaerobic process carried out only by the *Euryarchaeota*. All microorganisms capable of using methane (methanotrophs) were thought to be aerobic. However, for many years, geochemists had noticed that methane diffusing upwards from deep sediments often disappeared before it reached zones where contact with oxygen could occur. Sulfate is the only possible electron acceptor that could account for anaerobic methane oxidation in these anoxic sediments. Stable isotope methods combined with 16S rRNA gene sequencing and fluorescent *in situ* hybridization (FISH) showed that a consortium of *Archaea* and sulfate-reducing *Bacteria* (SRB) belonging to the *Desulfosarcina* and *Desulfococcus* genera is responsible for this reaction, although subsequent studies indicate that the SRB are more diverse than originally thought. This syntrophic consortium is illustrated in *Figure 5.2* and *Plate 5.1*. Together, the partners carry out the reaction

$$CH_4 + SO_4^{2-} \rightarrow HCO_3^- + HS^- + H_2O$$

The anaerobic methanotrophic *Archaea* seem to be related to the *Methanosarcinales*. The biochemistry and thermodynamics of this methanotrophic process have not been fully resolved, but it is believed to function as a reversal of methanogenesis. Genomic analyses of anaerobic archaeal methanotrophs supports

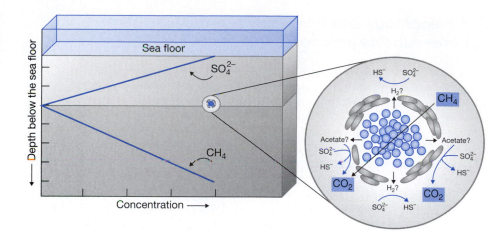

Figure 5.2 Model showing anaerobic oxidation of methane in marine sediments. The metabolic consortium contains about 100 methane-oxidizing archaeal cells surrounded by sulfate-reducing bacteria (SRB). See *Plate 5.1* for an image of the metabolic consortium revealed by use of FISH. (Reprinted from DeLong [2000] by permission of Macmillan Publishers Ltd., © Nature 2000.)

this conclusion. The sulfide and bicarbonate products of the reaction are very important. These consortia contribute to the construction of reefs of precipitated carbonate at methane seeps, especially in the Black Sea. The sulfide can be used in chemosynthetic symbioses (see Chapter 10) involving tubeworms, clams, or giant sulfide-oxidizing bacteria, which form thick carpets on the seabed above the methanotrophic consortia. Elsewhere, extensive microbial growth above methane seeps may sustain deep-sea coral communities. This type of methanotrophic metabolism may have evolved very early on Earth, possibly predating the development of O_2 in the atmosphere from oxygenic photosynthesis (about 2.2 billion years ago).

Thermococcus and *Pyrococcus* are hyperthermophiles found at hydrothermal vents

These genera of *Euryarchaeota* have optimum growth temperatures in excess of 80°C. Their phylogenetic position as a deeply branching group resembles that of the hyperthermophilic genera *Aquifex* and *Thermotoga* in *Bacteria* (*Figure 4.1*). It is interesting that extreme thermophiles in both domains branch very near the root of the tree, and this is taken as evidence that these organisms most closely resemble those that evolved first in the hotter conditions of the early Earth. Indeed, some scientists suggest that life may have evolved in submarine hydrothermal vents or subsurface rocks. The adaptation of hyperthermophiles to life at high temperatures is discussed on p. 84.

Thermococcus celer forms highly motile spherical cells about 0.8 μm in diameter. It is an obligately anaerobic chemoorganotroph, using complex substrates such as proteins and carbohydrates, with sulfur as the electron acceptor. It has an optimum growth temperature of 80°C. *Pyrococcus furiosus* has similar properties to *Thermococcus*, but has an optimum growth temperature of 100°C and a maximum of 106°C. Both organisms have been investigated extensively because of their biotechnological potential, and their thermostable enzymes are used in the PCR (see p. 42). The genomes of several *Pyrococcus* species have been compared, revealing evidence of significant gene rearrangements in their recent evolution.

ⓘ METHANOTROPHIC ARCHAEA FIX NITROGEN FOR THEIR SYNTROPHIC PARTNERS

As discussed in Chapter 9, there is uncertainty about the balance of the marine nitrogen cycle, since it appears that more nitrogen is being used than is being made. Victoria Orphan and colleagues at the California Institute of Technology have recently discovered that the methanotrophic *Archaea* found in consortia with SRB at methane cold seeps are able to fix nitrogen. Dekas et al. (2009) incubated the archaeal–bacterial assemblages in the laboratory with $^{15}N_2$ and used nanoSIMS (see *Plate 5.1*) to show that the archaeal cells fixed nitrogen. As the focused ion beam "burned" through the consortium, it became clear that the labeled fixed nitrogen was mainly associated with the core containing the *Archaea*, but there was evidence of some exchange of fixed nitrogen to the SRB. Nitrogen fixation is very expensive in terms of energy consumption and the growth of these consortia is very slow—the doubling time is 3–6 months—but it seems to be advantageous for the *Archaea* to transfer fixed nitrogen to the SRB, which they need to utilize methane.

Archaeoglobus and *Ferroglobus* are hyperthermophilic sulfate-reducers and iron-oxidizers

Archaeoglobus spp. are also extreme thermophiles (optimum temperature 83°C) found in sediments of shallow hydrothermal vents and around undersea volcanoes. They are strictly anaerobic organisms, which couple the reduction of sulfate to the oxidation of H_2 and certain organic compounds, resulting in the production of H_2S. Although *Archaeoglobus* forms a distinct phylogenetic group in the *Euryarchaeota*, it makes some of the key coenzymes used in methanogenesis, and analysis of its genome sequence shows that it shares some genes with the methanogens. However, it lacks the gene for one of the key enzymes, methyl-CoM reductase, so the origin of the small amount of methane produced is unknown. *Archaeoglobus* also occurs in oil reservoirs and has caused problems with sulfide "souring" of crude oil extracted from the North Sea and Arctic oilfields. Members of the related genus *Ferroglobus*, also found in hydrothermal vents, do not reduce sulfate but are iron-oxidizing/nitrate-reducing chemolithoautotrophs.

Some *Euryarchaeota* exist in hypersaline environments

Extreme halophiles (family *Halobacteriaceae*) grow in concentrations of NaCl greater than 9% and many can grow in saturated NaCl solutions (35%). They are found in salt lakes such as the Great Salt Lake (Utah, USA) and the Dead Sea (Jordan Rift Valley); as well as in hypersaline anoxic basins, such as those in the Mediterranean Sea (see *Research Focus Box 1.2*). Extreme halophiles also occur in coastal regions in solar salterns, which are lagoons in which seawater is allowed to evaporate to collect sea salt. They operate as semicontinuous systems and maintain a fairly constant range of salinity throughout the year. Both conventional microbiological and 16S rRNA methods show that the overall diversity of microorganisms decreases as salinity increases. Up to about 11% NaCl, the range of bacteria is similar to that found in coastal seawater (most marine *Bacteria* are moderately halophilic) whilst *Archaea* are scarce. However, above 15% salinity, culturable members of the *Archaea* such as *Halorubrum*, *Halobacterium*, *Halococcus*, *Haloarcula*, and *Haloquadratum* become dominant, as well as *Archaea* with novel gene sequences. The halophilic *Archaea* occur as rods or cocci and in a variety of unusual shapes including flattened squares. They may contain very large plasmids constituting up to 30% of the genome. They are chemoorganotrophs that can use a wide range of compounds as sources of carbon and energy. The requirement for very high Na^+ concentrations is achieved by pumping large amounts of K^+ across the membrane, so that the internal osmotic pressure remains high and protects the cell from dehydration. The intracellular machinery of these organisms is adapted to high levels of salt, and so they are obligate extreme halophiles. Cells lyse in the absence of sufficient concentrations of Na^+, possibly because Na^+ stabilizes the high levels of negatively charged acidic amino acids in the cell wall.

Halobacterium salinarum and some other species use a membrane protein called bacteriorhodopsin to synthesize ATP using light energy. This compound is so called because it is structurally similar to the rhodopsin pigment in animal eyes. As noted in *Research Focus Box 3.1*, bacteriorhodopsin is complexed with a carotenoid pigment, retinal, which absorbs light and generates a proton motive force for the production of ATP. Light is not used for photosynthesis, but it provides energy for the proton pump required to pump Na^+ out of and K^+ into the cell and can give enough energy for a low metabolic activity when organic nutrients are scarce. *H. salinarum* also contains other types of rhodopsin. Halorhodopsin captures light energy used for pumping Cl^- into the cell to counterbalance K^+ transport. Two other rhodopsin molecules act as light sensors that affect flagellar rotation, enabling chemotaxis toward light.

(i) THE NAMING OF RHODOPSINS IS CONFUSING!

The extremely halophilic genus *Halobacterium* was given this name because at the time of its discovery it was considered to be a bacterium. Because we were ignorant then of the true differences between the domains that we now know as the *Archaea* and *Bacteria*, they were known as the *Archaebacteria* and *Eubacteria* respectively. Therefore, when the rhodopsin-like pigment of *Halobacterium* was discovered, it was called bacteriorhodopsin, to distinguish it from the rhodopsin found in animal eyes. Another type of rhodopsin was discovered later, associated with sequences of *Proteobacteria*, so it was given the name proteorhodopsin (see *Research Focus Box 3.1*). However, subsequent studies showed that this molecule is also present in marine planktonic *Archaea* as well as *Bacteria*. Unfortunately, these names are now so well established that we will have to live with the potential confusion.

Nanoarchaeum is an obligate parasite of another archaeon, *Igniococcus*

Nanoarchaeum equitans was discovered as a member of hyperthermophilic microbial communities near a submarine hot vent off Iceland. After anaerobic incubation at 90°C (using specialized high-temperature fermenters) in the presence of sulfide, H_2, and CO_2, groups of tiny cocci about 400 nm in diameter were seen attached to the surface of the larger cells of another archaeon, *Igniococcus hospitalis* (*Figure 5.3A*). *Nanoarchaeum* appears to be an obligate parasite of *Igniococcus*, although the relationship between the two cells is unclear. All attempts to grow *N. equitans* in pure culture have been unsuccessful, even using extracts of *I. hospitalis*. However, the growth rate of *I. hospitalis* does not seem to be affected by the presence of *N. equitans* cells, and it can grow in their absence. Growth of *I. hospitalis* is anaerobic and occurs through a novel carbon dioxide fixation pathway, using sulfur reduction with molecular hydrogen as an electron donor. Genome analysis shows that *N. equitans* has no genes to support the same chemolithoautotrophic physiology as its host. *Nanoarchaeum equitans* obtains energy and metabolites, including membrane lipids, amino acids, nucleotides, and cofactors, from *I. hospitalis*. *Igniococcus hospitalis* is the only archaeon known with an outer membrane separated from the cytoplasmic membrane by a wide periplasmic space. As shown in *Figure 5.3B*, *N. equitans* attaches to this outer membrane via fibrillar structures, and at these contact points the periplasmic space is greatly reduced, so that the inner and outer membranes are in contact with each other. It is thought that large vesicles and tubes emerge from the *I. hospitalis* cytoplasm for the transport of metabolites and proteins.

The *Crenarchaeota* include hyperthermophiles and psychrophiles

The phylum *Crenarchaeota* is phylogenetically distinct from the *Euryarchaeota* described above, although many have similar physiological properties including extreme hyperthermophily. Most cultured representatives are known from extensive study of terrestrial hot springs, but several other species occur in submarine hydrothermal vents. They use a wide range of electron donors and acceptors in metabolism and can be either chemoorganotrophic or chemoheterotrophic. Most are obligate anaerobes. One of the greatest surprises of recent years has been the discovery that, far from being restricted to high-temperature habitats, members of the *Crenarchaeota* are ubiquitous in the marine environment. The discovery of widespread archaeal gene sequences, including their presence in Antarctic waters and sea ice and the subsequent realization that the *Crenarchaeota* are among the most numerous life forms in deep ocean waters led to a radical reappraisal of their diversity and ecology.

Hyperthermophilic *Crenarchaeota* belong to the order *Desulfurococcales*

Several species isolated from shallow and deep-sea hydrothermal areas belong to the order *Desulfurococcales*. The type genus, *Desulfurococcus*, is an obligate anaerobe with coccoid cells which uses the following reaction for energy generation.

$$H_2 + S^0 \rightarrow H_2S$$

Cells of *Pyrodictium* spp. are disc-shaped and are connected into a mycelium-like layer attached to crystals of sulfur by very fine hollow tubules. Most *Pyrodictium* spp. are chemolithoautotrophs which gain energy by sulfide reduction. However, *P. abyssi* is a heterotroph that ferments peptides to CO_2, H_2, and fatty acids.

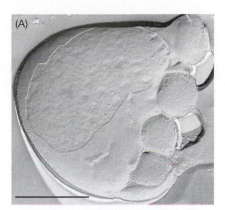

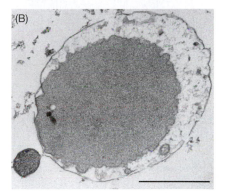

Figure 5.3 Electron micrographs showing larger cells of *Igniococcus hospitalis* with attached cells of *Nanoarchaeum equitans*.
(A) Scanning electron micrograph showing large cells of *Igniococcus* with several smaller cells of *Nanoarchaeum* attached.
(B) Transmission electron micrograph of ultrathin section showing close attachment of *Nanoarchaeum* to the outer membrane of *Igniococcus*. Bar = 1 μm. (Images courtesy of H. Huber, R. Rachel, B. Junglas, M.J. Hohn, C. Wimmer and K.O. Stetter, University of Regensburg, Germany. Reprinted from (A) Huber et al. [2002] and (B) Cerdeño-Tárraga [2009] by permission of Macmillan Publishers Ltd, © Nature.)

Pyrolobus fumarii is found in the walls of black smoker chimneys in hydrothermal vents and has an optimum growth temperature of 106°C. In this very extreme environment, it is probably a significant source of primary productivity. Cells are lobed cocci and the cell wall is composed of protein. It is a facultatively aerobic obligate chemolithotroph, using the reactions

$$4H_2 + NO_3^- + H^+ \rightarrow NH_4^+ + 2H_2O + OH^-$$

$$5H_2 + S_2O_3^{2-} \rightarrow 2H_2S + 3H_2O + 2e^-$$

$$H_2 + \tfrac{1}{2} O_2 \rightarrow H_2O$$

Igniococcus is a sulfide-reducing chemolithotroph with a very different structure to that of other *Archaea*. An outer membrane, reminiscent of that seen in Gram-negative *Bacteria*, surrounds the cell as a loose sac enclosing a very large periplasmic space containing vesicles, which may function in transport of nutrients (*Figure 5.3B*). As noted above, *I. hospitalis* was discovered to be the host of *Nanoarchaeum*.

Species of the genus *Pyrobaculum* show various modes of nutrition. Some species link sulfur reduction to anaerobic respiration of organic compounds, whilst others are anaerobic chemolithotrophic autotrophs using the reactions

$$H_2 + NO_3^- \rightarrow NO_2^- + H_2O \text{ or}$$

$$H_2 + 2Fe^{3+} \rightarrow 2Fe^{2+} + 2H^+$$

Staphylothermus marinus forms aggregates of cocci and is a chemoorganotroph, which, like *Pyrodictium abyssi*, ferments peptides to CO_2, H_2, and fatty acids. It is widely distributed in shallow and deep-sea hydrothermal systems and is a major decomposer of organic material. It is the largest known member of the *Archaea*. Although normally about 1 μm in diameter, it can form very large cells of up to 15 μm diameter in high-nutrient concentrations.

The psychrophilic marine *Crenarchaeota* are major members of the plankton

Analysis of ocean waters using primers for archaeal 16S rRNA genes led to some very unexpected results. Until the 1990s, crenarchaeotes were known only as extreme thermophiles, but since then several hundred sequences corresponding to this group have been found to be widespread in oligotrophic ocean waters, including the Antarctic and very deep waters. As shown in *Figure 5.4*, crenarchaeotes have been shown to comprise a large fraction of the picoplankton, especially in deeper waters. The total number of *Bacteria* and *Archaea* decreases with depth, from 10^5–10^6 ml^{-1} near the surface to 10^3–10^5 cells ml^{-1} below 1000 m. *Bacteria* are most prevalent in the upper 150 m of the ocean, but below this depth the fraction of *Archaea* equals or exceeds that of *Bacteria*. The pattern is consistent throughout the year. Combining the figures for cell density with oceanographic data for the volume of water at different depths indicates that the world's oceans contain approximately 1.3×10^{28} cells of *Archaea* and 3.1×10^{28} cells of *Bacteria*. About 1.0×10^{28} cells, that is 20% of all the picophytoplankton, is dominated by a single clade of the pelagic *Crenarchaeota*, suggesting that a common adaptive strategy has allowed them to radiate through the entire water column of the oceans. The physiology of these uncultured psychrophilic crenarchaeotes was completely unknown until very recently. As discussed in *Research Focus Box 5.1*, a combination of metabolic experiments and metagenomic evidence, together with study of the physiology of one cultured isolate (*Candidatus* Nitrosopumilus maritimus) showed that they are chemolithotrophic autotrophs that obtain energy from ammonia oxidation. Some may also grow mixotrophically.

(i) TINY CELLS ... TINY GENOME ... ANCIENT PARASITE?

When first described by Huber et al. (2002), the unusual rRNA sequence of *Nanoarchaeum equitans* indicated that it might be the sole representative of a new phylum, called the *Nanoarcheota*. However, it is now considered to be a rapidly evolving member of the *Euryarchaeota* related to *Thermococcus* (Brochier et al., 2005). The very small cell size and genome of *N. equitans* suggest that it may be close to the limits for cellular life. Sequencing of the genome (Waters et al., 2003) showed that it contains only 552 genes, about one third of which have no known homologs. Genes for many metabolic pathways are missing and *N. equitans* must derive nutrients from its *Igniococcus* host. Genes for many components of energy generation are also missing. Loss of genes during evolution is common in obligate parasites and symbionts (see *Research Focus Box 10.1*), but Waters et al. (2003) stated that *N. equitans* "does not fit the sterotype of a microbial parasite undergoing genomic degradation" and concluded that it is "genomically stable parasite that diverged anciently from the archaeal lineage." Recently, Das et al. (2006) compared the genome and proteome composition of *N. equitans* with those of other organisms and demonstrated dual adaptations for both host association and growth at extremely high temperatures. Similar sequences have been discovered in other thermal systems (Hohn et al., 2002). It is therefore possible that free-living species of *Nanoarchaeum* will eventually be discovered.

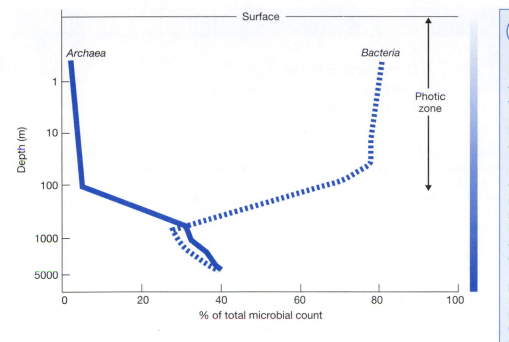

Figure 5.4 Distribution of *Bacteria* and *Archaea* in the ocean, showing mean annual depth profiles in the North Pacific subtropical gyre. Total cell abundance was measured using the 4'6'-diamido-2-phenylindole (DAPI) nucleic stain and *Bacteria* and *Archaea* measured using whole-cell rRNA-targeted fluorescent *in situ* hybridization with fluorescein-labeled polynucleotide probes. The counts of *Archaea* dominated the picoplankton below 1000 m and were almost entirely due to the crenarchaeote clade; counts of euryarchaeotes were a few percent of the total throughout the water column. (Redrawn from data in Karner et al. [2001]).

One species of crenarcheote, the uncultured *Candidatus* Cenarchaeum symbiosum, forms an association with the cold-water sponge *Axinella mexicana* and accounts for 65% of the microorganisms associated with the tissues of the sponge. It grows well at 10°C. Genomic analysis indicates that it also oxidizes ammonia and it may have a symbiotic function by removing nitrogenous wastes from the sponge. A number of other sponge species have now been found to harbor *Archaea* as symbionts, and it is likely that this is a very widespread phenomenon. This may have significant biotechnological potential, since sponges are rich sources of natural products. Evidence is accumulating that many of these compounds are actually synthesized by microorganisms colonizing the tissue, and isolation and characterization of these organisms could yield rich rewards.

Conclusions

Members of the domain *Archaea* exhibit a wide variety of morphology and metabolic types. In sediments, they play a major role in the carbon cycle in the production and oxidation of methane. At hydrothermal vents, chemolithotrophs and organotrophs are active at very high temperatures owing to the stability of their membranes, proteins, and intracellular constituents. One of the greatest surprises in the recent history of marine microbiology and oceanography is the realization that vast numbers of *Crenarchaeota* are active and productive in the deep sea, having the major role in the first steps of nitrification and fixation of carbon dioxide into cellular material.

? IS THERE AN UPPER TEMPERATURE LIMIT FOR LIFE?

Until recently, the "record holder" for growth at high temperatures was *Pyrolobus fumarii*. Kashefi and Lovley (2003) discovered a new isolate from a hydrothermal vent at the Juan de Fuca Ridge in the Pacific, which they nicknamed "strain 121" because of its ability to grow up to 121°C. This is noteworthy because this is the standard temperature that is used in an autoclave for sterilization at atmospheric pressure. It could also survive a temperature of 130°C for over 2 h. The isolate is phylogenetically related to both *Pyrolobus* and *Pyrodictium*. In a review of the discovery, Cowan (2003) cautions that exact measurement of growth and survival at these temperatures is difficult. Microscopic examination of the black smoker chimney from which the archaeon was isolated reveals diverse populations of microbes in cross-sections of the chimney material, indicating that microbes within even higher growth temperatures might exist. Research to investigate the basis of such extreme thermophily has not provided a clear explanation (see p. 84). Rather than depending entirely on the stability of proteins, as generally believed, Cowan suggests that the stability of small molecular-weight cofactors may be the most important factor. It is very likely that microbes that grow and survive even higher temperatures remain to be discovered. However, some essential biological molecules such as ATP cannot exist above about 150°C, so this may be the upper limit.

BOX 5.1   RESEARCH FOCUS

Deep sea FISHing

Genomic techniques reveal how *Archaea* contribute to carbon and nitrogen cycling in the oceans

Crenarchaeota are the most abundant microbial group in the deep ocean (see *Figure 5.4*). The possibility that this group is autotrophic was first indicated by isotope studies. Kuypers et al. (2001) analyzed sedimentary deposits containing "fossilized" *Archaea*—identified by the presence of characteristic membrane lipids), laid down about 112 million years ago. The organic matter contained a higher ratio of ^{12}C to ^{13}C, indicating that these *Archaea* had fixed CO_2 by enzymic processes (see p. 31 for an explanation of stable isotope methods). Wuchter et al. (2003) conducted *in situ* tracer experiments in seawater containing *Crenarchaeota* and confirmed that they are chemoautotrophic by showing the incorporation of large amounts of ^{13}C from bicarbonate into the distinctive crenarchaeotal membrane lipids. Herndl et al. (2005) investigated this process further by developing a highly sensitive CARD-FISH method in which oligonucleotide probes directed against specific archaeal groups were coupled with microautoradiography (see p. 54). To measure metabolic activity, seawater samples were incubated with radioactive [^{3}H]-leucine or [^{14}C]-bicarbonate. After collecting the cells on a membrane filter, the incorporation of the radioactive tracer into individual cells (tagged by the FISH probe) was visualized microscopically after overlaying with photographic emulsion. Herndl et al. (2005) concluded that *Crenarchaeota* in the oxygen-minimum layer of the ocean are metabolically active and that autotrophic growth makes a major contribution to productivity. However, a key question remained—what is the energy source that fuels this activity? In the first metagenomic study conducted in the Sargasso Sea by Venter et al. (2004), a gene encoding ammonia monoxygenase (AMO), the key enzyme in ammonia oxidation, was associated with archaeal genome fragments. The enzyme AMO is composed of three subunits, encoded by *amoA*, *amoB* and *amoC*, and archaeal homologs of all three genes have subsequently been discovered in metagenomic datasets.

Another breakthrough came when Könnecke et al. (2005) succeeded in cultivating a marine crenarchaeote for the first time. The organism, called *Candidatus* Nitrosopumilus maritimus strain SCM1, came from a marine aquarium and studies provided laboratory evidence for autotrophic oxidation of ammonia using nitrate. Martens-Habbena et al. (2009) measured the removal of ammonia by cultures of this organism, which grew with a doubling time of 26 h until the ammonia levels were depleted below the detection limit of 10 nM (about 100 times lower than the concentration needed by ammonia-oxidizing *Bacteria*). By measuring the kinetics of substrate binding, it was shown that SCM1 sustains high specific oxidation rates at ammonium concentrations found in the open oceans. SCM1 has a remarkably high specific affinity for ammonia, suggesting that organisms like this would successfully compete with heterotrophic bacterioplankton and phytoplankton.

Further evidence for the role of crenarcheaotal ammonia oxidation in marine systems was obtained in an enrichment experiment conducted by Wuchter et al. (2006), in which coastal sea water from the North Sea was kept in a mesocosm in darkness for 6 months without addition of any nutrients. During this time, the accumulation of characteristic archaeal membrane lipids was accompanied by the consumption of ammonia and production of nitrite. CARD-FISH showed that the water became enriched with crenarchaeotes with 16S rRNA sequences that were very similar to the cultured *Candidatus* N. maritimus. In addition, when the *amoA* gene was isolated from the archaeal cells, one dominant sequence closely matched those of *Cand.* N. maritimus and the archaeal *amoA* gene from the Sargasso Sea metagenome. Following analysis of seawater samples throughout the year and correlation with levels of expression of the *amoA* gene using quantitative PCR (Q-PCR, p. 44), the authors concluded that nitrification by *Crenarchaeota* in this coastal system may be more important than that carried out by the nitrifying *Bacteria*.

Given the abundance of crenarchaeotes in deep ocean waters, it is essential to know whether they are all capable of autotrophic nitrification, as this will affect our knowledge of carbon cycling as well as nitrogen budgets. The experiments of Herndl et al. (2005) showed that these *Archaea* take up amino acids, suggesting that heterotrophic or mixotrophic nutrition is occurring. Agogue et al. (2008) found that deep-sea crenarchaeotes from different oceanic regions varied in their nitrifying ability. Q-PCR analysis for the expression of *amoA* showed that the proportion of nitrifying deep-water crenarchaeotes was higher in subpolar regions and lower in subtropical regions. Elucidating the full ecological role of this group of *Archaea* is a major challenge for marine microbiologists.

References

Agogue, H., Brink, M., Dinasquet, J. and Herndl, G. J. (2008) Major gradients in putatively nitrifying and non-nitrifying Archaea in the deep North Atlantic. *Nature* 456: 788–791.

Brochier, C., Gribaldo, S., Zivanovic, Y., Confalonieri, F. and Forterre, P. (2005) Nanoarchaea: representatives of a novel archaeal phylum or a fast-evolving euryarchaeal lineage related to Thermococcales? *Genome Biol.* 6: R42, doi:10.1186/gb-2005-6-5-r42.

Cerdeño-Tárraga, A. M. (2009) Genome watch: what a scorcher! *Nat. Rev. Micro.* 7: 408–409.

Cowan, D. A. (2003) The upper temperature for life—where do we draw the line? *Trends Microbiol.* 12: 58–60.

Das, S., Paul, S., Bag, S. K. and Dutta, C. (2006) Analysis of *Nanoarchaeum equitans* genome and proteome composition: indications for hyperthermophilic and parasitic adaptation. *BMC Genomics* 7: 186.

Dekas, A. E., Poretsky, R. S. and Orphan, V. J. (2009) Deep-sea archaea fix and share nitrogen in methane-consuming microbial consortia. *Science* 326: 422–426.

DeLong, E. F. (2000) Resolving a methane mystery. *Nature* 407: 577–579.

DeLong, E.F., Wu, K.Y., Prezelin, B.B. and Jovine, R.V.M. (1994) High abundance of Archaea in Antarctic marine picoplankton. *Nature* 371: 695–697.

DeLong, E.F. (1992) Archaea in coastal marine environments. *Proc. Natl. Acad. Sci.* USA 89: 5685–5689.

Fuhrman, J.A., McCallum, K. and Davis, A.A. (1992) Novel major archaebacterial group from marine plankton. *Nature* 356: 148–149.

Fuhrman, J.A., Lee, S.H., Masuchi, Y., Davis, A.A. and Wilcox, R.M. (1994) Characterization of marine prokaryotic communities via DNA and RNA. *Microb. Ecol.* 28: 133–145.

Herndl, G. J., Reinthaler, T., Teira, E., van Aken, H., Veth, C., Pernthaler, A. and Pernthaler, J. (2005) Contribution of archaea to total prokaryotic production in the deep Atlantic Ocean. *Appl. Environ. Microbiol.* 71: 2303–2309.

Hohn, M. J., Hedlund, B. P. and Huber, H. (2002) Detection of 16S rDNA sequences representing the novel phylum "Nanoarchaeota": indication for a wide distribution in high temperature biotopes. *Syst. Appl. Microbiol.* 25: 551–554.

Huber, H., Hohn, M. J., Rachel, R., Fuchs, T., Wimmer, V. C. and Stetler, K. O. (2002) A new phylum of *Archaea* represented by a nanosized hyperthermophilic symbiont. *Nature* 417: 63–67.

Karner, M. B., DeLong, E. F. and Karl, D. M. (2000) Archaeal dominance in the mesopelagic zone of the Pacific Ocean. *Nature* 409: 507–510.

Kashefi, K. and Lovley, D. R. (2003) Extending the upper temperature limit for life. *Science* 301: 934.

Könnecke, M., Bernhard, A. E., de la Torre, J. R., Walker, C. B., Waterbury, J. B. and Stahl, D. A. (2005) Isolation of an autotrophic ammonia-oxidizing marine archaeon. *Nature* 437: 543–546.

Kuypers, M. M. M., Blokker, P., Erbacher, J., Kinkel, H., Pancost, R. D., Schouten, S. and Sinninghe Damste, J. S. (2001) Massive expansion of marine archaea during a mid-Cretaceous oceanic anoxic event. *Science* 293: 92–95.

Martens-Habbena, W., Berube, P. M., Urakawa, H., de la Torre, J. R. and Stahl, D. A. (2009) Ammonia oxidation kinetics determine niche separation of nitrifying *Archaea* and *Bacteria*. *Nature* 461: 976–979.

Venter, J. C., Remington, K., Heidelberg, J. F., et al. (2004) Environmental genome shotgun sequencing of the Sargasso Sea. *Science* 304: 58–60.

Waters, E., Hohn, M. J., Ahel, I., et al. (2003) The genome of *Nanoarchaeum equitans*: insights into early archaeal evolution and derived parasitism. *Proc. Natl Acad. Sci. USA* 100: 12984–12988.

Wuchter, C., Schouten, S., Boschker, H. T. S. and Sinninghe Damste, J. S. (2003) Bicarbonate uptake by marine *Crenarchaeota*. *FEMS Microbiol. Lett.* 219: 203–207.

Wuchter, C., Abbas, B., Coolen, M. J. L., et al. (2006) Archaeal nitrification in the ocean. *Proc. Natl Acad. Sci. USA* 103: 12317–12322.

Further reading

Amann, R., Jorgensen, B. B., Witte, U. and Pfannkuche, O. (2000) A marine microbial consortium apparently mediating anaerobic oxidation of methane. *Nature* 407: 623–626.

Cavicchioli, R. (2006) Cold-adapted archaea. *Nat. Rev. Microbiol.* 4: 331–343.

Cavicchioli, R. (ed.) (2007) Archaea: *Molecular and Cellular Biology*. ASM Press, Washington DC.

Danson, M. J. and Hough, D. W. (1998) Structure, function and stability of enzymes from the Archaea. *Trends Microbiol.* 6: 307–314.

Erguder, T. H., Boon, N., Wittebolle, L., Marzorati, M., Verstraete, W. and Erguder, T. H. (2009) Environmental factors shaping the ecological niches of ammonia-oxidizing archaea. *FEMS Microbiol. Rev.* 33: 855–869.

Friend, T. (2007) *The Third Domain*. National Academies Press, Washington DC.

Garrett, R. and Klenk, H.-P. (eds) (2007) Archaea: *Evolution, Physiology and Molecular Biology*. Blackwell, Oxford.

Knittel, K. and Boetius, A. (2009) Anaerobic oxidation of methane: progress with an unknown process. *Annu. Rev. Microbiol.* 63: 311–334.

Knittel, K., Lösekann, T., Boetius, A., Kort, R. and Amann, R. (2005) Diversity and distribution of methanotrophic archaea at cold seeps. *Appl. Environ. Microbiol.* 71: 467–479.

Makarova, K. S. and Koonin, E. V. (2005) Evolutionary and functional genomics of the *Archaea*. *Curr. Opin. Microbiol.* 8: 586–594.

Miroshnichenko, M. L. and Bonch-Osmolovskaya, E. A. (2006) Recent developments in the thermophilic microbiology of deep-sea hydrothermal vents. *Extremophiles* 10: 85–96.

Chapter 6

Marine Eukaryotic Microbes

The protists are highly diverse and play a major role in ocean food webs both as primary producers in the photic zone and as consumers in all parts of the water column and the benthos, providing a trophic link to zooplankton. New methods are revealing exciting new discoveries of hitherto unknown groups and their importance. The study of marine fungi (mycology) is considerably less developed than that of other microbes and it is commonly assumed that fungi do not play a significant role in marine ecosytems. This is probably a reasonable conclusion in pelagic environments, but is almost certainly an oversimplification in coastal habitats and sediments. This chapter reviews the biology of protists and fungi and highlights selected examples that have particular importance in marine ecology.

Key Concepts

- Protists are unicellular microbes that, in terms of diversity and cellular forms, dominate the eukaryotes.
- Classification of protists is complex and controversial, but major groups can be recognized by combining morphological and molecular information.
- Many planktonic protists are major producers of fixed carbon through photosynthesis.
- Heterotrophic protists provide trophic links in marine food webs via grazing.
- Blooms of some protists such as prymnesiophytes have major impacts on ocean–atmosphere interactions and affect global climate processes.
- Some protists form harmful blooms with deleterious consequences for marine life and human well-being.
- Application of environmental genomics has revealed unexpected diversity in very small plankton, with important consequences for marine ecology.
- Marine fungi are mainly important as degraders of organic matter in coastal environments.
- The protists and fungi include important pathogens of marine life.

The term "protist" is used to describe an extremely diverse collection of unicellular eukaryotic microbes

Today, the term "protist" is widely accepted as a term of convenience to describe unicellular eukaryotic microbes, many of which are major players in marine systems. As members of the domain *Eukarya*, protists possess a defined nucleus bounded by a double nuclear membrane, which is continuous with a membranous system of channels and vesicles called the endoplasmic reticulum. This is the site of fatty-acid synthesis and metabolism and is also lined with ribosomes responsible for protein synthesis. The Golgi apparatus (a series of flattened membrane vesicles) processes proteins for extracellular transport and is also responsible for the formation of lysosomes, membrane-bound vesicles containing digestive enzymes that fuse with vacuoles in the cell, either for the digestion of food in phagocytic vacuoles or for the recycling of damaged cell material.

Most protists demonstrate some form of sexual recombination as a regular component of their life cycle in which the fusion of the genomes of two cells and their nuclei is followed by meiosis. Unlike multicellular organisms, which are usually diploid (i.e. they contain two copies of each chromosome and reproduce by forming haploid gametes), protists vary greatly in whether the diploid or haploid state is the predominant phase in the life cycle.

The cytoskeleton is a system of hollow microtubules about 24 nm in diameter (composed of the protein tubulin) and microfilaments about 7 nm in diameter (composed of two actin monofilaments wound around each other). Protists often display a variety of other filaments. Microtubules often extend under the surface and provide the basic shape of the cell, or they may be grouped into bundles to support extensions of the cell such as pseudopodia. Microfilaments provide a strengthening framework for the cell. The cytoskeleton provides a mechanism for movement of intracellular structures within the cell, for example in the separation of chromosomes during nuclear division. Another function of the cytoskeleton, which is of particular importance in the marine amoebozoa, radiolarians, and foraminifera, is amoeboid movement due to changes in the cross-linking state of the protein actin. Rearrangement of the membrane also enables the process of phagocytosis, which brings food particles or prey organisms into the cell, enclosed within a food vacuole, which subsequently fuses with lysosomes.

A defining feature of almost all eukaryotes is the presence of mitochondria—organelles with an outer membrane and an extensively folded inner membrane. The nature of the infoldings, termed "cristae," is distinctive of different eukaryotic groups, and marine protists have either tubular cristae (as seen in alveolates and stramenopiles) or flattened cristae (as seen in euglenids and kinetoplastids). A few types of protists do not contain mitochondria. Most of these are animal parasites, but free-living protists lacking mitochondria may be important in marine anaerobic sediments, microbial mats, and particles.

Photosynthetic protists contain chloroplasts and are commonly referred to as algae. The term "microalgae" is often used to indicate unicellular algae of microscopic dimensions (see. p. 2) to distinguish them from larger multicellular types (seaweeds). Much evidence indicates that the chloroplasts evolved via primary endosymbiosis, in which an ancestral eukaryotic cell engulfed a cyanobacterium. Analysis of ribosomal RNA, the arrangement of the internal membranes, and the nature of photosynthetic pigments are used in classification of photosynthetic protists. Study of these features shows that chloroplasts have arisen on several occasions during evolution via secondary and tertiary symbiosis events.

Systems for the classification of eukaryotic microbes are still developing

There have been many attempts to develop a phylogenetic tree describing the relationships between the many different lineages that comprise the domain *Eukarya*. Traditional classification systems depended heavily on morphological characteristics, but the application of molecular methods has necessitated frequent major revisions in recent years and there is still disagreement among experts about the interpretation of data. Sequencing of 18S rRNA provided a basis for developing new phylogenetic trees, but, as noted in Chapter 1, reliance on a single gene can produce many anomalous results, and evidence of lateral gene transfer and secondary symbiosis events confounds the interpretation of relationships. By combining the sequences of various genes and amino-acid sequences of key proteins, together with morphological and biochemical evidence, consensus hypothetical trees similar to that shown in *Figure 6.1* have been developed. This indicates that evolutionary radiation occurred very early in the history of eukaryotes. Marine microbes appear in all of the five major "supergroups" envisaged by this model. Features of some of the most important groupings of marine eukaryotic microbes are described in the following sections. These are sometimes considered under general headings reflecting major ecological groupings rather than attempting to present them in a formal classification scheme.

Many protists possess flagella

Protists that possess one or more flagella for movement and feeding can be grouped under the general term "flagellates," although this characteristic has no phylogenetic significance and is found in about half of the major protistan phyla. Examples are shown in *Figure 6.2*. Eukaryotic flagella are composed of nine pairs of peripheral microtubules and two pairs of central microtubules. One pair of the tubules in each pair contains the protein dynein, which moves along an adjacent microtubule; rapid and repeated bending causes the flagellum to beat. The flagella are anchored in the cell via a basal body (kinetosome), which varies greatly in its ultrastructure and has been used in previous classification systems to distinguish different groups. For example, the group known as the heterokont algae now form part of the larger group known as the stramenopiles. The name heterokont refers to the presence of two different types of flagellum on each cell, which are inserted into the cell by a distinctive pattern of microtubule roots. The anterior or "tinsel" flagellum is covered with lateral bristles, which create a backwards current when it moves, pulling the cell forward. The posterior, or whiplash, flagellum is smaller and smooth; it is often inserted into a groove on the cell, or may be reduced to a basal body.

Since the introduction of epifluorescence microscopy and flow cytometry, heterotrophic and mixotrophic flagellates in the nanoplankton (2–20 μm) size range have been shown to be very abundant in the plankton and to be the major consumers of primary and secondary production in marine systems, consuming bacteria and small protists, which they encounter as they swim through the water and ingest via phagocytosis. This "grazing" leads to the recycling of nutrients since the flagellates are preyed upon by larger protists such as ciliates and dinoflagellates, providing a link to metazoan plankton in the food web (the microbial loop, discussed in Chapter 8). The nanoplanktonic flagellates are taxonomically very diverse, occurring in many distantly related different phyla. Their ecological and biogeochemical functions are probably also very diverse, and it is important to remember that grouping them as nanoplanktonic flagellates (sometimes

Figure 6.1 A hypothetical tree of eukaryotes. This is based on various types of data, including molecular phylogenies and other molecular characters, as well as morphological and biochemical evidence. Five "supergroups" are shown, each consisting of a diversity of eukaryotes, most of which are microbial. (Reprinted from Keeling et al. [2005] by permission of Elsevier.) Groups enclosed in boxes are discussed in the subsequent sections, although it should be noted that marine protists occur in numerous other groups.

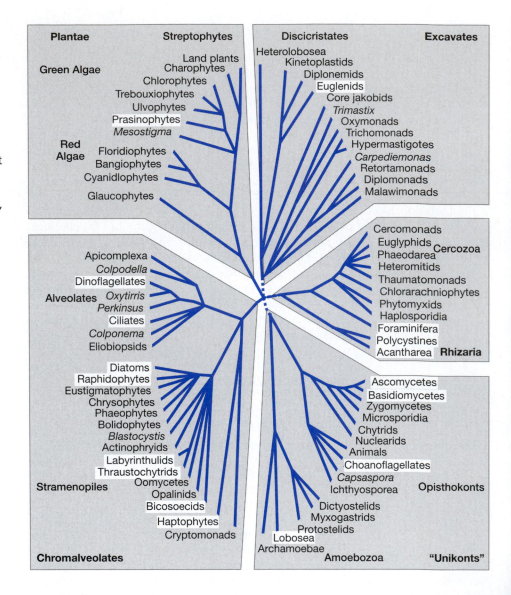

abbreviated to just "nanoflagellates") is only a convenience based on the use of the logarithmic scale for plankton size classes (*Table 1.2*) and "nano-" does not have the same meaning as its usual scientific use (1 nm = 10^{-9} m).

The euglenids may be phototrophic, heterotrophic, or mixotrophic

Study of the group known as the euglenids (see *Figure 6.2A*) provided one of the earliest challenges to traditional "zoo-" and "phyto-" based classification, since their nutrition may be phototrophic or heterotrophic, or combine features of both (mixotrophy). Thus, past taxonomic systems have placed these organisms in both the *Euglenozoa* and *Euglenophyta* by assuming that they are simple animals or plants, respectively.

Within this group, the kinetoplastids possess a unique structure termed the kinetoplast near the base of the flagellum, easily visible in the light microscope after applying DNA stains. This is a large, single mitochondrion containing a few interlocked large, circular chromosomes (maxicircles), which encode mitochondrial enzymes, plus thousands of smaller minicircles of DNA. The minicircles do

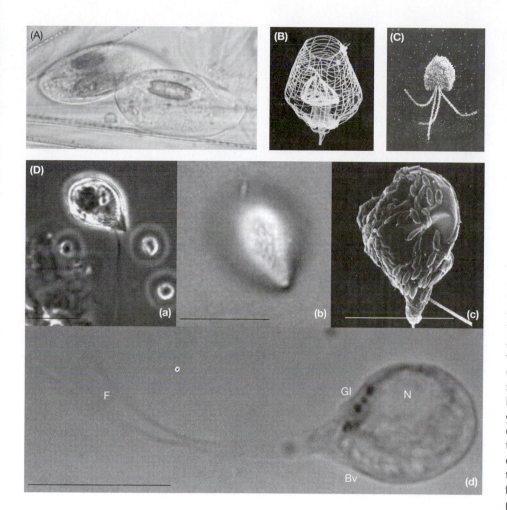

Figure 6.2 Examples of flagellated protists. (A) Light micrograph showing two euglenoid cells with ingested diatoms. (B) Transmission electron micrograph of a choanoflagellate (*Kakoeca antarctica*). (C) Scanning electron micrograph of the flagellated prasinophyte *Pyramimonas gelidicola*. (Images by Fiona Scott (A,B) and John van den Hoff (C), Electron Microscopy Unit, Australian Antarctic Division, © Commonwealth of Australia.) (D) Various aspects of *Telonema antarcticum* cells: (a) phase contrast light micrograph after Lugol fixation; (b) surface of living cell as seen by maximum resolution in the light microscope; (c) scanning electron micrograph shows that the cell surface is coated with angular paracrystalline objects in vesicles beneath the membrane; (d) light micrograph of fixed cell showing the nucleus (N), the lateral band of vesicles (Bv), stainable globules (Gl) and the two flagella (F) emerging diagonally, close to the protruding "tail" (a bacterium, out of focus, is located between them). Bars = 10 μm. (Reprinted from Klaveness et al. [2005] by permission of the International Union of Microbiological Societies.)

not appear to encode any complete genes, but are involved in post-transcriptional editing of mRNA. The bodonids (e.g. *Bodo* spp.) are kinetoplastids with oval cells about 4–10 μm long and are very common in coastal waters, often moving over surfaces using a trailing flagellum. A shorter flagellum propels water currents toward a cytopharynx lined with microtubules.

The bicosoecids are a group of highly active bacterivorous flagellates

The bicosoecids are stramenopiles occurring in the nanoplankton; whilst some genera have been known for many years, the most abundant types and ecologically important types have been described only recently. In particular, *Cafeteria roenbergensis* has now been found in all ocean waters examined, occurring at densities up to 10^5 per milliliter and has great importance as a bacterivore. Like the kinetoplastids, the bicosoecids have two flagella, one of which projects forwards to propel the cell in a spiral movement, whilst the other flagellum trails behind. When feeding, the flagellum beats about 40 times a second, creating a rapid water current that carries bacteria toward a "mouth" region at the base of the flagellum, where they are phagocytosed. In nonmotile species, the anterior flagellum attaches to the substrate. The cells have one nucleus and five mitochondria with tubular cristae, as found in all stramenopiles. Genome sequencing reveals that, unlike most eukaryotes, *Cafeteria* has a highly compact genome, with a very small amount of noncoding DNA. Another species, *Pseudobodo tremulans*, is found mainly in sediments.

The choanoflagellates have a unique feeding mechanism

The choanoflagellates in the opisthokont group are a distinctive group, which have a single flagellum that draws a current of water through a collar of 30–40 tentacle-like filaments around the top part of the cell. Bacteria are trapped and taken into food vacuoles in the cell. Some choanoflagellates produce a delicate basket-like shell around the cell (*Figure 6.2B*). About 100 species have been described from marine and estuarine waters throughout the world. The choanoflagellates resemble the feeding cells found in sponges and are of particular interest to evolutionary biologists as they are phylogenetically the group most closely related to the animals, as shown in *Figure 6.1*. Attached choanoflagellates can form dense colonies on surfaces, and the study of this process provides insights into the origins of multicellularity.

Dinoflagellates have critical roles in marine systems

The most distinctive feature of the dinoflagellates is the presence of a transverse flagellum that encircles the body in a groove, and a longitudinal flagellum that extends to the posterior of the cell. This gives rise to a distinctive spinning motion during swimming, from which the name dinoflagellate is derived (from the Greek *dinein*, "to whirl"). Another distinctive feature of their cell structure is the presence of a layer of vesicles (alveoli) underlying the cell membrane. In many species of dinoflagellates, these alveoli contain cellulose, which forms a protective system of plates that fits together like a suit of armor. There are numerous forms of such armored dinoflagellates, mainly in the genera *Dinophysis* (200 marine species) and *Protoperidinium* (280 marine species), often with unusual morphology with spines, spikes, or horn-like projections. The geometry of the arrangement of plates is an important factor in identification. Many other naked or unarmored dinoflagellates also occur, including those that resemble species in the genera *Gymnodinium*, *Gyrodinium*, and *Katodinium*. Most dinoflagellates

Figure 6.3 Scanning electron micrographs of dinoflagellates. (A) Zoospore of *Pfiesteria piscicida*, with its feeding peduncle extended. (Image courtesy of Howard Glasgow, North Carolina State University Center for Applied Aquatic Ecology). (B) *Protoperidinium incognitum*. (C) *Protoperidinium antarcticum*. (D) *Gonyaulax striata*. (Images (B–D) by Rick van den Enden, Electron Microscopy Unit, Australian Antarctic Division, © Commonwealth of Australia.)

are in the 20–200-μm microplankton size range; however, several mixotrophic genera have been identified that appear to be below 20 μm, whilst *Noctiluca* can reach a size of 2 mm.

At least 2000 species of dinoflagellates are known, of which about 90% are marine, with different species adapted to life in many habitats according to depth, temperature, and salinity. They occur as free-floating members of the plankton or attached to sediments, corals, seaweeds, and other sediments. Representative images are shown in *Figures 6.3* and *12.6*. Photosynthetic dinoflagellates contain chlorophylls *a* and *c*, plus carotenoids and xanthophylls, and make a significant contribution to carbon dioxide fixation and primary productivity in the oceans. The xanthophyll pigments confer the golden-brown color typical of many photosynthetic types and gives rise to the name zooxanthellae, used to describe the dinoflagellates that form important symbioses in invertebrates such as corals, anemones and clams (see Chapter 10). Endosymbioses also occur within other protists, such as ciliates, foraminiferans, and colonial radiolarians.

Although traditionally grouped with the algae or phytoplankton, heterotrophy occurs in about one half of known dinoflagellates and such species possess a variety of feeding mechanisms, including simple absorption of organic material, engulfment of other microbes by phagotrophy, or the presence of a tube that is used to suck out the contents of larger prey (*Figure 6.3A*). Such dinoflagellates can prey on bacteria and phytoplankton of all sizes, as well as zooplankton or even fish eggs and larvae; they therefore have major effects on food web structure. Nonpigmented, heterotrophic dinoflagellates often comprise more than half the biomass of the microzooplankton size class (20–200 μm) and are frequently the most important grazers of blooms of diatoms and other microplankton. Heterotrophic dinoflagellates also provide a major food source for copepods. Many dinoflagellates are mixotrophic and possess a number of very different forms in their life cycle. The most dramatic example of this occurs with the predatory dinoflagellate *Pfiesteria piscicida*, which was originally described as having more than 20 distinct morphological stages in the life cycle, ranging from photosynthetic to an "ambush predator" of fish. However, this interpretation has proved controversial (see *Research Focus Box 12.2*).

Some dinoflagellates are responsible for the formation of deleterious blooms, which may have damaging effects on marine life. Of most importance are the species that produce toxins that can cause illness and death of marine animals and humans; these are discussed in Chapters 12 and 13.

Dinoflagellates undertake diurnal vertical migration

In regions of the open ocean where there is little upwelling, most photosynthetic dinoflagellates are found in the photic zone during the day and migrate to deeper nutrient-rich waters at night. This vertical migration in the water column can exceed 30 m in each direction; this is a remarkable two million times the cell diameter. The onsets of both ascent and descent are regulated by a biological clock so that the cells anticipate sunrise and are in the optimum position to start photosynthesizing near the surface as soon as light is available. As light declines, they migrate to lower levels in the water column to take advantage of higher nutrient levels for the dark phase. During their daily migration, the cells encounter gradients of temperature, nutrients, and the amount and spectral quality of light; all of these can act as input signals for regulating the circadian rhythm.

Some dinoflagellates exhibit bioluminescence

About 2% of dinoflagellate species found in coastal waters are bioluminescent, with the best-known species being *Noctiluca scintillans* and *Lingulodinium polyedrum* (formerly *Gonyaulax polyedra*). They occur worldwide, but

ⓘ ACQUIRED PHOTOTROPHY— A KEY TO SUCCESS

Ultrastructural and molecular evidence provides strong support for the development of mixotrophy in protists that ingest their prey via phagocytosis, either by the maintenance of algal endosymbionts or the retention of algal plastids. Stoecker et al. (2009) emphasize that acquired phototrophy is usually overlooked in ecological and biogeochemical models, and understanding the process is important to explain how aquatic ecosystems respond to changing environments. Endosymbiosis, involving extensive gene transfer and reorganization of cellular functions, occurred very early in the evolution of eukaryotes, leading to organelles in several groups. However, acquired phototrophy can also be a facultative process involving transient retention of algae or their plastids. The heterotrophic dinoflagellate *Pfiesteria piscicida* can retain algal plastids for a few days, using photosynthesis as a source of energy and cellular material in a process known as kleptoplastidy (Lewitus et al. 1999). It is not known how the protist is able to digest most of the prey cell, but not digest the plastids. Many more persistent, facultative associations occur, especially in the dinoflagellates, ciliates, foraminifera, and radiolarians. Stoecker et al. (2009) argue that such facultative associations do not involve extensive adaptations to accommodate foreign cells or plastids. Such hosts receive the benefit of additional carbon with few "upkeep" costs. This might explain the development of features such as rapid motility, complex surface structures, or food storage systems.

? WHY DO DINOFLAGELLATES EMIT LIGHT?

There are two main hypotheses for the ecological function of luminescence in dinoflagellates. One idea is that mechanical stimulation occurs when a predator such as a copepod approaches a dinoflagellate cell—stimulating a brief, bright flash. This is thought to startle the predator and lead to its disorientation. An alternative possibility, called "the burglar alarm hypothesis," is that light produced by luminescent prey attracts grazing predators, which in turn sends a signal to larger predators, so that the grazers themselves become prey. Grazing on bioluminescent dinoflagellates will decrease if the risk of consuming them results in reduction of the net benefit that consumers receive. Thus, bioluminescence could reduce grazing pressure by reduction of feeding efficiency and the species as a whole could benefit, even though some mortality of individuals occurs. Various experimental approaches have been used to determine which of these alternative hypotheses is correct (e.g. Abrahams and Townsend, 1993), but there is no clear answer.

exceptionally high densities occur in certain tropical coastal waters and produce spectacular displays of "phosphorescence" when the surface of water is broken at night. Bioluminescence in dinoflagellates usually consists of brief flashes of blue-green light (wavelength about 475 nm), containing 10^8 photons and lasting about 0.01 s. *Lingulodinium* may also emit red light at 630–690 nm. The stimulus for light emission is deformation of the membrane due to shear forces, such as agitation of water by fish, breaking waves, or the wake of a boat. In the laboratory, bioluminescence can be stimulated by lowering the temperature or pH of the medium. The bioluminescent flash is preceded by an action potential, during which the inside of the membrane becomes negatively charged. This leads to acidification of vesicles in the vacuolar membrane containing the enzyme luciferase and the susbtrate luciferin. There are about 400 such vesicles (scintillons) in each cell and they occur as spherical evaginations of cytoplasm into the cell vacuole, each containing luciferin complexed with a special binding protein. A transient pH change results from the opening of membrane proton channels in the scintillons; this activates the reaction by release of the luciferin from its complexed state so that it can be oxidized in an ATP-mediated reaction similar to that in bacteria (*Figure 4.5*). Experimental studies of *Lingulodinium* cultured under various light conditions have revealed that bioluminescence is regulated by a circadian rhythm and peaks in the middle of the dark period. The circadian expression of bioluminescence involves the daily synthesis and destruction of the scintillons and component proteins; this is regulated at the translational level and may involve a clock-controlled repressor molecule that binds to mRNA.

The ciliates are voracious grazers of other protists and bacteria

This group is distinguished by the possession of organelles known as cilia in at least one stage in the life cycle. Representative examples are illustrated in *Figure 6.4*. Cilia have the same basic structure as flagella, but the cilia often cover the cell surface or are arranged in groups; they beat synchronously to provide movement or to create water currents to channel particulate food into the cell. At least 8000 species of ciliates are known—many of which are marine—and, although highly diverse, the group appears monophyletic in modern classification systems (see *Figure 6.1*). Marine ciliates are generally in the size range 15–80 μm, with some up to 200 μm. One group of ciliates, the tintinnids, produce a "house" called a lorica, which is constructed from protein, polysaccharides, and accumulated particulate debris collected from the water. As a result, tintinnid loricae are large enough to be collected in fine-mesh plankton nets and were, therefore, one of the first groups of marine ciliates to be studied. There has been intense interest in the ciliates in recent years because of growing recognition of the essential role that most types play in the microbial loop of ocean food webs by ingesting other small protists such as flagellates and bacteria and being preyed upon in turn by larger protists and mesozooplankton. Selective grazing on particular prey types is an important factor in structuring the composition of microbial communities in food webs. As a consequence, there have been numerous studies of the abundance and activity of marine ciliates, with a wide range of results. Typically, there are about 1–150 ciliates per milliliter in seawater, with the highest numbers in coastal waters; much higher concentrations occur in marine snow particles. Abundance varies greatly with water depth, temperature, and nutrient concentration. Water stratification during the different seasons is a major factor affecting ciliate numbers. Ciliates are also abundant in benthic sediments and microbial mats.

The most common marine ciliates are spherical, oval, or conical cells with a ring of cilia surrounding the cytostome ("mouth"), which they use to filter bacteria and small flagellates from the surrounding seawater. Upon entering the cytostome, ingested food particles are engulfed by phagosomes, which then fuse with lysosomes in the cytoplasm. The phagosomes become acidified and enzymes

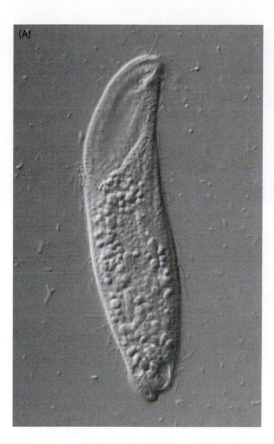

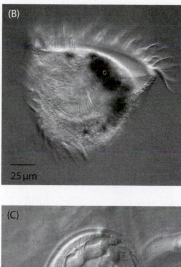

Figure 6.4 Light micrographs of representative marine ciliates. (A) *Acineria inurvata*. (B) *Gastrocirrhus monilifer*. (C) *Aspidisca leptaspis*. (Images by Bill Bourland, used by permission of David Patterson, MBL micro*scope project [http://starcentral.mbl.edu/microscope]).

contained in the lysosome lead to digestion. Nutrients pass into the cytoplasm and undigested waste material is egested. Although most genera are phagotrophic, some are strict photosynthetic autotrophs and have lost phagotrophic capability. The most important of these is *Mesodinium rubrum*, which can occur in large blooms in coastal waters and may make a sizable contribution to primary productivity. This ciliate contains functional photosynthetic cryptomonads as endosymbionts. There are many examples of mixotrophy arising from transient retention of ingested algal cells or retention of functioning plastids.

A distinctive feature of the ciliates is the possession of two types of nuclei. The larger macronucleus consists of multiple short pieces of DNA and is concerned largely with transcription of mRNA and growth processes. The smaller micronucleus is diploid and is responsible for an unusual type of sexual reproduction (conjugation), in which two cells fuse for a short period and exchange haploid nuclei derived from the micronuclei by meiosis. The macronuclei disappear during this process. The result of conjugation is that each partner ends up with one of its own haploid nuclei and one from its partner; these then fuse and the cells separate. In each cell, the diploid nucleus thus formed divides and differentiates into micro- and macronuclei.

The haptophytes (prymnesiophytes) are major components of ocean phytoplankton

The haptophytes, also known as prymnesiophytes, are very important members of the phytoplankton, some of which have profound influences on oceanic and atmospheric processes. There are about 500 living species in 50 genera, with many additional species identified from fossils. Most species are marine and occur worldwide, with the greatest diversity and abundance in tropical waters.

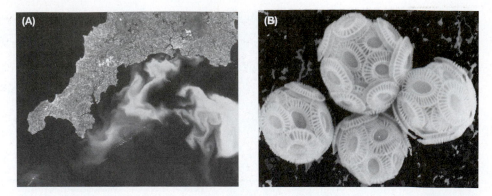

Figure 6.5 Bloom-forming prymnesiophytes. (A) Satellite image of *Emiliania huxleyi* bloom in the English Channel, south of Plymouth, UK. (Image courtesy of Plymouth Marine Laboratory Remote Sensing Group.) (B) Scanning electron microscope image of *E. huxleyi*. Each cell is ~5 μm in diameter and covered with about 30 coccoliths. (Image courtesy of Toby Colins and Willie Wilson, Marine Biological Association of the UK, Plymouth.)

They are flagellated unicells in one stage of their life cycle and contain a haptonema; this is a thin structure reminiscent of a flagellum, but with a different structure and unknown function. The haptonema is composed of six to seven microtubules in a ring or crescent, with a fold of endoplasmic reticulum extending out within the flagellum. A particular characteristic is the complex architecture and variety of shapes shown by the external scales or plates, formed within the Golgi apparatus, that cover the cell surface in many species. Haptophytes are photosynthetic and contain one or two plastids, together with the yellow-brown accessory pigments diadinoxanthin and fucoxanthin. They often have a complex life cycle, alternating between motile and nonmotile haploid and diploid stages.

The most studied group is the coccolithophorids, of which the best-known and most abundant example is *Emiliania huxleyi*, in which the scales are calcified and known as coccoliths. The typical cells of *E. huxleyi* are diploid, nonmotile, 4–5 μm in diameter and usually covered with about 30 coccoliths (*Figure 6.5B*). The haploid stage is thought to be in the form of motile swarmer cells with non-calcified scales; another nonmotile naked cell type may also occur as a mutant diploid stage. Each stage can reproduce vegetatively. The haploid stage seems to be very important in the resistance of *E. huxleyi* to viral infection, as discussed in *Research Focus Box 7.1*. Calcification takes place intracellularly in a specialized vesicle of the Golgi apparatus and depends on a high flux of inorganic carbon (in the form of bicarbonate) and calcium. The calcium binding protein is a key factor in the calcification process and the *gpa* gene that encodes this is often used as a genetic marker in studies of *E. huxleyi* community dynamics. Calcite (one of the crystalline forms of calcium carbonate) is precipitated onto the organic matrix of the vesicle and extruded to the surface to form a coccolith.

Each year, massive blooms of *E. huxleyi* occur in the upper photic layers of temperate and subtemperate waters. They are especially notable in the nutrient-rich coastal waters of northern Europe and Scandinavia where they are highly visible as milky-white areas in satellite images (*Figure 6.5A*), because of the reflective properties of free coccoliths released from the cells as they disintegrate. Under favorable conditions, individual blooms may cover 100000 km^2 and contain more than 10^{21} cells, accounting for almost 90% of all the phytoplankton in the upper water columns. The reflective "signature" of the coccoliths is unique and allows precise measurement of the fate of the bloom from satellite images. On a global scale, such dense blooms of *E. huxleyi* probably transiently cover 1 to 2 million km^2 in a typical year. *Emiliania huxleyi* is present at relatively constant, lower levels in many oligotrophic oceans, where they also contribute significantly to primary productivity.

Emiliania huxleyi has a global distribution and has been intensively studied in recent years because of its role in the global carbon cycle, being the largest global producer of calcium carbonate and, hence, a major sink for CO_2. However, there are several complications to the process and the exact contributions that *E. huxleyi* makes to carbon cycling and climate change are not clear. The light-scattering properties of the coccoliths may have a small albedo effect, reflecting solar radiation and helping to cool surface waters during a bloom. Calcification leads to rapid removal of inorganic carbon as the plates settle through the water column to form sediments. In fact, although *E. huxleyi* is the most dominant coccolithophore, other species with larger, heavier coccoliths may be more important in the flux of calcite. Various studies to estimate the global productivity of calcite in the modern ocean have been conducted, leading to a consensus value of about 1 Gt calcite carbon a year. Only about half of this amount will be precipitated to the ocean floor because in deep waters (over about 3000 m), changes in the water chemistry due to high pressure cause the calcite to redissolve. Fossils of the coccolithophorid group first appeared in the Jurassic period and reached their greatest abundance and diversity in the Late Cretaceous period, at the end of which a mass extinction of many genera occurred. Accumulation of coccolithophorid plates on the seabed contributes to the formation of ocean sediments and rocks such as the Mesozoic limestones and chalks. The famous White Cliffs of Dover in southern England are composed of a very fine-grained white chalk derived mainly from coccoliths.

Dissolved carbon exists in three forms in seawater—dissolved CO_2 gas, bicarbonate, and carbonate (see formula on p. 11)—and the relative proportions of each depend on pH, temperature, and pressure. When coccolithophore cells remove bicarbonate to form coccoliths, the pH drops and shifts more of the carbon into the dissolved CO_2 form. Calcification can be represented by the formula

$$Ca + 2HCO_3^- \rightleftharpoons CaCO_3 + H_2O + CO_2$$

Thus, coccolithophore blooms might increase global warming by causing release of carbon dioxide back to the atmosphere. However, this is complicated by the fact that the association of heavy coccoliths with marine snow particles and in the feces of grazing zooplankton could lead to more rapid settling of associated organic material, thus accelerating the sequestration of fixed carbon to the sediments. Understanding the fluxes involved is very important in view of rising atmospheric carbon dioxide levels and ocean acidification (see *Research Focus Box 1.1*), but despite many laboratory and mesocosm experiments, there are no clear answers. In addition, the role that *E. huxleyi* and other bloom-forming microalgae play in the production of dimethyl sulfide (DMS) may be highly significant in global climatic processes—this is discussed further in Chapter 9.

Two other prymnesiophytes, *Chrysochromulina* and *Phaeocystis*, have both been studied extensively because of their ability to produce large harmful blooms under certain conditions, especially in northern Europe and Scandinavia. Blooms are thought to be linked to climatic conditions and the input of pollutants from land runoff. *Phaeocystis globosa* aggregates into large colonies surrounded by polysaccharide mucilage. This is responsible for the foam that commonly affects seashores during summer, and excessive production of DMS (see p. 198) during the collapse of the blooms leads to sulfurous smells that beachgoers often confuse with sewage pollution. The mucilage may also clog aquaculture nets and causes mortalities of fish and shellfish. DMS is also suspected of affecting the migration behavior of fish, which seek to avoid it.

? WHY DO SOME COCCOLITHOPHORES CALCIFY THEIR SCALES?

One suggested role for the scales of prymnesiophytes is that of protection—perhaps they might have evolved to protect cells from mechanical stress, physical effects of temperature and radiation, or to deter grazing by zooplankton—especially if they are calcified. Although appealing, there is little experimental support for these roles. The cellular mechanism underlying the process of calcification in *Emiliania huxleyi* and related species is a subject of intense study by Colin Brownlee and colleagues at the Marine Biological Association laboratories in Plymouth, UK. Brownlee and Taylor (2004) evaluated evidence that the presence of coccoliths helps to regulate intracellular pH. An increase in hydrogen ions (lower pH) increases the availability of dissolved CO_2, in which form it is absorbed by the chloroplast for photosynthesis. The normal levels of CO_2 in seawater are thought to be well below the levels required for maximum activity of the CO_2-fixing enzyme RuBisCO (see p. 66). Thus, calcification may enhance the efficiency of photosynthesis. Current research is attempting to evaluate whether differences in calcification rates between strains of *E. huxleyi* results in improved growth of the cells. Such studies are particularly important in view of the effects of ocean acidification (see *Research Focus Box 1.1*).

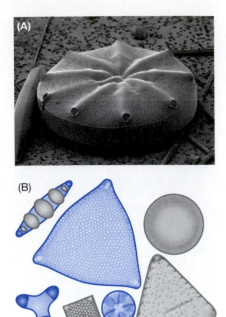

Figure 6.6 Representative marine diatoms. (A) Scanning electron micrograph of *Asteromphalus hookeri*. (Image by Rick van Enden; Electron Microscopy Unit, Australian Antarctic Division, © Commonwealth of Australia.) (B) Drawing based on light micrograph of acid cleaned diatoms to show their silica cell walls. (Image courtesy of the National Institute of Water & Atmospheric Research, New Zealand—Healey Collection [http://www.niwa.co.nz].)

Diatoms are extremely diverse and abundant primary producers in the oceans

The diatoms are one of most diverse groups of the protists, with in excess of 200 genera and about 100000 current species, with probably about the same number of fossil species. In modern classification schemes, diatoms are recognized as unicellular microbes belonging to the stramenopiles. The major pigments in this group are usually chlorophylls *a* and *c* and the carotenoid fucoxanthin. It is thought that the chloroplast in stramenopiles arose by a secondary endosymbiosis event about 500 million years ago, involving a red alga ancestor and a heterotrophic eukaryote. Several species of diatoms have now had their genome sequenced, revealing that genes from red algae have been transferred to the nucleus of diatoms—this provides an explanation for the unusual metabolic properties of diatoms. For example, unlike plants, diatoms have a complete urea cycle and generate energy from the breakdown of lipids. Carbohydrates are stored as a β-1,3-linked glucose polysaccharide.

For many years, diatoms were recognized as the dominant member of the marine phytoplankton and therefore attributed with the major role in primary productivity, but the recent recognition of other phototrophic microbes is changing this view. In some habitats—especially well-mixed coastal waters containing high levels of nutrients—this presumption is still valid, but we now realize that cyanobacteria and protists in the picoplankton size range dominate photosynthetic productivity in many parts of the world's oceans. In polar regions, diatoms play a critical role in the food chain, especially as their hybrid "phototroph–heterotroph" metabolism allows them to survive long periods of darkness before emerging as blooms in the spring. Like dinoflagellates, many open-ocean diatoms undertake a vertical migration from the nutrient-depleted photic zone to deeper waters in order to pick up nitrate and other nutrients. The complex ecological factors that determine community composition of the phytoplankton are discussed further in Chapter 8.

The most distinctive feature of diatoms is enclosure of the cell within a cell wall known as a frustule or test, composed of two overlapping plates (valves or thecae) of silicon dioxide (silica). Silica is synthesized intracellularly by the polymerization of silicic acid monomers. The two valves overlap rather like the lid and base of a petri dish. These glass structures form a variety of shapes of great architectural beauty and are highly distinctive for identification (*Figure 6.6*). Diatoms possess either radial or bilateral symmetry; these forms are termed "centric" or "pennate," respectively.

The nature of the frustule enclosing the diatom cell results in an interesting reproductive cycle. Reproduction is usually asexual with each daughter cell constructing a new theca within the old one, resulting in a progressive reduction in cell size with each division cycle. At a critical point—this is usually when the diatom is about one third of the original size—sexual reproduction occurs by formation of gametes via meiosis. These fuse to form a zygote, which increases in size and synthesizes new full-size thecae. Hence, sexual reproduction accomplishes genetic variability as well as allowing the cell line to regain maximum size before the next round of vegetative cell division.

Most diatoms are free-living in the plankton, but many attach to surfaces such as rocks, marine plants, mollusks, crustaceans, and larger animals. The skin of some whales has been shown to be covered with dense colonies of diatoms. Locomotion of diatoms occurs in contact with surfaces, probably via the action of minute filaments protruding through slits or pores in the frustule. Some diatoms have pairs of thin spines projecting from the ends of the cells, which link with those of other cells to form long chains, thereby increasing buoyancy and forming large mats.

Diatoms are largely responsible for the spring bloom along the continental shelf in temperate waters and for seasonal blooms in regions of nutrient upwelling. As with other algae, a key factor determining the size of blooms is the availability of nutrients, especially nitrogen and phosphorus, but the concentration of silica is also a limiting factor for diatoms and most will not grow at silica concentrations below about 2 μM. Some species make frustules with very high silica contents, and in some waters (e.g. the Antarctic circumpolar current) the shells are very resilient to dissolution as they settle through the water. Blooms are generally followed by the exhaustion of nutrients and aggregation and sinking of diatoms. The dynamics of diatom production and settlement are still poorly understood, but are important in understanding ocean biogeochemistry. *Research Focus Box 9.1* discusses open ocean experiments that have shown the importance of iron in stimulating diatom blooms.

Diatoms and their products—past and present—have many applications

Diatoms began to accumulate on the seabed about 100 million years ago and reached a peak of abundance in the middle part of the Cenozoic period. Settlement of dead diatoms over millennia led to thick deposits and formation of sedimentary rocks, which are mined as diatomaceous earth. This has many industrial uses such as filtration compounds, abrasives, insulating agents, pharmaceutical products, and insecticides. The small fraction of organic matter that escaped remineralization accumulated in sediments, leading to the formation of petroleum hydrocarbons. Today, diatoms are being exploited in biotechnology, including applications in nanotechnology and the production of biofuels (see Chapter 13).

Few diatoms are toxic, but an important exception is the genus *Pseudo-nitzschia*, which produces domoic acid. This is responsible for human illness associated with shellfish consumption and for mortalities in marine mammals and seabirds (p. 248).

Protists in the picoplankton size range are extremely widespread and diverse

Although some very small protists have been described for many years, their importance as components of the picoplankton has only been realized recently. The traditional logarithmic scale classification of plankton size groups classes the picoplankton range as 0.2–2 μm (*Table 1.2*), but owing to the sizes of filters used, some workers in this field consider cells up to 3 μm as "picoeukaryotes." As noted in the description of nanoplanktonic flagellates, it is important to remember that grouping organisms in this way is a pragmatic one, based solely on size, and does not imply that the organisms are related.

Phototrophic types belonging to the green algae (prasinophytes) and stramenopiles (haptophytes, prymnesiophytes) occur in the upper 100–200 m of all the world's oceans. Together with the cyanobacteria *Prochlorococus* and *Synechococcus* (see p. 113), the picophytoplankton may constitute up to 80% of the chlorophyll-containing biomass in subtropical oceans. Studies in the Pacific Ocean have shown that, although phototrophic picoeukaryotes comprise only a small proportion of the total phytoplankton cells, they account for the majority of the net carbon production. They are subject to heavy grazing pressure by other protists, possibly because they lack a cell wall. Thus, their role in marine food webs and the "biological pump" (see *Figure 8.1B*) may be much more important than previously realized because of the efficient capture of carbon dioxide and its rapid transfer to food webs via grazing.

(i) BACTERIA HELP TO DISSOLVE SHELLS AND RECYCLE SILICA

The availability of silicon is a critical factor in the growth of diatoms, which incorporate it into their shells. When diatoms die, they aggregate and sink through the water column. A fraction of the silica dissolves to form silicic acid, which is then available through upwelling to support more diatom growth. The input of silicic acid from terrestrial runoff is balanced by deposition in sediments, most of which occurs in the iron-limited areas of the ocean. Bacteria play a crucial role in the recycling of silicon by producing extracellular enzymes that break down polymeric substances in the cell wall. Bidle and Azam (1999, 2001) showed that if bacteria are removed from seawater or inhibited by antibiotics, the dissolution of silica from diatom debris is very low. Different species of diatoms vary in both the thickness of their silica shells and the degree to which they are protected by glycoprotein coats, and these factors determine the extent to which the shells of different species remain intact. Bidle et al. (2002) showed that temperature strongly affects the process of bacterial dissolution of silica, affecting the amount of organic material that remains attached to diatom shell debris. In colder waters, such as the polar regions, slow bacterial activity means that more carbon will be sequestered in sediments with sinking diatom shells. Understanding this process is important for modeling the role of the oceans in the removal of carbon dioxide from the atmosphere, and the effects of global warming on geochemical cycles.

The most studied prasinophyte phototroph is *Ostreococcus tauri*, which is the smallest eukaryotic organism known. Originally discovered and given its genome name because it was found in waters used for oyster (*Ostrea*) cultivation in France, *O. tauri* is now known to be ubiquitous in coastal waters and in the open ocean. The tiny cell (about 0.8 μm diameter) has no cell wall or flagella and contains the nucleus, only one mitochondrion, and one chloroplast, tightly packed within the cell membrane (see *Figure 1.4B*). Genome analysis has revealed that it has a small and compact genome with 20 linear chromosomes. It is thought that one chromosome may be a sex chromosome as it contains genetic information for meiosis, although sexual reproduction has not been observed. Genes on some chromosomes show little homology to other genes and appear to have been laterally transferred; they may encode surface proteins to protect the cell against predation. One of the most important findings of genome analysis is the presence of genes for C_4 photosynthesis. This route of CO_2 fixation may be more efficient than the normal C_3 route when dense populations of cells are competing for available CO_2. As with the cyanobacterium *Prochlorococcus* (see p. 113), comparative genome analysis shows that different ecotypes of *O. tauri* are adapted to different light and nutrient niches.

Another prasinophyte, *Micromonas pusilla* is a major component of the picoplanktonic community in several oceanic and coastal regions and shows summer blooms that may be terminated by viral infection. *Micromonas* is larger than *Ostreococcus* and is actively motile via a single flagellum, enabling it to swim toward light sources. Comparative genomic analysis of strains presumed to be of the same species, but isolated from different geographic regions, has shown remarkably large differences in genome sequences, with only about 90% of their genes shared.

Stramenopiles with small cells about 2–3 μm in diameter include *Aureococcus* spp., *Nannochloropsis* spp., and *Bolidomonas* spp. *Aureococcus anophagefferens* is notorious as a cause of regular "brown tides" in estuaries on the eastern US coast since the 1980s. It is present at low concentrations at all times and blooms to high concentrations under certain conditions, causing severe limitation of growth of seagrass (*Zostera marina*) beds, because of shading caused by the dense blooms. This affects larval recruitment of fish, scallops, and clams, with severe consequences for fisheries. *Aureococcus* is also suspected of producing a toxin that may affect ciliary function in bivalve mollusks, although this has not been isolated. *Nannochloropsis* has a very high content of polyunsaturated fatty acids, making it an important food source for fish larvae. This feature is being exploited in development of aquaculture feeds (see p. 312).

A very wide range of heterotrophic picoeukaryotes have been described; they seem to be abundant in all oceans at all depths examined, including in the vicinity of hydrothermal vents at densities ranging from 10^2 to 10^6 per milliliter. Many of the species described so far are stramenopiles, including *Paraphysomonas* and *Symbiomonas scintillans*. Despite its small size (~2 μm), *S. scintillans* contains intracellular bacteria within vesicles of the endoplasmic reticulum. Whether these are endosymbionts or prey that has somehow escaped digestion is unknown.

Raphidophytes are stramenopiles which may cause harmful blooms

Raphidophytes are photoautotrophs related to the diatoms and have large cells (50–100 μm) without cell walls; they possess two flagella. Blooms of species such as *Heterosigma akashiwo*, *Chatonnella* spp., and *Fibrocapsa* spp. have caused "red tides" responsible for extensive fish mortalities in Japan, Scotland, and Canada. They are thought to kill fish as a result of the production of a range of toxins, including reactive oxygen species, hemolytic substances, and neurotoxins.

Thraustochytrids and labyrinthulids play an important role in breakdown and absorption of organic matter

These two groups of closely related stramenopiles are characterized by the production of a "slime net"—a network of filaments produced as extensions of the cytoplasmic membrane that serve as tracks for the cells to glide along. They were originally grouped with slime molds in the fungi, but their heterokont flagella and mitochondrial structure mark them out as stramenopiles. They have been known for many years, but little is known of their ecology and importance in marine systems.

Several studies have shown that thraustochytrids are widely distributed in the plankton in coastal and open ocean, although they seem to occur at very low densities—perhaps only 1–100 per milliliter. Apart from bacterial and archaeal cells, these organisms are the only heterotrophic members of the plankton which feed by osmotic absorption of dissolved organic matter rather than by phagocytosis of cells or particles. The slime net contains extracellular enzymes that digest large polymeric organic compounds such as polysaccharides and proteins to smaller units for absorption. The slime net can penetrate particles, such as marine snow or clumps of decaying cells from algal blooms. These properties and the fact that their cells are many times bigger than bacterial or archaeal cells mean that, despite their low numbers, they may play a very significant role at the base of the food chain, especially since they are readily eaten by ciliates and amoebae. This link to higher trophic levels is particularly important because the thraustochytrids produce high levels of omega-3 polyunsaturated fatty acids, which are essential for the growth and reproduction of crustaceans and fish larvae. This feature is being exploited in aquaculture biotechnology. Thraustochytrids also degrade plant detritus such as algae and mangrove leaves, and a few species have been described as parasites of mollusks.

Labyrinthulids are very similar in structure to the thraustochytrids, but have not been found in plankton. Instead, they are usually associated with the surfaces of marine macroalgae, plants, or animals as parasites or commensals. One species, *Labyrinthula zosterae*, causes "wasting disease" of the seagrass *Zostera marina* (see p. 253).

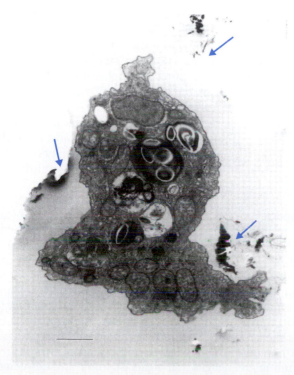

Figure 6.7 Transmission electron microscope image of an ultrathin section through a floc particle containing an *in situ* amoeba. Arrows indicate embedded particles, probably minerals. Scale bar = 1 μm. (Reprinted from Rogerson et al. [2003] by permission of Oxford University Press.)

Amoebozoa may be important grazers of bacteria associated with particles

A wide range of phylogenetically diverse protists demonstrate a crawling or amoeboid movement, through the use of pseudopodia. As seen in Figure 6.1, the Lobosea are recognized as a group of the Amoebozoa in the protistan phylogenetic tree. Planktonic amoebae are sparsely distributed in the open ocean water column, so their contribution to energy flow via bacterivory within pelagic food webs has largely been overlooked. However, they can reach high densities in more turbid estuarine and coastal waters and are particularly associated with surfaces such as flocs and marine snow particles, where the density can exceed that of the ciliated protists most commonly associated with bacterivory. Their plastic cell shape may facilitate removal of firmly attached bacteria by enabling them to penetrate crevices inaccessible to other grazers (see *Figure 6.7*), The unique ability of amoebae to graze floc-associated bacteria means that they may make a significant contribution to microbial loop processes (see p. 182).

Radiolarians and foraminifera have highly diverse morphologies with mineral shells

These two groups are classified in the Rhizaria in modern schemes (see *Figure 6.1*). They are characterized by an amoeboid body form, using pseudopodia for locomotion and feeding. The cells are usually less than 1 mm in size, but some species are the largest protists known, with diameters up to several centimeters.

Radiolarians have existed since the beginning of the Paleozoic era about 600 million years ago and produce a great diversity of beautiful, intricate shapes used in identification of more than 4000 species (*Figure 6.8*). Two closely related groups known as the polycystines and Acantharea are recognized. The polycystines are most important and form the majority of fossils. They are characterized by stiff, needle-like pseudopodia arranged in radial symmetry (from which they derive the name radiolarian) and internal skeletons made of silica. Larger species are often associated with surfaces and may contain algal symbionts that provide some nutrients to the cell, whilst the smaller types occur throughout the water column and in deep-sea sediments. Densities vary greatly, ranging from 10000 per cubic meter in some parts of the subtropical Pacific Ocean to less than 10 per cubic meter in the Sargasso Sea. The silica skeletons of these polycystine radiolarians deposit as microfossils and are second only to diatoms as a source of silica in sediments. The cell body consists of a central mass of cytoplasm surrounded by a capsular wall. This contains pores through which the cytoplasm extrudes into an extracapsular cytoplasm and forms the stiffened pseudopodia. The cytoplasm moves by streaming and captures other protists and small zooplankton, which are then surrounded and digested in food vacuoles. Hydrated, polymerized silicon dioxide (opal) is deposited within a framework of the cytoplasm. Reproduction occurs asexually by binary fission or sexually via the production of haploid gametes. The Acantharea produce skeletons of strontium sulfate (celestite) rather than silica.

About 275000 species of foraminifera have been documented, of which about 100000 species are living and the remainder are fossils. Foraminifers secrete shells (known as tests) usually composed of calcite, although some use the aragonite form of calcium carbonate. Most species are benthic. The single-celled organism forms a multichambered shell within a cytoplasmic envelope produced by pseudopodia; the shapes of these are highly distinctive for species identification. The chambers are connected by openings and have sealed pores that face the external environment, through which cytoplasmic spines stretch for long distances to form a net for the capture of prey, which can include bacteria, phytoplankton, and small metazoan animals. Many foraminifers contain other protists as endosymbionts, including green algae, red algae, diatoms, and dinoflagellates.

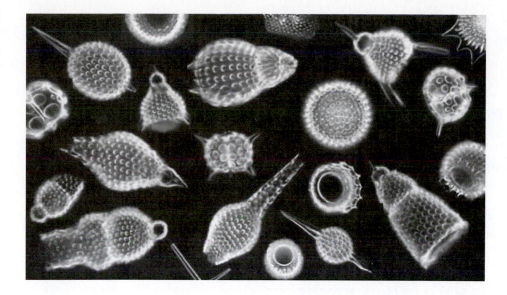

Figure 6.8 Radiolarian shells illuminated by dark field microscopy. (Image courtesy of Michael W. Davidson and the Florida State University.)

Foraminifers make up only a very small component of the plankton, but their discarded skeletons are the major source of calcareous deposits, accumulating as the "globigerian ooze" over vast areas of the ocean floor. Massive deposits accumulated during the Tertiary period, about 230 million years ago, and these have become uplifted over time and exposed as limestone beds in Europe, Asia, and Africa. The main application of the study of the many diverse forms of foraminifera is in biostratigraphy, a branch of geology which assigns relative ages of rock strata by recording fossil assemblages. This technique is widely used in oil exploration and drilling projects. Foraminifera are also used in reconstructing the history of oceanic conditions during geological eras.

Marine fungi are especially important in decomposition of complex materials in coastal habitats

The fungi are a very diverse group of organisms containing both microscopic forms (molds and yeasts) and large mushrooms and toadstools. Fungi are generally regarded as being primarily terrestrial organisms and have not received the depth of study afforded to some other groups in marine systems. Of nearly 100000 described species of fungi, only about 350 are recognized as truly marine species—these occur in the three major groups: *Ascomycetes*, *Basidiomycetes*, and *Zygomycetes*. However, many other fungi associated with freshwater and terrestrial habitats can grow in seawater, for example when decomposing detritus finds its way into the sea.

Fungi are heterotrophic and the majority are saprophytes playing a major role in decomposition of complex organic materials in the environment. The best-characterized marine fungi are those associated with decomposition of plant material (e.g. decaying wood, leaves, and intertidal grasses). Recent studies have shown that salt-marsh plants and tropical mangroves are particularly rich sources of marine fungi and it is likely that further investigation will reveal many more new species. These could have biotechnological potential as sources of enzymes and pharmaceuticals. Other fungal habitats include estuarine muds, the surface of algae, corals, sand, and the intestinal contents of animals. The role of fungi in decomposition of wood is particularly important, especially in low-oxygen environments such as estuarine muds and mangroves. These conditions inhibit wood-degrading invertebrates, and few bacteria are active in degradation of lignocellulose. As well as their importance in decomposition of plant detritus, mangrove roots, or tree branches washed downstream from rivers, fungi cause damage through attack of wooden structures such as piers and pilings. The enzymes involved in degradation of cellulose, hemicelluloses, and lignin have been characterized from a few types of marine fungi.

There has been considerable recent progress in the isolation and identification of marine yeasts (unicellular fungi). The isolation frequency of yeasts falls with

BOX 6.1   RESEARCH FOCUS

A hidden world of tiny eukaryotic plankton
New techniques reveal new roles for protists in marine ecosystems

Discovery of the picoplanktonic eukaryotes

Systematic study of marine protists is a well-established science with its roots in painstaking observations of marine life by scientists in the nineteenth and early twentieth centuries. Many thousands of species have been described and characterized using traditional methods of microscopy, taking advantage of major morphological differences between taxa. Sequencing of the 18S rRNA that characterizes the ribosomes of eukaryotes was applied extensively in the 1980s and 1990s and this, together with other genetic analyses, has helped to shape modern systems for classifying the protists within the eukaryotic domain (Adl et al., 2005; Keeling et al., 2005). However, the application of environmental genomic techniques to the study of protists in marine communities developed much more slowly than it did for bacteria and there were few published studies before 2000. In 2001, three separate studies published at about the same time revealed the discovery of a hitherto unknown extra level of diversity amongst the very small eukaryotes in the picoplankton size range. In these studies, DNA was extracted from filtered plankton cells and amplified via PCR using general primers for 18S rRNA genes, followed by construction of a clone library. Because eukaryotic plankton cells exist across such a broad range of sizes, from micrometers to millimeters, the method of specimen collection is very important. So, the researchers used a combination of filters and prefilters to trap cells of different size fractions and verified their findings by flow cytometry. The research teams produced similar results, showing a very great diversity of organisms in different regions and at different depths. Novel organisms were affiliated to many diverse groups, including prasinophytes, haptophytes, bolidophytes, and many types of alveolates related to dinoflagellates (Diez et al., 2001; López-García et al., 2001; Moon-van der Staay et al., 2001). This high diversity was greatly unexpected and challenged our understanding at the time of the role of picoeukaryotes in marine ecological processes (Worden and Not, 2008). Moreira and López-García (2002) concluded that a "large diversity of protists has escaped detection by classical methods ... with far-reaching implications for both evolution and ecology studies."

Subsequently, use of FISH, quantitative PCR, flow cytometry, and analysis of metagenomic data sets such as the GOS (see *Research Focus Box 4.1*) has revealed information about the numbers and activities of these organisms (Worden et al., 2006). Genomic approaches to investigate the diversity

and ecology of photosynthetic picoeukaryotes have generally been hampered by the fact that filter-fractionated samples in the GOS are dominated by bacterial and archaeal sequences. Also, clone libraries constructed from filtered seawater samples using universal eukaryotic 18S rRNA primers are strongly biased toward heterotrophs, especially alveolates and stramenopiles, despite the fact that autotrophic cells outnumber heterotrophs in the euphotic zone. To overcome this problem, Shi et al. (2009) developed a new approach based on sorting cells using a fluorescence activated cell sorter (see p. 34) before constructing clone libraries. This greatly increased the recovery of sequences from uncultivated, putative photosynthetic picoeukaryotes and revealed that community composition was highly variable. Many other methodological issues that may hamper interpretation of diversity studies are discussed by Worden and Not (2008).

Novel marine alveolates may be important parasites of other life

The dominance of novel lineages that group with marine alveolates (termed MALV) raised an obvious question—what do they do? Phylogenetic analysis by Guillou et al. (2008) showed that many MALV sequences group with a dinoflagellate order known as Syndiniales, a group known from previous culture-based studies to be composed entirely of parasites that infect a wide range of hosts including other protists, cnidarians, crustaceans, and fish. One genus, *Amoebophyra*, is known to infect many dinoflagellate species, including those responsible for harmful algal blooms (HABs) such as *Alexandrium* and *Gonyaulax*. The life cycle of *Amoebophyra* consists of alternation between a small, flagellated dinospore and an endoparasitic growing stage called the trophont, which eventually breaks the cell wall and is released as a swimming filament of cells that then differentiates into dinospores (Guillou et al., 2010). These authors argue that Syndiniales parasites may be important in controlling dinoflagellate populations and could be a factor in the emergence and decline of HABs. Furthermore, they speculate that the Syndiniales may have a key role in marine food webs by killing dinoflagellates and releasing dissolved organic matter into the microbial loop, as well as providing a trophic link to higher levels through grazing on the dinospores. Discovering the life cycles and host specificity of the new MALV organisms will reveal important new insights into community structure in the oceans.

increasing depth, and yeasts in the order *Ascomycetes* (e.g. *Candida, Debaryomyces, Kluyveromyces, Pichia,* and *Saccharomyces*) are more common in shallower water, whilst yeasts belonging to the *Basidiomycetes* are most common in deep waters (e.g. *Rhodotorula* has been isolated from the Marianas Trench at a depth of 11000 m). Marine yeasts have been investigated for their potential for use in aquaculture feeds (as a replacement for live larval food and as probiotic supplements) and for the production of fuel ethanol or single-cell protein from the breakdown of chitin in fisheries waste.

Fungi are also involved in symbiotic interactions with other organisms. Marine lichens are a familiar sight on the rock surfaces in intertidal zones. Lichens consist of an intimate mutualistic symbiotic association between a fungus and an alga or cyanobacterium. The fungus produces the mycelial structure that allows the lichen to attach firmly to rocks, and it also produces compatible solutes that help to retain water and inorganic nutrients from the atmosphere, receiving organic material from the photosynthesis of the algal partner. Marine fungi are also involved in symbiotic associations with salt-marsh plants, through the formation of mycorrhizae, in which the fungal mycelium assists the plant in obtaining nutrients from a wide area.

Some fungi are pathogens of marine animals, such as corals, crustaceans, mollusks, and fish; or of seaweeds, intertidal grasses, and mangrove roots. The skeleton of hard corals is usually penetrated by endolithic fungi, and fungal sequences occur in metagenomic studies of corals (see Chapter 10). These fungi infect the coral shortly after larval settlement, are usually considered parasitic, and are sometimes associated with disease.

Conclusions

This chapter has shown that marine protists are extremely diverse and have many important functions in ocean processes, as well as providing important clues to the history of our oceans through geological time. As well as their role in food webs and biogeochemical cycles, the activities of protists have many significant implications for human welfare, both negative (e.g. harmful or nuisance blooms) and positive (for biotechnological exploitation). The application of new techniques is increasing our recognition of the critical importance of protists and is propelling study of their diversity and properties to the "mainstream" of marine microbiology and oceanography.

References

Abrahams, M. and Townsend, L. (1993) Bioluminescence in dinoflagellates: a test of the burglar alarm hypothesis. *Ecology* 74: 258–260.

Adl, S. M., Simpson, A. G., Farmer, M. A., et al. (2005) The new higher level classification of eukaryotes with emphasis on the taxonomy of protists. *J. Eukaryot. Microbiol.* 52: 399–451.

Bidle, K. D. and Azam, F. (1999) Accelerated dissolution of diatom silica by marine bacterial assemblages. *Nature* 39: 508–512.

Bidle, K. D. and Azam, F. (2001) Bacterial control of silicon regeneration from diatom debris: significance of bacterial ectohydrolases and species identity. *Limnol. Oceanogr.* 46: 1606–1623.

Bidle, K. D., Manganelli, M. and Azam, F. (2002) Regulation of oceanic silicon and carbon preservation by temperature control on bacteria. *Science* 298: 1980–1984.

Brownlee, C. and Taylor, A. R. (2004) Calcification in coccolithophores: a cellular perspective. In: *Coccolithophores:*

From Molecular Processes to Global Impact (eds H. R. Thierstein and J. R. Young). Heidelberg, Springer-Verlag, pp. 31–50.

Diez, B., Pedros-Alio, C. and Massana, R. (2001) Study of genetic diversity of eukaryotic picoplankton in different oceanic regions by small-subunit rRNA gene cloning and sequencing. *Appl. Environ. Microbiol.* 67: 2932–2941.

Fenchel, T. and Patterson, D. J. (1988) *Cafeteria roenbergensis* nov. gen., nov. sp., a heterotrophic microflagellate from marine plankton. *Mar. Microb. Food Webs* 3: 9–19.

Guillou, L., Viprey, M., Chambouvet, A., et al. (2008) Widespread occurrence and genetic diversity of marine parasitoids belonging to Syndiniales (Alveolata). *Environ. Microbiol.* 10: 3349–3365.

Guillou, L., Alves-de-Souza, C., Siano, R. and Gonzalez, H. (2010) The ecological significance of small, eukaryotic parasites in marine ecosystems. *Microbiology Today*, accessed September 14, 2010, http://www.sgm.ac.uk.

Keeling, P. J., Burger, G., Durnford, D. G., et al. (2005) The tree of eukaryotes. *Trends Ecol. Evol.* 20: 670–676.

Klaveness, D., Shalchian-Tabrizi, K., Thomsen, H. A., Eikrem, W. and Jakobsen, K. S. (2005) *Telonema antarcticum* sp. nov., a common marine phagotrophic flagellate. *Int. J. Syst. Evol. Microbiol.* 55: 2595–2604.

Lewitus, A. J., Glasgow, H. B. and Burkholder, J. (1999) Kleptoplastidy in the toxic dinoflagellate *Pfiesteria piscicida* (Dinophyceae). *J. Phycol.* 35: 303–312.

López-García, P., Rodriguez-Valera, F., Pedros-Alio, C. and Moreira, D. (2001) Unexpected diversity of small eukaryotes in deep-sea Antarctic plankton. *Nature* 409: 603–607.

Moon-van der Staay, S., De Wachter, R. and Vaulot, D. (2001) Oceanic 18S rDNA sequences from picoplankton reveal unsuspected eukaryotic diversity. *Nature* 409: 607–610.

Moreira, D. and López-García, P. (2002) The molecular ecology of microbial eukaryotes unveils a hidden world. *Trends Microbiol.* 10: 31–38.

Rogerson, A., Anderson, O. R. and Vogel, C. (2003) Are planktonic naked amoebae predominately floc associated or free in the water column? *J. Plankton Res.* 25: 1359–1365.

Shi, X. L., Marie, D., Jardillier, L., Scanlan, D. J. and Vaulot, D. (2009) Groups without cultured representatives dominate eukaryotic picophytoplankton in the oligotrophic South East Pacific Ocean. *PLoS ONE* 4: e7657.

Stoecker, D. K., Johnson, M. D., de Vargas, C. and Not, F. (2009) Acquired phototrophy in aquatic protists. *Aquat. Microb. Ecol.* 57: 279–310.

Worden, A. Z. and Not, F. (2008) Ecology and diversity of picoeukaryotes. In: *Microbial Ecology of the Oceans* (ed. D. L. Kirchman). Wiley–Liss Inc., New York, pp. 159–205.

Worden, A. Z., Cuvelier, M. L. and Bartlett, D. H. (2006) In-depth analyses of marine microbial community genomics. *Trends Microbiol.* 14: 331–336.

Further reading

Anderson, O. R. (2001) Protozoa, radiolarians. In: *Encyclopedia of Ocean Sciences* (Ed. J. Steele, S. Thorpe and K. Turekian). Academic Press, New York, pp. 2315–2319.

Armbrust, E. V. (2009) The life of diatoms in the world's oceans. *Nature* 459: 185–192.

Caron, D. A., Worden, A. Z., Countway, P. D., Demir, E. and Heidelberg, K. B. (2008) Protists are microbes too: a perspective. *ISME J.* 3: 4–12.

Falkowski, P. G., Katz, M. E., Knoll, A. H., Quigg, A., Raven, J. A., Schofield, O. and Taylor, F. J. R. (2004) The evolution of modern eukaryotic phytoplankton. *Science* 305: 354–360.

Fay, S. A., Weber, M. X. and Lipps, J. H. (2009) The distribution of Symbiodinium diversity within individual host foraminifera. *Coral Reefs* 28: 717–726.

Fenchel, T. and Blackburn, N. (1999) Motile chemosensory behaviour of phagotrophic protists: mechanisms for and efficiency in congregating at food patches. *Protist* 150: 325–336.

Gast, R. J., Moran, D. M., Dennett, M. R. and Caron, D. A. (2007) Kleptoplasty in an Antarctic dinoflagellate: caught in evolutionary transition? *Environ. Microbiol.* 9: 39–45.

Gast, R. J., Sanders, R. W. and Caron, D. A. (2009) Ecological strategies of protists and their symbiotic relationships with prokaryotic microbes. *Trends Microbiol.* 17: 563–569.

Golubic, S., Radtke, G. and Campion-Alsumard, T. L. (2005) Endolithic fungi in marine ecosystems. *Trends Microbiol.* 13: 229–235.

Grossart, H. P., Levold, F., Allgaier, M., Simon, M. and Brinkhoff, T. (2005) Marine diatom species harbour distinct bacterial communities. *Environ. Microbiol.* 7: 860–873.

Haddock, S. H. D., Moline, M. A. and Case, J. F. (2010) Bioluminescence in the sea. *Annu. Rev. Mar. Sci.* 2: 443–493.

Hastings, J. W. (2007) The Gonyaulax clock at 50: translational control of circadian expression. *Cold Spring Harbor Symp. Quant. Biol.* 72: 141–144.

Hyde, K. D. (ed.) (2000) *Fungi in Marine Environments.* Fungal Diversity Press, Hong Kong.

Hyde, K. D., Jones, E. B. G., Leano, E., Pointing, S. B., Poonyth, A. D. and Vrijmoed, L. L. P. (1998) Role of fungi in marine ecosystems. *Biodivers. Conserv.* 7: 1147–1161.

Jezbera, J., Hornak, K. and Simek, K. (2005) Food selection by bacterivorous protists: insight from the analysis of the food vacuole content by means of fluorescence in situ hybridization. *FEMS Microbiol. Ecol.* 52: 351–363.

Moreira, D. and López-García, P. (2003) Are hydrothermal vents oases for parasitic protists? *Trends Parasitol.* 19: 556–558.

Murray, J. (2008) *Ecology and Applications of Benthic Foraminifera.* Cambridge, Cambridge University Press.

Pointing, S. B. and Hyde, K. D. (2000) Lignocellulose-degrading marine fungi. *Biofouling* 15: 221–229.

Pondaven, P., Gallinari, M., Chollet, S., Bucciarelli, E., Sarthou, G., Schultes, S. and Jean, F. (2007) Grazing-induced changes in cell wall silicification in a marine diatom. *Protist* 158: 21–28.

Raghukumar, S. (2008) Thraustochytrid marine protists: production of PUFAs and other emerging technologies. *Mar. Biotechnol.* 10: 631–640.

Rodriguez, F., Derelle, E., Guillou, L., Le Gall, F., Vaulot, D. and Moreau, H. (2005) Ecotype diversity in the marine picoeukaryote *Ostreococcus* (Chlorophyta, Prasinophyceae). *Environ. Microbiol.* 7: 853–859.

Schiebel, R. and Hemieben, C. (2001) Protozoa, plankton foraminifera. In: *Encyclopedia of Ocean Sciences* (eds J. Steele, S. Thorpe and K. Turekian). Academic Press, New York, pp. 2308–2314.

Schoemann, V., Becquevort, S., Stefels, J., Rousseau, V. and Lancelot, C. (2005) *Phaeocystis* blooms in the global ocean and their controlling mechanisms: a review. *J. Sea Res.* 53: 43–66.

Sherr, E. and Sherr, B. (2000) Marine microbes: an overview. In: *Microbial Ecology of the Oceans* (ed. D. L. Kirchman). Wiley–Liss Inc., New York, pp. 13–46.

Sherr, E. B. and Sherr, B. F. (2007) Heterotrophic dinoflagellates: a significant component of microzooplankton biomass and major grazers of diatoms in the sea. *Mar. Ecol. Prog. Ser.* 352: 187–197.

Sherr, B. F., Sherr, E. B., Caron, D. A., Vaulot, D. and Worden, A. Z. (2007) Oceanic protists. *Oceanography* 20: 130–134.

Smetacek, V. (1999) Diatoms and the ocean carbon cycle. *Protist* 150: 25–32.

Wootton, E. C., Zubkov, M. V., Jones, D. H., Jones, R. H., Martel, C. M., Thornton, C. A. and Roberts, E. C. (2007) Biochemical prey recognition by planktonic protozoa. *Environ. Microbiol.* 9: 216–222.

Chapter 7

Marine Viruses

The true abundance and ecological importance of viruses in marine ecosystems and global processes has only been realized in the last two decades. Recent discoveries have led to the emergence of virus ecology as one of the most exciting and fastest developing branches of marine science. The world's ocean is estimated to contain more than 10^{30} viruses—they are the smallest and most abundant members of marine ecosystems. All forms of cellular life—from bacteria to whales—are susceptible to viral infection. Through their interactions with all types of marine organisms, viruses play a critical role in the structuring of marine communities, in ocean processes, and in biogeochemical cycles. In this chapter, the focus is primarily on the properties and activities of marine viruses with respect to their interactions with other members of the plankton, especially bacteria, microalgae, and other protists. Viruses that infect other marine organisms, such as invertebrates, fish, marine vertebrates, or macroalgae are considered in Chapter 11. Human pathogenic viruses that may be introduced into the marine environment via pollution are considered in Chapter 12.

Key Concepts

- Viruses contain either DNA or RNA and are obligate intracellular parasites that take over the biosynthetic machinery of host cells.

- There are more than 10^{30} viruses in the oceans; marine viruses are extremely diverse in host range, size, structure, and genomic composition.

- Viruses that infect bacteria (phages) and other planktonic microbes are responsible for the turnover of microbial communities, with consequent major effects in nutrient cycles and biogeochemical processes.

- Viruses are responsible for structuring the diversity of microbial communities and manipulating life histories through coevolution with their hosts and widespread genetic exchange.

- Analysis of marine viral metagenomes (viromes) shows that groups of viral genotypes are globally distributed, although their relative abundance is affected by local selection factors.

- Studies of marine viromes are contributing to a reappraisal of the nature and evolution of viruses and their role in the living world.

Viruses are extremely diverse in structure and genetic composition

Viruses are noncellular biological entities. Virus particles, termed "virions," are composed of nucleic acid surrounded by a protein coat (capsid). A fundamental difference in the makeup of viruses compared with cells is that viruses contain only one type of nucleic acid, either DNA or RNA, whereas cells contain both. Since all forms of cellular life are susceptible to virus infection, we can reasonably speculate that every type of marine organism is host to at least one type of virus, whilst many organisms are known to host several different viruses. *Table 7.1* shows a list of representative marine virus families and their hosts; as can be seen, virions vary in size from less than 20 nm to more than 650 nm and exist in various morphological forms.

Various schemes have been developed to classify the different types of viruses. In the early days of virology in the mid-twentieth century, viruses were classified largely on the basis of their hosts (plants, animals, or humans). The structure and replication of viruses became the main criteria for classification as knowledge of these accumulated in the 1960s and 1970s. The Baltimore classification scheme devised in the 1970s divides viruses into seven groups, based on the nature of their genome and their replication strategy. Since the 1990s, the International Committee on Taxonomy of Viruses (ITCV) has developed rules for classifying and naming viruses using taxonomic divisions and binomial species names like those used for cellular organisms—the organization reflects the phylogenetic ancestry, structure, replication, hosts, and vectors of different viruses. Thus, at the present time, there are six orders (suffix *-virales*), 87 families (suffix *-viridae*), and about 350 genera (suffix *-virus*). This taxonomic system is well developed in the case of well-characterized plant and animal viruses, but is much less developed in the case of marine viruses. As with other microbes in which there is extensive genetic diversity, the concept of a "species" of virus is problematic. The ITCV provides a formal definition of viral species as a replicating lineage occupying a particular ecological niche. This requires that the virus can be cultured, but as this is not yet possible for the majority of marine viruses, the species epithet is rarely encountered. Individual viral isolates are usually referred to by strain numbers. For example, the abbreviations EhV-1 and EhV-86 are used to designate particular strains (genotypes) of the *Coccolithovirus* genus (family *Phycodnaviridae*) infecting the prymnesiophyte alga *Emiliania huxleyi*. Other virus strains may have informal names or acronyms reflecting their host; for example, cyanophage P60 is a member of the *Podoviridae* that infects the cyanobacterium *Synechococcus*; RDJLΦ1 is a member of the *Siphoviridae* that infects *Roseobacter denitrificans*; and HaRNAV is the acronym for *Heterosigma akashiwo* virus, formally classified as *Marnavirus* in the family *Marnaviridae*.

Viral genomes can be circular, linear, or segmented; with DNA or RNA that is either single- or double-stranded. A key factor in the replication strategy of single-stranded viruses—used in the Baltimore classification scheme—is whether the nucleic acid is positive or negative sense. Positive-sense RNA can serve as viral mRNA directly, whereas negative-sense RNA must be converted to a complementary mRNA by RNA-dependent RNA polymerase. The genome size of viruses varies greatly. The size of viral genomes is very variable, ranging from about 3200 bases in hepadnaviruses to more than 1.2 million base pairs in mimiviruses.

The virion contains identical protein subunits (capsomeres), which self-assemble to form the capsid, a shell-like structure surrounding the nucleic acid. The arrangement of the capsomeres and the overall morphology of the virion when visualized in the electron microscope is a key factor in the identification and classification of viruses. Helical viruses are rod shaped or filamentous, with a single type of capsomere arranged around a central cavity. Negative charges on the nucleic acid bind it to positive charges on the protein helix. Many viruses have an icosahedral structure—a symmetry that is the optimum way of forming a closed shell-like structure for identical subunits. Enveloped viruses are surrounded by an outer membrane containing lipids and carbohydrates derived from host and proteins encoded by the viral genome; the envelope is often involved in the infection of host cells by the virus. Many viruses have complex symmetry combining these different features. The best-known examples of this type are the tailed phages that have an icosahedral head—this contains the viral genome—bound to a helical tail with a base plate with protein fibers, which serves as a molecular syringe for delivery of the viral genome into the cell. An enormous variety of genomic structures can be seen among different viruses. Although we now know that there are millions of different virus types, only about 5000 have been described in detail.

Viruses are the most abundant biological entities in seawater

Early studies of free marine viruses in the plankton were made during the late 1970s and 1980s by direct observation of filtered seawater samples, using transmission electron microscopy (TEM). Samples may be concentrated onto grids by ultracentifrugation and negatively stained with uranyl acetate, or another similar heavy-metal stain. Strictly, we should refer to structures observed in this way as "virus-like particles" (VLPs), because TEM cannot indicate whether the particles contain genetic information and thus whether they are capable of infecting a host cell. Although technically straightforward, observing and enumerating viruses in seawater samples requires great care and control of variables such as the density of suspended particles. Although there were many technical difficulties with sample preparation in early studies, this method led to the gradual realization in the late 1980s that viruses are highly abundant in the marine environment. The TEM method is also very valuable because it shows the morphology of viruses; the majority of free VLPs observed in seawater are phages (see below) having capsids with pentagonal or hexagonal icosahedral three-dimensional symmetry. Both tailed and nontailed forms occur, and appendages such as capsid antennae or tail fibers can sometimes be seen (*Figure 7.1*). They vary in size, with most particles being 30–100 nm in diameter. Some very large virus structures up to 750 nm are also observed. The method of sample preparation has a great influence on the observed morphology, and it is generally easier to obtain details of morphology in viruses from laboratory-cultured hosts—in the few cases in which this is now possible—than in free VLPs observed directly in seawater samples.

Another approach developed for enumerating VLPs in sea water is an adaptation of epifluorescence microscopy, in which the sample is treated with a fluorochrome that binds to nucleic acid (see *Table 2.1*). Brightly fluorescent stains such as Yo-Pro-1 and SYBR Green® have been used successfully in many studies, with VLPs appearing as very small bright dots that can be distinguished from the larger stained cells of bacteria and other microscopic plankton (see *Plate 7.1*). This is a relatively rapid and inexpensive method, although caution is needed to avoid overestimating viral abundance because free nucleic acid bound to colloidal particles might appear as fluorescent dots; conversely, large VLPs might be confused with bacteria, resulting in an underestimation. This method may also result in underestimations of viral density because RNA viruses and single-stranded DNA viruses do not stain well with current methods. Since viruses are below the limits of resolution of the light microscope, fluorescence microscopy provides no information about the morphology of viruses—indeed the only reason that they can be seen with this method is because of the halo of bright of light emitted by the fluorochromes.

Based on numerous studies of seawater from various geographic sites and different depths using TEM and epifluorescence microscopy since the 1990s,

Table 7.1 Examples of viruses infecting marine organisms

Virus family	Morphology of virion	Size of virion (nm)[a]	Host(s)
Double-stranded DNA viruses			
Baculoviridae	Enveloped rods, some with tails	200–450 × 100–400	Crustaceans
Corticoviridae, Tectiviridae	Icosahedral with spikes	60–75	Bacteria
Herpesviridae	Pleomorphic, icosahedral, enveloped	150–200	Mollusks, fish, corals mammals, turtles
Iridoviridae	Round, icosahedral	190–200	Mollusks, fish
Lipothrixviridae	Thick rod with lipid coat	40 × 400	Archaea
Mimiviridae	Icosahedral with microtubule-like projections	650	Protists, corals (?), sponges (?)
Myoviridae	Polygonal head (icosahedral) with contractile tail (helical)	50–110 (head)	Bacteria
Nimaviridae	Enveloped, ovoid with tail-like appendage	120 × 275	Crustaceans
Papovaviridae	Round, icosahedral	40–50	Mollusks
Phycodnaviridae	Icosahedral	130–200	Algae
Podoviridae, Siphoviridae	Icosahedral with noncontractile tail	60 (head)	Bacteria
Single-stranded DNA viruses			
Microviridae	Icosahedral with spikes	25–27	Bacteria
Parvoviridae	Round, icosahedral	20	Crustaceans
Double-stranded RNA viruses			
Birnaviridae	Round, icosahedral	60	Mollusks, fish
Cystoviridae	Icosahedral with lipid coat	60–75	Bacteria
Reoviridae	Icosahedral, some with spikes	50–80	Crustaceans, mollusks, fish, protists
Totiviridae	Round, icosahedral	30–45	Protists
Single-stranded RNA viruses (positive sense)			
Caliciviridae	Round, icosahedral	35–40	Fish, mammals
Coronaviridae	Rod-shaped with projections	200 × 42	Crustaceans, fish, seabirds
Dicistroviridae	Round, icosahedral	30	Crustaceans
Leviviridae	Round, icosahedral	26	Bacteria
Marnaviridae	Round, icosahedral	25	Algae
Nodaviridae	Round, icosahedral	30	Fish
Picornaviridae	Round, icosahedral	27–30	Algae, crustaceans, thraustochytrids, other protists (?), mammals
Togaviridae	Round, with outer fringe	66	Fish
Single-stranded RNA viruses (negative sense)			
Bunyaviridae	Round, enveloped	80–120	Crustaceans
Orthomyxoviridae	Round, with spikes	80–120	Fish, mammals, seabirds
Paramyxoviridae	Various, mainly enveloped, filamentous	60–300 × 1000	Mammals
Rhabdoviridae	Bullet-shaped with projections	45–100 × 100–430	Fish

[a]For rod-shaped virions, dimensions are shown as diameter × length.

a consensus value for the density of viruses in seawater of about 10^7 per milliliter has emerged. However, there is considerable seasonal and geographic distribution, and it is hard to generalize to all depths and locations. In general, the viral abundance increases with the productivity of the system. In the open ocean, virus density declines rapidly below a depth of 250 m to a relatively constant value of about 10^6 per milliliter. Counts in highly productive coastal waters are usually higher than in the open ocean and are not so dependent on depth, with typical densities of 10^8 per milliliter. Viruses are also found in high densities in marine sediments and sea ice. The distribution of viruses in the water column generally mirrors the productivity and density of host populations of bacterioplankton as these are the most abundant hosts. As a rough approximation, there are about 10 times as many viruses as bacteria in most seawater samples. In the photic zone of very oligotrophic waters, cyanophages infecting photosynthetically active cyanobacteria such as *Prochlorococcus* and *Synechococcus* form the dominant group. The distribution of viruses infecting eukaryotic algae also generally mirrors the abundance of the population of algal hosts, but this is a small proportion of the total. Changes in the density of viral particles are very dynamic, and large fluctuations can occur over short timescales as a result of synchronized lysis of host cells and degradation of released virus particles. This emphasizes the important role of viruses as an active component of marine microbial communities, which will be discussed further in Chapter 8.

Recently, many experimental and field studies have used flow cytometry (see *Figure 2.3B*) as a high-throughput method to enumerate viruses alongside the community structure of their potential hosts in the plankton. It is possible to discriminate various host and viral populations based on their fluorescence and scatter signals after staining with fluorochromes such as SYBR Green®, and this can be applied to analysis of seasonal and spatial variation in viral dynamics in mesocosm and open-water studies. Recent improvements include the use of virus-specific probes for detection of infected cells.

Phages are viruses that infect bacterial and archaeal cells

When viruses that infect bacteria were discovered in 1915, they were called bacteriophages—literally "bacteria-eaters." Viruses that infect archaeal cells have also been discovered, but no specific term to denote them has yet been coined. Therefore, the abbreviated term "phage" is commonly used to indicate viruses infecting members of either domain. Prefixes indicating a particular host range are also used: terms such as "vibriophages" (infecting *Vibrio* spp.), "cyanophages" (infecting cyanobacteria), or "roseophages" (infecting members of the *Roseobacter* clade) are frequently used. The most commonly observed phage structures are tailed virions, comprising a head (nucleocapsid) connected to a tail structure used for adsorption to the cell and injection of the nucleic acid. These phages may be classified into a number of major groups, namely the *Myoviridae* (with a contractile tail; *Figure 7.1C*), the *Podoviridae* (with a very short tail; *Figure 7.1D*), and the *Siphoviridae* (with a long flexible tail). However, a wide range of other morphologies occur in marine phages, as shown in *Table 7.1*. To date only about 50 viruses infecting members of the *Archaea* have been described, mostly infecting extreme thermoacidophiles or halophiles. Phages have not yet been described for archaeal groups with major importance in sediments and ocean processes, such as the methanogens and crenarchaeotes (see Chapter 5).

Although the first marine phages were described in the 1950s, the significance of early findings was not appreciated and it was not until the late 1980s that serious attention was paid to this field. The relatively slow development of this area of research may be linked to the long-held fallacy that microbial populations in the oceans are insignificant and that, by association, viruses must also be unimportant. As discussed in Chapter 4, early marine microbiologists seriously

---

(i) "PLANET VIRUS"— 75 MILLION BLUE WHALES AND 10 MILLION LIGHT YEARS

The consensus estimates for viral density in different waters allow us to calculate the estimated total number of viral particles in the ocean. The total volume of the oceans is about 1.3×10^{21} liters, and assuming that the average abundance of viruses is 3×10^9 per liter, then the total number of viruses in the oceans is approximately 4×10^{30} (Suttle, 2005). This is about 10–15 times the total number of bacterial and archaeal cells, making viruses the most abundant biological entities in the ocean, comprising about 94% of all nucleic-acid-containing particles; although because of their small size, they only account for 5% of the biomass (Suttle, 2007). Nevertheless, Suttle calculates that the total carbon reservoir contained in marine viruses is 200 million tonnes (2×10^{11} kg)— the equivalent of 75 million blue whales—and that if the viruses were stretched end to end, they would span 10 million light years.

underestimated the abundance and diversity of bacteria because of reliance on inappropriate culture methods. The classical method of enumerating infective phages is based on the formation of plaques of lysis in lawns of susceptible hosts grown on agar plates (as shown in *Figure 7.2*) or by dilution in liquid media and enumeration by the most probable number method. In liquid culture, reduction in the turbidity of cultures can be measured as an indicator of cell lysis. These methods can be used with phages that infect easily cultivated heterotrophic and phototrophic bacteria and protists, but are not yet possible for the great majority of bacteria and archaea, which have not been cultured. It was not until reliable methods for direct observation and molecular analysis were developed in the study of both marine bacteria and their phages that it was realized that, far from being insignificant players in marine systems, phages play a critical and central role in ocean ecology and food webs.

The life cycle of phages shows a number of distinct stages

The first step in the life cycle of phages is adsorption to the host cell surface, which may be reversible, followed by irreversible binding to a specific receptor on the host cell surface. For those marine phages that have so far been propagated in cultivated bacteria, most show specificity for particular bacterial species

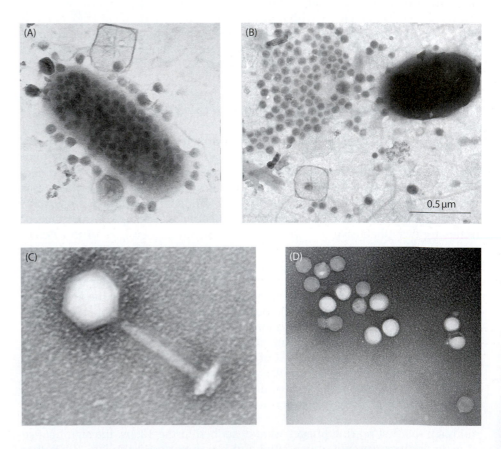

Figure 7.1 Transmission electron micrographs of phages. (A) Mature phages inside a bacterium before lysis. (B) Free viruses after lysis. (C) Myovirus infecting *Prochlorococcus*. (D) Podovirus infecting *Prochlorococcus*. (Images (A,B) courtesy of Mikal Heldal and Gunnar Bratbak, University of Bergen. (Images (C,D) © Chisholm Lab, Massachusetts Institute of Technology.)

Figure 7.2 The plaque assay for lytic phages. (A) The bacterial host is grown in broth culture and mixed with a sample containing the phage. This is resuspended in molten top agar, which contains a low concentration of agar so that the phage can diffuse through the gel to infect adjacent bacteria when poured onto a plate of agar medium containing appropriate nutrients for growth of the host. After incubation, clear areas of lysis appear as plaques in the lawn of bacterial growth. By plating different dilutions of the phage suspension, the density of infective phage can be calculated as plaque-forming units (PFU) per milliliter, analogous to the viable count of bacteria. Similar techniques can be used for the propagation of viruses infecting algae and other protists, providing suitable media for growth are available. (B) Photograph of culture plate showing plaque assay of a myovirus on a cyanobacterial host. (Image courtesy of Andrea Highfield, Marine Biological Association of the UK.)

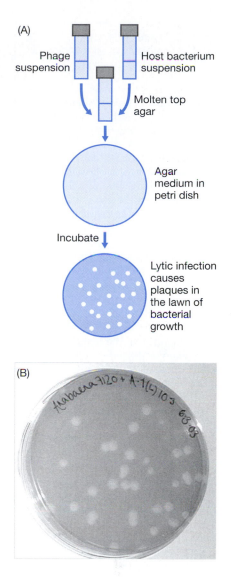

and sometimes for particular strains. These findings are similar to those seen in other well-known viruses and are usually thought to be due to the molecular specificity of virus receptors on the host cell surface, the presence of restriction enzymes, or compatibility of the replication processes. However, there are indications that these highly specific interactions may be something of an artifact introduced by the assay system *in vitro*, and some marine phages probably have a relatively broad host range in the natural environment. For example, some cyanophages have been shown to infect both *Synechococcus* and *Prochlorococcus* and some vibriophages infect several species of *Vibrio*. If correct, this has significant implications for the possibility of genetic exchange between different organisms and for the role of phages in determining bacterial community structure. Very little is known about the nature of the receptors for phage adsorption to marine bacteria. It is possible that broad host range phages target conserved amino-acid sequences of proteins on the cell surface of different bacterial types and that their tail fibers can recognize more than one type of receptor.

In most cases, enzymes in the tail or capsid of the phage attack the bacterial cell wall, forming a small pore through which the nucleic acid enters. The genetic material then remains in the cytoplasm or is integrated into the host cell genome. This is followed by expression of phage proteins, genome replication, and formation of the capsids and other parts of the virion. When assembly is complete, most phages cause lysis of the host cell by producing enzymes that damage the cytoplasmic membrane and hydrolyze the peptidoglycan in the cell wall.

The "burst size" of phage infection is a term used to describe the average number of progeny virus particles released from the host cell at the time of lysis. Knowing this value is important for modeling the dynamics of viral infection of host populations. The larger the burst size, the smaller the number of host cells that need to be lysed to support viral production. Burst size can be measured directly by observing visibly infected cells using TEM (see *Figure 7.1B*), or it can be calculated using models of virus: host ratios and theoretical rates of contact and infection required to maintain virus production. In some cases, lysis can also be induced artificially by the addition of the antibiotic streptomycin. There is a wide variation in results, depending on the location of the study and the methods used, but values in the range of 10–50 progeny viruses per cell are typical for *in situ* studies in natural communities. For those phages that can be propagated in laboratory cultures of bacteria, the burst sizes are usually much larger (average about 180) because *in vitro* grown bacteria are bigger and support greater virus densities. In *Synechococcus*, a large burst size of about 100–300 occurs.

Lysogeny occurs when the phage genome is integrated into the host genome

Virulent phages take over the host cell, replicate their nucleic acid and cause lysis of the host cell after assembly of the virus particles. However, another outcome is seen when phages known as temperate viruses infect the cell (*Figure 7.3*). Bacteria can enter into a symbiotic state in which the phage genome replicates along with the host DNA, but is not expressed. Often, the silent viral genome is stably integrated into the bacterial genome; this latent state is known as a prophage. Bacteria infected with these phages are known as lysogenic, because under certain conditions the bacteria spontaneously lyse and release infective virus particles. It is thought that about 1 in 10^5 lysogenic cells will spontaneously revert to the lytic cycle, so that there is a constant low-level production of temperate phages. Experimental conditions such as exposure to ultraviolet light, temperature shifts, treatment with antibiotics (mitomycin C is the most commonly used agent), or certain other chemicals will often induce a large proportion of cells to enter the lytic cycle. The genetic control of mechanisms that determine whether the phage enters the lytic or lysogenic cycle and the process by which expression of the prophage genes controlling the lytic pathway switch is repressed have been well studied in λ-bacteriophage of *Escherichia coli* and a few other examples. However, the molecular events in lysogeny in marine bacteria are largely unexplored and complete systems—temperate phage, lysogenic host, and uninfected host—have only been described for a few marine bacteria.

Integration and excision of the phage genome has very important evolutionary consequences because it provides a natural mechanism, known as specialized transduction, by which specific host genes can be transferred from one cell to another when part of the host DNA becomes incorporated into the mature virus particle. The presence of a prophage also confers resistance to infection of the host bacterium by viruses of the same type (phage immunity) because repressor proteins, which prevent replication of the prophage genome, also prevent replication of incoming genomes of the same virus. Prophage genes can also affect the phenotypic characteristics of the host cell in other ways, including general reproductive fitness or alteration of the phenotype through expression of different genes. An important example of phenotypic modification or lysogenic conversion is the role of phages in the acquisition of virulence factors enabling colonization and toxicity in *Vibrio cholerae*, a bacterium found in coastal and estuarine waters and responsible for the human disease cholera (see p. 260). Phage-mediated virulence may be very widespread in other *Vibrio* spp. (see p. 235) and other pathogenic bacteria.

Lysogenic immunity and phage conversion both carry a strong selective value for the host cell and could provide phages with a strategy to survive periods of low host density or low metabolic activity when the probability of encountering a host and initiating a lytic factor is low. Since many marine environments contain low levels of slow-growing bacterioplankton and because free viruses are inactivated quite quickly, virulent lytic phages could become rapidly depleted. For all these reasons, it would be reasonable to consider lysogeny to be a very common state of virus infections in nature. Indeed, some studies with cultivable bacteria have shown that lysogeny is more common in bacteria (up to 50%) in samples from offshore environments than nutrient-rich coastal environments, and in studies of deep-water communities the frequency of lysogenic cells appears to be negatively correlated with bacterial production. However, some other studies have produced conflicting results and we have no knowledge at all about the importance of lysogeny in microbes that have not yet been cultured. Another form of hidden infection may occur, in which the viral nucleic acid remains in the host cell for an extended period, but the lysis of the host cell is delayed. Such pseudolysogeny may be due to nutrient deprivation and low metabolism of the host cell and so might also be a common trend in oligotrophic marine waters. It

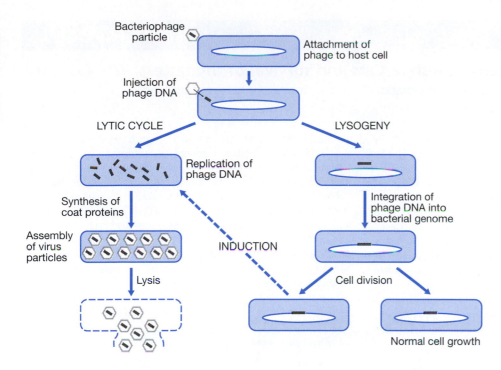

Figure 7.3 Possible outcomes of infection of a bacterial cell by a temperate DNA bacteriophage. If the phage DNA is incorporated and replicated with the host genome it is known as a prophage; the host cell is said to be lysogenic because it may be triggered to enter the lytic cycle under certain conditions.

is hoped that the application of new techniques will reveal the true importance of lysogeny and pseudolysogeny in marine virus–host interactions.

Large DNA viruses are important pathogens of planktonic protists

VLPs have been observed by TEM in many species of protists, although detailed study of host–virus interactions is so far restricted to a few cases. The *Phycodnaviridae* is a family of large DNA viruses that infect freshwater and marine algae including *Chlorella*, coccolithophores and other prymnesiophytes, prasinophytes, and raphidophytes. Phycodnaviruses infecting the coccolithophorid *Emiliania huxleyi* have major effects on the dynamics of *E. huxleyi* blooms (see *Figure 6.5*), with viral lysis leading to population crashes with major ecological and biogeochemical consequences, as discussed in *Research Focus Box 7.1*. Phycodnaviruses also regulate populations of other bloom-forming prymnesiophytes, including *Phaeocystis globosa* and *Chrysochromulina* spp. HaV is a phycodnavirus that infects the toxic bloom-forming raphidophyte *H. akashiwo*. Recently, the genome sequences of two phycodnaviruses infecting the smallest-known photosynthetic eukaryote, *Ostreococcus tauri*, have been described. This picoplanktonic prasinophyte alga is found throughout the oceanic euphotic zone and there are distinct ecotypes of this organism adapted to different light conditions. The ecological significance of the virus is as yet unclear, and it will be interesting to determine the specificity of *Ostreococcus* viruses and their coevolution with their hosts.

Evolutionary analysis of the genomes of several phycodnaviruses shows that they belong to a monophyletic group of large DNA viruses infecting eukaryotes—the nucleocytoplasmic large DNA viruses (NCLDVs). Other viruses within this group

BOX 7.1   RESEARCH FOCUS

Red Queens, Cheshire Cats and survival of the fattest
New insights into interactions between a virus and its algal host

"Boom and bust"

Natural blooms of *Emiliania huxleyi* develop annually in many coastal and ocean waters, before collapsing with the release of millions of coccoliths. The blooms that appear annually off the coast of Plymouth in the English Channel (see *Figure 6.5A*) have been investigated intensely in recent years, and the virus–host systems have been explored in the natural setting, in laboratory experiments, and in mesocosms (see *Figure 2.2*). Several different strains of the virus have been characterized from different locations (Schroeder et al., 2002; Wilson et al., 2002) and shown to possess some of the largest viral genomes known (>400 kbp; Wilson et al., 2005). There are differences between isolates of both the alga and the virus, with host succession appearing to show "kill the winner" dynamics (Thingstad, 2000). Studies of community dynamics and genetic diversity structure in mesocosms were carried out in different years using PCR-DGGE analysis of genes encoding the calcium-binding protein (GPA) in *E. huxleyi*, and the major capsid protein (MCP) gene in *Emiliania huxleyi* viruses (EhVs) (Martinez et al., 2007; Schroeder et al., 2003). This showed a periodic annual succession of identical *E. huxleyi* genotypes, which in turn determines the succession of viral genotypes. The dominant strains appear to survive between blooms. By monitoring the MCP and GPA genes every 4 h during a bloom, Sorensen et al. (2009) showed that the viral community was highly dynamic, with appearance and disappearance of different viral genotypes occurring at very short intervals.

An evolutionary "arms race"

These studies suggest that there is a constant coevolution of the host and virus—an "arms race" in which there is continual selection pressure for the host to evolve resistance and for the virus to evolve to overcome that resistance. In ecology, this is referred to as the "Red Queen Hypothesis," so named because in Lewis Carroll's *Alice in Wonderland: Through the Looking Glass*, the Red Queen says, "It takes all the running you can do, to keep in the same place." Until the study by Frada et al. (2008), all studies on *E. huxleyi* had focused on the diploid cells, which are coated in calcified coccoliths. Noncalcified motile haploid cells also occur, but were overlooked until flow cytometric analysis showed the appearance of smaller cells in populations of *E. huxleyi* undergoing viral lysis (Jacquet et al., 2002). Frada et al. (2008) showed that three strains of cultured diploid *E. huxleyi* were all sensitive to five EhV strains, whereas haploid stages were not infected. Electron microscopy and PCR amplification of the gene encoding MCP revealed that viruses did not adsorb to haploid cells, perhaps because a different cell surface structure prevents the enveloped virus from fusing with the cytoplasmic membrane (see *Figure 7.4A*). Additional experiments showed that the motile, haploid cells only appeared in the late stages of infected cultures, indicating that viral infection triggers meiosis. The

authors termed this strategy of escape from infection as the "Cheshire Cat" phenomenon—in reference to the fictional cat in *Alice*, which disappears and appears without its head—and discuss the implications of releasing host evolution from pathogen pressure, acting to counter the "Red Queen" coevolutionary arms race. This may explain why *E. huxleyi* can "afford" to show boom and bust dynamics—the Cheshire Cat strategy creates a reservoir of resistant haploid cells that may persist in the environment for some time, eventually mating to produce new *E. huxleyi* genotypes. Thus, selection for sexual reproduction as an antiviral mechanism helps to maintain high diversity that helps to buffer the effects of environmental change.

Sphingolipids and cellular suicide

The genome of EhV86 contains a cluster of seven genes encoding an almost complete sphingolipid biosynthesis pathway (SBP) that has been transferred during evolution from the host genome, with subsequent small modifications. One of the key products of the SBP is ceramide, a membrane sphingolipid that is known to be an important signaling molecule which regulates the cell cycle, differentiation, and programmed cell death (PCD; Lane, 2008). Allen et al. (2006) showed that the viral SBP genes are expressed during infection. Bidle et al. (2007) showed that the SBP activates metacaspases and that inhibition of these enzymes prevents viral production because activity of some of the viral proteins appears to depend on specific cleavage. Perhaps PCD evolved as a defense mechanism, which serves to limit viral spread within a host population, but has been subverted by the pathogen? Vardi et al. (2009) showed the accumulation of unique glycosphingolipids in infected cells, which seem to act as a trigger for release of virions. These authors also suggest that release of the glycosphingolipids acts to initiate PCD in nearby *E. huxleyi* cells, leading to the rapid termination of a bloom. Pagarete et al. (2009) designed host-specific and virus-specific primers for use in quantitative PCR (see p. 44) to follow the transcription of SBP genes, showing that the virus induces a switch from the normal host pathway to a massive increase in transcription of the viral genes as the infection progresses. They speculate that this may be another facet of the Red Queen versus Cheshire Cat scenario. Perhaps the host sphingolipids are involved in triggering meiosis to generate haploid cells, and the viruses counter this by producing slightly different specific sphingolipids that prevent the host from switching on this defense mechanism. Another theory is that the sphingolipids assemble in the cell membrane to provide focal points (lipid rafts) that facilitate the release of the virions through budding. The authors conclude their paper by providing a twist on the phrase so often associated with Darwin's evolutionary ideas: "in the case of the *E. huxleyi*/coccolithovirus sytem ... the survival of the fittest may actually be dependent on survival of the fattest."

include poxviruses, iridoviruses, and mimiviruses, including a range of plant and animal pathogens. These viruses replicate inside the nucleus and/or cytoplasm and their large genomes encode many genes of unknown function. Unlike many viruses, NCLDVs seem to be able to carry out several biosynthetic pathways independent of the host cell's replication machinery. Genome analysis of the NCLDVs shows that they have a very ancient origin and coevolution with their hosts— divergence of the major families of NCLDVs appears to have occurred prior to the radiation of the major eukaryotic lineages 2–3 billion years ago. As discussed in *Research Focus Box 7.2*, remarkable new insights into the origin and evolution of these viruses and their importance in marine ecosystems have followed the recent description of a giant virus called mimivirus and the discovery that related viruses seem to be very abundant in marine ecosystems as possible pathogens of eukaryotic plankton.

The mechanism of infection of host cells varies amongst the NCLDVs, with the mechanisms of entry and exit showing large variations. Among the phycodnaviruses, *Chlorovirus* possesses a lipid-containing membrane (underlying the capsid), which fuses with the host cell membrane after enzymic digestion of the algal cell wall, followed by injection of the viral DNA and associated proteins. *Phaeovirus* also fuses with the host cell, in this case infecting the spores or gametes, which lack a cell wall. Recently, it has been discovered that the *E. huxleyi* virus infects cells using a mechanism that resembles that seen in enveloped animal viruses, involving entry of the entire nucleocapsid into the cytoplasm following either fusion of the envelope with the cell membrane or an endocytotic process, as shown in *Figure 7.4*. Together with genetic evidence, this suggests that the coccolithoviruses should be reclassified into a new family.

Photosynthetic protists are also infected by RNA viruses

It is generally assumed that DNA viruses are the most dominant members of the virioplankton, but this assumption may be influenced by methodological constraints because RNA viruses with small genomes are more difficult to observe and enumerate using electron microscopy or flow cytometry. Nevertheless, RNA viruses appear to be very diverse, and numerous ecologically important examples occur in the major virus groups, as shown in *Table 7.1*. PCR-based screening of seawater samples using conserved sequences of the RNA-dependent RNA polymerase gene has demonstrated that distinct groups of picorna-like viruses are abundant, widespread, and persistent. Metagenomic analysis based on reverse-transcribed shotgun sequencing has shown that most sequences are unrelated to known RNA sequences. Many RNA viruses are well known because of their economic importance as pathogens of marine fish and shellfish, and, until recently, there has been little information about their role in marine ecology as pathogens of planktonic microbes. Only one RNA phage infecting bacteria is known, but several RNA viruses infecting protists have now been described. A single-stranded (ss)RNA virus called HaRNAV in the family *Reoviridae* infects the raphidophyte *H. akashiwo* and has been investigated extensively in view of the importance of controlling blooms of this organism, which is responsible for large fish kills. Electron microscopy of infected cells shows that they may contain as many as 2×10^4 viral particles. Genome sequencing led to the definition of a new family called the *Marnaviridae*. These viruses seem to be widespread and may play an important role in population dynamics of other marine protists. A double-stranded (ds)RNA virus has been isolated from the prasinophyte *Micromonas pusilla*.

Diatoms are of major importance in ocean food chains, being one of the main components of the phytoplankton responsible for primary productivity, especially in coastal waters and at high latitudes. One might imagine that the silica frustule would protect diatoms from viral infection, but viruses have also recently been identified as lytic agents of bloom-forming diatoms. Small viruses

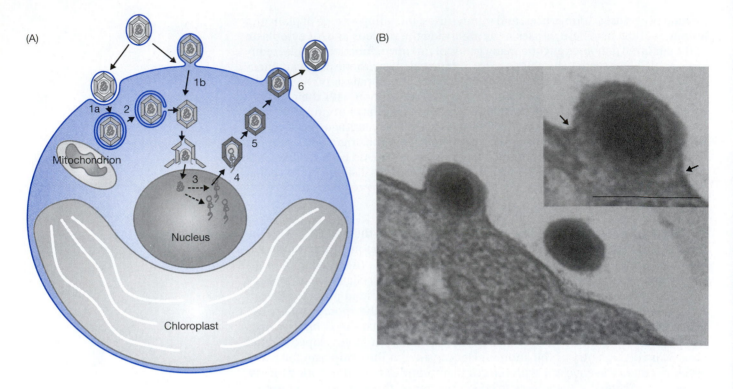

Figure 7.4 Stages in the proposed life cycle of an *Emiliania huxleyi* virus (EhV86). (A) Entry may occur by either: (1a) endocytosis and internalization within a vacuole, with (2) subsequent release; or (1b) fusion of the envelope with the cell membrane. (3) Release of viral genome and expression of early genes in the nucleus. (4) Expression of mid-late genes and capsid assembly in the cytoplasm. (5) Transport of early assembled virions to the plasma membrane. (6) Release by budding. (B) Electron micrograph showing release of a viral particle via a budding process. Bar = 100 nm. (Reproduced from Mackinder et al. [2009] by permission of the authors and Society of General Microbiology.)

presumably enter via pores in the frustule. Novel ssRNA viruses were isolated from *Rhizosolenia setigera*, in temperate coastal waters. Another ssRNA virus has also been described in *Chaetoceros* spp., where infection leads to a burst size of more than 10^4 virions per cell. As infection of diatoms is likely to have a major impact on primary production and ocean processes, further study of the role of viruses in population dynamics is necessary. For example, we do not know if viral infection is involved in the collapse of diatom blooms, which is a major contributor to the formation of marine snow and the cycling of silica.

Dinoflagellates form another major group of the phytoplankton of special importance in coastal waters, with many forming harmful blooms. Two distinct types of ssRNA virus (HcRNAV) and an icosahedral dsDNA virus (HcV) have now been described from *Heteroplasma circularisquama*. There appears to be a clear correlation between density of HcRNAV and host population dynamics. Again, the burst size of infected cells seems to be very large.

The role of viruses as pathogens of heterotrophic protists remains unclear

Densely packed small icosahedral VLPs have been observed in myxosporean parasites of fish, and putative viruses have been described as pathogens of sarcodines, rhizopods, and radiolarians. VLPs have also been observed at high densities in heterotrophic and mixotrophic protists in Arctic sea-ice brines. A novel icosahedral ssRNA virus (possibly a picornavirus) that affects the thraustochytrid protist *Schizochytrium* sp. has been described; since thraustochytrids have special importance in marine ecology because of their high content of omega-3 polyunsaturated fatty acids, viral infection of this group could be very significant in ocean food webs.

The heterotrophic nanoflagellates *Bodo* sp. and *Cafeteria roenebergensis* have both been shown to be infected by large dsDNA viruses, which lyse the cells and cause a population crash in culture. Although uncharacterized, based on size and morphology, these viruses may be related to the NCLDVs, for which gene sequences are highly abundant in marine metagenomes (see *Research Focus Box 7.2*).

Loss of infectivity of viruses arises from irreparable damage to the nucleic acid or protein capsid

Usually, a virus will lose its infectivity before showing obvious signs of degradation. However, since most marine viruses are studied by microscopy or flow cytometry, the term "virus decay" reflects the observation of a decline in numbers of VLPs over time in the absence of new viral production. Many of the studies of virus inactivation in water were originally carried out in connection with the health hazards associated with sewage-associated viruses (such as enteroviruses or coliphages; p. 291) in waters for swimming or cultivation of shellfish. Subsequently, the results of these studies have been applied to the population dynamics of autochthonous marine viruses. A wide range of physical, chemical, and biological factors can influence virus infectivity. Different studies have produced various estimates of decay rates, but a value of about 1% per hour is typical in natural seawater kept in the dark. Visible light and ultraviolet irradiation are by far the most important factors influencing survival, and in full-strength sunlight the decay rate may increase to 3–10% per hour, and in some circumstances can be as high as 80%. Light will have its greatest effect in the upper part of the water column, but is probably still effective down to about 200 m in clear ocean water. Even in very turbid coastal waters, viral inactivation by light can be observed down to several meters. Such high rates of inactivation would lead to the conclusion that there are no or very few infective viruses in the top layer of water. However, this conclusion is incorrect, probably because repair of UV-induced damage can occur by mechanisms encoded either by the host or the virus. Another important factor in decay is the presence of particulate matter and enzymes such as proteases and nucleases produced by bacteria and other members of the plankton. Here, there are complex interactions, because adsorption of viruses to particles can also afford some protection. Virus inactivation does not proceed at a constant rate and it appears that there is variation in the resistance of viral particles to damaging effects, presumably because of minor imperfections in the capsid. Thus, over time, inactivation will lead to a slowly decaying low level of infective particles.

Measurement of virus production rates is important for assessing the role of virus-induced mortality

Many studies in viral ecology have attempted to measure the effect of viruses on microbial mortality and the rate of production of new virus particles. This enables estimates of the proportion of primary and secondary production that is "turned over" by viral lysis. It is possible to use filtration and high-speed centrifugation to obtain a pellet of planktonic microbes, which can be embedded in resin and sectioned for TEM. By examining such samples from various locations, it has been found that about 1–4% of microbial cells contain mature, fully assembled VLPs. Since VLPs can only be seen within infected host cells in the final stages of the lytic cycle of infection—this stage usually represents about a quarter of the life cycle—it is possible to estimate the total proportion of plankton that are infected at any one time. Another method used in early studies of virus production was to measure the incorporation of radioactively labeled precursors such as ^{3}H-thymidine, ^{14}C-leucine, or ^{32}P-phosphate into virus particles, which can be separated for cells and cell debris by filtration. More recently, a virus dilution (reduction) method has been developed and applied in numerous studies. In this method, free virus particles are removed from a water sample by filtration and the number of new particles occurring in the water over time is measured by epifluorescence microscopy or flow cytometry. After adjustments for the relative abundance of host cells in the sample, the production rate and the original percentage of infected cells can be calculated. Each of these methods has advantages and disadvantages and no single method gives precise estimates of virus-mediated mortality. Nevertheless, despite variability depending on the method, location, and time of sampling, the unequivocal conclusion from various studies is that viral mortality has a highly significant impact on mortality of microbes that is at least as significant as grazing by protists and zooplankton. A consensus value is that about 20–40% of marine bacteria are killed each day by viral action. In the case of algae, laboratory experiments and mesocosm studies have indicated that viral infection can account for up to 100% of mortality of bloom-forming algae such as *E. huxleyi*. Using modified dilution methods to distinguish the effects of mortality due to viruses and protistan grazing, viral mortality of *Micromonas pusilla* has been estimated at 9–25% standing stock per day and in this case it is likely that there is a more stable coexistence of viruses with their algal hosts. As with bacteria, precise values for the effects of viral mortality on natural populations of algae are hard to obtain, but there is no doubt that it has a very significant impact on primary production.

Viral mortality "lubricates" the biological pump

Viral infection of heterotrophic and autotrophic bacteria, archaea, algae, and other protists seems to be the most important factor influencing nutrient cycles in the oceans, because it leads to the release of massive amounts of organic material and essential elements from these microbes into the dissolved organic pool from where it reabsorbed and metabolized by heterotrophs. Because the contents of cells lysed by viruses are rich in nitrogen and phosphorus, this "viral shunt" speeds up the recycling of nutrients, enhances the rate of microbial respiration, and reduces the amount of organic material available to higher trophic levels through protistan grazing in the microbial loop (see *Figure 8.2*). Furthermore, the release of high-molecular-weight polymeric substances from viral lysis contributes to the aggregation of algal flocs and marine snow (see *Figure 1.9*). Thus, viral lysis has complex effects on the export of carbon compounds to the ocean floor via the biological pump). The importance of these processes is considered further in the Chapter 8.

Viral mortality plays a major role in structuring diversity of microbial communities

As well as influencing microbial diversity by horizontal gene transfer and genetic rearrangements between host and virus, viruses influence host diversity at ecological scales. The discovery in the 1990s that phages are so numerous and responsible for high levels of bacterial mortality led to the development of models to explain how this might be important in influencing community structure. These approaches have subsequently extended to interactions between viruses and algae. As in all predator–prey relationships, the probability of an encounter is more likely at high host densities because increased contact rates between a virus and its host are determined by their relative abundances. Also, some viruses may be very specific, and small changes in the surface receptors or replication

---

(i) THE DEEP IMPACT OF THE VIRAL SHUNT

Biogeochemical cycles in deep-sea ecosystems—which cover about 65% of the Earth's surface—are determined by the activities of bacteria and archaea that largely depend on organic matter exported from the ocean surface. Microbial life is abundant in deep-sea sediments and constitutes a major fraction of the total microbial carbon on Earth, but most of this carbon appears to be unused by higher trophic levels. This paradox is explained by the discovery by Danovaro et al. (2008) that viruses are the main agents of mortality of bacteria and archaea in deep-sea sediments. They studied viral activity at 232 sites at different locations and depths—showing that over 80% of the heterotrophic production is shunted by viral lysis into labile dissolved organic matter, producing between 0.37 and 0.63 gigatonnes of carbon a year. Viruses therefore play a major role in deep-sea processes by injecting huge amounts of organic material into the ecosystem—stimulating microbial metabolism and accelerating biogeochemical processes.

processes of the hosts could make them resistant to infection by a particular virus genotype. (But note the earlier discussion suggesting that some phages probably have a broad host range.) The density dependence and host range factors mean that viruses should preferentially infect the most common hosts: abundant hosts are more susceptible and rare hosts are less susceptible. Therefore, viruses may control excessive proliferation of hosts that have an advantage in nutrient acquisition and growth, encouraging a high diversity in the host population so that less competitive but virus-resistant microbes survive. This has been developed into a mathematical model termed the "kill the winner hypothesis."

Marine viruses show enormous genetic diversity

As with other microbial groups, the application of culture-independent methods has led to major advances in our understanding of the level of diversity of marine viruses. One of the problems in the study of viral diversity is that there is no universal marker like the ribosomal RNA genes used for other microbes. However, it is becoming possible to identify representative "signature" genes that are sufficiently conserved to be used as markers of particular groups. For example, extensive studies of the diversity of viruses have been made by analysis of variation in the sequences of structural proteins, such as the g20 capsid protein gene in cyanophages or the major capsid protein (MCP) in phycodnaviruses. Primers for parts of the DNA or RNA polymerase genes can be used in PCR amplification reactions for phycodnaviruses or picornaviruses, respectively. Such approaches can be used for identifying and "fingerprinting" uncultured viruses in environmental samples, for example by PCR-DGGE or TGGE analysis (see p. 48) or construction of clone libraries and RFLP analysis (see p. 45). These fingerprinting methods give an indication of the richness of virus genotypes within a sample, but are unreliable indicators of the abundance of different genotypes due to variations in the efficiency of amplification. However, quantitative PCR (see p. 44) has given some promising results for community analysis. These studies show that there is a very high level of diversity of single genes, even within these restricted viral groups.

It is important to recognize that analysis of diversity based on single genes—as previously discussed in the case of bacterial and protistan diversity—can give misleading results. One way to overcome this is to use microarrays (see p. 54), which can be used to screen for the presence or absence of specific genes in particular virus strains. Pulsed field gel electrophoresis (PFGE; see p. 45) can also be used to separate large DNA fragments; with careful control of the concentration steps, it can provide quantitative estimates of the abundance of viral genomes of different sizes. If specific oligonucleotide probes are available for a virus, these can be applied to Southern blots of the PFGE gels to detect particular viruses. Numerous studies using conserved gene sequences have shown that viral diversity in the marine environment is very high and largely uncharacterized.

Since the first genomic sequences of the phages MS2 (RNA) and phage Φ-X174 (DNA) were published in 1976/1977, about 3000 viral genomes—including representatives of all the major virus families—have been sequenced. The ever-declining cost and rapidity of sequencing (see p. 46) means that it is becoming possible to compare genomes of closely related viruses to reveal hidden diversity and to explain the evolution of viruses. Recently, further revelations about the extent of viral diversity have emerged from shotgun sequencing and metagenomic studies.

Viromes are reservoirs of genetic diversity and exchange

Besides affecting microbial population dynamics, viruses influence diversity in the marine environment as a result of genetic exchange. Analysis of viral metagenomes (viromes) has unexpectedly revealed numerous genes involved in metabolic pathways, indicating that they are a reservoir of genetic information, which

? DO "KILL THE WINNER" DYNAMICS OPERATE IN NATURE?

The "kill the winner" hypothesis was developed by Thingstad (2000) as a mathematical model in the context of an idealized food web incorporating bacteria, protists, and grazing protists based on traditional host–parasite dynamics. The "winner" was conceived as the most active bacterial population, not necessarily the most abundant. Laboratory experiments involving enrichment or depletion of viruses have provided some support for the hypothesis, with strong effects on particular taxa of bacteria (e.g. Hewson and Furhman, 2006), but effects on overall bacterial community composition are less clear-cut, possibly because of interactive effects of viral lysis and protistan grazing (e.g. Winter et al., 2005). Resistance may also incur a cost—changes in surface protein or polysaccharide receptors might make a strain less competitive if these molecules are involved in uptake of nutrients or metabolism. By contrast, very strong evidence for "kill the winner" dynamics has emerged from studies of the effects of viruses during the collapse of algal phytoplankton blooms. The role of viruses in the collapse of blooms of the coccolithophore *Emiliania huxleyi* has been investigated extensively in mesocosm and open-ocean studies, as discussed in *Research Focus Box 7.2*. Winter et al. (2010) review the current status of the hypothesis and set out challenges that need to be overcome for a thorough test—they emphasize the need for further development in the light of evidence, stressing that investigators should not "fall victim to the idea that the model is set in stone."

? HOW DIVERSE ARE MARINE VIRUSES?

The application of metagenomics to study virus diversity is providing new insights into their geographical distribution and role in ecosystem processes. Early application of this approach showed that a 200-liter sample of coastal marine water contained about 5000 genotypes (Breitbart et al., 2002) and a 1-kg marine sediment sample contained possibly as many as one million genotypes (Breitbart et al., 2004). Angly et al. (2006) report on the analysis by high-throughput 454-pyrosequencing (see p. 46) of 184 water samples collected from 68 sites over 10 years from four regions. Global diversity was very high, with an estimated several hundred thousand genotypes. More than 91% of the DNA sequences found did not match sequences already in genomic databases. Using statistical approaches to analyze the distribution of viral sequences, Angly et al. (2006) found regional variations. For example, prophage-like sequences were most dominant in the Arctic whilst cyanophages and a newly discovered clade of single-stranded DNA phages were most common in the Sargasso Sea. Most of the differences between viral assemblages could be explained by variation in the occurrence of the most common viral genotypes. Together, these results indicate that phages are globally distributed and that local environmental conditions enrich for certain viral types through selective pressure. Breitbart and Rohwer (2005) propose a model in which only the most abundant viruses are active, complemented by a large "seed-bank" of inactive viruses for potential recruitment.

is important in the evolution and adaptation of their hosts to different ecological niches. Lysogeny of bacteria by temperate phages may lead to the introduction and expression of new genes into a host and excision of a prophage can lead to transduction of genes to new hosts. In addition, a process known as generalized transduction can occur, in which the enzymes responsible for packaging viral DNA into the capsid may mistakenly incorporate host DNA. These virions are defective and cannot induce a lytic infection, but DNA can be passed from one host to another and may recombine with the DNA of the recipient host. We have known for some time that viruses are vectors for horizontal gene transfer, but we now know that this has powerful effects on the evolution of both the microbial hosts and the viruses. Genetic information is moved by viruses from organism to organism and throughout the biosphere. As discussed in *Research Focus Box 7.1*, genes encoding entire metabolic pathways may move between viruses and their hosts.

Unlike defective phages, which are "genuine" virions that have mistakenly packaged host DNA, another class of virus-like entities called gene transfer agents (GTAs) has recently been discovered in *Rhodobacter* and some other members of the *Alphaproteobacteria*. Some authors also refer to these as generalized transducing agents, but GTAs are different to the defective viruses described above because they seem to function only to transfer random fragments of DNA between cells, they are smaller than phages (about 30–40 nm head diameter), and they carry less DNA (about 4 kb) than would be required to encode the protein components of the particle. There are no negative effects associated with gene transfer to the recipient. It appears that some species of bacteria may produce these GTA particles as a dedicated mechanism of gene transfer and it has been suggested that they are cellular structures akin to flagella or pili. The gene cluster encoding the GTA is widespread in the *Alphaproteobacteria* and may have evolved from a defective prophage—through loss of replication, regulatory, and lysis genes—long before the divergence of the major phylogenetic groups of bacteria.

Conclusions

This chapter has hopefully revealed some of the excitement and rapid pace at which the field of marine virology is moving. It has shown how viruses are emerging to take "center stage" in marine ecology and biogeochemical processes, through their effects on plankton composition and production. On a more fundamental level, recent discoveries revealed by genetic and genomic analysis of individual viruses have revealed that viruses are a reservoir of diversity and play a major role in evolution through gene transfer. Analysis of marine metagenomes is revealing the existence of giant viruses and viruses that appear to infect other viruses; discoveries that mean that many biologists will need to reevaluate completely their perceptions of the nature and evolution of viruses. Viruses feature again in the next three chapters, which include further discussion of their activities in nutrient cycles, in specific biogeochemical cycles, and as disease agents of marine organisms other than microbes.

BOX 7.2   RESEARCH FOCUS

Did viruses precede cells in the "primordial soup"?
Studies of giant viruses and marine metagenomes lead to new ideas about the evolution of life

Mimivirus—bigger and more complex than some cellular organisms

Mimivirus was isolated from the protist *Acanthamoeba* in a cooling water tower in 1992 during a search for pathogenic *Legionella* and was originally thought to be an unidentified bacterium. For many years, the isolate remained in the freezer of the Bradford Public Health Laboratory, until it was reexamined by scientists at the Université de la Méditerranée in Marseilles, who realized that it was a giant, double-stranded DNA virus with an icosahedral capsid. La Scola et al. (2003) gave it the name mimivirus (derived from mimicking microbe) and placed it in the new family *Mimiviridae*. With 650-nm diameter particles and a 1.2-Mbp genome encoding nearly 1000 genes (Legendre et al., 2010), mimivirus is larger and genetically more complex than some bacteria such as *Rickettsia* and *Candidatus* Pelagibacter, and the archaeon *Nanoarchaeum*. The virus has sufficient information to allow it to carry out most, but not all, parts of its life cycle. It possesses some genes for energy production but is dependent on the host ribosomes for translation of its mRNA into proteins (Suzan-Monti et al., 2006). The life cycle thus resembles that of *Rickettsia*, but is clearly viral in nature as it does not replicate through cell division, but by assembly of preformed units. Other giant viruses within this family have since been discovered and nicknamed "giruses" by some workers.

Did the eukaryotic cell nucleus evolve from a virus?

Most DNA viruses of eukaryotes insert their DNA into the nucleus of host cells, where it is replicated. However, after infection of the *Acanthamoeba*, mimivirus enters an eclipse period, during which host cells are instructed to build a large organelle-like "virion factory" in which metabolic processes leading to the synthesis of new virus particles are coordinated. The function of these intracytoplasmic virion factories has been compared with that of a cell nucleus because they both contain the apparatus for DNA replication and transcription, but lack the ability for energy production and mRNA translation. Bell (2001) had previously proposed that the eukaryotic nucleus may have evolved from an infection of an ancestral bacterial cell by a virus, and his controversial viral eukaryogenesis hypothesis has recently been revamped in the light of the mimivirus discovery. Bell (2009) suggests that "the first eukaryotic cell was a multimember consortium consisting of a viral ancestor of the nucleus, an archaeal ancestor of the eukaryotic cytoplasm, and a bacterial ancestor of the mitochondria." Other evidence supports the concept that DNA viruses are the origin of replication proteins in eukaryotic cells (Villareal and DeFilippis, 2000).

The origin of viruses revisited

One view for the origin of viruses is that they developed from degeneration of cells that were intracellular parasites of other cells; another is that they have evolved within cells via the escape of nucleic acids that have become partially independent of their cellular origin as "selfish" genetic entities. Koonin et al. (2006) review the evidence for an alternative scenario—that viruses developed before the evolution of cellular organisms. Interestingly, this idea had been first put forward in the 1920s by Felix d'Herelle following his discovery of phages, and by J. B. S. Haldane in his 1928 treatise *The Origin of Life*. Koonin et al. (2006) show how analysis of genes with key roles in viral replication and capsid formation can be used to identify viral hallmark genes that are shared by apparently unrelated viruses, but never found in cellular organisms (except as proviruses integrated into the genome). They use this to construct a concept of an ancient "Virus World" with extensive gene mixing that predated the emergence of cellular organisms. Claverie (2006) proposes a model in which the initial step was infection of an RNA-based cell to form a primitive nucleus, which becomes transformed to a DNA-based cell because of the selective advantages of DNA biochemistry. He proposes that new pre-eukaryotic viruses were created by rapid reassortment of genes from the viral and cellular pools before the evolution of a stable eukaryotic cell with a fully developed DNA nuclear genome. Forterre (2006) suggested that the three domains of life originated from different RNA cell precursors containing DNA viral genomes. These ideas have provoked much controversy and alternative explanations have been proposed (e.g. Moreira and López-García, 2009).

NCLDV-related and "virophage" sequences are abundant in marine metagenomes

After the discovery of mimivirus, reexamination of Sargasso Sea and GOS metagenomes (see p. 100) revealed very large numbers of sequences homologous to NCLDV genes, and found that *Mimiviridae* are the second most abundant group after phages (Ghedin and Claverie, 2005), suggesting that they commonly infect plankton. Further analysis showed multiple sequences of specific genes that clustered with four of the six NCLDV families. Because these metagenomes were derived from filters designed to capture particles between 0.1 and 0.8 µm, Koonin and Yutin (2010) conclude that these genes have been recovered either from free giant viruses of similar size to mimivirus or from infected heterotrophic and mixotrophic picoeukaryotes (see p. 145). An alternative possibility is that protists may harbor latent viruses in a mechanism similar to temperate phages. Even more surprising was the discovery of genes related to the Sputnik virus (La Scola et al., 2008), a very small icosahedral virus that has been designated a virophage because it prevents the replication of mamavirus—another member of the *Mimiviridae* found in amoebae—by infecting the virus factory (Desnues and Raoult, 2010). Unraveling the ecological importance of these findings and their significance for viral evolution will be a fascinating challenge in the coming years.

References

Allen, M. J., Forster, T., Schroeder, D. C., Hall, M., Roy, D., Ghazal, P. and Wilson, W. H. (2006) Locus-specific gene expression pattern suggests a unique propagation strategy for a giant algal virus. *J. Virol.* 80: 7699–7705.

Angly, F. E., Felts, B., Breitbart, M., et al. (2006) The marine viromes of four oceanic regions. *PLOS Biol.* 4: 2121–2131.

Bell, P. J. L. (2001) Viral eukaryogenesis: Was the ancestor of the nucleus a complex DNA virus? *J. of Molec. Evol.* 53: 251–256.

Bell, P. J. L. (2009) The viral eukaryogenesis hypothesis: a key role for viruses in the emergence of eukaryotes from a prokaryotic world environment. *Ann. NY Acad. Sci.* 1178: 91–105.

Bidle, K. D., Haramaty, L., Ramos, J. B. E. and Falkowski, P. (2007) Viral activation and recruitment of metacaspases in the unicellular coccolithophore, *Emiliania huxleyi. Proc. Natl Acad. Sci. USA* 104: 6049–6054.

Breitbart, M. and Rohwer, F. (2005) Here a virus, there a virus, everywhere the same virus? *Trends Microbiol.* 13: 278–284.

Breitbart, M., Felts, B., Kelley, S., Mahaffey, J. M., Nulton, J., Salamon, P. and Rowher, F. (2004) Diversity and population structure of a near-shore marine-sediment viral community. *Proc. Roy. Soc. B. Biol. Sci.* 271: 565–574.

Breitbart, M., Salamon, P., Andresen, B., Mahaffey, J. M., Segall, A. M., Mead, D., Azam, F. and Rohwer, F. (2002) Genomic analysis of uncultured marine viral communities. *Proc. Natl Acad. Sci. USA* 99: 14250–14255.

Claverie, J. M. (2006) Viruses take center stage in cellular evolution. *Gen. Biol.* 7: 110.

Danovaro, R., Dell'Anno, A., Corinaldesi, C., Magagnini, M., Noble, R., Tamburini, C. and Weinbauer, M. (2008) Major viral impact on the functioning of benthic deep-sea ecosystems. *Nature* 454: 1084–1087.

Desnues, C. and Raoult, D. (2010) Inside the lifestyle of the virophage. *Intervirology* 53: 293–303.

Forterre, P. (2006) Three RNA cells for ribosomal lineages and three DNA viruses to replicate their genomes: a hypothesis for the origin of cellular domain. *Proc. Natl Acad. Sci. USA* 103: 3669–3674.

Forterre, P. (2010) Giant viruses: conflicts in revisiting the virus concept. *Intervirology* 53: 362–378.

Frada, M., Probert, I., Allen, M. J., Wilson, W. H. and de Vargas, C. (2008) The "Cheshire Cat" escape strategy of the coccolithophore *Emiliania huxleyi* in response to viral infection. *Proc. Natl Acad. Sci. USA* 105: 15944–15949.

Ghedin, E. and Claverie, J. M. (2005) Mimivirus relatives in the Sargasso sea. *Virol. J.* 2.

Hewson, I. and Fuhrman, J. A. (2006) Viral impacts upon marine bacterioplankton assemblage structure. *J. Mar. Biol. Assoc. UK* 86: 577–589.

Jacquet, S., Heldal, M., Iglesias-Rodriguez, D., Larsen, A., Wilson, W. and Bratbak, G. (2002) Flow cytometric analysis of an *Emiliania huxleyi* bloom terminated by viral infection. *Aquat. Microb. Ecol.* 27: 111–124.

Koonin, E. V., Senkevich, T. G., Dolja, V. V. (2006) The ancient virus world and evolution of cells. *Biol. Direct* 1: 29.

Koonin, E. V. and Yutin, N. (2010) Origin and evolution of eukaryotic large nucleo-cytoplasmic DNA viruses. *Intervirology* 53: 284–292.

Lane, N. (2008) Origins of death. *Nature* 453: 583–585.

La Scola, B., Audic, S., Robert, C., et al. (2003) A giant virus in amoebae. *Science* 299: 2033.

La Scola, B., Desnues, C., Pagnier, I., et al. (2008) The virophage as a unique parasite of the giant mimivirus. *Nature* 455: 100–104.

Legendre, M., Audic, S., Poirot, O., et al. (2010) mRNA deep sequencing reveals 75 new genes and a complex transcriptional landscape in mimivirus. *Genome Res.* 20: 664–674.

Lindell, D., Sullivan, M. B., Johnson, Z. I., Tolonen, A. C., Rohwer, F. and Chisholm, S. W. (2004) Transfer of photosynthesis genes to and from *Prochlorococcus* viruses. *Proc. Natl Acad. Sci. USA* 101: 11013–11018.

Lindell, D., Jaffe, J. D., Johnson, Z. I., Church, G. M. and Chisholm, S. W. (2005) Photosynthesis genes in marine viruses yield proteins during host infection. *Nature* 438: 86–89.

Mackinder, L. C. M., Worthy, C. A., Biggi, G., et al. (2009) A unicellular algal virus, *Emiliania huxleyi* virus 86, exploits an animal-like infection strategy. *J. Gen. Virol.* 90: 2306–2316.

Mann, N. H., Cook, A., Millard, A., Bailey, S. and Clokie, M. (2003) Marine ecosystems: bacterial photosynthesis genes in a virus. *Nature* 424: 741.

Martinez, J. M., Schroeder, D. C., Larsen, A., Bratbak, G. and Wilson, W. H. (2007) Molecular dynamics of *Emiliania huxleyi* and cooccurring viruses during two separate mesocosm studies. *Appl. Environ. Microbiol.* 73: 554–562.

Moreira, D. and López-García, P. (2009) Ten reasons to exclude viruses from the tree of life. *Nat. Rev. Microbiol.* 7: 306–311.

Pagarete, A., Allen, M. J., Wilson, W. H., Kimmance, S. A. and de Vargas, C. (2009) Host–virus shift of the sphingolipid pathway along an *Emiliania huxleyi* bloom: survival of the fattest. *Environ. Microbiol.* 11: 2840–2848.

Patel, A., Noble, R. T., Steele, J. A., Schwalbach, M. S., Hewson, I. and Fuhrman, J. A. (2007) Virus and prokaryote enumeration from planktonic aquatic environments by epifluorescence microscopy with SYBR Green I. *Nat. Protoc.* 2: 269–276.

Schroeder, D. C., Oke, J., Malin, G. and Wilson, W. H. (2002) *Coccolithovirus (Phycodnaviridae)*: Characterisation of a new large dsDNA algal virus that infects *Emiliania huxleyi. Arch. Virol.* 147: 1685–1698.

Schroeder, D. C., Oke, J., Hall, M., Malin, G. and Wilson, W. H. (2003) Virus succession observed during an *Emiliania huxleyi* bloom. *Appl. Environ. Microbiol.* 69: 2484–2490.

Sorensen, G., Baker, A. C., Hall, M. J., Munn, C. B. and Schroeder, D. C. (2009) Novel virus dynamics in an *Emiliania huxleyi* bloom. *J. Plankton Res.* 31: 787–791.

Sullivan, M. B., Lindell, D., Lee, J. A., Thompson, L. R., Bielawski, J. P. and Chisholm, S. W. (2006) Prevalence and evolution of core photosystem II genes in marine cyanobacterial viruses and their hosts. *PLOS Biol.* 4: 1344–1357.

Suttle, C. A. (2005) Viruses in the sea. *Nature* 437: 359–361.

Suttle, C. A. (2007) Marine viruses— major players in the global ecosystem. *Nat. Rev. Microbiol.* 5: 801–812.

Suzan-Monti, M., La Scola, B. and Raoult, D. (2006) Genomic and evolutionary aspects of Mimivirus. *Virus Res.* 117: 145–155.

Thingstad, T. F. (2000) Elements of a theory for the mechanisms controlling abundance, diversity, and biogeochemical role of lytic bacterial viruses in aquatic systems. *Limnol. Oceanogr.* 45: 1320–1328.

Vardi, A., Van Mooy, B. A. S., Fredricks, H. F., Popendorf, K. J., Ossolinski, J. E., Haramaty, L. and Bidle, K. D. (2009) Viral glycosphingolipids induce lytic infection and cell death in marine phytoplankton. *Science* 326.

Villarreal, L. P. (2004) Are viruses alive? *Sci. Am.* 291: 100–105.

Villarreal, L. P. and DeFilippis, V. R. (2000) A hypothesis for DNA viruses as the origin of eukaryotic replication proteins. *J. Virol.* 74: 7079–7084.

Villarreal, L. P. and Witzany, G. (2010) Viruses are essential agents within the roots and stem of the tree of life. *J. Theor. Biol.* 262: 698–710.

Wilson, W. H., Tarran, G. A., Schroeder, D., Cox, M., Oke, J. and Malin, G. (2002) Isolation of viruses responsible for the demise of an *Emiliania huxleyi* bloom in the English Channel. *J. Mar. Biol. Assoc. UK* 82: 369–377.

Wilson, W. H., Allen, M. J., Schroeder, D. C., et al. (2005) Complete genome sequence and lytic phase transcription profile of a *Coccolithovirus. Science* 309: 1090–1092.

Winter, C., Smit, A., Herndl, G. J. and Weinbauer, M. G. (2005) Linking bacterial richness with viral abundance and prokaryotic activity. *Limnol. Oceanogr.* 50: 968–977.

Winter, C., Bouvier, T., Weinbauer, M. G. and Thingstad, T. F. (2010) Trade-offs between competition and defense specialists among unicellular planktonic organisms: the "killing the winner" hypothesis revisited. *Microbiol. Mol. Biol. Rev.* 74: 42–57.

Further reading

Allen, M. J. and Wilson, W. H. (2008) Aquatic virus diversity accessed through omic techniques: a route map to function. *Curr. Opin. Microbiol.* 11: 226–232.

Bouvier, T. and del Giorgio, P. A. (2007) Key role of selective viral-induced mortality in determining marine bacterial community composition. *Environ. Microbiol.* 9: 287–297.

Brussaard, C. P. D. (2004) Viral control of phytoplankton populations—a review. *J. Eukaryot. Microbiol.* 51: 125–138.

Fuhrman, J. A. and Schwalbach, M. (2003) Viral influence on aquatic bacterial communities. *Biol. Bull.* 204: 192–195.

Kristensen, D. M., Mushegian, A. R., Dolja, V. V. and Koonin, E. V. (2010) New dimensions of the virus world discovered through metagenomics. *Trends Microbiol.* 18: 11–19.

Lang, A. S. and Beatty, J. T. (2007) Importance of widespread gene transfer agent genes in alpha-proteobacteria. *Trends Microbiol.* 15: 54–62.

Lang, A. S., Rise, M. L., Culley, A. I. and Steward, G. F. (2009) RNA viruses in the sea. *FEMS Microbiol. Rev.* 33: 295–323.

Munn, C. B. (2006) Viruses as pathogens of marine organisms—from bacteria to whales. *J. Mar. Biol. Assoc. UK* 86: 1–15.

Nagasaki, K. (2008) Dinoflagellates, diatoms, and their viruses. *J. Microbiol.* 46: 235–243.

Paul, J. H. (2008) Prophages in marine bacteria: dangerous molecular time bombs or the key to survival in the seas? *ISME J.* 2: 579–589.

Rohwer, F. and Thurber, R. V. (2009) Viruses manipulate the marine environment. *Nature* 459: 207–212.

Van Etten, J. L., Graves, M. V., Muller, D. G., Boland, W. and Delaroque, N. (2002) *Phycodnaviridae*—large DNA algal viruses. *Arch. Virol.* 147: 1479–1516.

Weinbauer, M. G. (2004) Ecology of prokaryotic viruses. *FEMS Microbiol. Rev.* 28: 127–181.

Weinbauer, M. G. and Rassoulzadegan, F. (2004) Are viruses driving microbial diversification and diversity? *Environ. Microbiol.* 6: 1–11.

Wilhelm, S. W., Weinbauer, M. G. and Suttle, C. A. (eds) (2010) *Manual of Aquatic Viral Ecology.* American Society of Limnology and Oceanography, Waco, TX. doi:10.4319/mave.2010.978-0-9845591-0-7.

Chapter 8

Microbes in Ocean Processes—Carbon Cycling

Preceding chapters have discussed the activities of individual types of marine microbes in major processes such as photosynthesis, chemolithotrophy, breakdown of organic material, and oxidation–reduction transformations of major elements. These processes occur in seawater and sediments, as well as in specialized habitats such as hydrothermal vents, cold seeps, and epi- or endobiotic associations. In this chapter and in Chapter 9, the focus is largely on the role of planktonic microbes in processes in the upper layers of the pelagic zone and emphasizes the overall picture of carbon cycling and its global biogeochemical significance. This exciting area of work, in which the activities of microbiologists and chemical and physical oceanographers come together, has led to spectacular paradigm shifts in our view of the importance of microbes in ocean processes.

Key Concepts

- Microbes play a central role in the biological pump for the cycling of carbon in the oceans, both as primary producers and in the retention of dissolved nutrients and remineralization of carbon through the activity of heterotrophs (the microbial loop).

- Marine phytoplankton are responsible for about half of the global CO_2 fixation, with cyanobacteria and picoeukaryotes having a dominant role in oligotrophic gyres.

- Primary production depends on the availability of light and nutrients and shows large variations in different coastal and oceanic regions.

- Ocean experiments have confirmed that low concentrations of iron limit photosynthesis, but proposals to use iron fertilization are controversial and these ocean experiments have provided little evidence of effective sequestration of carbon to deep waters.

- Grazing by protists and filter feeding by tunicates results in the transfer of fixed carbon to different trophic levels.

- Numerous factors determine the amount of fixed carbon that is transferred to higher trophic levels (link) or reaches the seafloor sediments (sink).

- Viral lysis of microbes catalyzes nutrient regeneration in the upper ocean by diverting the flow of carbon from the food chain into a semi-closed cycle of bacterial uptake and release of organic matter.

Development of the microbial loop concept transformed our understanding of ocean processes

Space does not permit a full historical treatment of the development of modern ideas about carbon cycling and the ocean food web, but it is instructive to outline some of the most important highlights from the past 30 years. Until the mid-1970s, marine bacteria were regarded as of little importance other than as decomposers of detritus. The classic view of trophic interactions in the oceans was of a simple food chain, in which primary production is due mainly to algae like diatoms and dinoflagellates that are large enough to be trapped in traditional plankton nets. A simple food chain was envisaged, in which these algae are consumed by copepods (a diverse group of small crustaceans), which in turn are consumed by larger zooplankton, eventually reaching fish at the end of the food chain. In this scenario, phytoplankton would be consumed as rapidly as it is produced, with all the primary production going through herbivorous zooplankton—bacteria did not feature at all in this food chain. Estimates of bacterial abundance were orders of magnitude lower than is now known to be the case, fitting with the prevalent view at that time that most of the oceans are biological "deserts" with low nutrient fluxes, low biomass of phytoplankton, and low productivity. Until the early 1980s, the perceived small populations of bacteria were thought to be largely dormant and this perspective only changed as a result of the application of new techniques, as discussed in Chapter 2. The development of ultraclean analytical techniques, measurement of the incorporation of radio-labeled precursors, and the measurement of ATP levels led to the realization that metabolic activities and productivity in seawater are much higher than previously envisaged. The development of controlled pore-size filters and epifluorescence light microscopy in combination with fluorescent DNA stains revealed that the oceans are, in fact, teeming with microorganisms—typically 10^6 to 10^7 per milliliter—in the picoplankton size class (i.e. <2 μm). Epifluorescence microscopy also led to the discovery of *Synechococcus*, a previously overlooked cyanobacterium now recognized as one of the major contributors to primary productivity in a large part of the oceans. These new methods also revealed the importance of the consumption of bacteria by small phagotrophic protists, providing evidence of a direct link between bacterial production and higher components of the food web.

A paper by Lawrence Pomeroy of the University of Georgia, entitled "The ocean's food web: a changing paradigm" (Pomeroy, 1974), is widely credited as being one of the most significant advances in our thinking about the role of microbes in the movement of energy and nutrients in marine systems. The main arguments in this paper were: (i) that the main primary producers in the oceans are nanoplankton (small phototrophs <60 μm in size) rather than the "net" phytoplankton previously recognized; (ii) that microbes are responsible for most of the metabolic activity in seawater; and (iii) that dissolved and particulate organic matter form an important source of nutrients consumed by heterotrophic microbes in the marine food web.

Evidence about the role of the heterotrophic bacterioplankton steadily accumulated until a series of seminal papers were published in the early 1980s by Farooq Azam and coworkers at the Scripps Oceanographic Institute, California. They developed the concept of the "microbial loop" to explain the flow and cycling of dissolved organic material (DOM) in the oceans. DOM is a term used to describe dissolved monomeric, oligomeric, and polymeric compounds plus colloids and small cell fragments. It is important to remember that there is no clear cutoff between DOM and particulate organic matter (POM) and the distinction is purely empirical, reflecting the size of filters used in sample preparation (see p. 14). The original microbial loop concept has been continuously developed as a result of new discoveries made possible by technical advances. For example, the development of flow cytometers that could be used on research vessels led to the discovery of photosynthetic *Prochlorococcus*. This is now known to be one of the major primary producers in the oceans, and possibly the most abundant photosynthetic organism on Earth, yet it escaped detection until 1988. Since the late 1980s, the large biomass and diversity of marine microbes has been revealed through the application of molecular techniques. The previously unimagined importance of *Archaea*, picoeukaryotes, and mixotrophs as components of the plankton has emerged in the last few years, whilst metagenomics has led to the discovery of new organisms, new energy-generating mechanisms, and the unraveling of metabolic processes in organisms that have not yet been cultured. Perhaps the most significant recent addition to the microbial loop concept is recognition of the role of viruses in controlling community structure in the plankton and shunting nutrients from organisms into DOM.

The fate of carbon dominates consideration of the microbial ecology of the oceans

The unique chemical properties of carbon mean that it is present in many different inorganic and organic forms. The major reservoir of carbon occurs as carbonate minerals in sedimentary rocks in the Earth's crust and in organic compounds such as coal, oil, and natural gas. The turnover of this carbon by natural geological processes is very slow, occurring over timescales of millennia. However, in the last few hundred years, human activities have accelerated the flux of carbon to the biosphere because of the use of fossil fuels and changes in land use.

The oceans (excluding sediments) are the largest reservoir of biologically active carbon, containing about 4×10^{13} tonnes of carbon—47 times more than the atmosphere and 23 times more than the terrestrial biosphere. Gaseous CO_2 is highly soluble in water and forms carbonic acid (H_2CO_3). At the current natural pH of seawater (7.8–8.2), this mostly dissociates rapidly to form bicarbonate (see p. 11 and *Research Focus Box 1.1*). Absorption of CO_2 at the interface between ocean and atmosphere and its transfer to deeper layers is driven to a large extent by physical processes as circulation of water occurs as a result of turbulence created by temperature gradients, surface winds, and the Coriolis effect caused by the Earth's rotation. Toward the poles, cold water with a high density sinks vertically to a depth of 2000–4000 m and then distributes through the ocean basins. The sinking mass of water displaces deep ocean water to the surface through upwelling. As the water warms, CO_2 escapes to the atmosphere again. This model constitutes the so-called solubility pump for CO_2 circulation in the oceans (*Figure 8.1A*; see also *Figure 1.7*).

In addition to purely physical factors, a biological pump is responsible for massive redistribution of carbon in the oceans (*Figure 8.1B*). Fixation of CO_2 by phytoplankton leads to incorporation of carbon into cellular material. Phytoplankton cells may be consumed by zooplankton or fish, so that some of the carbon enters the food chain directly. A small, but significant, proportion of cellular carbon is released as DOM from phytoplankton cells by exudation, and large amounts of DOM are produced when organisms die and break up. Polymeric substances and cell debris aggregate to form POM (see *Figure 1.9*). DOM and POM act as a source of energy, carbon, and other nutrients for heterotrophic organisms via the microbial loop. On average, about half of the organic carbon fixed by phytoplankton passes through the microbial loop. Heterotrophic microbes provide a *sink* for fixed carbon through assimilation of DOM and remineralization to CO_2, as well as a *link* to higher trophic levels, as discussed below.

Marine phytoplankton are responsible for about half of the global CO₂ fixation

Primary production can be defined in terms of the total amount of CO_2 fixed into cellular material during daylight (gross primary production) or the fraction

WHY WAS USE OF THE TERM "MICROBIAL LOOP" SO IMPORTANT?

The development of the microbial loop concept represents a true paradigm shift, as thinking about the role of microbes in mineral cycling and food webs underwent a sudden and dramatic change. The term "microbial loop" was originally coined in the paper "The ecological role of water-column microbes in the sea" by Azam et al. (1983). The paper arose from a conference on biological oceanography and connected discoveries made during the preceding decade by several marine biologists. The paper has had one of the greatest impacts of all papers in the aquatic sciences, with more than 2160 citations to date (October 2010). Twenty-five years after publication, one of the original authors on the paper reflects on why it became so successful (Fenchel, 2008). Tom Fenchel points out that the paper "only summarised the previous work of some of the co-authors and that of many other colleagues, and the general idea of the existence of a microbial loop had been expressed more or less explicitly in several papers preceding the 1983 paper." He attributes the impact of the paper to the fact that the simple descriptive name "microbial loop" was used for the first time—suddenly clarifying and connecting the various studies and catalyzing further research.

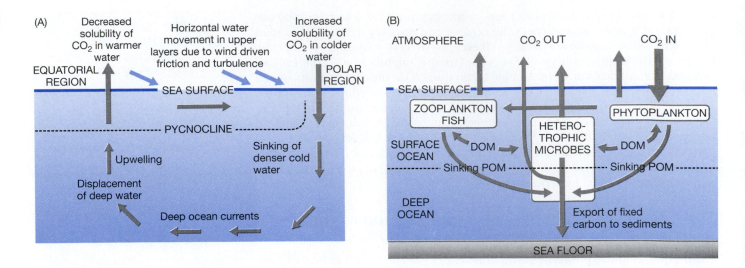

Figure 8.1 Carbon cycling processes in the oceans. (A) The solubility pump. Arrows show the circulation of water and CO₂. The dotted line represents the boundary between the upper ocean layer and deeper waters, caused by differences in temperature and density. (B) The biological pump. Arrows show the movement of CO₂ and fixed carbon compounds. DOM, dissolved organic matter; POM, particulate organic matter. The dotted line represents a depth of about 150–200 m.

that remains after losses due to respiration by the phytoplankton, which occurs both day and night (net primary production). Another concept, net community production, has been introduced to encompass the effects of respiration by heterotrophic organisms. Oceanographers also use various other ways of measuring productivity in terms of new, export, and regenerated production. Both the biomass of phytoplankton and the rate of gross and net production are affected by a range of factors, which vary greatly over both long-term and short-term timescales, owing to the interplay of hydrographic conditions (currents, upwelling, and diffusion) with light and nutrient availability. Temperature increases both photosynthetic and respiratory rates; but this effect is complicated in the oceans because, although most upwelling waters have a low temperature, the stimulation of photosynthetic rates by their increased nutrient levels overcomes the inhibitory effects of cooling.

Quantifying photosynthetic productivity has been a central question in biological oceanography for more than a century. Various methods are available and these may produce different results depending on the parameter measured and the timescale employed. The most common laboratory experiments measure incorporation of radioactively labeled bicarbonate ($H^{14}CO_3^-$) into organic material, comparing the effects of incubation in light and dark conditions. Oxygen concentration can be measured by high-precision titration, and changes indicate O_2 evolution by oxygenic photosynthesis and O_2 consumption by respiration. Alternatively, specialized equipment can be used to follow $^{18}O_2$ evolution from $H_2^{18}O$, indicating photosynthesis without interference from respiratory processes. Clearly, the discovery of widespread anoxygenic phototrophic bacteria (*Research Focus Box 3.1*) means that earlier assumptions based on the use of O_2 evolution need to be reconsidered. All methods have certain drawbacks, and it is necessary to make allowances for isotope effects and differences in the photosynthetic quotient (ratio of HCO_3^- assimilated to O_2 generated), which varies according to species and nutrient availability. It is important to recognize that growth of phytoplankton is not a simple process concerned only with these processes; growth also depends on assimilation of other nutrients and many metabolic transformations to produce macromolecules in the cell. Measuring the physiological rates of photosynthesis in the natural situation is difficult. Most experiments are conducted by enclosing seawater containing natural communities in bottles, which are then held at various depths in the field. Photosynthetic rates are measured over time under different environmental conditions, including light. Bottle experiments obviously have inherent limitations because nutrients are quickly depleted in the absence of diffusion and artifacts can be introduced during deployment and retrieval of samples, by contamination or by alteration of the spectral qualities of light entering the containers. Mesocosm experiments in large enclosures containing several cubic meters of sea water can overcome some of these problems (see p. 29).

The penetration of water by light of different wavelengths limits photosynthesis to the upper 100–200 m of clear ocean waters, and considerably less in the presence of suspended material. Different phototrophic organisms are adapted to use light of different wavelengths by the use of different types of chlorophyll and ancillary pigments, and this affects their distribution in the water column. Indeed, the traditional definition of the limit of the euphotic zone as the depth at which light intensity is reduced to 1% of its surface level is no longer valid, since some types of *Prochlorococcus* are photosynthetic at light levels less than 0.1% of surface irradiance. Very high irradiances can inhibit photosynthesis, especially due to damage to the photosystem by ultraviolet light, and organisms possess a number of photoprotective mechanisms to overcome these effects (p. 87). The water depth at which the light intensity is just sufficient to balance the incorporation of CO_2 by photosynthesis and its loss due to respiration is known as the compensation depth. This varies according to species composition of the phytoplankton, geographic region, season, light penetration during the day, and nutrient availability. Clouds and atmospheric dust have a marked effect on light penetration, and the density of phytoplankton itself is a major factor. Usually, maximum primary production does not occur at the very surface of the ocean, but several meters deeper where light and nutrient levels are optimal.

Table 8.1 Estimates of primary productivity of the oceans

	Primary production ($\times 10^{15}$g carbon fixed)	% of total
(a) Annual production in different oceans		
Pacific	19.7	42.8
Atlantic	14.5	31.5
Indian	8.0	17.3
Antarctic	2.9	6.3
Arctic	0.4	0.9
Mediterranean	0.6	1.2
Total global annual production	46.1	100.0
(b) Seasonal estimates of global ocean productivity		
March–May	10.9	23.0
June–August	13.0	28.2
September–November	12.3	26.7
December–February	11.3	22.1
Total global annual production	47.5	100.0

Estimated from SeaWiFS (Sea-viewing Wide Field-of-View Sensor) remote sensing data using a vertically generalized production model. Note: there is a slight discrepancy in the total productivity estimates in (a) and (b). (Adapted from data in the ICMS Ocean Productivity Study [2000]; see *Plate 8.1* for a map of annual global productivity.)

NOT ALL CO$_2$ FIXATION OCCURS BY OXYGENIC PHOTOSYNTHESIS

Although photosynthesis by the phytoplankton in the upper layers of the oceans is by far the most important source of marine primary production, the reader should bear in mind the significant local importance of primary production by photoautotrophs and chemolithotrophs in benthic microbial mats, symbiotic interactions, vents, and seeps, as discussed in other chapters. Also, anoxygenic phototrophs are thought to be responsible for about 5% of primary production (see p. 62). Furthermore, we have recently discovered that about one-fifth of all the picoplankton in the deep ocean are chemolithotrophic crenarchaeotes, fixing CO_2 autotrophically using energy from ammonia oxidation (see *Research Focus Box 5.1*). At present we have very little idea of the contribution that archaeal production in the deep sea makes to the overall carbon budget of the oceans, but Herndl et al. (2005) concluded that planktonic archaea are actively growing in the dark ocean (although at lower growth rates than bacteria) and estimated that archaeal production in the mesopelagic and bathypelagic north Atlantic Ocean accounts for 10–20% of production.

Satellite observation of ocean color has had a major impact on the study of variability in phytoplanktonic communities over long timescales (see *Plate 8.1*). Scientists at Plymouth Marine Laboratory and Laboratoire d'Océanologie et de Géoscience, Wimereux, are developing new approaches to remote sensing which permit differentiation of size classes among the phytoplankton, based on the models that allow estimation of photosynthetic rates relevant to cell sizes (Hirata et al., 2008). This reveals differences in geographic distribution and the relative contribution of different groups to the total production. Microphytoplankton (larger diatoms and dinoflagellates) are abundant mainly at mid to high latitudes and are distributed patchily. Nanophytoplankton (smaller diatoms and dinoflagellates) are moderately abundant globally, but concentrations are higher at equatorial and mid to high latitudes and relatively low in subtropical gyres; whilst picophytoplankton (cyanobacteria and picoeukaryotes) mainly dominate oligotrophic gyres. Detailed analysis of monthly and annual variations in distributions of different size classes and correlation with other oceanographic parameters will reveal how phytoplankton biomass and production varies over different periods and allow predictions about the impact of climate change.

Since the distribution of primary production is highly dynamic and influenced by many factors, reliable estimates of productivity throughout the oceans have only been possible since the advent of remote sensing using the spectral scanners on satellites. By measuring the different absorbance properties of chlorophyll *a* and other photosynthetic pigments, it is possible to map phytoplankton biomass on a daily basis (cloud cover permitting). Mathematical models linking chlorophyll measurements, photosynthesis rates, and irradiance are used to extrapolate experimental values to different depths and to estimate the effects of diurnal variation.

The total net annual primary production of the world's oceans is close to 5×10^{16} g carbon fixed, which is very similar to the total productivity of all terrestrial ecosystems. However, on an area basis the annual global marine productivity is about 50 g carbon fixed per square meter, which is only one third that of production on land. This discrepancy is due to the lower utilization of solar radiation by ocean phytoplankton than terrestrial plants, largely because of nutrient limitation and the effect of suspended particles (including the phytoplankton cells themselves) in absorbing light.

As well as light, photosynthetic activity depends on the availability of nutrients

As shown in *Table 8.1* there is great variation in the distribution of primary productivity in the world's oceans and shelf seas. The most productive regions are upwelling regions associated with currents found along the west coasts of continents; namely, the Canary Current (off Northwest Africa), the Benguela Current (off southern Africa), the California Current (off California and Oregon), the Humboldt Current (off Peru and Chile), and the Somali Current (off Western India). All of these currents (see *Figure 1.6*) support major fisheries. Around the equator, winds and the Coriolis effect generate a divergence current resulting in upwelling of denser, nutrient-rich water. There is also large-scale upwelling in the Southern Ocean resulting from upwelling to replace a northward flow of water driven by strong winds. The least productive areas are the central gyres of the ocean basins.

Although, productivity on a global basis is fairly constant across the year (*Table 8.1b*), it is highly seasonal in some parts of the world. High latitude temperate regions (especially the north Atlantic) characteristically show a spring bloom. Strong mixing occurs during the cold, dark, windy winter bringing nutrients to the surface layers and promoting a rapid increase in photosynthesis as light levels increase during spring. Increased stratification and nutrient depletion lead to reduction in productivity during summer, even though light levels are at their highest. During autumn, mixing occurs again and a small secondary peak of production results. In tropical seas, seasonal effects are much less pronounced, except where there is a seasonal upwelling of nutrients, such as that occurring in the Arabian Sea. Around coastlines, photosynthetic activity is generally high owing to the input of nutrients from rivers and wind-blown dust and because of the activity of seaweeds, seagrasses, and salt-marsh plants, which contribute about 5% of total global production. Tropical coral reefs are locally very productive because of the dense concentrations of photosynthetic dinoflagellates in many corals, but because they occupy such a small area they probably contribute less than 1% to global primary production. Carbon is rarely a limiting factor in productivity, since HCO_3^- is abundant in seawater, but nitrogen, phosphorus, silica, trace metals, and some trace organic compounds such as B vitamins all have the potential to be a limiting nutrient, thus affecting production. The availability of iron has special significance, as discussed in *Research Focus Box 8.1* and in Chapter 9.

The importance of various components of the microbial loop varies according to circumstances

As noted earlier, the classic view of marine trophic interactions was a simple pyramidal food chain, summarized as microalgae (mainly diatoms) → copepods → fish and whales. Development of the microbial loop concept has placed much greater emphasis on the importance of DOM and the activity of bacteria, protists, and viruses, culminating in our modern view of the ocean food web represented schematically in *Figure 8.2*. This shows that there are trophic interactions at multiple levels, with microbial processes occupying center stage. However, elements of the classic model are still valid and the relative importance of the various components varies according to the circumstances. In weakly stratified water masses, with a high degree of mixing and turbulence—such as polar and coastal temperate seas during the spring bloom—the food web is often based strongly on the dominance of organisms in the microplankton size class or greater (i.e. diatoms, dinoflagellates, prymnesiophytes, and copepod grazers). By contrast, the large areas of the ocean in the tropics and subtropics that have highly stratified waters with constant low nutrient levels have food webs that are strongly dependent on microbial loop processes dominated by the activities of nanoplankton (small protists), picoplankton (bacteria and very small protists), and viruses. As a generalization, the relative influence of the pico- or nanoplankton size classes of microbes is greatest in oligotrophic waters, because conditions of low-nutrient flux exert positive selection for small size (see p. 5). The relative importance of activity of larger particle-feeding protists, zooplankton, and fish becomes progressively greater as nutrient levels in the water column increase. A useful analogy may be made between the behavior of food webs in nature and the behavior of laboratory cultures. Food webs dominated by microbial loops tend to behave like chemostat (steady state) cultures. Despite rapid multiplication rates of some of the component organisms, species abundance and the concomitant geochemical processes in the oligotrophic ocean gyres

Figure 8.2 Schematic representation of the food web in the sea, showing the central importance of dissolved organic material (DOM) and the microbial loop. The left side of the figure represents the "classic" food chain, whilst the right side represents microbial loop processes. Solid arrows show the main routes for transfer of organic material. Dotted lines indicate the release of cellular material as DOM, mainly by viral lysis of plankton; some compounds are also released as exudates or by cell breakage during feeding by zooplankton. Low-molecular-weight DOM is absorbed directly by heterotrophs, whilst high-molecular-weight DOM is degraded by bacterial ectoenzyme activity before absorption. Dashed lines indicate the formation of particulate organic material (POM) formed from aggregation of fecal pellets, mucus and cellular debris; POM, which is more recalcitrant to breakdown, will be exported to deep water and sediments. Mucus net feeders such as larvaceans and salps can feed directly on very small microbes. For clarity, not all trophic connections and groups of organisms are shown and images are illustrative only, and not to scale.

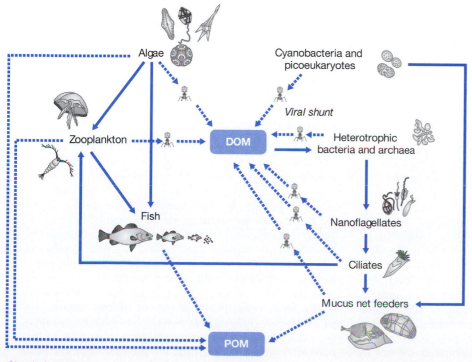

Classic food chain   Microbial loop

Fertilizing the oceans
Controversy surrounds plans to boost phytoplankton with iron

The iron hypothesis

The idea that iron deficiency limits phytoplankton growth in many parts of the ocean is largely credited to John Martin of the Moss Landing Marine Laboratory, California. In the 1990s, Martin proposed that climatic changes in the Earth's history (glacial and interglacial periods) were caused by changing patterns of deposition of iron-rich dust, resulting in modification of CO_2 exchange between the atmosphere and oceans. He developed the ultraclean analytical techniques necessary to study trace metals and used bottle experiments to show that iron addition caused rapid growth of phytoplankton in water from parts of the ocean that are high in nutrients but low in chlorophyll (HNLC; see p. 188). Although many other researchers were involved in the research in this field, Martin is most associated with the "iron hypothesis" because of his famous and controversial quip at a lecture: "Give me half a tanker of iron and I will give you an ice age." Amidst much controversy, Martin planned experiments to enrich areas of the ocean with iron, but he died shortly before his hypothesis could be tested. The first experiment (IronEx I) was conducted in 1993 in the equatorial Pacific off the Galapagos Islands, by addition of iron sulfate dispersed in the wake of a ship's propeller as it sailed over the area. In this patch, the surface of the sea turned from blue to green—the idea that addition of iron would stimulate phytoplankton growth in the ocean had been vindicated. Oceanographers developed plans for more experiments.

Iron enrichment experiments—what have they shown?

At the time of writing (October 2010), 12 mesoscale open-ocean iron enrichment experiments have been conducted during research cruises—four in the northwest Pacific, two in the equatorial Pacific, and six in the Southern Ocean (reviewed by Boyd et al., 2007). The Southern Ocean is thought to play an especially important role in the global carbon cycle because of large-scale export of fixed carbon to deep waters. Owing to its distance from continental landmasses, the Southern Ocean receives little dust and hence has very low iron levels, resulting in large reservoirs of unused macronutrients in the upper and deep mixed layers. Most experiments employed similar methodology, in which hundreds or thousands of kilos of iron sulfate plus an inert tracer (sulfur hexafluoride) were added over patches of ocean a few kilometers in diameter. Over periods of a few days to weeks, scientists on board research vessels measured biomass and diversity of phytoplankton, together with a wide range of biochemical, geochemical, hydrographic, and physical parameters. In all but one of the experiments, phytoplankton blooms dominated by diatoms have been stimulated with up to 15-fold increases in the chlorophyll content of surface waters, providing strong evidence that iron is a growth-limiting factor. However, estimates of the amount of carbon drawn down into deeper water as blooms sink from the surface have been variable and much lower than predicted from models based on laboratory scale experiments. Recycling of fixed carbon by microbial loop processes may return CO_2 to the atmosphere after a short period. The experiments have provided little evidence that long-term sequestration would occur. Watson et al. (2008) review the limitations of these first-generation experiments and conclude that the timescale and spatial extent of the experiments conducted to date have been severely hampered by difficulties of logistics and expense associated with carrying out this type of work—large scientific teams are needed for shipboard experiments, many miles from land. Smetacek and Naqvi (2008) note that the total ship time dedicated to these experiments over 10 years is only a total of 11 months, compared with more than 2500 months to other oceanographic research cruises. These authors argue that the iron enrichment experiments have provided much useful information about ocean processes, but the design and timescale of the experiments were simply not appropriate to answer the question whether iron fertilization can be a successful strategy for carbon sequestration. They propose that the next generation of experiments need to be conducted on a very large scale—over hundreds of kilometers—with observations using a wide range of techniques made over periods of years rather than days or weeks, and incorporated into new oceanographic models. In addition, experiments in semi-closed areas of water formed by eddies, and new approaches to the method of release of iron compounds are needed to provide better evidence for the effect of iron fertilization on carbon sequestration.

What are the risks?

To be effective, iron fertilization requires a deliberate alteration of marine ecology, by stimulating blooms of relatively large phytoplankton that are usually not abundant. Opponents argue that this could shift the food web in unpredictable ways and distort biogeochemical and ecological relationships throughout the system (Strong et al., 2009). Our experience of the effects of eutrophication in the creation of hypoxic "dead zones" (see p. 185) suggests that similar effects might occur with large-scale iron fertilization. Another possibility is that iron fertilization could stimulate the growth of toxic algae. Trick et al. (2010) used *in situ* measurements and shipboard culture experiments with seawater collected from an iron enrichment research site off Alaska to demonstrate that oceanic species of the diatom *Pseudo-nitzschia* produce the toxin domoic acid (see p. 248). They concluded that large-scale ocean iron

BOX 8.1   RESEARCH FOCUS

enrichment could potentially lead to levels of domoic acid sufficient to cause mortalities in seabirds and mammals or necessitate the closure of fisheries to protect public health.

Priming the biological pump—commerce and controversy

Early models based on the potential of fertilization of low-productivity regions were promoted enthusiastically by some scientists and entrepreneurs. They suggested that the technique might accelerate the removal of billions of tons of carbon dioxide from the atmosphere—sequestering it as fixed carbon compounds in deep waters and helping to control the effects of global warming—as well as having potential benefits for enhancement of fisheries. Several companies have been formed to exploit the technology and some continue to promote plans for large-scale trials. Following the development of the Kyoto Protocol for reduction of CO_2 emissions at the United Nations Framework Convention on Climate Change conference in 1997, the concept of a global market in the trading of "carbon offset credits" emerged. Subsequent treaties on climate change imposed by governments have placed strict caps and taxes on emissions, encouraging entrepreneurs to establish companies with the aim of profiting from ocean fertilization whilst mitigating the consistent rise in atmospheric CO_2 levels. Such iron fertilization takes place in international waters and is subject to treaties and laws administered by the London Convention through the International Maritime Organization. In 2007, they ruled that that commercial iron "dumping" and carbon trading based on ocean iron fertilization should be prohibited unless research provides the scientific foundation to evaluate risks and benefits (see Buessler et al., 2008). This moratorium was subsequently supported by nearly 200 countries at the United Nations Convention on Biological Diversity in 2008. The Royal Society (2009) considered the feasibility of iron fertilization as part of a wider review of planetary scale geoengineering approaches to control increased CO_2 levels and global warming. They concluded that to have even a modest impact, we would need to remove several gigatonnes of CO_2 per year, maintained over decades and more probably centuries. Theoretically—based on Redfield ratios—one atom of Fe could stimulate

production of 100000 organic carbon atoms, but much of this would be remineralized by microbial loop processes. Verifying the overall effect of fertilizing a patch of ocean on CO_2 sequestration is very difficult because stimulating phytoplankton growth in one area might remove nutrients like nitrogen or phosphorus from surrounding areas. It would be very difficult to audit carbon removal, as required by a trading scheme.

Scientific opinion about iron fertilization remains very divided among the world's leading oceanographers. For example, Victor Smetacek from the Alfred Wegener Institute for Polar and Marine Research (Germany) proposes that continued research is necessary to study the relationship between phytoplankton growth and the effects of grazing and breakdown of biomass, before we can determine whether significant amounts of CO_2 are removed from the atmosphere and whether large-scale iron fertilization could be useful (Smetacek and Naqvi, 2008). In 2009, he led the LOHAFEX experiment in the Southern Ocean with researchers from India's National Institute of Oceanography. The expedition attracted extreme opposition from other scientists and environmental lobbyists, who argued that it contravened the moratorium. For a short period, the German government put a stop to the venture before allowing it to proceed. LOHAFEX was conducted in a mesoscale eddy along the Polar Front Antarctic circumpolar current. The phytoplankton here comprise more fast-growing coastal species that are expected to respond more vigorously to iron input than the slower-growing oceanic diatom species studied in previous fertilization experiments (Smetacek and Naqvi, 2010). The results have been interpreted as dampening hopes on the potential of the Southern Ocean to sequester significant amounts of CO_2 because low levels of silicic acid limited the growth of diatoms, and other phytoplankton were quickly removed by zooplankton grazing. Strong et al. (2009) argue that it is "time to move on," because further iron enrichment experimentation "will not resolve any remaining debate about the risks of iron fertilization for geoengineering … and is both unnecessary and potentially counterproductive, because it diverts scientific resources and encourages what we see as inappropriate commercial interest."

seem remarkably stable if viewed over fairly long timescales. By contrast, food chains dominated by microalgae and copepods typical of spring bloom events have dynamics that resemble batch laboratory cultures. Biomass and photosynthetic activity increase rapidly and then decline owing to nutrient depletion and the pressure of copepod grazing, leading to a greater export of photosynthetically fixed carbon to deeper waters through the sedimentation of zooplankton fecal pellets and cell debris. Certain hydrographic, nutritional, and climatic conditions can lead to the very sudden development of massive, dense blooms, such as dinoflagellates or diatoms and prymnesiophytes, which then disappear equally suddenly.

The microbial loop results in retention of dissolved nutrients

About 50% of the daily net production from photosynthesis enters the ocean system as DOM, which supports the growth of heterotrophic bacteria via the microbial loop, resulting in greater retention of dissolved nutrients in the upper layers of the ocean. Some DOM is formed from direct extracellular release of carbohydrates, amino acids, lipids, and organic acids from phytoplankton cells as they grow. The amounts released by this route are very variable, but are generally highest in the most photosynthetically active regions in high light intensities. Extracellular release might result from overproduction of photosynthate when CO_2 fixation provides more organic material than can be incorporated into growing cells because of nutrient limitation. There is probably also a constant leakage of low-molecular-weight compounds across cell membranes owing to the steep concentration gradient caused by very low solute concentrations in seawater. Under some circumstances, phytoplankton cells can lose 5–10% of cell mass a day. A large fraction of DOM is released as dimethylsulfoniopropionate (DMSP)—no other single compound contributes as much to the DOM pool—and the formation and fate of this compound is discussed in the next chapter. A large amount of DOM and POM is released by protistan grazing on plankton. Partially digested particles of prey, unabsorbed small molecules, and digestive enzymes are released as colloidal material during the egestion process, when food vacuoles fuse with the external membrane. DOM and POM are also released during grazing as a result of "sloppy feeding," when phytoplankton cells are broken apart by the mouthparts of zooplankton such as copepods. However, it is now recognized that by far the most important source of DOM is the lysis of phytoplankton and heterotrophic microbes by viruses. For example, up to 80% of the total photosynthetic production can be released as DOM during the collapse of phytoplankton blooms.

The fecal pellets of zooplankton and fish, together with polymers and debris from lysed cells, contribute to the formation of aggregates, and these settle to deeper waters as marine snow. Thus, some of the carbon fixed in the upper layers is removed to deeper waters where the turnover time increases greatly. As it sinks, some of the carbon in POM is remineralized to CO_2 by the respiratory activities of microbes, zooplankton, or fish following ingestion of particles. As shown in *Figure 1.9*, viral lysis and the activity of extracellular enzymes around marine-snow particles leads to a substantial release of DOM from particles as they sink.

Some DOM is converted by purely chemical mechanisms into humic substances; these are complex polymers that resist biodegradation and form a reservoir of refractory organic matter, which eventually sinks to the ocean floor. Large amounts of carbon—possibly about one quarter of net production—are also removed to the ocean floor as $CaCO_3$ in the skeletons of plankton such as coccolithophores and foraminifera. In waters deeper than 3000 m, some of this $CaCO_3$ redissolves to HCO_3^-, but in shallower waters the skeletons of these planktonic forms lead to the formation of calcareous sediments. The vertical movement of organic matter is accentuated by the daily migration of plankton over hundreds of meters in the water column. Together, all the processes described constitute the biological pump depicted in *Figure 8.1B*, which results in the transport of CO_2 from the atmosphere to deep ocean waters.

Despite many advances in recent years, we still do not have reliable estimates of the amounts of carbon exported from the upper ocean and the effects of mixing and water flow in its redistribution. Nor do we have a clear idea of the quantitative role of bacteria in carbon flux in the oceans. Because bacterial growth efficiency is generally low—that is, large nutrient and energy inputs result in low yield of biomass—their overall role in carbon flux is probably as a sink rather than a link to higher trophic levels.

Ingestion of bacteria by protists plays a key role in the microbial loop

The consumption of heterotrophic bacteria by protistan microbes (Chapter 6) helps to explain the regulation of bacterial biomass at near-constant levels and is critical to the ocean food web. Bacterivory constitutes a mechanism by which nutrients contained in very small bacterial cells are made available to larger planktonic organisms. In addition to its importance as a "top-down" control on bacterial production involved in heterotrophic recycling, grazing of cyanobacteria by protists has a direct influence on primary production. It should be noted that, for many species of bacterivorous protists, bacteria are not the sole diet. Feeding experiments and analysis of the food vacuole contents of protists show that many can feed on larger photosynthetic and heterotrophic protists as well as inanimate organic particles (by phagotrophy) or dissolved compounds (by absorption). The most active bacterivorous protists are flagellates in the nanoplankton size class (2–20 μm), and most are less than 5 μm. Flagellated protists, including dinoflagellates, cryptomonads, euglenoids, and ciliates in the microplankton (20–200 μm) class are also active bacterivores. These organisms feed by the generation of water currents with cilia and flagella and they can process hundreds of thousands of body volumes per hour. Larger protists such as radiolarians and foraminifera are also bacterivorous, but their importance is less well known.

Some metazoan zooplankton can also consume bacteria directly. The larvae and juvenile stages of copepods have been shown to consume labeled bacteria (1–5 μm) in experimental studies, but they are possibly less efficient grazers of the very small bacteria that dominate pelagic systems. Tunicates such as larvaceans and salps trap bacteria in fine mesh structures constructed of gelatinous mucus, and the discarded structures and fecal pellets make a major contribution to the formation of marine snow. Indeed, since many pelagic bacteria associate with particles of marine snow rather than exist as freely suspended forms, they may be consumed by a wide range of larger metazoan zooplankton and small fish, which would not be able to feed directly on individual bacteria.

An additional key component of the microbial loop is the release of DOM through protistan grazing. Large amounts of DOM—up to 25% of ingested prey carbon—are egested by protists in the form of fecal pellets, and some of this is readily metabolized and enters the DOM pool for further recycling by bacteria, whilst the remainder enters the "sink" of long-term refractory material. The relatively high respiration and excretion rates of protistan grazers mean that the cycling of "lost" photosynthetic products through the microbial loop is rather inefficient.

Detailed measurements of grazing rates are technically difficult. In one type of experimental study, the uptake of labeled bacteria and their accumulation in food vacuoles is followed using microscopy (fluorescently labeled prey) or radioactive tracer methods (^{3}H– or ^{14}C– labeled prey). Another approach is the dilution method, in which a dilution series of seawater is prepared, so that the natural growth rate of the prey bacteria is unaltered, whilst the predator:prey ratio is reduced. The rate of change in bacterial density is then monitored during incubation at different dilution levels. Separation of bacteria and their grazers in different size classes may also be achieved by filtration. Using such techniques, many studies have attempted to evaluate the impact of protistan grazing on the density, dynamics, and structure of bacterial populations in the oceans. In low-productivity waters, grazing seems to be the dominant factor in balancing bacterial production. However, in higher-productivity waters viral lysis will be favored because of the higher bacterial population densities, as discussed below. Other factors such as removal by benthic filter-feeding animals may also be important in controlling bacterial numbers in coastal regions and estuaries.

Up to a certain limit, which depends on their own size, protistan grazers show a preference for larger prey as food. Therefore, one consequence of grazing pressure

? DO SALPS FAST-TRACK CARBON EXPORT?

Salps are small, pelagic tunicates that pump large volumes of seawater through their body for movement by jet propulsion and trap food particles in a sticky mucus net that is rolled up and passed to the digestive tract. Until recently, it was assumed that salps only feed on organisms larger than 1.5 μm, since this is the pore size in the net. However, Sutherland et al. (2010) developed a mathematical model predicting that salps encounter submicron particles at higher rates than larger particles. This was confirmed by experiments in which salps were provided with fluorescent microspheres of different sizes (0.5, 1 and 3 μm). When 0.5-μm microspheres were used at densities of 10^5 particles ml^{-1} (a value approaching the density of picoplankton cells in the ocean), up to 80% of them were trapped. Calculations indicate that salps could fulfill all their nutritional requirements by this highly efficient uptake of very small microbes. Madin et al. (2006) showed that periodic swarms of one salp species off the northeast US coast consume up to 74% of phytoplankton. Because salps migrate to a depth of 600–800 m at night, fecal pellets and material from dead salps are quickly sequestered as a long-lived pool in deep waters and sediments. Fecal pellets contain partially digested prey and are therefore rich in C, N, P, and other nutrients. The discovery that salps can feed on very small picoplankton, which includes primary producers, as well as heterotrophic bacteria and picoeukaryotes, means that large populations of salps might divert a substantial fraction of the carbon and energy from primary production away from heterotrophic remineralization by transferring it rapidly to deep waters. There have been proposals to develop a network of wave-driven pumps to transport nutrient-rich water from the deep ocean, in order to stimulate production in the surface waters and encourage growth of salps (Kithil, 2006). Like the proposals for iron fertilization, it will be difficult to prove that such a scheme will be effective in sequestering significant quantities of CO_2, unless we can undertake large-scale experiments over a long period.

is to encourage the domination of microbial assemblages by small cells. As discussed on p. 5, most bacteria in oligotrophic pelagic environments are less than 0.6 μm, with many less than 0.3 μm, and many of the smallest bacteria may be in a state of metabolic maintenance rather than active cell division. Thus, the interesting concept arises that protistan grazers are preferentially removing the larger, actively growing and dividing bacteria, leaving a large stock of smaller bacteria that are growing slowly, if at all. However, selective bacterivory of larger, active bacteria seems to stimulate the growth of other bacteria by the release of regenerated nutrients such as ammonium and phosphate. Use of high-resolution video techniques shows that, as well as chance encounters, flagellates may actively select their prey. Species-specific differences in the processing of food particles explain the coexistence of various bacterivorous nanoflagellates in the 3–5-μm size range and indicate the existence of specific predation pressure on different bacteria. Motility, surface characteristics, and toxicity can all affect the outcome of bacterium–protist interactions at the various stages of capture, ingestion and digestion.

Bacteria are also susceptible to predation by other bacteria. The only well-studied example is *Bdellovibrio* (p. 108), but it is likely that many other types remain to be discovered; these could have considerable ecological importance in nutrient dynamics in marine systems.

The "viral shunt" catalyzes nutrient regeneration in the upper ocean

When the microbial loop was first conceived, we had little idea of the abundance and activity of marine viruses. Now, we know that viral lysis of primary producers and heterotrophic organisms act as a short circuit that disrupts the flow of nutrients into higher trophic levels—this has been termed the "viral loop" or the "viral shunt," as shown in *Figure 8.3*. Viral lysis leads to the release of cell contents, much of which enters the DOM pool and is readily recycled by heterotrophs. However, cell fragments and high-molecular-weight components may be more recalcitrant to breakdown. For example, phage lysis of bacteria leads to release of cell envelope fragments containing embedded bacterial outer-membrane proteins, which are relatively resistant to proteolytic degradation. Some components of algal cells lysed by viruses are also very refractory. The quantity and dynamics of these less labile components resulting from viral lysis are not yet fully known. Mathematical models can be constructed to compare the effects of different

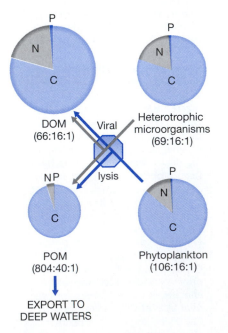

DOM
(66:16:1)

Viral

Heterotrophic
microorganisms
(69:16:1)

NP   lysis   P

POM
(804:40:1)

Phytoplankton
(106:16:1)

EXPORT TO
DEEP WATERS

Figure 8.3 Schematic representation of the viral shunt. The viral shunt moves material from heterotrophs (*gray arrows*) and photoautotrophs (*blue arrows*) into particulate organic matter (POM) and dissolved organic matter (DOM). In this process there is a stoichiometric effect, such that the chemical composition of the POM and DOM pools are not necessarily the same as the composition of the organisms from which the material was derived. Highly labile materials, such as amino acids and nucleic acids, tend to be recycled in the photic zone, whereas more recalcitrant carbon-rich material, such as that found in cell walls, is probably exported to deeper waters. Thus, the material that is exported to deeper waters by the viral shunt is probably more carbon rich than the material from which it was derived. This would increase the efficiency of the biological pump. The numbers in parentheses are the estimated ratios of carbon:nitrogen:phosphorus (in atoms). (Redrawn from Suttle [2007] by permission of Macmillan Publishers Ltd, © Nature.)

levels of viral mortality on nutrient budgets. At a level of 50% bacterial mortality from viruses, the overall level of bacterial production and respiration rate is increased by about a third, compared with that occurring with zero mortality due to viruses. In these models, a high level of protistan grazing leads to carbon input to the higher trophic levels of the food chain (animals), whereas a high level of viral lysis diverts the flow of carbon from the food chain into a semi-closed cycle of bacterial uptake and release of organic matter. Because cell fragments, viruses, and dissolved substances do not sink—unless they aggregate into larger particles—viral lysis has the effect of maintaining carbon and inorganic nutrients such as nitrogen, phosphorus, and iron in the upper levels of the ocean. Viral lysis also contributes to the microscale heterogeneity of seawater through the release of polymeric substances and the dissolution of material from marine snow.

As discussed in Chapter 7, both lytic and lysogenic cycles of infection occur when phages infect marine bacteria. Although the lysogenic state is common and has important consequences for genetic transfer, induction in natural marine systems seems to be relatively low, so the majority of viral infection seems to occur via encounter of bacteria with active virions that initiate the lytic cycle. This process is highly dynamic and is affected by the virus:host ratio, the rate of viral replication, the burst size of virus progeny, and the rates of viral decay. As discussed in Chapter 7, estimation of the impact of viral lysis on bacterial and algal populations produces very variable results, depending on the techniques used, but there is no doubt that viruses are responsible for a significant turnover—perhaps 20–40% per day overall—of ocean microbes. As noted above, viral lysis is relatively more important in eutrophic coastal waters, whereas protistan grazing is more important in oligotrophic ocean waters.

Eutrophication of coastal waters affects microbial activity

Nutrient enrichment of estuaries and coastal waters is a growing problem as a result of anthropogenic sources of pollutants such as runoff from heavily fertilized land, sewage, and animal wastes from agriculture and aquaculture. The impact of eutrophication depends on the source, nature, and level of nutrient inputs, as well as hydrographic factors (especially tidal flushing and mixing) and other physical factors (especially light and temperature). Increased nutrient loading, especially nitrates, stimulates phytoplankton growth beyond the point at which it is controlled by zooplankton grazing. For example, massive cyanobacterial blooms (e.g. *Nodularia*, *Microcystis*, and *Oscillatoria*) occur regularly in the Baltic Sea, and eutrophication is probably a major factor in the increased occurrence of harmful dinoflagellate blooms (see p. 268). Excessive growth of macroalgae also occurs frequently. Active microbial loop processes convert excess primary production, but these too may be overwhelmed, and large amounts of decaying detritus and particles of organic material sink toward the seafloor. Bacterial decomposition leads to a heavy demand for oxygen and the overlying water column may become hypoxic or anoxic, resulting in mass mortality of benthic animals and fish.

Conclusions

This chapter has shown how the integration of discoveries by marine microbiologists with those of physical oceanographers and geochemists has revolutionized our understanding of ocean processes. As atmospheric CO_2 levels and ocean acidification continue to rise, further research to study the flux of carbon is necessary in order to determine the balance of net export of carbon to sediments and regeneration of CO_2 by bacterial respiration. This is especially important in consideration of plans to mitigate climate change by CO_2 sequestration in ocean systems.

? WILL BACTERIAL ACTIVITY IN "DEAD ZONES" ACCELERATE CLIMATE CHANGE?

The number of hypoxic "dead zones" in shallow coastal waters has approximately doubled since the 1960s, largely as a result of the input of excess nutrients, especially nitrate and phosphorus from fertilizers, which enter coastal waters from terrestrial runoff and stimulate phytoplankton blooms. Ironically, plans to reduce the use of fossil fuels by production of ethanol from crops such as corn may increase the spread of dead zones because of increased use of fertilizers (Potera, 2008). Bacterial decomposition results in depletion of oxygen, producing hypoxic levels that are too low to support fish and many invertebrate animals (Diaz and Rosesenberg, 2008). Even slightly reduced oxygen levels may disrupt food chains or cause stress-induced disease in economically important fish and shellfish. As well as these local effects, the expansion of dead zones may have serious consequences for climate change. Heterotrophic denitrification leads to the production of the greenhouse gas nitrous oxide, and Codispoti (2010) comments that bacterial decomposition and denitrification in shallow waters, fuelled by copious supplies of decaying phytoplankton, may lead to a large increase in the rate of nitrous oxide production. Elevated atmospheric concentrations of the gas could further exacerbate the effects of global warming and contribute to ozone loss, causing an increase in exposure to harmful ultraviolet radiation.

References

Azam, F., Fenchel, T., Field, J. G., Gray, J. S., Meyer-Reil, L. A. and Thingstad, F. (1983) The ecological role of water-column microbes in the sea. *Mar. Ecol. Prog. Ser.* 10: 257–263.

Boyd, P. W., Jickells, T., Law, C. S., et al. (2007) Mesoscale iron enrichment experiments 1993–2005: synthesis and future directions. *Science* 315: 612–617.

Buesseler, K. O., Doney, S. C., Karl, D. M., et al. (2008) Ocean iron fertilization moving forward in a sea of uncertainty. *Science* 319: 162.

Codispoti, L. A. (2010) Interesting times for marine N_2O. *Science* 327: 1339–1340.

Diaz, R. J. and Rosenberg, R. (2008) Spreading dead zones and consequences for marine ecosystems. *Science* 321: 926–929.

Fenchel, T. (2008) The microbial loop—25 years later. *J. Exp. Mar. Biol. Ecol.* 366: 99–103.

Herndl, G. J., Reinthaler, T., Teira, E., van Aken, H., Veth, C., Pernthaler, A. and Pernthaler, J. (2005) Contribution of *Archaea* to total prokaryotic production in the ocean. *Appl. Environ. Microbiol.* 71: 2303–2309.

Hirata, T., Aiken, J., Hardman-Mountford, N., Smyth, T. J. and Barlow, R. G. (2008) An absorption model to determine phytoplankton size classes from satellite ocean colour. *Remote Sensing Environ.* 112: 3153–3159.

ICMS Ocean Productivity Study (2000) Rutgers, The State University of New Jersey Institute of Marine and Coastal Sciences. http://marine.rutgers.edu/opp/.

Kithil, P. W. (2006) Are salps a silver bullet against global warming and ocean acidification? American Geophysical Union, Fall Meeting 2006: abstract OS21A-1571.

Madin, L. P., Kremer, P., Wiebe, P. H., Purcell, J. E., Horgan, E. H. and Nemazie, D. A. (2006) Periodic swarms of the salp *Salpa aspera* in the slope water off the NE United States: biovolume, vertical migration, grazing, and vertical flux. *Deep-Sea Res. I: Oceanogr. Res. Pap.* 53: 804–819.

Pomeroy, L. R. (1974) The ocean's food web: a changing paradigm. *Bioscience* 24: 409–504.

Potera, C. (2008) Fuels: corn ethanol goal revives dead zone concerns. *Environ. Health Perspect.* 116: A242–243.

Royal Society (2009) *Geoengineering the Climate: Science, Governance and Uncertainty*, accessed August 16, 2010, http://royalsociety.org/Geoengineering-the-climate/.

Smetacek, V. and Naqvi, S. W. A. (2008) The next generation of iron fertilization experiments in the Southern Ocean. *Philos. Transact. A Math. Phys. Eng. Sci.* 366: 3947–3967.

Smetacek, V. and Naqvi, S. W. A. (2010) The expedition of the research vessel "Polarstern" to the Antarctic in 2009. Alfred-Wegener-Institut für Polar- und Meeresforschung, accessed August 16, 2010, hdl:10013/epic.35169.d001.

Strong, A., Chisholm, S., Miller, C. and Cullen, J. (2009) Ocean fertilization: time to move on. *Nature* 461: 347–348.

Sutherland, K. R., Madin, L. P. and Stocker, R. (2010) Filtration of submicrometer particles by pelagic tunicates. *Proc. Natl Acad. Sci. USA* 107: 15129–15134.

Suttle, C. A. (2007) Marine viruses—major players in the global ecosystem. *Nat. Rev. Microbiol.* 5: 801–812.

Trick, C. G., Bill, B. D., Cochlan, W. P., Wells, M. L., Trainer, V. L. and Pickell, L. D. (2010) Iron enrichment stimulates toxic diatom production in high-nitrate, low-chlorophyll areas. *Proc. Natl Acad. Sci. USA*, doi: 10.1073/pnas.0910579107.

Watson, A. J., Boyd, P. W., Turner, S. M., Jickells, T. D. and Liss, P. S. (2008) Designing the next generation of ocean iron fertilization experiments. *Mar. Ecol. Prog. Ser.* 364: 303–309.

Further reading

Azam, F. and Malfatti, F. (2007) Microbial structuring of marine ecosystems. *Nat. Rev. Microbiol.* 5: 782–791.

Carlson, C. A., del Giorgio, P. A. and Herndl, G. J. (2007) Microbes and the dissipation of energy and respiration: from cells to ecosystems. *Oceanography* 20: 89–100.

Cotner, J. B. and Biddanda, B. A. (2002) Small players, large role: microbial influence on biogeochemical processes in pelagic aquatic ecosytems. *Ecosystems* 5: 105–121.

Falkowski, P. G. (2002) The ocean's invisible forest. *Sci. Am.* 287: 54–61.

Jiao, N., Herndl, G. J., Hansell, D. A., et al. (2010) Microbial production of recalcitrant dissolved organic matter: long-term carbon storage in the global ocean. *Nat. Rev. Microbiol.* 8: 593–599.

Jürgens, K. and Massana, R. (2008) Protistan grazing on marine bacterioplankton. In: *Microbial Ecology of the Oceans*, 2nd Edn (ed. D. L. Kirchman). Wiley, Hoboken, NJ, pp. 383–342.

Karl, D. M. (2002) Nutrient dynamics in the deep blue sea. *Trends Microbiol.* 10: 410–418.

Kolber, Z. (2007) Energy cycle in the ocean: powering the microbial world. *Oceanography* 20: 79–88.

Middelboe, M. (2008) Microbial disease in the sea: effects of viruses on marine carbon and nutrient cycling. In: *Infectious Disease Ecology: Effects of Ecosystems on Disease and of Disease on Ecosystems* (Eds R. S. Ostfeld, F. Keesing and V. T. Eviner). Princeton University Press, Princeton, NJ, pp. 242–259.

Moran, M. A. and Miller, W. L. (2007) Resourceful heterotrophs make the most of light in the coastal ocean. *Nat. Rev. Microbiol.* 5: 792–800.

Nagata, T. (2008) Organic matter-bacteria interactions in seawater. In: *Microbial Ecology of the Oceans*, 2nd Edn (ed. D. L. Kirchman). Wiley, Hoboken, NJ, pp. 207–242.

Pernthaler, J. (2005) Predation on prokaryotes in the water column and its ecological implications. *Nat. Rev. Microbiol.* 3: 537–546.

Pernthaler, J. and Amann, R. (2005) Role and fate of heterotrophic microbes in pelagic habitats: focus on populations. *Microbiol. Mol. Biol. Rev.* 69: 440–461.

Pomeroy, L. R., Williams, P. J. L., Azam, F. and Hobbie, J. E. (2007) The microbial loop. *Oceanography* 20: 28–33.

Robinson, C. (2008) Heterotrophic bacterial respiration. In: *Microbial Ecology of the Oceans*, 2nd Edn (ed. D. L. Kirchman). Wiley, Hoboken, NJ, pp. 299–327.

Sherr, E. and Sherr, B. (2008) Understanding roles of microbes in marine pelagic food webs: a brief history. In: *Microbial Ecology of the Oceans*, 2nd Edn (ed. D. L. Kirchman). Wiley, Hoboken, NJ, pp. 27–44.

Wilhelm, S. W. and Suttle, C. A. (1999) Viruses and nutrient cycles in the sea—viruses play critical roles in the structure and function of aquatic food webs. *Bioscience* 49: 781–788.

Chapter 9

Microbes in Ocean Processes—Nitrogen, Sulfur, Iron, and Phosphorus Cycling

This chapter builds on the knowledge of ecophysiological processes and transformations by the different microbial groups introduced in Chapters 3–5. In the first section, the role of nutrients as limiting factors in ocean productivity is considered. Then, the significance of nitrogen cycling in ocean processes is discussed with particular emphasis on recent discoveries of new nitrogen-fixing organisms, anaerobic ammonia oxidation, and the role of archaea in nitrification, which require major revision of conventional ideas. Sulfur cycling is considered in the third section, mainly from the perspective of the transformations of organic sulfur compounds, the implications for climate control, and the structuring of microbial communities. The final section gives a brief overview of recent developments in our understanding of the uptake of inorganic and organic phosphorus. In all cases, major advances in our understanding have occurred in the past few years by the integration of biogeochemical studies with genomic and metagenomic approaches.

Key Concepts

- Earlier concepts of single elements acting as limiting nutrients have been revised owing to recognition of more complex co-limitation and effects on different community members.

- Estimates of nitrogen fixation in the oceans have increased and the process may be a greater contributor to the nitrogen cycle than previously thought, especially in the tropical and subtropical oceans, as a result largely of the activities of *Trichodesmium* and a range of other newly discovered cyanobacteria.

- The relative importance of the output of nitrogen to the atmosphere by heterotrophic denitrifiers and the newly discovered autotrophic anammox bacteria is an area of active research.

- Algae produce the organic sulfur compound dimethylsulfoniopropionate (DMSP), which is used by numerous heterotrophic bacteria; products from its breakdown affect community structure and ocean ecology and contribute to climatic effects.

- Marine microbes have diverse adaptations to phosphorus limitation, which are important factors in microbial evolution and community dynamics.

? WHAT IS THE REDFIELD RATIO?

During a series of cruises from Woods Hole Oceanographic Institution, Alfred Redfield analyzed thousands of marine samples from different ocean regions. In a keynote paper published in 1934, he reported the empirical observation that the elemental ratios of phytoplankton and dead organic matter in the oceans are very similar throughout the world. The C:N:P ratio became known as the Redfield ratio, averaging approximately 106:16:1 in atomic composition. Subsequently, Redfield developed the concept that the N:P ratio of 16:1 found in the interior of all ocean basins is controlled by the biological activity of phytoplankton, which release N and P to the environment when they break down. Many subsequent studies have confirmed Redfield's hypothesis; but although the Redfield ratio is very stable in the deep ocean, the N:P ratio can vary considerably when different species of phytoplankton grow with limiting concentrations of these nutrients. Arrigo (2005) emphasizes that the N:P Redfield ratio of 16:1 is a general average for various types of phytoplankton growing with different life-cycle strategies under various conditions; N:P ratios vary in different taxonomic groups according to their evolutionary history and are affected by the proportion of ribosomes, enzymes, and pigments within cells. These factors determine whether the organisms are adapted to sustained growth under low-resource conditions, or respond to high-nutrient concentrations with rapid growth (blooming).

NUTRIENT LIMITATION

Key elements may act as limiting nutrients for different groups of microbes

Nitrogen, phosphorus, silica, and trace metals (especially iron) all have the potential to be limiting nutrients, thus affecting productivity. The concept of a single limiting nutrient originates from Liebig's Law of the Minimum, a principle that holds that the single chemical factor in shortest supply will act to limit chemical reactions and, by inference, biological growth, including plankton. In the last few years, there has been a radical reappraisal of nutrient cycling in the oceans, and this simple view has been replaced by the realization that, in some circumstances, limitation may occur by multiple nutrients (or other resources such as light). Such co-limitation can operate at the level of individual cells, at the population level, or at the community level. Different species of plankton may be limited by different nutrients and may be affected differently at different stages of their life cycle. Furthermore, the use of these and other nutrients is closely linked in mixed natural communities, so it is difficult to untangle these complex interactions.

Productivity of surface waters shows marked geographical variations

The idea that available nitrogen is the most important limiting nutrient in the oceans arose largely because most of the early studies of biological oceanography were carried out in temperate North Atlantic waters. During the spring bloom, dominated by diatoms, investigators observed that the increase in phytoplankton biomass and concentration of nitrate were inversely proportional. Phytoplankton growth in seawater increased with experimental addition of ammonium, but not of phosphate. In addition, if the concentrations of phosphate and nitrate in surface seawater are compared as they are utilized, some phosphate remains even when the nitrate levels reach zero. At the onset of the spring bloom in temperate waters, phytoplankton biomass—as indicated by chlorophyll concentrations—is uniform throughout the mixed layer, but as stratification develops, the maximum chlorophyll levels occur in deeper waters where nitrate levels are higher. New production is that portion of primary production dependent on new sources of nitrogen to the photic zone, including nitrate from the deep layers, nitrogen fixed by diazotrophs, and terrestrial runoff. Regenerated production is that portion of primary production supported by nitrogen recycled by heterotrophic decomposition in the photic zone.

This view of limiting nutrients has been challenged by detailed study of other ecosystems. In particular, remote sensing has confirmed the existence of high-nutrient, low-chlorophyll (HNLC) expanses of the oceans, especially in the eastern equatorial and subarctic regions of the Pacific and the Southern Oceans. These areas occupy about 20% of the surface area of the oceans and have high (>2 μM) concentrations of nitrate, but support some of the lowest levels of phytoplankton biomass (<0.5 μg chlorophyll per liter). As discussed below, there is growing evidence that this anomaly is largely explained by low levels of iron in HNLC regions.

The oligotrophic gyres occupy 70% of the oceans' area and are characterized as low-nutrient, low-chlorophyll (LNLC) biomes, whilst high-nutrient, high-chlorophyll (HNHC) and low-nutrient, high-chlorophyll (LNHC) regions are typical of coastal waters—occupying about 5% of the area each. In some areas, such as the eastern Mediterranean, phosphorus appears to be the key limiting nutrient, whilst silicon availability can be limiting for diatom growth in coastal waters and may lead to harmful algal blooms (see p. 268).

Ocean microbes require iron

Iron is an essential component of many enzymes and electron transport proteins, but is present in extremely low concentrations in seawater (see p. 71). Nitrogen fixation and photosynthesis are especially dependent on iron because the key components in these processes—the nitrogenase enzyme complex and the photosystem-cytochrome complex respectively—rely on iron-containing proteins. Over 99% of "dissolved" iron is tightly bound to organic compounds; it occurs mainly as colloidal $Fe(OH)_3$, which is very insoluble, precipitates rapidly and adsorbs to organic particles (*Figure 9.1*). Microbes need special mechanisms to acquire this iron and some bacteria achieve this by production of siderophores (see p. 71), which bring iron into the cell via binding to surface receptors. Some bacteria utilize iron bound to siderophores produced by other species. Photochemical reactions with α-hydroxy acid-containing siderophores (such as aquachelins) may lead to an increased bioavailability of the siderophore-complexed iron. Although some novel mechanisms for iron acquisition have been identified in some members of the *Gammaproteobacteria*, their general significance for marine bacteria is not clear, and almost nothing is known about iron uptake in archaea. Eukaryotic phytoplankton do not appear to synthesize siderophores, but can take up iron via a cell-surface ferrireductase enzyme that liberates iron bound to organic compounds such as porphyrins, which are released from cells through zooplankton grazing and viral lysis. Phagotrophic eukaryotes such as flagellates can acquire iron from ingested bacteria. Depending on the chemical nature of available iron complexes, microbes could therefore be competing for iron and the outcome will affect the separation of ecological niches and community composition. This will have consequences for the fate of carbon in iron-limited waters (see *Research Focus Box 8.1*). Siderophore-bound iron may be the major source of the element in regions dominated by cyanobacterial photosynthesis and microbial loops (such as oligotrophic tropical and subtropical oceans), whilst porphyrin-complexed iron may be the major source in coastal regions dominated by diatoms and zooplankton grazing. Another important factor may be the production of storage proteins to sequester scarce elements like iron within cells. We know that some bacteria produce ferritin-like compounds to store iron when it is more abundant than is needed for immediate use, but these compounds have not yet been detected in aquatic bacteria. Microalgae are known to synthesize phytochelatins, which can store a range of trace metals, but these do not seem to sequester iron. An interesting observation is that domoic acid—a toxin produced by the diatom *Pseudo-nitzschia* (p. 248)—binds iron and is produced in greater amounts when cultures are grown with high iron concentrations.

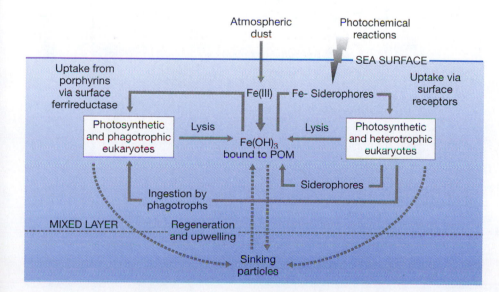

Figure 9.1 Iron cycling in the oceans. Arrows show the transfer of iron and the acquisition mechanisms employed by marine microbes. POM, particulate organic matter.

Low concentrations of iron have been proposed to explain the paradox of the low productivity of HNLC regions. As noted above, these oceanic regions show a much lower rate of photosynthesis than would be expected from the availability of nitrate. The major source of iron is terrestrial, and it follows that most coastal regions receive regular input from rivers, runoff from weathering of rocks, and wind-borne transport of dust. The extent of this input depends on the geology of the land, and some upwelling coastal regions can be iron-deficient. By contrast, the HNLC regions occur at great distances from the continents and will only receive iron inputs via wind-blown dust or from remobilization of iron in sediments and upwelling of deep waters. Iron concentrations in surface water (<200 m) decrease with distance from land, although deep water (>500 m) in open oceans contains about 10 times the level near the surface. This vertical profile resembles that of major nutrients and occurs because biological uptake in the upper (photic) water column is followed by heterotrophic microbial processes of regeneration of particulate organic matter (POM) as it settles into deeper water. Some of the experimental evidence in support of the iron-limitation theory is discussed in *Research Focus Box 8.1*, together with the controversial proposals to use iron fertilization of the oceans as a way of removing excess carbon dioxide from the atmosphere as a countermeasure for global warming.

THE NITROGEN CYCLE

Major shifts in our understanding of the marine nitrogen cycle are in progress

Nitrogen is a key element that constitutes about 12% by dry weight of all living cells because of its role in the structure of proteins, nucleic acids, and several other cell materials. In the environment, nitrogen exists in various oxidation states from –3 in ammonium (NH_4^+) to +5 in nitrate (NO_3^-); the role of microbes in these oxidation–reduction transformations was introduced in Chapter 3. Whilst the principal aspects of the nitrogen cycle have been understood for more than 100 years, major changes in our understanding of how it functions in the marine environment have resulted from recent discoveries of previously unknown microbial processes, leading to the model shown in *Figure 9.2*. One of the biggest challenges facing oceanographers is determining a budget for the input and output of nitrogen via nitrogen fixation and nitrate reduction processes, respectively.

Figure 9.2 Microbial transformations in the marine nitrogen cycle. (Reprinted, with minor modifications, from Francis et al. [2007] by permission of Macmillan Publishers Ltd.) Key to functional genes discussed in the text: *amo*, ammonia mono-oxygenase; *hao*, bacterial hydroxylamine oxido-reductase; (?), unknown gene/enzyme in ammonia-oxidizing archaea; *nif*, nitrogenase; *nir*, nitrite reductase; *nor*, nitric oxide reductase. PON, particulate organic nitrogen; DNRA, dissimilatory nitrate reduction to ammonium. Figures in blue boxes show annual input and output of nitrogen in teragrams (Tg), using data from Capone (2008); but note that the balance of the nitrogen budget is the focus of current research (see *Research Focus Box 9.1*) and this model does not include the recently discovered contribution of nitrogen fixation by consortia of methanotrophic archaea and bacteria in sediments (see p. 125).

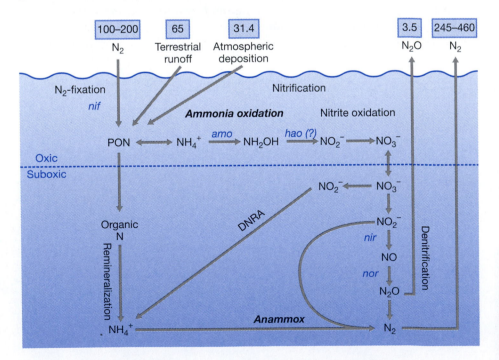

New nitrogen-fixers have been discovered recently

Nitrogen gas (triple-bonded N_2) forms 78% of the modern atmosphere and must be fixed in a reduced state into organic material before it can used in cellular processes. Only a small number of microbes (diazotrophs) are capable of carrying out this transformation (see *Figure 3.5*). For many years, the rate of nitrogen fixation in the surface waters of the oceans was thought to be negligible compared with other sources of nitrogen such as upwelling of nitrate from deep waters; it was also thought to be only a small fraction of the nitrogen fixed in soils on land. It has always been a mystery that there seem to be so few ocean organisms that can exploit the abundant supplies of gaseous nitrogen under the evolutionary pressure of severe shortages of dissolved inorganic nitrogen compounds, although it must be recognized that nitrogen fixation is energetically expensive and depends on a large use of ATP. However, biogeochemical evidence indicates that diazotrophy in the large portion of the biosphere occupied by the tropical and subtropical oceans may be much more important than previously realized. Although there are still many uncertainties, oceanic nitrogen-fixation rates are now estimated at between 100 and 200 Tg per year—a figure close to that believed to occur in soils. In the 1960s, classical microbiological and physiological techniques using isotope tracer techniques led to the suggestion that the large filamentous cyanobacterium *Trichodesmium*—highly abundant as colonial blooms in tropical seas (see *Figure 4.10D*)—could carry out nitrogen fixation. For some time, this was discounted because it was not clear how *Trichodesmium* could fix nitrogen without forming heterocysts, structures known to protect the nitrogenase enzyme from oxygen (see p. 67). However, use of a sensitive acetylene reduction assay technique that could be used in the field proved that *Trichodesmium* does make a significant contribution to the global nitrogen fixation budget. A major breakthrough in this area has come from the application of molecular biological approaches to detect the presence of nitrogenase genes in water samples and the discovery of new groups of very small, abundant unicellular diazotrophic cyanobacteria occurring as free-living forms and in association with diatoms, tintinnids and radiolarians. Other groups of *Alpha-* and *Gammaproteobacteria* have also been identified as diazotrophs, but little is known about their activity. Nitrogen fixation is also thought to be carried out by anaerobic bacteria that inhabit the gut of copepods; as these are highly abundant members of the zooplankton, this could be a significant source of fixed nitrogen into food webs. There appear to be many complex factors, including temperature and the availability of either phosphorus or iron as limiting nutrients, which determine the geographical and seasonal distribution of the various types of diazotrophic bacteria. This means that incorporation of these new findings into biogeochemical models is still very difficult.

Fixed nitrogen is returned to the inorganic pool by ammonification and nitrification

When organisms die, the amino groups of proteins and nucleotides are returned to the inorganic pool via the process of ammonification, carried out by many types of bacteria. Ammonium is rapidly oxidized to nitrate by chemolithotrophic bacterial and archaeal nitrifiers. There is now considerable evidence showing that ammonia-oxidizing archaea are highly diverse, abundant, and active in areas critical for global nitrogen cycling, including the base of the photic zone, suboxic waters, and coastal sediments. As a group, the marine *Crenarchaeota* seem to have a high degree of metabolic versatility, and in addition to their role in the nitrogen cycle, the possibility that they contribute to the carbon cycle via autotrophy means that measurement of their activities in different geographical regions and at different depths is a high priority for research. This is discussed in *Research Focus Box 5.1*.

Because it is highly reduced, ammonium can be easily assimilated by many microbes and used to synthesize amino acids, which are then transformed to

DIFFERENCES IN NITRATE USAGE BY *PROCHLOROCOCCUS*

Some phytoplankton such as *Synechococcus* diatoms respond with rapid growth to high nitrate concentrations found at high latitudes and in coastal and open-ocean upwelling, leading to the formation of seasonal or transient blooms. However, when the genome sequences of cultured strains of high- and low-light ecotypes of *Prochlorococcus* were studied (see p. 114), it was found that they lacked genes for assimilatory nitrate reductase, leading to the assumption that *Prochlorococcus* depends entirely on regenerated nitrogen and can grow only by using ammonium ion. This is very significant because this cyanobacterium is the dominant member of the phytoplankton in subtropical and tropical oceans. However, by analysis of the GOS metagenome (see p. 100), Martiny et al. (2009) showed that there are microdiverse lineages within *Prochlorococcus* with genes required for nitrite and nitrate assimilation in both ecotypes. However, the currently cultured members of these ecotypes, used for genome sequencing, do not have these genes. The GOS data also indicate that nitrate assimilation genes are more prevalent in some regions than others, and Martiny and colleagues speculate that this is related to nitrogen availability. There may be selection of lineages with the ability to acquire nitrate in regions with low N concentrations, whereas in high N regions, this may be outweighed by the advantage of possessing a smaller genome. This finding has major implications for understanding the role of *Prochlorococcus* in productivity and export of carbon from the upper ocean.

other compounds via transamination reactions. This is a less energy-demanding process than the uptake of nitrate, which requires large amounts of NADPH and ATP because of its high oxidation state.

Denitrification and anammox reactions return nitrogen to its elemental form

Denitrification is an anaerobic respiratory pathway in which fixed nitrogen is removed from the ocean, usually in the form of nitrate, returning nitrogen to the atmosphere and completing the cycle shown in *Figure 9.2*. Nitrous oxide and nitrite can also accumulate under certain conditions. Although the denitrification process only occurs under low-oxygen conditions, potential dentrifiers belonging to a very wide diversity of heterotrophic microbes can be identified in almost all environments in both oxic and anoxic waters and sediments, and many of these organisms can also grow aerobically using oxygen as an alternative electron acceptor.

Since the 1960s, measurements of the levels of ammonium in anoxic basins have indicated that there are much lower levels of ammonium than might be expected to accumulate during the intense remineralization (either as aerobic or N-dependent respiration) supposed to occur there, suggesting that there is some unknown oxidation mechanism for its removal by anaerobic microbial activity. The discovery of anammox (anaerobic ammonia oxidation) explains this deficit and has necessitated a major reevaluation of the ocean nitrogen cycle. Only a very narrow phylogenetic group of bacteria (*Planctomycetes*; p. 117) are thought to be responsible for this process, which depends on a specialized cellular structure with a membrane that protects the cytoplasm from hydrazine, a highly reactive intermediate formed during the oxidation of ammonia to nitrogen using nitrite as the electron acceptor. Anammox bacteria grow extremely slowly—perhaps doubling only every 14 days or so—and many oceanographers believed that they would be unlikely to thrive in natural environments. However, several studies carried out since 2003 have indicated that, in sediments and in some anoxic basins, anammox is responsible for at least as much nitrogen loss as occurs by denitrification. It is possible that nitrate reduction and anammox are coupled and the nitrite generated by denitrifying bacteria might then serve as the electron acceptor in the anammox reaction.

These processes occur in the oxygen-minimum zones (OMZs). About half of the global extent of OMZs corresponds with zones of nitrogen loss due to denitrification and anammox. A suboxic layer (usually defined as <20 µM dissolved oxygen) typically occurs in all oceans at intermediate depths (usually 1000–1500 m). This usually corresponds to the pycnocline, where there is change in density and accumulation of high levels of particulate organic matter, resulting in a zone of intense heterotrophic bacterial activity which depletes oxygen. OMZs usually occur at much shallower depths (typically 50–200 m) and have extremely low oxygen concentrations at their core—often about 50 times lower than those at the "classical" oxygen minimum. Defining the extent of the OMZs depends on the definition adopted; if defined by an oxygen concentration below 5 µM, they occupy about 0.05% of the volume of the world's oceans, whereas if defined at the suboxic concentration of below 20 µM, they occupy about 1% of the volume. The major permanent open ocean OMZs—in which extremely low oxygen levels and extensive loss of nitrogen occur—are the eastern tropical North Pacific, the eastern South Pacific near the equator and Chile and Peru, and the Arabian Sea (shown in *Plate 9.1*). Recently, another permanent OMZ has been identified in deeper waters of the eastern subtropical North Pacific (25°N–52°N) off the west coast of North America. In addition, OMZs form during the winter in higher latitudes, such as the western Bering Sea and the Gulf of Alaska. Oxygen concentration is not uniform throughout the OMZ. There is a core region—occupying about 10% of the volume of the whole OMZ—with the lowest oxygen concentration (between 1 and <20 µM), as well as broader suboxic zones. There is usually a narrow oxycline (about 10–20 m thick) at the top and bottom of the OMZ.

Microbial processes in sediments are a major contributor to nitrogen cycling

Nitrogen-containing POM that reaches the seafloor may be permanently buried, whilst a range of anaerobic heterotrophs are responsible for production of ammonium ion which is available for regeneration by other organisms, or may be coupled to nitrate production in the oxic layers of sediments. POM exported to sediments in deep waters usually has a much higher ratio of carbon to nitrogen than plankton in the upper ocean because of the extensive recycling of labile nitrogen-rich compounds in the microbial loop and viral processes (see *Figure 8.3*). Nitrogenous compounds reaching the seafloor are predominantly proteins, chitin, and other polymers with a relatively high molecular weight which have escaped degradation by bacterial enzymes during their descent to the seafloor. Organisms in the sediment may also absorb nitrate, and this process is especially important in coastal waters receiving high levels of nitrate from agricultural run-off. As noted on p. 101, in some regions nitrate can serve as an electron acceptor for bacteria such as *Thioploca* and *Thiomargarita*, which form dense colonies underneath anoxic waters with high levels of nitrate from natural upwelling, and form filaments that stretch up to collect nitrate which is then reduced to ammonium using hydrogen sulfide as electron donor. Sediments on the seafloor are usually anoxic below the top few millimeters of the surface and are therefore a major site for the anammox and denitrification processes previously described for the loss of fixed nitrogen. Nitrogen cycling also occurs on the surface of sediments in coastal and shelf seas in relatively shallow waters, which permit the penetration of sufficient light for establishment of communities of microalgae and cyanobacteria. Although nitrate is the main source of nitrogen for these communities, some nitrogen fixation may take place within microbial mats. Benthic nitrogen fixation by symbiotic or closely coupled diatoms, cyanobacteria, and other diazotrophs helps to explain the high productivity of some benthic systems, such as seagrass beds and coral reefs (see p. 68).

THE SULFUR CYCLE

The oceans contain large quantities of sulfur—an essential element for life

Sulfur is an essential element for all organisms, constituting about 1% of cellular mass, mainly in the amino acids methionine and cysteine and in various coenzymes and metalloproteins. The oceans and sediments contain a high concentration of inorganic sulfur compounds in the various oxidation states. The most important oxidation states are –2 (sulfhydryl R-SH and sulfide HS$^-$), 0 (sulfur S^0), and +6 (sulfate SO$_4{}^{2-}$). These are transformed by microbes and also by some abiotic processes. The oceans are the largest reservoir of sulfur (in the form of sulfate) in the biosphere. As noted on p. 69, it is necessary to distinguish between dissimilative and assimilative sulfate reduction. The dissimilative process—for energy generation—leads to the production of hydrogen sulfide and is restricted to the sulfate-reducing bacteria (including many members of the *Deltaproteobacteria*), which are a key part of the community in anaerobic sediments. By contrast, many protists, bacteria, and archaea can assimilate sulfide produced during ATP-driven reduction of sulfate and incorporate it into organic sulfur compounds.

Metabolism of organic sulfur compounds is especially important in surface waters

Organisms need sulfur for the synthesis of sulfur-containing amino acids and other essential compounds. Despite the abundance of sulfate in seawater, it seems that most marine microbes obtain sulfur from recycling of organic sulfur compounds produced by phytoplankton, especially DMSP, [(CH$_3$)$_2$S$^+$CH$_2$CH$_2$COO$^-$]

BOX 9.1   RESEARCH FOCUS

Fixing the nitrogen cycle

Recent research leads to a major re-evaluation of the input and output of nitrogen to the oceans

Is there a feedback mechanism to balance the marine nitrogen cycle?

Douglas Capone of the University of Southern California, one of the leading experts on marine nitrogen fixation, provides values of 100–200 Tg for the annual global marine input of nitrogen by diazotrophy and 400 Tg for nitrogen lost by denitrification and anammox (Capone, 2008). He argues that this imbalance would lead to the oceans being depleted of nitrate within a few thousand years; but this is not supported by geochemical evidence. Is there really such an imbalance, or is there some "homeostatic" mechanism that couples the processes that add or remove nitrogen? Deutsch et al. (2007) determined nitrogen fixation rates in the world's oceans by use of an ocean circulation model to assess the impact on deviations of nitrate and phosphate concentrations from the 16:1 Redfield ratio. They found that in upwelling waters in contact with oxygen-minimum zones (OMZs), the N:P ratio is less than 16:1. They suggested that, as surface waters flow offshore, the "excess" phosphorus is removed by nitrogen fixers, which presumably extract phosphorus without absorbing nitrogen compounds. Thus, it is possible that nitrogen-fixation rates are highest in the Pacific Ocean and other areas downstream of the OMZs (see *Plate 9.1*), although this has not been confirmed. Excess phosphate declines toward zero (i.e. Redfield ratio approximates to 16:1) in the subtropical gyre of the North Pacific, throughout the North Atlantic, and in the southern subtropical gyres of all three ocean basins. Thus, the model predicts that nitrogen fixation is focused in the low-latitude ocean and is the main source of nitrogen sustaining biological carbon export in these regions. Supporting this idea, these phosphate-enriched regions coincide with the geographical distribution of the diazotroph *Trichodesmium* observed by satellite. This research supports the idea that there is a negative feedback mechanism, so that externally driven changes in the oceanic nitrogen levels are counteracted by decreased or increased rates of nitrogen fixation, respectively. It seems that either we are overestimating the loss of fixed nitrogen or missing a major additional input of nitrogen to the oceans.

"Nifty" techniques reveal that small cyanobacteria have a major role in nitrogen fixation

The work of Jonathan Zehr and colleagues at the University of California–Santa Cruz has had a large impact on our knowledge of marine nitrogen fixation. They exploited the fact that the nitrogenase enzyme contains two proteins that are highly conserved in a wide range of bacteria. PCR primers that recognize conserved sequences of the *nifH* gene (which encodes the iron protein component of nitrogenase reductase) were used to amplify the gene from samples of picoplankton from the Atlantic and Pacific Oceans. Amplified DNA fragments were cloned and sequenced and a variety of *nif* sequences were identified from several diverse bacterial lineages (Zehr et al., 2000). The most abundant cyanobacterial genera, *Synechococcus* and *Prochlorococcus*, do not fix nitrogen, but cyanobacteria in the 3–10-μm diameter size range were show to be diazotrophs. Zehr et al. (2001) showed that nitrogenase genes from the subtropical north Pacific Ocean gyre were closely related to genes from the unicellular cyanobacterium *Crocosphaera watsonii*. Similar gene sequences have since been discovered in many other regions. Montoya et al. (2004) measured high rates of nitrogen fixation down to at least 100 m in tropical Pacific waters and calculated that these organisms contribute at least 10% of total oceanic new production. Studies of the distribution of these cyanobacteria are now in progress (Moisander et al., 2010). One active nitrogen-fixing cyanobacterium (strain UCYN-A) has defied all cultivation attempts, but reconstruction shows that the genome size is only 1.4 Mb, with no photosystem II, no TCA cycle and no RuBisCO (Tripp et al., 2010). Strain UCNY-A seems to be an obligate photoheterotroph that relies on an exogenous source of nutrients for the large amounts of energy needed for nitrogen fixation. It is possible that this photofermentative metabolism could have interesting biotechnological potential.

Nitrogen loss—a tale of two oceans?

A major reevaluation of the marine nitrogen cycle has become necessary with the discovery of the anammox reaction for oxidation of ammonium to dinitrogen in OMZs. What proportion of the nitrogen loss is due to this mechanism, compared with heterotrophic denitrification? Is the relative importance the same in all OMZs? One reason why it is important to determine the relative contribution of these two pathways is because production of the powerful greenhouse gas nitrous oxide is an intermediate in denitrification, but not in anammox. Also, the two processes are likely to be regulated by different factors and may be affected differently by climate change. Experts disagree on these matters, and some of the controversy is evident from two major studies published in 2009.

A major revision of the nitrogen cycle was proposed by Phyllis Lam and coworkers at the Max Planck Institute in Bremen, Germany (Lam et al., 2009). They studied the OMZ of the eastern tropical South Pacific (ETSP) that extends for 10000 km off the coast of Peru (see *Plate 9.1*). Using genome information from the anammox bacterium *Candidatus* Scalindula (see p. 69), primers for the anammox-specific nitrite reductase gene *nirS* were used in quantitative RT-PCR (see p. 44), showing expression of this gene.

BOX 9.1   RESEARCH FOCUS

This result confirmed other studies that showed abundance of the anammox bacteria in this OMZ (Hamersley et al., 2007). The rates of reduction of nitrate and nitrite and the rate of ammonium oxidation were measured using stable-isotope pairing techniques, in which ^{15}N tracer molecules are added to water samples. Chemical transformations result in products with different ratios of ^{15}N and ^{14}N (see Steingruber et al., 2001). These assays were conducted alongside RT-PCR of the relevant functional genes. Results showed that anammox is the major pathway by which nitrogen is lost in this system and that it is coupled directly to various aerobic and anaerobic transformations of nitrogen compounds. Reduction of nitrate provides nitrite for the anammox reaction, whilst dissimilatory nitrate reduction to ammonium (DNRA) in the core OMZ provides a substantial part of the ammonium requirement for anammox (see Figure 9.2), with the remainder coming from remineralization of organic nitrogen coupled to nitrate reduction. Lam et al. (2009) concluded that heterotrophic denitrification is of minor importance and that remineralization of organic matter by anammox plays the dominant role in nitrogen loss. Thus, calculations based on nitrate deficits alone may lead to the apparent discrepancies in the nitrogen budget, discussed above.

In another paper published shortly afterwards, Bess Ward and colleagues at Princeton University, New Jersey, confirmed that anammox is dominant in the ETSP, but came to very different conclusions about the nitrogen cycle in the Arabian Sea. Here, use of two different isotope incubation methods showed that heterotrophic denitrification was the dominant process—responsible for 87–99% of the total nitrogen production (Ward et al., 2009). Molecular studies using different nirS primers to those used by Lam's group showed that diverse denitrifying bacteria were present at much higher levels than anammox bacteria in both study regions. Ward et al. (2009) suggested that denitrification is limited by the availability of organic carbon in the OMZ of the ETSP, but not in the Arabian Sea. Another paper from this group (Bulow et al., 2010) confirms their conclusion that denitrification exceeds anammox as a pathway for nitrogen loss in the Arabian Sea OMZ. A major difference between the two studies was that Lam and colleagues measured gene expression, whereas Ward and colleagues only measured gene abundance. It is possible that denitrifier nirS genes are more abundant than anammox nirS at the gene level. Only anammox nirS genes were found to be consistently and actively expressed, but denitrifier genes were only sporadically expressed—consistent with rate measurements; furthermore, since the primers used by Ward's group would exclude the predominant anammox bacterium Candidatus Scalindua arabica, anammox bacterial abundance would be underestimated. Here, there are frequent pulses of organic matter and Ward and colleagues suggest that the lack of correlation between bacterial abundance and the rates of denitrification in the two OMZs sites is due to the contrasting lifestyles of the denitrifying and anammox bacteria. Denitrifying bacteria are diverse opportunistic heterotrophs that can respond rapidly to episodic supplies of organic matter, whereas anammox bacteria are more constrained, slow-growing chemolithotrophic autotrophs. However, the Bremen group also conducted experiments in the Arabian Sea and concluded that most nitrogen loss was attributed to anammox, and denitrification activities were sporadic at best (Phyllis Lam, personal communication). Since a large proportion of the global nitrogen loss occurs in the Arabian Sea, it is important to resolve the different interpretations of these results. Clearly, the methods and timing of experimental measurements could lead to very different conclusions about the importance of different processes, which can only be resolved by long-term studies at sea. Sadly, further scientific work in the Arabian Sea is likely to be hampered by the serious problem of piracy and kidnapping off the coast of Somalia.

produced from the amino acid methionine by many algae. Some of the bloom-forming prymnesiophytes (e.g. *Phaeocystis globosa*), coccolithophorids (e.g. *Emiliania huxleyi*), and dinoflagellates are particularly potent producers of the compound, with up to 300 mM intracellular concentrations. DMSP is believed to be accumulated at high intracellular concentrations as a compatible solute that provides protection against osmotic stress. It may also have cryprotectant, antioxidant, and defense functions. An alternative view is that it has evolved as a metabolic relief mechanism when algae are undergoing unbalanced growth, in order to eliminate excess reducing power and reduced sulfur. There is some evidence that synthesis of DMSP (and that of another compatible solute, glycine betaine) is regulated by nitrogen levels. Some DMSP is exuded naturally from healthy algae, but this probably represents only 1–10% of the cellular contents—most DMSP enters the water column as result of disruption of the cells by zooplankton grazing or by viral lysis. The annual global production of sulfur by DMSP production and release from phytoplankton has been estimated at over 6×10^7 Gt. DMSP is the main vehicle for the cycling of organic sulfur compounds in the oceans. The high carbon content of DMSP means that it also plays a very important role in the flow of carbon in the marine food web—it is probably the most important single organic compound transferring fixed carbon to the microbial loop processes (see *Figure 8.2*).

Many algae also incorporate large amounts of sulfur into sulfated polysaccharides such as mucus and cell-wall components. These compounds have important functions in defense against grazing and the sequestration of metals. Because they are quite recalcitrant to microbial attack, these compounds form a significant component of marine snow aggregates.

A fraction of DMSP production leads to release of the gas dimethyl sulfide (DMS)

The fate of DMSP is a very important factor in marine ecology and ocean–atmosphere interactions. When DMS was first discovered, it was assumed that phytoplankton produce the gas directly, but we know that it is generated by the action of a group of DMSP lyase enzymes, which are widespread in many marine microbes and break down DMSP into DMS and acrylate (*Figure 9.3B*). DMSP-producing algae often contain DMSP lyase themselves, but this is usually in a separate cellular compartment, so it is thought that the enzyme only comes into contact with the substrate when cells are broken apart by grazing or viral lysis.

DMS is highly volatile and some of it escapes to the atmosphere, where it leads to the formation of sulfates and other oxidized products, which act as nuclei for water vapor, causing the formation of clouds. DMS, therefore, has an albedo function, affecting temperature and light availability to surface waters. Because DMS production was thought to be directly related to phytoplankton growth and therefore was controlled by light and nutrient availability, a hypothesis was developed in which DMS production by algae provided a feedback mechanism that regulated the climate. This idea is one of the real-life "planks" for the controversial "Gaia Hypothesis" developed by James Lovelock and Lynn Margulis, in which the biological and physical components of Earth interact to maintain the climatic and biogeochemical conditions of the planet. As shown in *Figure 9.3A*, one version of the model envisages a decline in phytoplankton growth as cloud increases owing to a rise in the level of DMS until cloud formation diminishes again. As shown in *Figure 9.3B*, we now know that most of the DMSP is broken down by microbial activity; 80–90% is demethylated and enters the microbial loop, whilst 10–20% is metabolized by the DddD cotransferase or degraded by DMSP lyases to DMS and acrylate, which are also metabolized by bacteria. Only DMSP lyase activity leads to production of DMS and only 1–2% of the sulfur in DMSP is thought to enter the atmosphere as DMS. Although the amount of DMS transferred to the atmosphere is very significant (15–33 Mt a year) it is much less than previously thought, necessitating revision of ocean–atmosphere climate models. As shown in *Figure 9.3B*, demethylation and demethiolation reactions generate products that are readily assimilated by microbes. Recent research on this topic is discussed in *Research Focus Box 9.2*.

Microbial sulfate reduction and sulfide oxidation occur in sediments, vents, and seeps

When organic matter reaches the seafloor and is incorporated into sediments, conditions quickly become anoxic. As noted on p. 109, a specialized group of sulfur- and sulfate-reducing bacteria belonging to the *Deltaproteobacteria* use organic compounds or hydrogen as electron donors for the reduction of sulfate to sulfide. In deep sediments, archaea carry out anaerobic oxidation of methane coupled to sulfate reduction (see *Figure 5.2*). Sulfide is in turn metabolized anaerobically by bacteria such as *Thioploca*, *Thiothrix*, and *Thiomargarita*, which transport nitrate into the sediment for use as an electron acceptor. Bacteria such as *Beggiatoa*, *Thiobacillus*, and *Thiovulum* metabolize sulfide aerobically in the

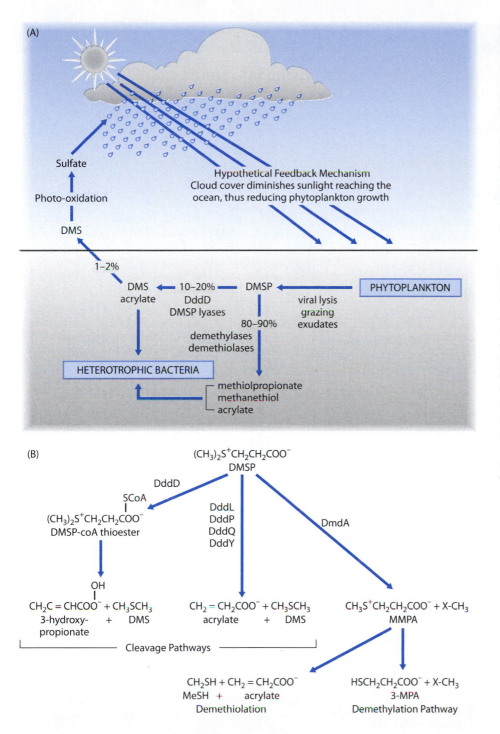

Figure 9.3 Microbial transformations and the fates of dimethylsulfoniopropionate (DMSP) in the ocean, resulting in production of dimethyl sulfide (DMS) and products that are assimilated by different bacterial groups for cell growth. (A) Representation of the hypothetical mechanism by which release of DMS gas to the atmosphere results in the formation of cloud condensation nuclei, producing cloud albedo that may constitute a feedback process that regulates phytoplankton growth. Figures show the estimated percentage of DMSP transformed at each stage. (B) Principal transformations of DMS by microbial activity. Arrows are labeled with the principal enzymes known for the cleavage and demethylation pathways. DddD, a class III acyl CoA transferase enzyme, which is thought to perform the initial step in DMS release, as an alternative to DMSP lyase; MeSH, methanethiol; MMPA, 3-methiolpropionate; MPA, 3-mercaptopropionate; X-CH3, unidentified intermediate with terminal methyl group.

upper part of sediments or microbial mats. High concentrations of hydrogen sulfide occurring at hydrothermal vents and cold seeps result in abundant chemolithotrophic sulfide-oxidizers, both as free-living forms and in symbiosis with animals (see Chapter 10).

THE PHOSPHORUS CYCLE

Phosphorus is often a limiting or colimiting nutrient

Phosphorus is an essential element for all life, being a key constituent of nucleic acids and lipids and a crucial component of ATP in energy transfer reactions. Inorganic phosphorus also has a key role in photosynthesis. In the oceans, there

BOX 9.2   RESEARCH FOCUS

DMSP, DMS, and marine microbial communities
Transformations of organic sulfur compounds are emerging as major drivers of ocean processes

Enzymes that liberate DMS from DMSP are highly diverse and widely distributed

Bacteria that catabolize dimethylsulfoniopropionate (DMSP) have been isolated from several environments, including corals, algal blooms, copepods, and intertidal salt marshes (Yoch, 2002). Recent work by Andy Johnston and colleagues at the University of East Anglia, UK, has revealed that the nature of the enzymes that degrade DMSP varies greatly among different microbes. Culture plates containing DMSP were used to isolate bacteria that can utilize the compound, and genes were cloned into *Escherichia coli*. It was simple to determine whether DMSP-degrading genes were expressed in the recombinants by sniffing the *E. coli* cultures for the characteristic smell of dimethyl sulfide (DMS), then selecting cultures for confirmation by gas chromatography. The first enzyme characterized (DddD) was identified in the bacterium *Marinomonas* found in a salt-marsh plant and shown to be a transferase that releases DMS via an unknown CoA intermediate. Homologs of this enzyme have subsequently been identified in other members of the *Gammaproteobacteria* and in the alga *Emiliania huxleyi*. Four other gene families encode DMSP lyases called DddL, DddP, DddQ, and DddY. These all produce DMS and acrylate, but they seem to be completely unrelated as they have almost no sequence similarity and differ in their requirement for metal cofactors. The L, P, and Q enzymes occur mainly in members of the marine alphaproteobacterial order *Rhodobacterales*, but the P type also occurs in ascomycete fungi such as *Fusarium* and *Aspergillus* (Kirkwood et al., 2010a; Todd et al., 2009). The Y and D enzymes seem to be most important in bacteria that depend on DMSP for a large part of their carbon sources, occurring in high DMSP habitats such as coral reefs and mudflats. Analysis of the GOS metagenomes (see *Research Focus Box 4.1*) shows that sequences homologous to the genes *dddP* and *ddQ* are widespread. In another study, Curson et al. (2009) isolated strains of *Pseudomonas* and *Psychrobacter* that grew on DMSP as its sole carbon source from the gut of herring, due to the production of an inducible DMSP lyase encoded by *dddD*. This is the first demonstration of DMSP-degrading bacteria and their relevant genes in the gut microbiota of a vertebrate. Other studies have shown that a large proportion of the carbon in DMSP ingested by other planktivorous fish is assimilated into fish tissue with very little release of DMS, suggesting that other transformations may be carried out by the microbial community in the fish gut. The most obvious explanation for this remarkable distribution of the *ddd* genes is that they have been transferred by extensive lateral gene transfer (Kirkwood et al., 2010b).

Demethylation of DMSP diverts the production of DMS

It has been known since the early 1990s that DMSP is readily degraded by chemoheterotrophic bacteria. DMSP is estimated to supply up to 15% of the total bacterial carbon demand and nearly all of the bacterial demand in the surface waters of the ocean—no other single compound contributes as much carbon to the DOM pool (Yoch, 2002). However, the gene *dmdA* responsible for the demethylating enzyme that diverts primary production away from DMS production has only been recognized recently, following work by the group of Mary Ann Moran at the University of Georgia. Howard et al. (2006) identified genes for a glycine cleavage T-family protein with DMSP methyltransferase activity in the highly abundant *Roseobacter* and SAR11 clades in the *Alphaproteobacteria*. The enzymes from cultures of *Silicibacter pomeroyi* and *Candidatus* Pelagibacter ubique (the cultured representative of the abundant SAR11 clade; see p. 37) have been cloned and expressed in *E. coli*, in order to study the kinetics of demethylation (Reisch et al., 2008). An extended analysis of GOS data showed that more than half of the bacteria in marine surface waters have the genetic capability to demethylate DMSP. Howard et al. (2008) found 1701 homologs of the gene *dmdA*, which they calculated as sufficient for 58% of sampled cells to participate in the demethylation step. They also enumerated gene homologs encoding proteorhodopsin (p. 64); based on these analyses, they concluded that DMSP-demethylating bacteria with photoheterotrophic potential are the most abundant type of small bacterioplankton in the surface ocean. Also, a new cluster of *dmdA* sequences were affiliated with the *Gammaproteobacteria*. Genetic analysis provided strong evidence for extensive lateral gene transfer as a mechanism for the wide distribution of this mechanism, as is the case with *ddd* genes. Moran's group is developing ocean monitoring assays to measure abundance and expression of DMSP degrading genes, which may resolve some of the uncertainties about the role of marine bacterioplankton in the hypothetical feedback mechanism for global temperature regulation. For example, Vila-Costa et al. (2010) used metatranscriptomics (see p. 52) to show that enrichment of surface water with DMSP led to an increased expression of genes supporting heterotrophic activity and depletion of those genes associated with phototrophy, suggesting a very rapid shift in the use of DMSP by a variety of marine bacteria.

The role of DMSP and DMS in coral biology

As discussed in Chapter 10, corals are holobionts consisting of complex communities of the coral animal and various microbes—*Symbiodinium* dinoflagellates (zooxanthellae), other protists, bacteria, archaea, fungi, and viruses. *Symbiodinium* has been found to produce high intracellular levels of DMSP, and the very high densities of these organisms occurring within the tissue of many corals means that the concentration of DMSP produced by corals may be a

BOX 9.2   **RESEARCH FOCUS**

highly significant component of the global sulfur cycle, despite the restricted distribution of reefs. Release of large quantities of DMSP in coral mucus is likely to provide a major link to microbial loop processes in the nutrient-deficient waters inhabited by most tropical corals. This has been investigated on the Great Barrier Reef by scientists from the Australian Institute of Marine Science. Raina et al. (2009) used enrichment culture to isolate bacteria that use DMSP as their sole carbon source by placing fragments of coral tissue and mucus from two species of corals in vials of minimal media, to which DMSP was added. Cultures were plated onto agar minimal medium supplemented with DMSP, DMS, or acrylic acid, and breakdown of these compounds was monitored by gas chromatography or nuclear magnetic resonance. A variety of alpha- and gammaproteobacterial species capable of degrading DMSP were isolated. Of particular interest was the dominance of *Spongiobacter*, suggesting that it is a major player in the microbial sulfur cycle associated with *Acropora* corals. This correlates with observations of Bourne et al. (2008) showing the decline of *Spongiobacter* in corals during bleaching (when loss of *Symbiodinium* would result in lower DMSP production) and its reappearance as the corals recover (leading to a resumption of DMSP production). Raina et al. (2010) undertook bioinformatic analysis of an extensive collection of marine metagenomes and showed that sequences corresponding to the genes encoding DMSP degradation are present in 17% of the datasets, including metagenomes from corals. The *dmdA* gene was consistently found in coastal water samples, consistent with the idea that DMSP demethylation is an important source of sulfur and carbon for microbial communities in coral reefs. Many of these genes were associated with phage sequences, and Raina and colleagues suggest that they may have originated from host bacteria and are carried by phages as additional information to ensure effective propagation, as occurs in viruses infecting cyanobacteria and *E. huxleyi* (see *Research Focus Box 7.2*).

Raina et al. (2010) suggest that sulfur cycling plays a major role in structuring the microbial communities associated with the coral holobiont.

Does the breakdown of DMSP have a defense function?

The breakdown of DMSP to DMS and acrylate may be important as a deterrent against grazing by protists. Wolfe et al. (1997) examined grazing on the alga *E. huxleyi* by the dinoflagellate *Oxyrrhis marina*, concluding that lysis of the ingested prey resulted in mixing of the enzyme DMSP lyase and its substrate DMSP. When different protistan grazers were presented with a mixture of algal strains possessing high and low levels of DMSP lyase, they preferentially selected strains with low enzyme activity. The highly concentrated acrylate that is produced in this reaction was thought to act as a deterrent in view of its toxic effects. However, Strom et al. (2003) concluded that grazers were responding to preingestion cues, such as dissolved chemicals or compounds on the cell surface, rather than to toxic signals received after ingestion. They showed that dissolved DMSP itself inhibits grazing and suggested that this may have been strong selective pressure in the evolution of high levels of DMSP production in bloom-forming algae such as prymnesiophytes like *E. huxleyi*.

Evans et al. (2006) investigated the phenomenon of DMS production during viral lysis of *E. huxleyi* following an earlier report that strains of *E. huxleyi* with high DMSP lyase activity were resistant to *Coccolithovirus* infection (Schroeder et al., 2002). They showed that both DMS and acrylic acid affected viral infectivity and concluded that the DMSP system functions as an antiviral defense, protecting remaining members of the population from infection. It appears that DMSP has a range of functions; the factors affecting evolution of this system in different phytoplankton groups are discussed by Simó (2001).

are various forms of dissolved inorganic phosphorus (DIP), of which orthophosphate PO_4^{2-} is the most abundant. A large fraction of the phosphorus in surface oligotrophic waters is present as dissolved organic phosphorus (DOP) compounds. The sole input and output of phosphorus to the oceans is via terrestrial runoff and burial in sediments, respectively. Thus, biogeochemical cycling of phosphorus is closely linked with carbon flux and the availability of phosphorus has a major limiting or colimiting effect on oceanic primary production rates and microbial community composition. This phosphorus limitation has been studied extensively in oligotrophic regions of the north Pacific, western north Atlantic (Sargasso Sea), and the Mediterranean Sea, where analyses show that DIP is turned over very rapidly. There is some evidence that climate change over the last few decades is leading to stratification of the subtropical Pacific gyre, resulting in an increase in nitrogen fixation and changing the system toward limitation by phosphorus. The organic and inorganic forms of phosphorus are transformed continuously by microbial activity. In the surface ocean, phytoplankton assimilate DIP and incorporate it into cellular material, which is ingested by zooplankton or enters the microbial loop via exudation or cell lysis, where it is taken up and recycled by heterotrophic bacteria. Heterotrophic and autotrophic microbes compete for available phosphorus, which affects productivity. Recycling of DOP

? DOES PLANKTON MIGRATION AFFECT NUTRIENT DISTRIBUTION?

Some diatoms such as *Rhizosolenia* and the cyanobacterium *Trichodesmium* form large mats that can migrate across physical mixing barriers between the surface and deep pools of nutrients (80–100 m) in order to assimilate nitrate, before returning to shallower water with optimum light levels for photosynthesis. This process depends on the regulation of cell buoyancy by gas vesicles. The daily migration of zooplankton between the surface and deeper waters may also result in a two-way flux of nutrients, the net effect of which will depend on the relative N:P balance at different depths. Passive upward flux of low-density, lipid-rich material containing high levels of phosphorus could also be important. The effect of these processes affecting nitrogen and phosphorus must now be considered alongside the purely physical processes of upwelling and turbulent mixing, but at present we lack a clear picture of the importance of vertical zooplankton migration in supplying nutrients to the phytoplankton.

occurs throughout the water column and regeneration of inorganic phosphorus occurs in deeper waters. Although phosphate seems to be the preferred source of phosphorus for phytoplankton, they can also hydrolyze organic phosphorus compounds when phosphate does not meet their demands.

Marine microbes are adapted to low and variable levels of phosphorus

As discussed in Chapter 3, planktonic bacteria demonstrate one of two "lifestyles." Many are oligotrophic and are adapted to subsist on a relatively fixed diet of constant low levels of nutrients. This is often reflected in very small cell sizes and streamlined genomes, as exemplified by the photoautotroph *Prochlorococcus* and the chemoheterotroph *Candidatus* Pelagibacter ubique. Others are copiotrophic or opportunistic and are able to respond to inputs of increased levels of nutrients because they have larger genomes and more metabolic versatility. Such organisms may respond to nutrient starvation by initiating a program of changes in cellular composition and activity. Small cells with reduced genomes contain lower amounts of DNA and phospholipids (owing to less cell membrane content). Since these compounds incorporate large amounts of phosphorus, small cells require less and are at an advantage in conditions of phosphorus limitation. In fact, phytoplankton (including larger eukaryotic cells) growing in low-phosphorus environments seem to reduce demand further by substituting phospholipids in their membranes with lipids containing sulfur and sugars.

The mechanisms of uptake of phosphorus compounds have been extensively studied in laboratory cultures of bacteria like *Escherichia coli*, but less so in marine bacteria. Homologs of PstS, the high-affinity phosphate-binding protein in *E. coli*, are produced by *Synechococcus*, *Prochlorococcus*, and *Trichodesmium*, and multiple copies of the gene are present in the genome sequences of cultured strains. These genes are also abundant in marine metagenomes. High-affinity phosphate-uptake genes are also present in the genomes of viruses infecting cyanobacteria and *E. huxleyi*, and their expression is upregulated under conditions of phosphate deficiency—presumably increased phosphate uptake ensures efficient viral replication. Other phosphate transporter systems have also been identified in marine algae. Cell lysis leads to the release of enzymes such as nucleases, lipases, and esterases that degrade organic compounds to regenerate phosphate.

To obtain phosphorus from organic phosphorus compounds, bacterial cells must hydrolyze them to orthophosphate. In *E. coli* the well-studied Pho regulon contains a number of genes associated with regulation, uptake, and hydrolysis of phosphorus compounds. Pho gene homologs are abundant in genomes and metagenomes of marine bacteria, which appear to possess several different mechanisms for transporting and hydrolyzing phosphate esters. The most common type of enzyme is alkaline phosphatase, whose expression is repressed by high phosphorus levels and is often used as a marker for phosphate stress. These enzymes are grouped into a number of families, including PhoA, PhoD, and PhoX that differ in their substrate specificity, cellular location, and requirement for metals as cofactors. Based on bioinformatic analysis of the GOS metagenome, it seems that a large proportion of coastal and oceanic bacteria contain phosphatases within the cytoplasm, suggesting that small DOP compounds are carried across the cytoplasmic membrane, providing carbon (and possibly nitrogen) to the cell, as well as phosphorus.

The phosphonates are another group of organic phosphorus compound whose contributions to marine nutrient cycles have been relatively little studied until recently. Phosphonates are characterized by a stable C–P bond, rather than the C–O–P bond found in esters, and they are generally very resistant to chemical and thermal decomposition. Naturally occurring phosphonates are produced by a range of organisms as membrane components and therefore are found in the

DOP pool, but interest in these compounds has increased following the use of synthetic (xenobiotic) phosphonates used as herbicides, insecticides, detergent additives, and other applications. Tens of thousands of tons of these long-lived compounds are released each year into the environment. Analysis of metagenomic databases for phosphonate utilization genes (*phn*) has revealed that they are widespread and abundant among diverse bacterial phyla that are common in marine bacterioplankton, including *Proteobacteria*, *Planctomycetes*, and *Cyanobacteria*. *In situ* studies have shown that expression of these genes and phosphonate degradation occurs in phosphate-limited regions. Since up to 25% of DOP may occur as phosphonates, these molecules appear to be a significant phosphorus source for marine microbes.

Conclusions

This chapter has illustrated how recent discoveries have led to dramatic shifts in our thinking about the role of microbes in the major nutrient cycles and their consequences for global ocean processes. It is important that we have a clear understanding of underlying mechanisms because of the increased effects of anthropogenic inputs on the composition of our atmosphere and oceans. Increasing concentrations of CO_2 and sulfur from the use of fossil fuel and industrial activity, together with input of nitrogen and phosphorus nutrients from agriculture and aquaculture will cause shifts in microbial community composition and alter the balance of marine microbial nutrient cycles. Indeed, major shifts in the nitrogen and sulfur nutrient cycles might catalyze further climatic changes in a runaway effect, for example through increased output of greenhouse gases like nitrous oxide or depletion of the albedo effect due to changes in DMS emissions. However, predicting the outcomes of these changes is very difficult unless we have a clear understanding of underlying processes (see *Research Focus Box 1.1* for a discussion of ocean acidification). Metagenomic analyses are providing a huge amount of information, leading to new discoveries about hitherto unknown processes. New tools for applying this knowledge to *in situ* observations (e.g. NanoSIMS, see p. 31) are being developed, and the next decade will undoubtedly see many more surprises.

References

Arrigo, K. R. (2005) Marine microorganisms and global nutrient cycles. *Nature* 437: 349–355.

Bourne, D., Iida, Y., Uthicke, S. and Smith-Keune, C. (2008) Changes in coral-associated microbial communities during a bleaching event. *ISME J.* 2: 350–363.

Bulow, S. E., Rich, J. J., Naik, H. S., Pratihary, A. K. and Ward, B. B. (2010) Denitrification exceeds anammox as a nitrogen loss pathway in the Arabian Sea oxygen minimum zone. *Deep-Sea Res. I Oceanogr. Res. Pap.* 57: 384–393.

Capone, D. G. (2008) The marine nitrogen cycle. *Microbe* 3: 186–192.

Curson, A. R. J., Sullivan, M. J., Todd, J. D. and Johnston, A. W. B. (2009) Identification of genes for dimethyl sulfide production in bacteria in the gut of Atlantic herring (*Clupea harengus*). *ISME J.* 4: 144–146.

Debose, J. L., Lema, S. C. and Nevitt, G. A. (2008) Dimethylsulfoniopropionate as a foraging cue for reef fishes. *Science* 319: 1356.

Deutsch, C., Sarmiento, J. L., Sigman, D. M., Gruber, N. and Dunne, J. P. (2007) Spatial coupling of nitrogen inputs and losses in the ocean. *Nature* 445: 163–167.

Evans, C., Malin, G., Wilson, W. H. and Liss, P. S. (2006) Infectious titers of *Emiliania huxleyi* virus 86 are reduced by exposure to millimolar dimethyl sulfide and acrylic acid. *Limnol. Oceanogr.* 51: 2468–2471.

Francis, C. A., Beman, J. M. and Kuypers, M. M. M. (2007) New processes and players in the nitrogen cycle: the microbial ecology of anaerobic and archaeal ammonia oxidation. *ISME J.* 1: 19–27.

Hamersley, M. R., Lavik, G., Woebken, D., et al. (2007) Anaerobic ammonium oxidation in the Peruvian oxygen minimum zone. *Limnol. Oceanogr.* 52: 923–933.

Howard, E. C., Henriksen, J. R., Buchan, A., et al. (2006) Bacterial taxa that limit sulfur flux from the ocean. *Science* 314: 649–652.

Howard, E. C., Sun, S. L., Biers, E. J. and Moran, M. A. (2008) Abundant and diverse bacteria involved in DMSP degradation in marine surface waters. *Environ. Microbiol.* 10: 2397–2410.

Kirkwood, M., Le Brun, N. E., Todd, J. D. and Johnston, A. W. B. (2010a) The *dddP* gene of *Roseovarius nubinhibens* encodes a novel lyase that cleaves dimethylsulfoniopropionate into acrylate plus dimethyl sulfide. *Microbiology* 156: 1900–1906.

Kirkwood, M., Todd, J. D., Rypien, K. L. and Johnston, A. W. B. (2010b) The opportunistic coral pathogen *Aspergillus sydowii* contains dddP and makes dimethyl sulfide from dimethylsulfoniopropionate. *ISME J.* 4: 147–150.

Lam, P., Lavik, G., Jensen, M. M., et al. (2009) Revising the nitrogen cycle in the Peruvian oxygen minimum zone. *Proc. Natl Acad. Sci. USA* 106: 4752–4757.

Martiny, A. C., Kathuria, S. and Berube, P. M. (2009) Widespread metabolic potential for nitrite and nitrate assimilation among *Prochlorococcus* ecotypes. *Proc. Natl Acad. Sci. USA* 106: 10787–10792.

Moisander, P. H., Beinart, R. A., Hewson, I., et al. (2010) Unicellular cyanobacterial distributions broaden the oceanic N_2 fixation domain. *Science* 327: 1512–1514.

Montoya, J. P., Holl, C. M., Zehr, J. P., Hansen, A., Villareal, T. A. and Capone, D. G. (2004) High rates of N_2 fixation by unicellular diazotrophs in the oligotrophic Pacific Ocean. *Nature* 430: 1027–1031.

Nevitt, G. A. and Bonadonna, F. (2005) Seeing the world through the nose of a bird: new developments in the sensory ecology of procellariiform seabirds. *Mar. Ecol. Prog. Ser.* 287: 292–295.

Nevitt, G. A., Losekoot, M. and Weimerskirch, H. (2008) Evidence for olfactory search in wandering albatross, *Diomedea exulans*. *Proc. Natl Acad. Sci. USA* 105: 4576–4581.

Raina, J. B., Tapiolas, D., Willis, B. L. and Bourne, D. G. (2009) Coral-associated bacteria and their role in the biogeochemical cycling of sulfur. *Appl. Environ. Microbiol.* 75: 3492–3501.

Raina, J. B., Dinsdale, E. A., Willis, B. L. and Bourne, D. G. (2010) Do the organic sulfur compounds DMSP and DMS drive coral microbial associations? *Trends Microbiol.* 18: 101–108.

Reisch, C. R., Moran, M. A. and Whitman, W. B. (2008) Dimethylsulfoniopropionate-dependent demethylase (DmdA) from *Pelagibacter ubique* and *Silicibacter pomeroyi*. *J. Bacteriol.* 190: 8018–8024.

Schroeder, D. C., Oke, J., Malin, G. and Wilson, W. H. (2002) *Coccolithovirus* (*Phycodnaviridae*): Characterisation of a new large dsDNA algal virus that infects *Emiliania huxleyi*. *Arch. Virol.* 147: 1685–1698.

Simó, R. (2001) Production of atmospheric sulfur by oceanic plankton: biogeochemical, ecological and evolutionary links. *Trends Ecol. Evol.* 16: 287–294.

Steingruber, S. M., Friedrich, J., Gachter, R. and Wehrli, B. (2001) Measurement of denitrification in sediments with the 15N isotope pairing technique. *Appl. Environ. Microbiol.* 67: 3771–3778.

Strom, S., Wolfe, G., Holmes, J., Stecher, H., Shimeneck, C., Lambert, S. and Moreno, E. (2003) Chemical defense in the microplankton I: Feeding and growth rates of heterotrophic protists on the DMS-producing phytoplankter *Emiliania huxleyi*. *Limnol. Oceanogr.* 48: 217–229.

Todd, J. D., Curson, A. R. J., Dupont, C. L., Nicholson, P. and Johnston, A. W. B. (2009) The *dddP* gene, encoding a novel enzyme that converts dimethylsulfoniopropionate into dimethyl sulfide, is widespread in ocean metagenomes and marine bacteria and also occurs in some Ascomycete fungi. *Environ. Microbiol.* 11: 1376–1385.

Tripp, H. J., Bench, S. R., Turk, K. A., et al. (2010) Metabolic streamlining in an open-ocean nitrogen-fixing cyanobacterium. *Nature* 464: 90–94.

Vila-Costa, M., Rinta-Kanto, J. M., Sun, S., Sharma, S., Poretsky, R. and Moran, M. A. (2010) Transcriptomic analysis of a marine bacterial community enriched with dimethylsulfoniopropionate. *ISME J.*, doi:10.1038/ismej.2010.62.

Ward, B. B., Devol, A. H., Rich, J. J., et al. (2009) Denitrification as the dominant nitrogen loss process in the Arabian Sea. *Nature* 461: 78–81.

Wolfe, G. V., Steinke, M. and Kirst, G. O. (1997) Grazing-activated chemical defence in a unicellular marine alga. *Nature* 387: 894–897.

Yoch, D. C. (2002) Dimethylsulfoniopropionate: its sources, role in the marine food web, and biological degradation to dimethylsulfide. *Appl. Environ. Microbiol.* 68: 5804–5815.

Zehr, J. P., Carpenter, E. J. and Villareal, T. A. (2000) New perspectives on nitrogen-fixing microorganisms in tropical and subtropical oceans. *Trends Microbiol.* 8: 68–73.

Zehr, J. P., Waterbury, J. B., Turner, P. J., et al. (2001) Unicellular cyanobacteria fix N_2 in the subtropical North Pacific Ocean. *Nature* 412: 635–638.

Further reading

Chistoserdova, L., Kalyuzhnaya, M. G. and Lidstrom, M. E. (2009) The expanding world of methylotrophic metabolism. *Annu. Rev. Microbiol.* 63: 477–499.

Dyhrman, S. T., Ammerman, J. W. and Van Mooy, B. A. S. (2007) Microbes and the marine phosphorus cycle. *Oceanography* 20: 110–116.

Erguder, T. H., Boon, N., Wittebolle, L., Marzorati, M. and Verstraete, W. (2009) Environmental factors shaping the ecological niches of ammonia-oxidizing archaea. *FEMS Microbiol. Rev.* 33: 855–869.

Gilbert, J. A., Thomas, S., Cooley, N. A., et al. (2009) Potential for phosphonoacetate utilization by marine bacteria in temperate coastal waters. *Environ. Microbiol.* 11: 111–125.

Hutchins, D. A., Mulhollan, M. R. and Fu, F. (2009) Nutrient cycles and marine microbes in a CO_2-enriched ocean. *Oceanography* 22: 129–145.

Martinez, A., Tyson, G. W. and DeLong, E. F. (2010) Widespread known and novel phosphonate utilization pathways in marine bacteria revealed by functional screening and metagenomic analyses. *Environ. Microbiol.* 12: 222–238.

Paulmier, A. and Ruiz-Pino, D. (2009) Oxygen minimum zones (OMZs) in the modern ocean. *Prog. Oceanogr.* 80: 113–128.

Quinn, J. P., Kulakova, A. N., Cooley, N. A. and McGrath, J. W. (2007) New ways to break an old bond: the bacterial carbon–phosphorus hydrolases and their role in biogeochemical phosphorus cycling. *Environ. Microbiol.* 9: 2392–2400.

Van Mooy, B. A. S., Fredricks, H. F., Pedler, B. E., et al. (2009) Phytoplankton in the ocean use non-phosphorus lipids in response to phosphorus scarcity. *Nature* 458: 69–72.

Chapter 10
Symbiotic Associations

In the broadest definition, the term symbiosis (literally "living together") is used to describe any close, long-term relationship between two different organisms, which can range from commensalism (a loose association in which one partner gains benefit but does no harm to the host) to parasitism (in which the parasite benefits at the expense of the host). In common usage, the term symbiosis usually refers to a mutualistic relationship, in which both partners benefit through increased fitness in evolution. We can classify the relationship further in terms of the degree of "intimacy" of the association. Endosymbionts live inside the host cells, usually within specialized cells; these microbes show special adaptations to intracellular existence. Ectosymbionts (or episymbionts) are extracellular microbes that colonize surfaces, including infoldings or chambers such as the exoskeleton of crustaceans or intestinal tract of other animals. Symbiotic microbes may also have a free-living stage, but many have lost the ability to exist independently. This chapter provides examples of mutualistic and commensal interactions that illustrate the importance of microbial symbiosis in marine ecology.

Key Concepts

- Many marine invertebrates obtain nutrition from photosynthetic microbes within their tissues.
- Animals found in hydrothermal vents, cold seeps, and many other habitats depend on chemosynthetic microbes, which use reduced sulfur compounds or methane to fuel autotrophic metabolism.
- Coevolution of hosts and their symbionts has led to complex adaptations to maximize benefits of the relationship and in some cases to obligate interdependence.
- Environmental stress may lead to breakdown of the symbiotic relationship.
- Hosts pass on symbionts to the next generation by horizontal or vertical transmission, in which the symbionts are acquired from the environment or inherited from the parent host, respectively.
- Bioluminescence in some fish and invertebrates is produced by bacterial symbionts.
- Mixotrophic metabolism in protists often arises from symbiosis.
- Study of symbiosis has important applications in biotechnology and medicine.

Zooxanthellae and other photosynthetic endosymbionts are vital for the nutrition of many marine animals

Biologists first recognized associations between marine invertebrates and photosynthetic microbes more than a century ago and introduced several descriptive terms at this time, based largely on the coloration of the tissue due to pigments from the symbionts. The best-known and most important group is the zooxanthellae—the term refers to the golden-brown color—which are dinoflagellates occurring in a wide range of corals, anemones, and some mollusks and sponges. Zoochlorellae—having a green color—are members of the Chlorophyceae, occurring mainly in sponges, coelenterates and flatworms, whilst zoocyanellae—with a blue-green color—are cyanobacteria (including *Prochloron*) found in some sea squirts and mollusks. Many tropical sponges also rely on cyanobacteria for more than half of their energy requirements.

A variety of dinoflagellate types belonging to the genus *Symbiodinium* have been recognized as zooxanthellae in association with a wide range of host species. Until the 1970s, all zooxanthellae were considered members of the single species *Symbiodinium microadriaticum*. Molecular analysis has revealed a very high level of diversity among different isolates of *Symbiodinium*. The genus can be divided into eight major groupings (clades) based on sequence analysis of 18S rRNA genes (see p. 40). Members of the same host species often harbor the same *Symbiodinium* clade, but there are now many examples of animals with more than one clade. Multiple clade types seem rarer in Pacific corals. However, more discriminating methods based on gene sequencing of the variable internal transcribed (ITS) region show that the zooxanthellae may have significant intraclade diversity, and there are several hundred genetically distinct types, which might be considered separate species. Analysis of full genome sequences will reveal more insights into the diversity of zooxanthellae.

In their hosts, the *Symbiodinium* usually form coccoid cells surrounded by a cellulose cell wall, within a host vacuole called a symbiosome; however, in clams the zooxanthellae lie extracellularly in specialized channels in the gill mantle. Many symbiotic dinoflagellates can be isolated and maintained in culture given suitable conditions. Most investigators have concluded that the morphology (e.g. the loss of flagella) and life cycle of the free-living and symbiotic forms are fundamentally different, but few details are known.

The dinoflagellates carry out photosynthesis, harvesting light energy via a complex of chlorophyll *a*, chlorophyll c_2 and the protein peridinin. Light energy is used to fix CO_2 which occurs mainly via the C3 pathway (Calvin–Benson cycle) using RuBisCO (see p. 66). This can be shown by enzyme assays and by incubating with radiolabeled bicarbonate and measuring the incorporation of ^{14}C into the tissues. Some species may also use a C4 route employing phospho-enol pyruvate carboxylase as the key enzyme. The animal host clearly derives nutritional benefit from the relationship in the form of photosynthetically fixed carbon compounds, although views differ on the exact contribution that each partner makes to the acquisition of other nutrients—especially nitrogen and phosphorus. Radiolabeling experiments show that the zooxanthellae release a high proportion of photosynthetically fixed carbon as small molecules, including glycerol, glucose and organic acids. It is possible that all the essential amino acids are provided by the zooxanthellae. In return, the host is presumed to provide certain key nutrients and suitable environmental conditions for the dinoflagellate symbionts, including protection from predators. The density of zooxanthellae varies widely in different hosts; high numbers are often associated with a reduced dependence on feeding by capture of plankton, manifested by smaller tentacles or reduced digestive systems.

We know little about how these intimate relationships came about, but available evidence suggests that they evolved in the mid-Triassic period (about 250 million

? WHAT'S IN IT FOR THE ZOOXANTHELLAE?

Free-living dinoflagellates "make big sacrifices"—loss of cell wall and flagella, altered metabolism, and reproduction—when they become zooxanthellae. The benefits for the host are obvious, but what benefits the zooxanthellae receive in return is less clear. The most likely explanation is that they have stable access to a favorable light regime and receive a reliable source of nitrogen in the form of ammonium compounds produced by the host and other associated microbes. The *Symbiodinium* population is dynamic and can be released in large numbers from their hosts. Many hosts acquire zooxanthellae from the environment at the larval stage, whilst others transmit them vertically in the eggs. The discovery that corals can associate with different genetic types of *Symbiodinium*, as well as a diverse range of other microbes, complicates earlier ideas about the pairwise coevolution of host and microbe. Development of models to explain the evolution of such relationships is a major goal of theoretical biology.

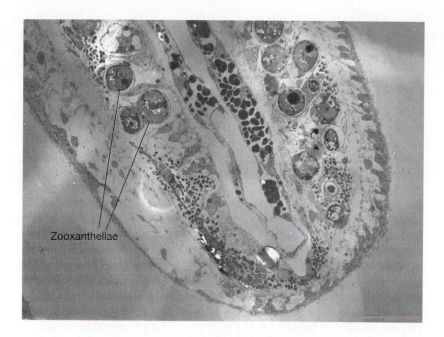

Zooxanthellae

Figure 10.1 Zooxanthellae in corals. Transmission electron micrograph of section of a coral tentacle. (Image courtesy of Simon Davy, Victoria University of Wellington.)

years ago). In all of the animal phyla that harbor dinoflagellates, the final step in digestion is intracellular. Cells of the animal's digestive tract may have retained algal cells that were resistant to digestion, and subsequent evolution "refined" this association. The importance of the association to the host is emphasized by anatomical and behavioral adaptations observed in the various animal groups, which have evolved in order to expose the zooxanthellae to light.

Coral bleaching occurs due to the breakdown of the symbiosis between zooxanthellae and their host

Zooxanthellae are very widespread and of major importance in the nutrition of corals, especially in tropical waters. In the group known as hermatypic corals, the individual animals (polyps) usually associate into colonies and secrete a calcium carbonate skeleton, which leads to the development of coral reefs. Usually, the zooxanthellae occur in the gastroderm (innermost cell layer) of the coral tissue, as shown in *Figure 10.1*.

Reef-building corals lay down their skeleton over a very fine organic matrix and field observations show that a high content and activity of zooxanthellae is essential for the rapid building of the reef structure. Under optimum growth conditions, zooxanthellae may produce up to 100 times more carbon than they need for their own growth and reproduction. Most of this excess is transferred to the coral, where it is mainly respired. Excess carbon is also secreted in the mucus, which provides a major source of nutrients sustaining the microbial loop and benthic processes, explaining why coral reefs are so productive. The coral produces signal molecules, which alter the control of photosynthesis and stimulate the release of organic compounds from the zooxanthellae by altering membrane permeability. When photosynthesis is inhibited, for example by restriction of light, the uptake of calcium and subsequent secretion of calcium carbonate is reduced. Excretion of lipids by the zooxanthellae is also important in construction of the skeleton.

The importance of zooxanthellae to tropical corals is shown dramatically by the phenomenon of coral bleaching. Bleaching occurs either when the host loses its symbiotic zooxanthellae or when their photosynthetic ability is reduced through loss of pigment (*Plate 10.1*). Loss of zooxanthellae does not necessarily lead to immediate death of the corals; however, the health of the colony is severely impaired and the corals may become susceptible to disease or overgrowth by

Since 1998, several particularly severe mass bleaching events have affected tropical reefs in all the world's oceans, occurring in years when average sea surface temperatures have been just 1°C or so above normal. The thermal tolerance of most corals is very narrow. Also, reefs may not be able to grow quickly enough to maintain optimum light levels as sea levels rise and the decrease in ocean pH will reduce their ability to calcify. Some coral biologists predict that major reef systems will be irreversibly damaged during this century (Kleypas and Eakin, 2007). Hoegh-Guldberg et al. (2009) conclude that "by the time atmospheric CO_2 concentrations rise above 450 ppm, most coral reef ecosystems will be functionally gone, and the structures they build will begin to erode. Such losses will destabilize the safety and livelihoods of millions of people." As well as temperature-induced bleaching, disturbance of reef ecology by overfishing and nutrient inputs can lead to a vicious cycle of colonization by algae, excessive microbial growth, and the complete demise of reefs (see p. 224). Rohwer and Youle (2010) argue that these local stressors provide the greatest current threat to corals and that we need to provide immediate and complete protection from overfishing and nutrient pollution to selected reefs in regions where they have the greatest potential to survive. They remind us that corals have survived previous major environmental upheavals over millennia and inject a note of optimism with the view that "The possibility of adaptation by the holobiont, thanks to its algal, viral and microbial partners, is a source of hope."

algae, especially when high temperatures prevail in successive years. Many factors affect the severity and duration of bleaching, including nutrients and pollution. There is general agreement that environmental stress, especially elevated sea temperature and high solar irradiation, is the major trigger for bleaching, but the cellular mechanisms by which the host recognizes that its zooxanthellae are impaired and eliminates them are still unclear. Possible mechanisms include: expulsion of zooxanthellae by exocytosis; induction of apoptosis, in which programmed cell death is initiated; and symbiophagy, in which the vacuole containing the zooxanthellae is transformed into a digestive organelle. It is likely that different environmental factors trigger different physiological mechanisms of bleaching. Infection by bacteria and viruses is also implicated as the cause of some types of coral bleaching, as discussed in *Research Focus Box 11.1*.

Some corals seem to be able to associate with different clades of zooxanthellae, and different genotypes with a greater tolerance to temperature may recolonize the tissue after bleaching. This "adaptive bleaching hypothesis" was first proposed by Buddemeier and Fautin in 1993, and there is now considerable experimental evidence to support many aspects of the concept. For example, aquarium experiments have shown that *Symbiodinium* clade D has a better quantum yield of photosynthesis than other clades, but only at elevated temperatures. Thus, the physiological cost of harboring a thermotolerant clade may override the apparent benefits. Field experiments support the idea that corals may acquire more thermotolerant zooxanthellae (*Figure 10.2*).

Although zooxanthellae are vitally important in the reef-building corals, it is important to bear in mind that most corals supplement their nutrition by feeding on zooplankton via their tentacles, and some do not contain symbionts at all (this is referred to as azooxanthellate). For example, an azooxanthellate variant of the Mediterranean coral *Oculina patagonica* occurs in undersea caves. In particular, thousands of species of slow-growing azooxanthellate cold-water corals, such as *Lophelia* spp., have recently been found in deep waters, hundreds of meters below the photic zone.

Scleractinian corals are multipartner symbiotic systems (holobionts)

In addition to the zooxanthellae discussed earlier, the tissue of healthy corals and their secreted mucus layer support a diverse and dynamic community of other microbes, including bacteria, archaea, fungi, and viruses. These contribute to coral nutrition and play important roles in maintaining coral health through inhibition of pathogens and possible removal of toxic oxygen species produced during photosynthesis. Although bacteria had been isolated from corals since the 1970s, rapid advances in this field did not take off until 2002, since when culture-independent methods have revealed a high diversity of bacteria associated with tropical corals, with similar communities in corals of the same species from geographically very different regions. Recent metagenomic analyses have confirmed a very high diversity of metabolic functions associated with coral microbes, including nitrogen cycling, sulfur cycling, photosynthesis, breakdown of complex proteins, and polysaccharides. Genes for stress response and virulence, in particular DNA repair and antibiotic resistance, also seem to be important. The chemical properties of coral mucus, as well as intercellular signaling and antagonistic interactions between the microbes themselves, seem to be important factors in the establishment of specific microbiota, although the protective function can be disturbed, leading to disease (see *Research Focus Box 11.1*). Of particular significance is the discovery that a diverse group of resident nitrogen-fixing bacteria (including *Cyanobacteria* and vibrio-like *Gammaproteobacteria*) are present in many corals. Provision of nitrate to the zooxanthellae may help to explain the high productivity of corals in the apparently nitrogen-limited environments that they inhabit. The role of coral bacteria in the global cycling of sulfur has also

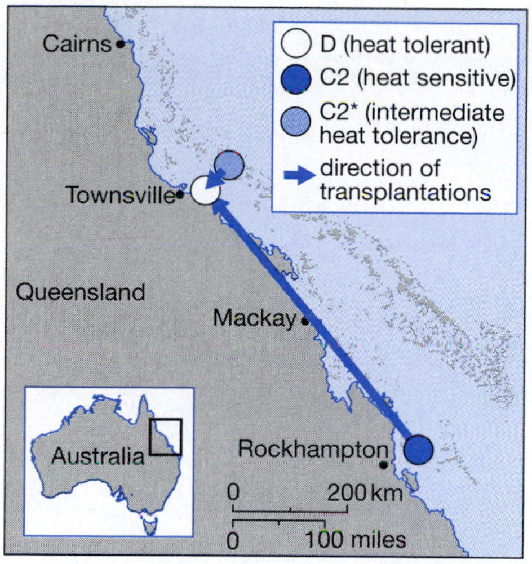

Figure 10.2 Transplantation experiments on the Great Barrier Reef have shown that when corals were moved from "cooler" to "warmer" (plus ~1°C) spots on the reef, there was a shift in the dominant *Symbiodinium* type to the more thermotolerant clade D, leading to resistance to bleaching. (Reproduced from Berkelmans and van Oppen [2006] by permission of the Royal Society.)

recently been shown to be significant (see *Research Focus Box 9.2*). Coral-associated archaea seem to be especially important in recycling waste products in the coral holobiont, by nitrification and denitrification of ammonia in the mucus layer. Study of cold-water, deep-sea corals such as *Lophelia* spp. is much more difficult, but it appears that they may also have specific microbiota, although these are very different, owing to their adaptation to cold, dark environments at high pressure.

The role of viruses in the health of corals is also unclear; we are conditioned to think of viruses only as disease agents but some recent studies suggest that they could have a mutualistic role.

Photosynthetic zooxanthellae boost the growth of giant clams in nutrient-poor waters

Anyone who has dived or snorkeled on the Great Barrier Reef or Pacific islands will have marveled at the size and beautiful colors of the shell mantles of the giant clams such as *Tridacna gigas*, which can reach enormous sizes, up to 300 kg (*Plate 10.2*). How can these animals grow so rapidly when the waters they inhabit are very poor in nutrients? The siphon tissue around the mantle is packed with endosymbiotic zooxanthellae. In this case, there are additional pigments of many different colors that provide protection to the zooxanthellae from the high ultraviolet irradiation in the clear waters that the clams inhabit. Recent studies of the importance of light in the symbiotic relationship have shown that, under the right experimental conditions, the zooxanthellae can provide all of the clam's carbon requirements through the release of small molecules such as glycerol and organic acids. However, as in many corals, acquisition of nutrients by heterotrophic feeding is also important. Tridacnid clams have particularly efficient filter-feeding mechanisms and can extract significant quantities of plankton from the water, even though the plankton is in low concentrations because of the oligotrophic nature of the environment. Thus, it seems that under natural conditions, the clams rely on both autotrophic and heterotrophic sources for their nutrition, with between 35% and 70% of the carbon requirements coming from photosynthetic activity of the zooxanthellae.

Worms and clams at hydrothermal vents obtain nutrition from chemosynthetic bacterial endosymbionts

Chapter 1 introduced the dense undersea communities of giant tubeworms, bivalve mollusks, and other creatures growing around deep-sea hydrothermal vents (*Plate 10.3*), whose discovery in 1977 was such a surprise. How can such

---

ⓘ CLAMS BUILT FOR SUNBATHING AND UPSIDE-DOWN JELLYFISH

Tridacna has evolved remarkable anatomical and behavioral modifications to optimize the benefits from their photosynthetic symbionts. This type of clam is unusual among bivalve mollusks because it lives above the sediment. During evolution, the orientation of the internal organs have twisted through 180°, allowing the siphon to be at the top of the body so that the maximum surface area colonized by the zooxanthellae is exposed to the light. Photosynthesis is further enhanced by specialized hyaline structures, which focus light onto the zooxanthellae. Another interesting evolutionary adaptation to derive maximum benefit from photosynthesis occurs in the jellyfish *Cassiopea xamachana*, which usually lies inverted on the seabed in shallow water, exposing the zooxanthellae to light.

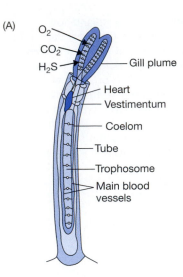

(A)

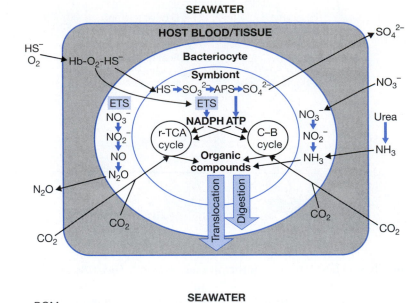

(B)

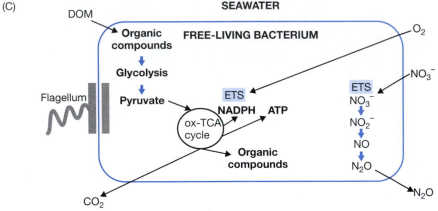

(C)

Figure 10.3 Schematic diagrams illustrating symbiosis between the tube worm *Riftia pachyptila* and a chemosynthetic sulfur-oxidizing bacterium. (A) Illustration of the tube worm. An extensive capillary system transports O_2, H_2S and CO_2 bound to hemoglobin (Hb) to the trophosome, which consists of host cells (bacteriocytes) containing endosymbiotic bacteria. (B) Endosymbiotic bacteria oxidize sulfide via the APS pathway (see *Figure 3.3*), yielding ATP and NADPH via an electron transport system (ETS). CO_2 fixation occurs mainly via the Calvin–Benson (C–B) and reductive tricarboxylic acid (r-TCA) cycles (see *Figure 3.4*). Transfer of organic matter from symbionts to host occurs via both translocation of simple organic compounds and digestion of symbiont cells. Nitrogen for the symbiont's biosynthesis is derived from urea excreted by host cells, or from reduction of nitrate absorbed from seawater, probably in the periplasm. (C) In their free-living state, the bacterial symbionts of *Riftia* have the genomic potential to live heterotrophically by absorbing amino acids from dissolved organic matter (DOM) in seawater. These may be respired via glycolysis and the oxidative (ox-)TCA cycle. [Data from Stewart et al. (2005) and Robidart et al. (2008).]

a highly productive ecosystem be sustained? Vent habitats are completely dark, under enormous pressure, and at such great depth that organic material from the upper layers of water settles too slowly to support this level of growth. The idea developed that the nutrition of these animals depends on thioautotrophic bacteria carrying out CO_2 fixation using reduced sulfur compounds as electron donors for the production of ATP (see p. 63).

By measuring the ratios of different isotopes of carbon, it was found that the cellular material of the bivalve mollusks (clams) does not have its ultimate origin in photosynthesis. The ratios of ^{13}C to ^{12}C reflect the efficiency with which different enzymes deal with the different isotopes. The ratio found in these animals is well outside the value of photosynthetically fixed carbon, proving that they are not feeding on material derived from CO_2 fixed in the upper photic zone. At first, investigators thought that the vent animals feed by filtration of chemoautotrophic bacteria that exist in the water around the vents, concentrated by local warm-water currents created by the vent activity. This idea became unlikely when equipment for bringing intact animals to the surface was developed—it was discovered that the filter-feeding mechanism and digestive tract of the clams found at the vents was greatly reduced.

With investigation of the giant tubeworm *Riftia pachyptila*, new questions arose about how this dense animal community obtains its nutrients. *Riftia* has no gut, so it was assumed at first that it must feed by simply absorbing nutrients from the surrounding water. However, the rate of uptake of organic material is insufficient to explain the extremely high metabolic rates of these animals. The cavity of the worm is filled by an organ—the trophosome—containing granules of elemental sulfur and supplied with many blood vessels. Electron microscopy revealed the presence of large numbers of structures strongly resembling bacterial cells, with the trophosome tissue containing a large amount of lipopolysaccharide—indicating that it is packed with Gram-negative bacteria as endosymbionts—and enzymes associated with sulfur metabolism, including ATP-reductase and ATP-sulfurylase. The presence of high levels of RuBisCO provided further proof of endogenous autotrophic metabolism. This is a truly remarkable animal–bacterium symbiosis; there are about 10^9 bacteria per gram of tissue, occupying up to a quarter of the volume of the trophosome. The endosymbionts have been identified by 16S rRNA gene sequences as members of the *Gammaproteobacteria*, but all attempts to culture them from trophosome tissue have failed. They use the oxidation of reduced sulfur compounds as a source of energy for the fixation of CO_2 into organic material, which is transferred across the host membrane surrounding the bacteria to the animal tissue for growth. *Riftia* has evolved into a truly autotrophic animal, with sophisticated mechanisms to ensure the acquisition of its symbionts and optimum conditions for their maintenance (see *Figure 10.3* and *Research Focus Box 10.1*).

In the vent clam *Calyptogena magnifica*, microscopy reveals that the symbiotic bacteria are contained in large cells that are exposed to the seawater on one side and the blood supply on the other, although the anatomical adaptations to ensure efficient transport of nutrients to the symbionts are not as sophisticated as those of *Riftia*. Once again, enzyme assays show the thioautotrophic nature of the bacteria. The vent mussel *Bathymodiolus thermophilus* harbors intracellular thioautotrophic bacteria in its gills; here, the physiological adaptations to control the nutrient environment of the symbionts are less well developed than in other vent animals.

Chemosynthetic symbionts are widely distributed in marine invertebrates

After the discovery of chemosynthetic symbionts in vent animals, biologists realized that the phenomenon is very widespread and occurs in many other habitats and hundreds of animal species in at least seven phyla. In addition to sulfide-based metabolism, methanotrophic symbioses have now been described in a range of worms and mollusks, following the discovery of huge communities of mussels and tubeworms in methane-rich "cold seeps" in the Gulf of Mexico and elsewhere. The bacteria contain densely packed lamellae and use methane as a source of both carbon and energy (see p. 70). Using enzyme assays, 16S rRNA gene sequencing, and fluorescent *in situ* hybridization (FISH), it has been shown that mussels such as *Bathymodiolus* host two distinct populations of chemoautotrophic bacteria, one of which oxidizes sulfur and the other oxidizes methane. The ratio of the two types of endosymbiont within the host reflects the relative abundance of sulfide and methane in the site at which the animal is growing. The mussels also have a filter-feeding apparatus and probably rely on a mixture of autotrophic and heterotrophic feeding. This, combined with their greater mobility, explains the fact that these mussels can tolerate a wider range of habitats at the hydrothermal vent sites. Snails in the genus *Ifremeria*, found at hydrothermal vents, are also thought to harbor both methanotrophic and thiotrophic bacteria, whilst the "scaly foot" snail *Crysomallon squamiferum* has populations of several distinct bacteria potentially involved in nutrition and protection of the host (see *Plate 10.6*).

Perhaps the most remarkable example of multiple symbiosis discovered so far occurs in the small oligochaete worms (*Olavius algarvensis* and related species), which contain both sulfate-reducing and sulfur-oxidizing bacteria, members of the *Deltaproteobacteria* and *Gammaproteobacteria* respectively. These tiny

? COULD SYMBIOSIS IN MARINE ANIMALS HELP IN SPACE EXPLORATION?

The discovery that many species of animals have evolved coordinated interactions with different microbial partners could provide a model for studies of self-sustaining systems, such as those that may be required to support interplanetary space travel. In *Olavius*, sulfate-reducing and sulfate-oxidizing bacteria feed each other and fix carbon dioxide for the production of organic compounds. The symbiotic bacteria also detoxify waste products produced by the host and recycle nitrogen within the system. Although some external nutrients must be provided, there is very extensive recycling within the system. Perhaps, manipulation and bioengineering to scale up such systems could help to develop almost self-sufficient biospheres.

BOX 10.1   RESEARCH FOCUS

Evolving together to thrive in a hellish environment
Genomic techniques reveal the interdependence of hydrothermal vent animals and
their chemosynthetic bacterial partners

Many of the animals found at hydrothermal vents are sustained by internal chemosynthetic symbiotic bacteria, whilst subsequent research has revealed that such associations are very widespread (Dubilier et al., 2008). Many of these animals have highly reduced digestive systems and rely entirely on their microbial symbionts for nutrition. How do the animals acquire their specific microbial partners during development? How have these associations evolved? Answers to these questions are emerging from the application of genomic and proteomic techniques.

Complete genome sequences are now known for the bacterial intracellular symbionts that colonize specialized cells within the gills of the vesicomyid clams *Calyptogena okutanii* and *Calyptogena magnifica* found at hydrothermal vents (Kuwahara et al., 2007; Newton et al., 2007). These *Gammaproteobacteria* have very small genomes (1.0–1.2 Mb) encoding only about 1000 genes. Thus, in common with many other endosymbionts and parasites, they would not be expected to survive outside of their host. Genome analysis shows that the major chemoautotrophic pathways are present, as well as routes for the production of vitamins, cofactors, and amino acids required by the clam. However, many of the genes associated with a free-living lifestyle—encoding processes such as motility, DNA repair mechanisms, and stress responses—are missing. In *Calyptogena* spp., the symbiotic bacteria are predominantly passed from one generation to the next via vertical transmission in the eggs. Hurtado et al. (2003) showed that DNA sequences in the symbionts are closely coupled with the DNA in host mitochondria (which are only inherited in the cytoplasm of the eggs), confirming that transmission is vertical. The association between these clams and their microbial symbionts is relatively young in phylogenetic terms—less than 50 million years—so perhaps we are witnessing an early stage in the evolution of this symbiosis toward full enslavement of the bacteria to become organelles through genome reduction (Moya et al., 2008), as we believe occurred in the evolution of mitochondria and chloroplasts.

By contrast, the symbionts found in vestimentiferan tubeworms are acquired from the environment at an early stage of larval development. Hydrothermal vents are very ephemeral and the success of these tubeworms depends on larval dispersal over great distances. Since the adult worms depend absolutely on the symbionts for their nutrition, it seems a risky strategy for the larvae not to contain symbionts. However, horizontal transmission may offer more opportunities for the host to acquire specific types of bacteria that might be best adapted to local conditions. How do the bacteria colonize the worms? To investigate this, Nussbaumer et al. (2006) constructed specially designed

chambers with grooved plates that were placed at a hydrothermal vent using the submersible *Alvin*. When collected a year later, the tiny (200 μm to 2 mm) and very fragile bodies of newly settled larvae and juvenile animals of three different species (*Riftia pachyptila*, *Tevnia jerichonana*, and *Oasisia alvinae*) were examined microscopically. Symbiotic bacteria were detected using symbiont-specific FISH probes (see p. 52). When they first settle, the larvae have an obvious digestive tract, including a mouth and anus. The symbiotic bacteria can be detected when the animals reach a length of about 250 μm. However, even though the larvae feed actively on microbes, the FISH micrographs showed that the symbionts enter via the skin (where they become enclosed in vacuoles), not via the mouth. Fewer than 20 bacteria are needed for infection, but these rapidly migrate through the layers of host tissue, multiply, and initiate differentiation of the mesodermal tissue. Their presence then provokes massive apoptosis (programmed cell death) of the epidermis, muscles, and mesoderm, followed by trophosome development.

Harmer et al. (2008) filtered large volumes of sea water from a vent site and used another type of collection device placed at various distances from the tube worm colonies. Free-living bacteria corresponding to the symbiont phylotype (again detected using 16S rRNA gene probes with FISH) were found as far as 100 m away from the vent source. The methods used in this study would not enable measurements of the numbers of free-living bacteria, so it is not known if there is any mechanism by which the environment is seeded from the host, such as occurs during the daily shedding of *Vibrio fischeri* by *Euprymna scolopes* (see p. 216) and in other systems (Polz et al., 2000). Although all attempts to culture the *Riftia* symbiont have failed, the high densities of an apparently pure population of bacteria within the trophosome enabled Robidart et al. (2008) to separate bacteria from host tissue using density gradient centrifugation and to construct a composite metagenome of the population. The findings explain why the bacterium is successful in the two very different environments (within host cells and free-living). The authors recognized this dual nature by naming the symbiont *Candidatus* Endoriftia persephone (from the goddess in Greek mythology, who was the goddess of fertility as well as queen of the underworld). Autotrophic metabolism in the host depends on CO_2 fixation by both the Calvin–Benson cycle and the reverse TCA cycle (*Figure 10.3*), but the bacterium also has a wide suite of genes necessary for heterotrophic metabolism. Further innovative experimental approaches will be needed to determine the relative contributions of autotrophic and heterotrophic metabolism in the free-living state.

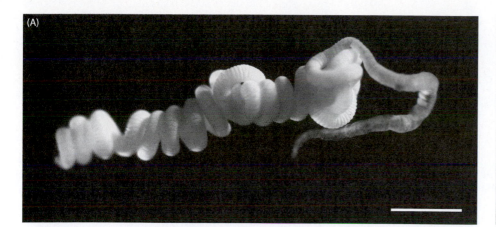

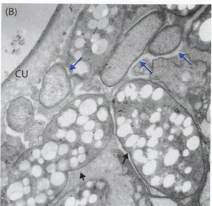

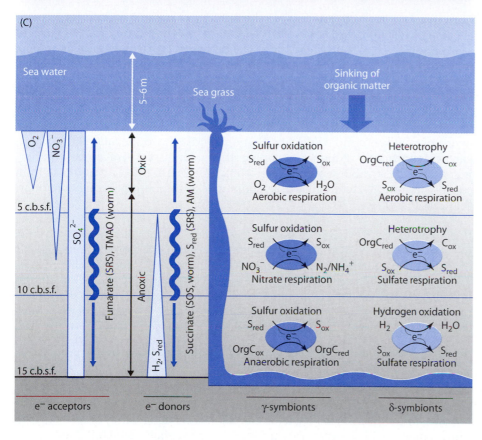

Figure 10.4 Bacterial endosymbiosis in *Olavius algarvensis*. (A) Photograph of the worm isolated from sediment. Scale bar = 1 mm. (Image courtesy of Christian Lott and Nicole Dubilier, Max Planck Institute for Marine Microbiology, Bremen, Germany). (B) Electron micrograph of tissue section showing endosymbionts; blue arrows indicate small sulfate-reducing bacteria; black arrows indicate large sulfide-oxidizing bacteria. The pale areas within the large cells are sulfur globules. CU, cuticle. Scale bar = 1μm. (Reprinted from Dubilier et al. [2001] by permission of Macmillan Publishers Ltd.) (C) Model showing availability of electron acceptors and donors in sediment layers. External sources are shown in the triangles, internal sources are shown next to the worms. AM, anaerobic metabolites; c.b.s.f., centimeters below sea floor; OrgC, organic compounds; S_{ox}, oxidized sulfur compounds; SOS, sulfur-oxidizing symbionts; SRS, sulfate-reducing symbionts; S_{red}, reduced sulfur compounds; TMAO, trimethylamine N-oxide. (Reprinted from Woyke et al. [2006] by permission of Macmillan Publishers Ltd.)

worms, only 0.1–0.2 mm in diameter and a few millimeters long, are completely devoid of a gut system and rely entirely on the production of nutrients by the thioautotrophic bacteria. Unlike the previous examples, the bacteria are not contained within specialized cells, but occur extracellularly just below the cuticle of the worm (*Figure 10.4*). The bacteria show cooperative metabolism, known as syntrophy. The deltaproteobacterial symbionts reduce sulfate absorbed by the worm from the sediment to sulfide, which is used by the gammaproteobacterial symbionts as an electron donor for the autotrophic fixation of CO_2. Because the worm's habitat has very low external concentrations of sulfide, the internal generation of sulfide by the deltaproteobacterial symbionts is essential for autrophy by the gammaproteobacterial symbionts. When first described, it was thought that the worm contained one population of each type, but metagenomic analysis has revealed that there are two genetically different types of symbionts affiliated with *Deltaproteobacteria* and two types affiliated with *Gammaproteobacteria*. Some individuals also contain a spirochaete. Analysis of the metagenome shows that one of the symbionts possesses an alternative mechanism for production of electrons for CO_2 fixation, using nitrate or fumarate in the absence of O_2. The deltaproteobacterial symbionts possess genes for CO_2 fixation via the reductive tricarboxylic acid (TCA) cycle (see p. 66), as well as numerous genes for heterotrophic metabolism of organic compounds in aerobic conditions. Thus, as the worm moves through the oxic and anoxic regions of the coastal sediments that it inhabits, electron donors will be available at all levels and the symbiotic consortium provides the worm with an optimal energy supply. Furthermore, the symbionts also recycle the host's waste products (ammonia and urea), which explains why the worms can live with a much-reduced excretory system. Examination of related species from other parts of the world reveals similar symbiotic consortia.

Animals colonizing whale falls depend on autotrophic and heterotrophic symbionts

Large whales weigh 30–160 tons, so the amount of organic carbon from a single carcass reaching the seafloor can be the equivalent of thousands of years of marine snow falling from the surface. The occasional but huge pulse of organic matter to the ocean floor supports diverse animal communities. After initial removal of soft tissue by sharks, hagfish, crabs, and other scavengers, followed by colonization of the surrounding sediments, the whale skeleton can support a rich population of animals sustained by thioautotrophic microbes oxidizing sulfur obtained from the bones. Many of these associations resemble those found at hydrothermal vents and cold seeps. Some scientists have speculated that whale carcasses provide "evolutionary stepping stones" for dispersal of animals harboring autotrophic symbionts (such as vesicomyid clams) as they have radiated from shallow- to deep-water habitats, but some specialist researchers do not support this idea.

In 2002, scientists studying the decaying carcass of a gray whale, which had sunk to a depth of nearly 3000 m off the coast of California, discovered a new genus of gutless polychaete worms (*Osedax*; Latin for "bone devourer"). The animals covered the skeleton, penetrating the bones with a root system developed from an enlarged egg sac and richly supplied with blood vessels (*Plate 10.4*). In electron micrographs, large bacterial cells were observed inside vacuoles in the host cells; 16S rRNA gene sequencing showed these to be very different to the bacteria found in other gutless worms. The main type belongs to the *Oceanspirillales* group; these bacteria are heterotrophs and provide nutrition for the worm; probably by breakdown of collagen, cholesterol, and lipids, which are present in large amounts in the whale bone. Study of *Osedax* is at an early stage, but it seems likely that the host digests the bacteriocytes rather than relying on translocation of nutrients from the bacteria. Other types of bacteria may also be present. The animals are biologically very unusual since the males are contained within the female body; little is known about the nutrition of the males. The wide geographic distribution of *Osedax* poses something of a mystery, given the rarity of carcasses of whales since their massive depletion by whaling in the nineteenth century. However, a large whale carcass should sustain a community for many years and, based on calculation of natural mortality rates, significant deposition of carcasses could still occur, especially along migration routes. Thus, whale falls may still play a major role in the ecology of deep-sea benthic processes, sustained by long-term microbial activity.

Some hydrothermal vent animals have dense populations of bacteria on their surface

When hydrothermal vents on the mid-Atlantic Ridge were first investigated in 1985, large populations of a type of shrimp (*Rimicaris exoculata*) were discovered, but no evidence of endosymbionts was found. These animals have specially adapted mouthparts and an enlarged gill chamber, containing large populations

IT'S NOT EASY TO SINK A WHALE

Although some knowledge of the biology of whale falls came from chance discovery of carcasses, most of our knowledge of the succession of events in the decay of whale skeletons comes from manipulative experiments. Craig Smith (University of Hawaii) and colleagues have towed carcasses of beached whales to drop sites (mapped precisely with GPS coordinates) and sunk them to the seafloor. Apart from the logistic problems of arranging transport and support, sinking a whale carcass bloated with gases after many days of towing behind a ship is a difficult and unpleasant task. Manned submersibles and ROVs are used for long-term video recording and sampling of the carcass over many years.

(up to 10^7) of thioautotrophic bacteria. The same phylotype of bacteria is also present in large numbers on and around the chimney, but the shrimp population seems to obtain most of its nutrition by feeding on the episymbionts that it has "cultivated" in its gill chamber. The shrimps provide the optimum concentrations of sulfide and oxygen by scraping metal sulfides from the chimney wall and fanning water currents over the bacteria. Unlike the thioautotrophs previously described, the shrimp episymbionts are members of the *Epsilonproteobacteria*.

Another vent animal that is the subject of intensive research is *Alvinella pompejana*, an oligochaete worm about 9 cm long and 2 cm wide, which lives in dense masses in the walls of black smoker chimneys on the East Pacific vents. This animal is known as the "Pompeii worm" in recognition of the association with volcanic activity and its remarkable temperature gradient across its body; the worm's posterior end is in galleries within the chimney wall (~85°C), while the anterior end projects into the surrounding cooler water (~20°C). Hair-like projections secreted from mucous glands on the dorsal surface are covered in episymbiotic bacteria, forming a dense fleece-like structure. Recent metagenomic studies show that the episymbionts are affiliated with *Epsilonproteobacteria* and form a multispecies biofilm. They are chemoautotrophic, possessing genes for sulfide oxidation and denitrification, and fix CO_2 via the reductive TCA cycle. It is not clear whether the host obtains nutritional benefit, but it is possible that the worms eat the bacteria from the back of neighboring worms as a source of vitamins or amino acids. The episymbionts may also detoxify metals and shield the host from fluctuations in temperature. The bacterial proteins predicted from the metagenomic sequences appear to have special adaptations to the variable environment—temperature and chemical composition—that the worm inhabits.

Some fish and invertebrates use bacteria to make light

Another type of cooperative association exists between bacteria and animals, in which the activities of the bacteria confer a behavioral and ecological benefit to the host, rather than a direct nutritional one. Bioluminescence is the emission of blue or green light from oxygen-utilizing reactions via the luciferase enzymes (see Figure 4.5). Bioluminescence occurs very widely in the oceans, particularly in animals that inhabit deep waters where no sunlight penetrates or in animals that are active in shallow waters at night. Bioluminescent animals use light emission for avoidance or escape from predators, attracting prey, or as a means of communication (such as mate recognition). Animals often possess complex structures such as lenses, filters, and shutters to control or modify the light emitted. In most bioluminescent fish and invertebrates the light is produced within specialized cells of the animal. However, in some animals symbiotic bacteria living within specialized structures produce the light. Bioluminescence seems to have multiple evolutionary origins, as there is a very wide diversity of enzymes and substrates for the bioluminescent reaction mechanism. Apart from light emission, the only common feature is the requirement for oxygen, and it is possible that bioluminescence evolved originally as an antioxidative mechanism.

Several groups of fish and invertebrates are colonized by bioluminescent bacteria. In many cases, these bacteria can be isolated and will emit light in culture, but some have not yet been cultured and have been identified only using gene sequencing. Another clue to the bacterial origin of bioluminescence is that bacteria usually emit light continuously over a period, whereas light produced by eukaryotic enzymes tends to occur as brief flashes. Final proof of the role of bacteria in light emission by animals is the use of an assay for luciferase in the presence of reduced flavin mononucleotide, as this reaction is unique to bacterial bioluminescence. Bacterial bioluminescence occurs in a wide range of phylogenetically distinct animals, and the associations range from relatively unspecialized facultative colonization of the intestinal tract or organs derived from it, through to obligate interactions with specialized external light organs.

EPISYMBIOTIC BACTERIA MAKE THE "YETI CRAB" HAIRY

A new species of crab was discovered in 2005 at a hydrothermal vent in the South Pacific Ocean. The legs of the animal are covered in silky fur-like projections (setae) that in turn are covered with clusters of filamentous bacteria (*Plate 10.5*). The crab thus appears to be extraordinarily hairy and has been nicknamed the "Yeti crab," although it is actually a species of squat lobster, *Kiwa hirsuta*. Molecular phylogenetic analyses revealed the setae-associated bacteria to be dominated by members of the *Epsilonproteobacteria*, *Gammaproteobacteria*, and *Bacteroidetes*. Key enzymes for carbon fixation and sulfur cycling have been detected (Goffredi et al., 2008). It has been suggested that the bacteria may be harvested by the crabs for nutrition, or they may play a role in detoxification of minerals, as in *Alvinella*. One intriguing puzzle is how the bacteria are transferred during the process of molting and loss of the exoskeleton.

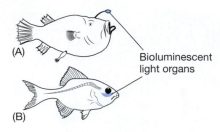

Bioluminescent light organs

Figure 10.5 Location of external light organs harboring bioluminescent bacteria. (A) Anglerfish. (B) Flashlight fish. See also *Plate 10.7*.

The bacteria associated with bioluminescence have probably evolved via adaptive radiation with fish and invertebrate hosts over the last few hundred million years. Bioluminescent bacteria often occur as members of the commensal gut microbiota in fish, and luminescence in the feces may encourage ingestion by other animals, facilitating their transmission to new hosts.

The commonest culturable types of bacteria are members of the *Vibrionaceae*, including *Photobacterium phosphoreum*, *Photobacterium leiognathi*, and *Vibrio fischeri* (recently reclassified as *Aliivibrio fischeri*; see p. 104). The bacteria grow with much lower generation times in the light organ than they are capable of in culture and it seems that some mechanism of the animal host controls this. It is in the host's interest to maximize bioluminescence but minimize diversion of nutrients to the bacteria; restriction of oxygen supply or iron limitation may be important factors in this process. In all cases, the light organs contain dense communities of extracellular bacteria in tubules that communicate with the intestine or the external environment. Release of bacteria is an important component in the control of the bacterial population.

Figure 10.5 illustrates the location of external light organs in the flashlight fishes and deep-sea anglerfishes, which in both cases contain symbionts that have not yet been isolated in culture. Bacterial light organs in flashlight fishes (members of the family *Anomalopidae*) are the largest in any fish relative to body size. The light organs in these strictly nocturnal tropical reef fish are located below the eyes. In very dark conditions, the forward illumination is bright enough to allow the fish to seek out prey, and the fish can control the light emission in order to communicate with other members of the species (possibly for mating). Anglerfishes, which belong to nine families in the suborder *Ceratoidei*, are usually solitary animals found in the deep sea. They have a small light organ (esca) at the end of a projection from the head (see *Plate 10.7*). This acts as a lure to attract prey near to the jaws, and light emission is controlled by the supply of oxygenated blood to the esca.

The bobtail squid uses bacterial bioluminescence for camouflage

The Hawaiian bobtail squid (*Euprymna scolopes*) inhabits shallow reefs, hiding during the day and feeding at night. The squid emit light from their ventral surface, adjusted to match the intensity of moonlight. This counter-illumination camouflage helps them to be less visible to predators from below. A highly specific association with a certain strain of bioluminescent *V. fischeri* is responsible. Newly hatched squid acquire this particular vibrio exclusively from the surrounding seawater within a few hours, even though it is present at very low densities—just one to a few hundred cells per milliliter—and many other species of bacteria are present. The squid flush seawater in and out of their body during respiration and movement, and this brings bacteria into contact with the nascent light organ (*Figure 10.6*). To reach the crypts of the light organ, bacteria must swim through a duct lined with ciliated cells and covered with mucus. In the crypt spaces, the bacteria encounter macrophage-like cells that have an immune surveillance function. The macrophages engulf unwanted bacteria and some of the *V. fischeri*, probably serving to keep the symbionts from overgrowing. Most of the symbiotic bacteria colonize the microvilli of the crypt epithelial cells and this step involves adhesion of the bacteria to cell surface receptors. After adhesion, the bacteria grow rapidly (about three doublings per hour) for the first 10–12 h until they reach a density of about 10^{11} bacteria per milliliter of crypt fluid. Once this critical density is reached, an autoinducer of bioluminescence accumulates, leading to initiation of light emission via quorum-sensing regulation of the *lux* operon (see *Figure 3.8*). Once established, the bacteria grow more slowly (about 0.2 doublings per hour). Each morning, in response to daylight, the squid squeezes out the contents of the light organ—this is a thick paste of

---

ⓘ BACTERIA HELP FEMALE ANGLERFISHES FIND THEIR MATE

Anglerfishes show an unusual sexual dimorphism. The males are dwarfs that reach sexual maturity soon after the larval stage and they probably never feed. Only the females possess the bioluminescent organ. The males have large eyes and a large olfactory organ, so it is assumed that they locate the females by a combination of looking for light from the flashing esca and detection of pheromones. The male bites the underside of the female; his body then fuses with hers and all organs apart from the testis degenerate. The female now carries a supply of sperm for fertilization of her eggs.

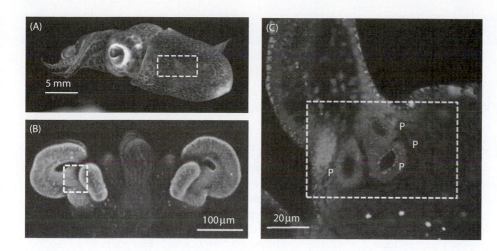

Figure 10.6 The light organ of *Euprymna scolopes*. (A) An adult squid (box shows location of the light organ). (B) Confocal micrograph showing the complex ciliated epithelium on the surface of the juvenile squid. (C) Confocal micrograph showing the tiny pores (P) through which *Vibrio fischeri* enters the host. (Reprinted from Nyholm and McFall-Ngai [2004] by permission of Macmillan Publishers Ltd.)

mucus, bacteria, and macrophage-like cells—so that about 90% of the bacteria are expelled. This behavior of the squid ensures that it maintains a "fresh" active culture of *V. fischeri* that will build up to the high density required for biolumi-nescence when the squid emerges from its hiding place at night. It also provides regular seeding of the environment with the specific strain of *V. fischeri*, which ensures horizontal transmission of the symbiont to juvenile squid. Study of this remarkable association is revealing many important findings about the "dia-logue" between an animal host and its bacterial partner, which has great signifi-cance in consideration of the evolution of symbiotic and pathogenic associations (see *Research Focus Box 10.2*).

Endosymbionts of bryozoans produce compounds that protect the host from predation

Bryozoans are tiny colonial animals that resemble moss coating the surface of rocks. One species, *Bugula neritina*, has been the subject of interest because it produces complex polyketides (bryostatins), which may be effective in the treat-ment of some cancers (see p. 312). We know now that an endosymbiotic bac-terium synthesizes bryostatin and the genes responsible have been identified. The bryostatins coat the surface of the larvae at high concentrations and seem to function in protecting the larvae from predation, as they are unpalatable to fish. They may also play a role in attracting conspecific larvae to settle to form colo-nies. The bacterium has never been cultivated, but genetic analysis shows that it is a gammaproteobacterium, and it has been designated *Candidatus* Endobu-gula sertula.

Sponges contain dense communities of specific microbes

Sponges (Phylum *Porifera*) are the oldest group of multicellular animals, com-prising more than 6000 species in a wide variety of tropical, temperate, and cold-water marine habitats, with a smaller number of freshwater examples. They have a simple three-layered body structure with an internal system of pores and chan-nels lined by flagellated cells (choanocytes) that pump water through the body. They feed by filtering bacteria and microalgae from the surrounding water and digesting them with specialized phagocyte cells. Sponges filter huge volumes of water to concentrate food particles from the surrounding water—a modest sized sponge filters tens of thousands of liters of water a day—and they can remove over 90% of the bacteria in the inhalant water. The sponge body (mesohyl) is a gelatinous matrix strengthened by fibers and spicules of calcium carbonate or silica. In most sponges, the mesohyl is packed with a diverse community of many different microbes, which can constitute up to half of the volume of the sponge (*Figure 10.7*). Most microbes in the mesohyl are extracellular, but intracellular

BOX 10.2   RESEARCH FOCUS

A perfect partnership?

Genomic techniques reveal secrets of the *Euprymna scolopes–Vibrio fischeri* symbiosis

The pioneering work in the study of this intriguing interaction has been carried out since the 1990s by Ned Ruby, who works on the bacterium, and Margaret McFall-Ngai, who studies the developmental biology of the host. In addition, their many graduate students have gone on to form their own research teams. This collaboration has yielded many remarkable results, and numerous research papers have been published on diverse aspects of the host–symbiont interactions. Of the many different facets of the symbiosis, two questions of special interest are considered here: first, how does the host associate so specifically with particular strains of *V. fischeri*; and second, what is the mechanism that leads to the daily expulsion of most of the bacteria in the light organ, requiring subsequent regrowth?

Visick and Skoufos (2001) found one type of mutant strain that was unable to colonize the light organ because it lacks a protein termed RscS (regulator of symbiotic colonization-sensor), which has a sequence similarity to the sensor kinase component of two-component regulators (see p. 76). Such sensor kinases recognize external signals and relay information to regulatory proteins that increase or decrease transcription of specific genes. Elucidation of the full genome sequence of *V. fischeri* (Ruby et al., 2005) has also revealed a large number of putative response regulator genes, and mutation studies showed that several of these are important in different stages of the colonization process. Yip et al. (2006) showed that RscS is a master regulator that activates the expression of large clusters of *syp* genes associated with polysaccharide synthesis, biofilm formation, and aggregation outside the squid light organ in the early stages of colonization. Hussa et al. (2008) identified the response regulator protein (SypG) and showed that it is part of a complex network of interconnected regulatory systems controlling colonization processes. Yildiz and Visick (2009) reviewed the complex interactions between genes for structural (flagella, pili, polysaccharides) and regulatory (two-component regulators, quorum sensing, and other signaling) processes in the formation of biofilms in *V. fischeri* and other *Vibrio* spp. Recently, Mandel et al. (2009) made the remarkable discovery that a single gene encoding the RscS regulator is responsible for the specificity seen in this interaction. They compared the complete genome sequence of a *V. fischeri* strain, MJ11, isolated from the bioluminescent Japanese pineconefish (*Monocentris japonica*) with that of strain ES114, isolated from *E. scolopes*. MJ11 is unable to colonize squid larvae, even if present at many times the density of ES114. The genomes have a highly similar content and gene order, although there appear to be differences in about 300 proteins. Mandel et al. (2009) showed that the gene *rscS* is present in the genome of all strains of *Vibrio* isolated from squid, but not in the genome of *V. fischeri* isolated from the pineconefish (although some strains have a comparable gene with a very divergent sequence). After transferring the *rscS* gene to the pineconefish strain MJ119, this strain developed the ability to complete the initial stages of infection of the squid. Phylogenetic analysis showed an important role for RscS in the evolution of the squid symbiosis.

Another feature of the genome sequence is its close resemblance to that of pathogenic vibrios, including *V. cholerae*, *V. parahaemolyticus*, and *V. vulnificus* (see Chapter 12). Numerous genes of *V. fischeri* were recognized to have similarities to those associated with pathogenicity in these bacteria, such as genes for toxins and surface-associated virulence factors, prompting Ruby et al. (2005) to conclude that "effectors may result in either a beneficial or a pathogenic outcome, depending on the host species or tissue location colonized." Nyholm et al. (2009) have found that host hemocytes react differently over time to *V. fischeri* through a process that resembles the immune tolerance seen in chronic human infections. It is clear that the adult squid host "tolerates" the presence of *V. fischeri* for a while, but the daily expulsion of the bacteria from the light organ indicates that a process akin to that seen in pathogenic infections takes over. By using microarray technology (see p. 54), Chun et al. (2008) were able to determine which genes of the squid host are regulated (up or down). Gene expression in squid exposed to mutants of different types showed that the host response depended on signals associated with presence of the bacteria, detection of bioluminescence, and production of quorum-sensing autoinducers. More recent studies have shown that hundreds of changes in gene expression of the symbiont and its host occur throughout the different stages of the daily cycle (Wier et al., 2010). A key set of virulence-like bacterial genes seems to be switched on just before dawn, and this correlates with the appearance in the host tissue of blebbing and effacement of the light organ crypt epithelial surface, very similar to what is seen during infection of the human gut by some pathogens. The bacteria vary the expression of genes associated with utilization of different substrates throughout the diel cycle. After the dawn expulsion, the remaining bacteria regrow, upregulating genes for the anaerobic metabolism of glycerol derived from the host membranes. After 12 h, genes associated with fermentation of chitin are upregulated. Unraveling the mechanisms underlying these processes may have important spin-offs for medicine, because of parallels with the behavior of bacteria in the human intestine.

bacteria and microalgae are sometimes found. Interest in the study of the microbial content of sponge tissue is largely driven by the interest in natural products such as antibiotics and antitumor compounds, which they often contain. Only a tiny fraction of sponge microbes have been cultured, but suspicion that the microbes (rather than the host's own metabolism) produce a wide range of bioactive compounds has been confirmed by recognition of microbial genes for their biosynthesis. This offers new possibilities for biotechnological exploitation, which is discussed in Chapter 13.

In addition to the biotechnological focus, the interaction between microbes and sponges is inherently very interesting and intriguing from a biological and evolutionary perspective. In the past few years, there has been intensive application of molecular methods to investigate sponge microbes, and more than 2000 16S rRNA gene sequences have been described. Sponge microbes are very diverse including *Bacteria* from 15 phyla; *Archaea*, in particular members of the *Crenarchaeota*; and various types of dinoflagellates, diatoms and yeasts. Remarkably, there appear to be clusters of closely related sponge-specific microbes that have not been found elsewhere, including a new phylum that is apparently unique to sponges (*Poribacteria*). A large number of studies have now provided compelling evidence that sponges contain widespread specific microbial communities. Even distantly related sponges isolated from a wide geographic range have been found to contain overlapping microbial communities. One explanation for this is that specific microbes became associated with sponges very early in their evolution, more than 600 million years ago, and remained associated with the sponges as they underwent evolutionary radiation. This hypothesis requires that the microbes are passed vertically from one generation to the next, and there is now considerable evidence (largely from microscopic studies) that this occurs during sexual reproduction, with bacteria observed in the eggs, embryos, or larvae of many species. During asexual reproduction, bacteria could be passed via tissue buds.

The almost ubiquitous occurrence of symbiotic microbes and their probable presence throughout evolution of sponges indicates strongly that these are mutualistic associations. As noted earlier, some sponges rely on photosynthetic microbes for their nutrition, but most sponges are heterotrophic, consuming microbes harvested from the seawater. The extent to which bacteria within the mesohyl contribute to carbon requirements or the provision of specific organic nutrients (e.g. by translocation of nutrients, as occurs in tubeworms) is unclear. Some species of sponges growing at methane seeps contain methanotrophic symbionts, which undoubtedly provide nutritional benefit to their host. In some sponges, ingestion and breakdown of the bacteria in the mesohyl has been observed, leading some scientists to suggest that sponges cultivate bacteria as a food source. Bacteria certainly seem to play a key role in nitrogen cycling within sponges, particularly by removal of toxic ammonia by denitrification. Nitrogen fixation by cyanobacteria may also be important. The production of large amounts of mucus

? DO ASCIDIANS OBTAIN BENEFIT FROM AEROBIC ANOXYGENIC PHOTOTROPHS?

In their adult stage, ascidians (commonly known as sea squirts) are colonial sessile filter feeding invertebrates, which are characterized by a tough polysaccharide coat or "tunic." The adult stage of some species harbors the photosynthetic cyanobacterium *Prochloron*, and cells of the symbionts are transmitted vertically during late stages of embryonic development of the larvae (Hirose and Hirose, 2007). Recent studies by Martinez-Garcia et al. (2007) using DNA fingerprinting and FISH analysis showed that some species of ascidians harbor a diverse community of bacteria, including some types that carry out aerobic anoxygenic phototrophy (AAnP). As discussed on p. 62, this is a recently discovered process carried out by many planktonic bacteria, especially those in the *Roseobacter* clade, using light as an energy source to supplement heterotrophic metabolism. Martinez-Garcia and colleagues showed that the *pufM* gene, characteristic of AAnP, is expressed in the tunic tissues and appears to be widespread in ascidians within the photic zone. This unexpected finding indicates a new type of light-based symbiosis very different from the oxygen-generating activities of zooxanthellae and symbiotic cyanobacteria. It will be interesting to find out whether the host derives nutritional benefit from the association and, if so, the mechanisms involved.

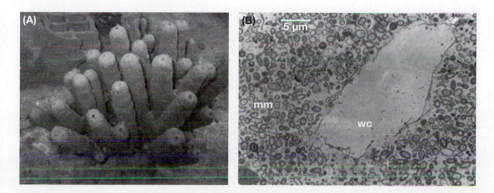

Figure 10.7 The Mediterranean sponge *Aplysina aerophoba*. (A) Underwater photograph. (B) Transmission electron micrograph showing the microorganism-filled mesohyl matrix (mm) and a water channel (wc). (Photography by K. Bayer and M. Wehrl, Würzburg, Germany, respectively). (Reprinted from Taylor et al. [2007] by permission of Macmillan Publishers Ltd.)

? COULD SPONGES ACQUIRE RARE BACTERIA FROM THE ENVIRONMENT?

Although the idea of a very ancient symbiosis with specific microbes is favored by most sponge microbiologists, alternative explanations involving acquisition of some organisms from the environment are possible. One of the main reasons given in favor of specific associations is that many types of sponge microbe can never be detected in the environment. However, because a sponge can filter tens of thousands of liters each day, even if a very rare bacterium was present at only one cell per liter—and therefore unlikely to be detected by commonly used methods such as PCR amplification of genes in water samples—a sponge could still acquire tens of thousands of this type every day.

by the mesohyl bacteria is also thought to contribute to the structural integrity of the tissue. The most important beneficial function is probably the production of bioactive compounds that may protect the sponge against the harmful effects of ultraviolet radiation or oxidative stress, or act as defense compounds to prevent predation.

Given that sponges filter-feed on microbes, how is it that most of the bacteria can live within the tissues, adjacent to phagocytic cells? It is likely that the symbionts possess surface properties such as production of grazing deterrents, altered cell walls, or slime capsules that protect them from recognition and ingestion by the phagocytes. Some insights into this come from examining the genome of *Cenarchaeum symbiosum*, which is the dominant symbiont of the sponge *Axinella mexicana*. Numerous genes not found in free-living relatives, encoding cell surface, regulatory functions, and defense mechanisms, are present. As yet, firm conclusions about the origin of the association between sponges and their resident microbes are not possible. Further application of genomics and other advanced molecular techniques, such as those carried out with symbionts of vent animals (see *Research Focus Box 10.1*), will aid our understanding of the specificity and interdependence of the host–symbiont interactions, which is such a fascinating aspect of sponge microbiology.

Some protists with endosymbionts can switch from heterotrophic to phototrophic metabolism

Study of marine protists provides powerful evidence in support of the endosymbiosis theory for the evolution of eukaryotic cells, discussed in Chapter 1. Many unicellular protists form associations with other species of protists, or with bacteria and archaea; these range from loose associations, through intracellular endosymbionts, to fully integrated organelles. This is well illustrated by the phenomenon of mixotrophy, a term used to describe the ability of protists to use both heterotrophic and phototrophic modes of nutrition. Many species of phytoplankton carry out photosynthesis, but can also ingest particulate food (phagotrophy). This is particularly common in the dinoflagellates, where over 20% of species have been shown to be mixotrophic. Many small mixotrophic algae can ingest bacteria, whilst larger forms can also ingest other microalgae, ciliates, and other prey. Different stages in the life cycle of some dinoflagellates may show a spectrum from fully photosynthetic, through mixotrophic, to fully phagotrophic lifestyles. Phagotrophy clearly provides nutrients in the form of carbon to supplement photosynthesis and other essential substances such as nitrogen, phosphorus, and iron. Mixotrophs have sophisticated mechanisms to regulate their metabolism. For example, when high densities of bacteria are present and/ or light levels are low, photosynthesis may be switched off completely and the protist relies entirely on ingested bacteria as food. When densities drop below a certain level, the photosynthesis machinery is activated. Protists will grow at different rates depending on the mix of nutrition employed at the time.

Viruses may help a sea slug to use "stolen" chloroplasts for photosynthesis

There are many examples of the ingestion of photosynthetic microbes by "normally" heterotrophic protists, in which the chloroplast organelle is retained by the cell and becomes "enslaved" for its benefit. This process, known as "kleptoplastidy," is widespread in ciliates, foraminiferans, and dinoflagellates (see p. 139). Clearly, the integration of the metabolic control of organelles with that of the cytoplasm requires complex adaptations, and many protozoa retain ingested chloroplasts for a few days before they are digested or eliminated. True endosymbionts must avoid digestion by the host. This can occur either by physical separation of the food vacuoles and symbiont-capturing vacuoles, or by alteration of the phagosome membrane so that lysosomes containing digestive enzymes do not fuse with the phagosome. New insight has come from the observation that *Elysia chlorotica*, a species of sea slug, can acquire chloroplasts from ingesting its algal prey; the chloroplasts are retained in cells of the animal's digestive tract and are fully functional in photosynthesis. The chloroplasts do not contain all the genetic information needed to carry out photosynthesis and normally rely on nuclear genes in the alga. A gene for a key protein needed for photosynthesis is transferred to the sea-slug genome and expressed by the animal cells. Viruses can be observed in the stolen chloroplasts, the nuclei, and the cytoplasm, and they are probably responsible for the transfer of genes, as well as causing the death of snails after they have laid their eggs.

Conclusions

Since the discovery of hydrothermal vent communities in 1977 and realization that the animals found there depend for their nutrition on endosymbiotic bacteria, knowledge of the range and diversity of mutualistic associations between microbes and animals has expanded enormously. The examples provided here should convince the reader that such symbioses, far from being an exceptional occurrence in specialized habitats, occur in diverse habitats and in many phyla from simple protists to vertebrates. Frequently, multiple microbial partners are involved. As the use of genomic techniques to explore these relationships is expanded, we are gaining an insight into the molecular basis of these interactions and the evolutionary adaptations of both the hosts and their microbial partners.

References

Berkelmans, R. and Van Oppen, M. J. H. (2006) The role of zooxanthellae in the thermal tolerance of corals: a "nugget of hope" for coral reefs in an era of climate change. *Proc. Roy. Soc. B Biol. Sci.* 273: 2305–2312.

Chun, C. K., Troll, J. V., Koroleva, I., et al. (2008) Effects of colonization, luminescence, and autoinducer on host transcription during development of the squid–vibrio association. *Proc. Natl Acad. Sci. USA* 105: 11323–11328.

Dubilier, N., Mulders, C., Ferdelman, T., et al. (2001) Endosymbiotic sulphate-reducing and sulphide-oxidizing bacteria in an oligochaete worm. *Nature* 411: 298–302.

Dubilier, N., Bergin, C. and Lott, C. (2008) Symbiotic diversity in marine animals: the art of harnessing chemosynthesis. *Nat. Rev. Microbiol.* 6: 725–740.

Goffredi, S. K., Jones, W. J., Erhlich, H., Springer, A. and Vrijenhoek, R. C. (2008) Epibiotic bacteria associated with the recently discovered Yeti crab, *Kiwa hirsuta*. *Environ. Microbiol.* 10: 2623–2634.

Harmer, T. L., Rotjan, R. D., Nussbaumer, A. D., Bright, M., Ng, A. W., DeChaine, E. G. and Cavanaugh, C. M. (2008) Free-living tube worm endosymbionts found at deep-sea vents. *Appl. Environ. Microbiol.* 74: 3895–3898.

Hirose, E. and Hirose, M. (2006) Morphological process of vertical transmission of photosymbionts in the colonial ascidian *Trididemnum miniatum* Kott, 1977. *Mar. Biol.* 150: 359–367.

Hoegh-Guldberg, O., Hughes, T., Anthony, K., Caldeira, K., Hatziolos, M. and Kleypas, J. (2009) Coral reefs and rapid climate change: Impacts, risks and implications for tropical societies. *IOP Conf. Ser.: Earth Environ. Sci.* 6: 302004

Hurtado, L. A., Mateos, M., Lutz, R. A. and Vrijenhoek, R. C. (2003) Coupling of bacterial endosymbiont and host mitochondrial genomes in the hydrothermal vent clam *Calyptogena magnifica*. *Appl. Environ. Microbiol.* 69: 2058–2064.

Hussa, E. A., Darnell, C. L. and Visick, K. L. (2008) RscS functions upstream of SypG to control the *syp* locus and biofilm formation in *Vibrio fischeri*. *J. Bacteriol.* 190: 4576–4583.

Kleypas, J. A. and Eakin, C. M. (2007) Scientists' perceptions of threats to coral reefs: results of a survey of coral reef researchers. *Bull. Mar. Sci.* 80: 419–436.

Kuwahara, H., Yoshida, T., Takaki, Y., et al. (2007) Reduced genome of the thioautotrophic intracellular symbiont in a deep-sea clam, *Calyptogena okutanii*. *Curr. Biol.* 17: 881–886.

Mandel, M. J., Wollenberg, M. S., Stabb, E. V., Visick, K. L. and Ruby, E. G. (2009) A single regulatory gene is sufficient to alter bacterial host range. *Nature* 458: 215–2U7.

Martinez-Garcia, M., Diaz-Valdes, M., Wanner, G., Ramos-Espla, A. and Anton, J. (2007) Microbial community associated with the colonial ascidian *Cystodytes dellechiajei*. *Environ. Microbiol.* 9: 521–534.

Moya, A., Pereto, J., Gil, R. and Latorre, A. (2008) Learning how to live together: genomic insights into prokaryote–animal symbioses. *Nat. Rev. Genet.* 9: 218–229.

Newton, I. L. G., Woyke, T., Auchtung, T. A., et al. (2007) The *Calyptogena magnifica* chemoautotrophic symbiont genome. *Science* 315: 998–1000.

Nussbaumer, A. D., Fisher, C. R. and Bright, M. (2006) Horizontal endosymbiont transmission in hydrothermal vent tubeworms. *Nature* 441: 345–348.

Nyholm, S. V. and McFall-Ngai, M. J. (2004) The winnowing: establishing the squid–vibrio symbiosis. *Nat. Rev. Microbiol.* 2: 632–642.

Nyholm, S. V., Stewart, J. J., Ruby, E. G. and McFall-Ngai, M. J. (2009) Recognition between symbiotic *Vibrio fischeri* and the haemocytes of *Euprymna scolopes*. *Environ. Microbiol.* 11: 483–493.

Polz, M. F., Ott, J. A., Bright, M. and Cavanaugh, C. M. (2000) When bacteria hitch a ride. *ASM News* 66: 531–539.

Robidart, J. C., Bench, S. R., Feldman, R. A., et al. (2008) Metabolic versatility of the *Riftia pachyptila* endosymbiont revealed through metagenomics. *Environ. Microbiol.* 10: 727–737.

Rowher, F. and Youle, M. (2010) *Coral Reefs in the Microbial Seas*. Plaid Press, San Diego, CA.

Ruby, E. G., Urbanowski, M., Campbell, J., et al. (2005) Complete genome sequence of *Vibrio fischeri*: a symbiotic bacterium with pathogenic congeners. *Proc. Natl Acad. Sci. USA* 102: 3004–3009.

Stewart, F. J., Newton, I. L. G. and Cavanaugh, C. M. (2005) Chemosynthetic endosymbioses: adaptations to oxic-anoxic interfaces. *Trends Microbiol.* 13: 439–448.

Suzuki, Y., Kopp, R. E., Kogure, T., et al. (2006) Sclerite formation in the hydrothermal-vent "scaly-foot" gastropod—possible control of iron sulfide biomineralization by the animal. *Earth Planet. Sci. Lett.* 242: 39–50.

Taylor, M. W., Radax, R., Steger, D. and Wagner, M. (2007) Sponge-associated microorganisms: evolution, ecology, and biotechnological potential. *Microbiol. Mol. Biol. Rev.* 71: 295–347.

Visick, K. L. and Skoufos, L. M. (2001) Two-component sensor required for normal symbiotic colonization of *Euprymna scolopes* by *Vibrio fischeri*. *J. Bacteriol.* 183: 835–842.

Wier, A. M., Nyholm, S. V., Mandel, M. J., et al. (2010) Transcriptional patterns in both host and bacterium underlie a daily rhythm of anatomical and metabolic change in a beneficial symbiosis. *Proc. Natl Acad. Sci. USA* 107: 2259–2264.

Woyke, T., Teeling, H., Ivanova, N. N., et al. (2006) Symbiosis insights through metagenomic analysis of a metabolic consortium. *Nature* 443: 950–955.

Yildiz, F. H. and Visick, K. L. (2009) Vibrio biofilms: so much the same yet so different. *Trends Microbiol.* 17: 109–118.

Yip, E. S., Geszvain, K., DeLoney-Marino, C. R. and Visick, K. L. (2006) The symbiosis regulator RscS controls the *syp* gene locus, biofilm formation and symbiotic aggregation by *Vibrio fischeri*. *Mol. Microbiol.* 62: 1586–1600.

Further reading

Books and general reviews

Bright, M. and Bulgheresi, S. (2010) A complex journey: transmission of microbial symbionts. *Nat. Rev. Microbiol.* 8: 218–230.

Moya, A., Pereto, J., Gil, R., and Latorre, A. (2008) Learning how to live together: genomic insights into prokaryote–animal symbioses. *Nat. Rev. Genet.* 9: 218–229.

White, J. and Torres, M. (2009) *Defensive Mutualism in Microbial Symbiosis*. CRC Press, Boca Raton, FL.

Zooxanthellae and the coral holobiont

Mouchka, M. E., Hewson, I. and Harvell, C. D. (2010) Coral-associated bacterial assemblages: current knowledge and the potential for climate-driven impacts. *Integr. Comp. Biol.* 50: 662–674.

Raina, J.-B., Tapiolas, D., Willis, B. L. and Borne, D. G. (2009) Coral-associated bacteria and their role in the biogeochemical cycling of sulfur. *Appl. Environ. Microbiol.* 74: 3492–3501.

Rosenberg, E. and Loya, Y. (2004) *Coral Health and Disease*. Springer, Berlin.

Rosenberg, E., Koren, O., Reshef, L., Efrony, R. and Zilber-Rosenberg, I. (2007) The role of microorganisms in coral health, disease and evolution. *Nat. Rev. Microbiol.* 5: 355–362.

Siboni, N., Ben-Dov, E., Sivan, A. and Kushmaro, A. (2009) Global distribution and diversity of coral-associated *Archaea* and their possible role in the coral holobiont nitrogen cycle. *Environ. Microbiol.* 10: 2979–2990.

Stat, M., Carter, D. and Hoegh-Guldberg, O. (2006) The evolutionary history of *Symbiodinium* and scleractinian hosts—symbiosis, diversity, and the effect of climate change. *Perspect. Plant Ecol. Evol. Syst.* 8: 23–43.

Van Oppen, M. J. H., Leong, J. A. and Gates, R. D. (2009) Coral-virus interactions: a double-edged sword? *Symbiosis* 47: 1–8.

Yellowlees, D., Rees, T. A. V. and Leggat, W. (2008) Metabolic interactions between algal symbionts and invertebrate hosts. *Plant Cell Environ.* 31: 679–694.

Hydrothermal vent symbioses

Goffredi, S. K., Waren, A., Orphan, V. J., Van Dover, C. L. and Vrijenhoek, R. C. (2004) Novel forms of structural integration between microbes and a hydrothermal vent gastropod from the Indian Ocean. *Appl. Environ. Microbiol.* 70: 3082–3090.

Grzymski, J. J., Murray, A. E., Campbell, B. J., et al. (2008) Metagenome analysis of an extreme microbial symbiosis reveals eurythermal adaptation and metabolic flexibility. *Proc. Natl Acad. Sci. USA* 105: 17516–17521.

Whale falls and *Osedax*

Goffredi, S. K., Orphan, V. J., Rouse, G. W., Jahnke, L., Embaye, T., Turk, K., Lee, R. and Vrijenhoek, R. C. (2005) Evolutionary innovation: a bone-eating marine symbiosis. *Environ. Microbiol.* 7: 1369–1378.

Goffredi, S. K., Johnson, S. B. and Vrijenhoek, R. C. (2007) Genetic diversity and potential function of microbial symbionts associated with newly discovered species of *Osedax* polychaete worms. *Appl. Environ. Microbiol.* 73: 2314–2323.

Smith, C. R. and Baco, A. R. (2003) Ecology of whale falls at the deep-sea floor. *Oceanogr. Mar. Biol. Ann. Rev.* 41: 311–354.

Bioluminescent bacteria

Dunlap, P. V. and Kita-Tsukamoto, K. (2006) Luminous bacteria. In: *The Prokaryotes, a Handbook on the Biology of Bacteria*, 3rd Edn, Vol. 2 (eds M. Dworkin, S. Falkow, E. Rosenberg, K. H. Schleifer, and E. Stackebrandt). Springer, New York, pp. 863–892.

Symbiosis in bryozoans

Davidson, S. K., Allen, S. W., Lim, G. E., Anderson, C. M. and Haygood, M. G. (2001) Evidence for the biosynthesis of bryostatins by the bacterial symbiont "*Candidatus* Endobugula sertula" of the bryozoan *Bugula neritina*. *Appl. Environ. Microbiol.* 67: 4531–4537.

Sharp, K. H., Davidson, S. K. and Haygood, M. G. (2007) Localization of "*Candidatus* Endobugula sertula" and the bryostatins throughout the life cycle of the bryozoan *Bugula neritina*. *ISME J.* 1: 693–702.

Sponge microbiology

Steger, D., Ettinger-Epstein, P., Whalan, S., Hentschel, U., de Nys, R., Wagner, M. and Taylor, M. W. (2008) Diversity and mode of transmission of ammonia-oxidizing archaea in marine sponges. *Environ. Microbiol.* 10: 1087–1094.

Vogel, G. (2008) The inner lives of sponges. *Science* 320: 1028–1030.

Mixotrophy and acquired photosynthesis

Moorthi, S., Caron, D. A., Gast, R. J. and Sanders, R. W. (2009) Mixotrophy: a widespread and important ecological strategy for planktonic and sea-ice nanoflagellates in the Ross Sea, Antarctica. *Aquat. Microb. Ecol.* 54: 269–277.

Pierce, S. K., Maugel, T. K., Rumpho, M. E., Hanten, J. J. and Mondy, W. L. (1999) Annual viral expression in a sea slug population: life cycle control and symbiotic chloroplast maintenance. *Biol. Bull.* 197: 1–6.

Rumpho, M. E., Worful, J. M., Lee, J., et al. (2008) Horizontal gene transfer of the algal nuclear gene *psbO* to the photosynthetic sea slug *Elysia chlorotica. Proc. Natl Acad. Sci. USA* 105: 17867–17871.

Chapter 11

Microbial Diseases of Marine Organisms

Marine biologists have become increasingly aware of the effects of microbial diseases on populations and communities of marine organisms, which can sometimes result in major changes at the ecosystem level. Infectious diseases caused by viruses, bacteria, fungi, and protists have been characterized in marine mammals, fish, and turtles, as well as in a wide range of invertebrates, including mollusks, crustaceans, sponges, and corals. Infections of multicellular algae and seagrasses also have important ecological and economic effects. Toxins produced by diatoms and dinoflagellates may also cause disease. This chapter explores the various types of microbial diseases and their importance in natural ecosystems and aquaculture, with special consideration of the impact of climate change, pollution, and other anthropogenic factors. It is divided into three sections, dealing with diseases of invertebrates, vertebrates, algae and seagrasses, respectively. Biotechnological aspects of the control of disease are considered in Chapter 14.

Key Concepts

- Incidence of disease in many groups of organisms has increased in natural marine ecosystems in the past few decades.

- The development of disease depends on complex interactions between the host, the pathogen, and environmental factors (*Figure 11.1*); it is not possible to define single causative agents for many marine diseases.

- Global climate change and other anthropogenic factors appear to have led to increased disease in some groups of marine organisms.

- Development of intensive marine aquaculture has led to increased levels of bacterial and viral diseases in mollusks, crustaceans, fish, and algae, with some major economic effects in the industry as well as impacts on natural ecosystems.

- Bacteria (especially vibrios) are strongly associated with diseases of corals, mollusks, crustaceans, and fish.

- Viruses are responsible for major epizootics in mollusks, crustaceans, fish, reptiles, and mammals, some of which seriously affect natural populations.

Figure 11.1 The disease process in marine organisms depends on complex interactions of the host, pathogen, and environmental factors.

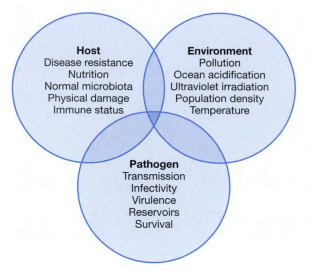

Host
Disease resistance
Nutrition
Normal microbiota
Physical damage
Immune status

Environment
Pollution
Ocean acidification
Ultraviolet irradiation
Population density
Temperature

Pathogen
Transmission
Infectivity
Virulence
Reservoirs
Survival

? ARE DISEASES IN MARINE ECOSYSTEMS REALLY INCREASING?

Many studies suggest that disease rates in marine organisms have increased dramatically over the past few decades, and global climate change is often cited as one of the main causes. Do these reports represent a true increase in disease rates, or is it simply that more scientists are conducting studies in this field? To answer this question, Ward and Lafferty (2004) conducted an online search of 5900 journals published between 1970 and 2001 for reports of disease in nine taxonomic groups. The authors used a sophisticated normalization method to determine the proportion of disease reports about a given taxonomic group relative to the total number of reports about any subject in that group. The data were also adjusted to remove the effect of multiple reports of the same disease outbreak or the effect of particularly prolific authors investigating disease. Reports of disease were increased for all groups over this period. However, after normalization to remove the skewing effects, Ward and Lafferty (2004) concluded that there was a clear increase in disease among turtles, corals, mammals, urchins, and mollusks, but no significant trends for seagrasses, decapods, and sharks/rays. Reports of diseases in fishes actually decreased, which they suggested was caused by drastic reductions in population density due to overfishing, presenting fewer opportunities for transmitting infection.

DISEASES OF INVERTEBRATES

Diseases of invertebrates have major ecological and economic impact

The marine invertebrates comprise hundreds of thousands of species. Although microbial diseases have been described in only a tiny fraction of these species, a huge bank of knowledge has been collected by zoologists, especially for the many protozoan parasites that infect invertebrates. Rather less is known about the impact of bacterial and viral diseases in the ecology of marine invertebrates, except in a few situations in which our attention has been drawn to animals of particular interest to humans. Thus, over many years we have built up knowledge of microbial diseases of invertebrate animals used for food; these are mainly molluskan and crustacean shellfish, either harvested from the sea or cultured in traditional aquaculture systems. Here, a number of microbial diseases that affect fishery production have been described. In the last few decades, the intensification of shellfish culture has increased rapidly, accompanied by a dramatic rise in disease incidence, with consequent economic effects due to lost production.

Another area that has expanded greatly is the recent attention given to diseases of corals. Besides the contribution of direct damage by mining, boating and diving activities, dynamite fishing, and pollution to the serious degradation of tropical coral reefs, we have recently become aware of the role of microbes in bleaching and diseases of corals. In addition to the obvious concerns about the loss of biodiversity and the ecological impact of the demise of coral reefs, the economic impact due to adverse effects on fisheries and tourism is also very significant. This section of the chapter therefore discusses some of the main bacterial and viral diseases of corals, sponges, mollusks, and crustaceans.

Infectious diseases of corals have emerged as a major threat to their survival

As discussed in Chapter 10, many of the world's tropical and subtropical reefs are severely at risk. Since the 1970s, there has been increasing recognition of the importance of coral diseases associated with microbial infection as a factor in this decline. Coral diseases were first studied intensively in the Caribbean, where successive outbreaks of disease seem to have contributed to the decline of the major reef-building species, especially *Acropora*, resulting in major ecosystem changes. Diseases have now been described in all the major reef systems of the world, involving several hundred different species of coral. In some cases, such emerging diseases reach epidemic status and dramatically alter both the abundance of specific corals and the overall diversity of reefs. With "emerging diseases," it is always difficult to decide whether there is a genuine increase in incidence, or whether it simply reflects more intensive observation. In the case of corals, there is no doubt that there has been recent intense interest in monitoring the health of the world's reefs, and there is now a large community of subaqua diving scientists, which partly explains the increased reporting of disease. However, there is a general consensus that disease is increasing as a result of recent changes to the marine environment caused by anthropogenic effects such a pollution and global climate change. If, as predicted, the effects of global warming escalate during the twenty-first century, it is vital that we develop a better understanding of the role of microbes in the health of corals.

One of the most difficult problems in evaluating the importance of coral diseases is that the early literature contains many poorly documented reports of disorders for which details of pathology and evidence of specific causation is incomplete. Diagnosis of disease in corals is primarily based on characteristics such as the loss of tissue, change in color, and exposure of coral skeleton. These limited macroscopic descriptions lead to imprecise names for the conditions—for example, the terms "white plague," "white disease," "white pox," and "white band" have all been used to represent different syndromes—and this results in inaccurate subsequent diagnosis. It is often difficult to meet Koch's postulates, which are the usual criteria needed to prove the etiology of disease (see information box on p. 232). As shown in *Table 11.1*, a small number of diseases have been well characterized, with strong evidence of association with specific bacteria or fungi. It can be seen that *Vibrio* spp. are very dominant, being associated with five of the nine conditions listed in *Table 11.1*. However, for most diseases, knowledge of the mechanisms of pathogenesis and the host responses remains very limited and many coral biologists remain skeptical about the role of microbes as primary causative agents of disease in uncompromised hosts (see *Research Focus Box 11.1*). Examples of some major diseases of corals are illustrated in *Plate 11.1*.

Table 11.1 Diseases of corals for which there is a strong association with microbial infection

Disease	Location	Principal host(s)	Pathogen(s)
Aspergillosis	CS	*Gorgonia ventalina* and other octocorals	*Aspergillus sydowii*
Black band disease	CS, IP	Wide range of corals	Mixed consortium; *Phormidium corallyticum*, *Desulfovibrio* spp. and *Beggiatoa* spp. are key components; *Vibrio* spp. may be associated
Bleaching	RS	*Oculina patagonica*	*Vibrio shiloi*[a]
Bleaching and tissue lysis ("white syndrome" in IP)	RS, IP	*Pocillopora damicornis, Montipora, Acropora, Pachyseris speciosa*	*Vibrio coralliilyticus* and related *Vibrio* spp.
White band	CS	*Acropora*	*Vibrio carchariae* (= *V. harveyi*) and other *Vibrio* spp.
White plague	RS	*Favia favus*	*Thallassomonas loyana*
White plague type II	CS	*Dichocenia* and other scleractinian corals	*Aurantimonas coralicida*
White pox	CS	*Acropora palmata*	*Serratia marcescens*
Yellow blotch/band	CS, IP	*Acropora, Montastrea, Diploastrea, Fungia*	Consortium of *Vibrio* spp.: *V. rotiferianus, V. harveyi, V. alginolyticus, V. proteolyticus*

[a]The original name *V. shiloi* continues to be used in most references, although the correct taxonomic designation was amended to *V. shiloni* and subsequently shown to be a junior synonym of *V. mediterranei* (see *Table 4.1*). CS, Caribbean Sea; IP, Indo-Pacific Ocean; RS, Red Sea.

The fungus *Aspergillus sydowii* caused a mass mortality of sea fans in the Caribbean Sea

Beginning in 1995, one of the most devastating epizootics yet seen in the marine environment occurred in sea fans (*Gorgonia ventalina*) in the Caribbean Sea. A new fungal species, *Aspergillus sydowii*, was identified as the causative agent responsible for massive tissue destruction. It was shown to be transmissible through contact with healthy corals. Since 1998, the severity of the disease has declined and it now occurs only sporadically, although many of the largest sea fans have been killed, especially in the Florida Keys. Investigations reveal that there is a complex relationship between temperature, growth of the pathogen, and host resistance due to production of antifungal compounds. The host attempts to limit spread of the pathogen by encapsulation of necrotic tissue, resulting in the characteristic purple lesions (*Plate 11.1F*). Other microbes associated with surface mucus also differ between healthy and diseased sea fans and between healthy and diseased parts of the same individual. *Aspergillus sydowii* has also been suspected as the causative agent of the virtual eradication of sea urchins (*Diadema antillarum*) from parts of the Caribbean Sea in the 1980s, although there is no direct evidence for this. There is recent evidence that the fungus can be isolated from apparently healthy *G. ventalina* and other species of octocorals, suggesting that it is an opportunist pathogen that causes disease only under certain conditions.

Black band disease of corals is a long-established disease of corals worldwide

Black band disease (BBD) takes its name from the characteristic black band (from about 10 mm to several centimeters wide), which is a thick microbial mat that migrates over the coral colony by as much as 2 cm a day (*Plate 11.1A*). Healthy coral tissue is killed and the band moves on, leaving the exposed skeleton, which usually becomes overgrown by turf algae. BBD usually occurs in the summer when water temperatures rise. It can arise on apparently pristine reefs, but is more often associated with reefs receiving sewage and runoff from the coast. BBD can have severe effects because it often attacks large corals such as *Montastrea* spp., which are important in building the framework of reefs. Infection can be transmitted via contact, and damaged colonies are more susceptible. Since its first description in the Caribbean in the 1970s, BBD has been observed in a wide range of coral species throughout the world. Microscopic examination of diseased tissue consistently shows the presence of large, gliding, filamentous cyanobacteria, together with numerous other heterotrophic bacteria including sulfide-oxidizing bacteria (especially *Beggiatoa* spp.) and sulfate-reducing bacteria (especially *Desulfovibrio* spp.), which are responsible for the characteristic black pigment. Several species of cyanobacteria, including *Phormidium corallyticum*, *Leptolyngbya* spp., *Geitlerinema* spp., and *Pseudoscillatoria coralii* have been associated with the disease. Production of microcystin toxins has been suggested by some to have role in the disease process, but this is now considered unlikely. It is generally assumed that BBD is caused by this mixed microbial consortium due to anoxic and sulfide-rich conditions at the base of the band, resulting in death of the coral tissue. Recent use of PCR amplification of 16S rRNA genes separated by denaturing gradient gel electrophoresis (DGGE; see p. 48) has shown marked differences between the communities inhabiting healthy tissue, diseased tissue, and the overlying seawater. Different investigators have found rather different results with regard to the presence of sequences from cyanobacteria and sewage-associated bacteria (from terrestrial runoff) in healthy and diseased BBD tissue, although sequences corresponding to sulfate-reducing bacteria (SRB) are consistently found, as predicted from cultural and metabolic studies. *Vibrio* spp. have also been recovered from some BBD infections, so the etiology of BBD remains unclear. A similar condition called red band occasionally affects hard star and brain corals in the Caribbean and Great Barrier

Reef. It seems to involve similar cyanobacteria and other members of the microbial consortium; the reasons for difference in pigmentation are not clear.

White plague and white pox are major diseases affecting Caribbean reefs

White plague disease (WPD) of corals was first described as a disease of large encrusting and branching corals in the 1970s and so far has been observed only in the Florida Keys in the Caribbean. It starts at the base of the coral and rapidly progresses upwards, with the destruction of tissue. There appear to be two forms of the disease, termed WPD-I and WPD-II, with the latter being more virulent. A *Sphingomonas* sp. was originally suspected as the causative agent of WPD-II, but this was subsequently shown to be incorrect, and the pathogen is now recognized as *Aurantimonas coralicida*. Unlike BBD, there seems to be a sharply defined line between healthy and diseased tissue, suggesting that toxins or enzymes are excreted by the advancing microbial population.

Since the mid-1990s, the disease termed "white pox" has had a devastating effect on the ecology of the shallow-water elkhorn coral, *Acropora palmata*, with losses of over 70% in living cover in the Florida Keys. Rapid tissue loss resulting in patches of coral skeleton occurs in the summer and disease signs (*Plate 11.1C*) are very different from the tissue loss occurring due to bleaching or predation by snails; the disease spreads to neighboring colonies, implicating an infectious agent. Using a combination of cultural and molecular techniques, the causative agent was identified as *Serratia marcescens*. Since this bacterium is commonly associated with the human gut, it was suggested that the infection originated from sewage pollution. Other human enteric bacteria can be isolated from the mucus of corals in the Keys; it is possible that the emergence of white pox is linked to expansion of coastal dwellings in the region and the widespread use of septic tanks, which release sewage runoff to the sea. Koch's postulates were fulfilled in experimental infection, but a very high inoculum was required, raising some doubts about the ecological significance of the conclusions. Fish guts are also known to harbor *S. marcescens*, so this may provide an alternative explanation. The ability to use specific components of the coral mucus may depend on enzymatic properties of the pathogen.

Extensive tissue necrosis of corals may involve bacteria and protistan parasites

Rapid tissue necrosis (RTN) is a common scourge of aquarium corals of many types, resulting in death within a few days. Vibrios including *Vibrio harveyi* and *Vibrio alginolyticus* have been suspected as being responsible for the disease, but results are inconclusive, and another hypothesis is that RTN is initiated either by a virus or by an immune-like cross-reaction between corals in close contact in closed aquarium systems, with subsequent bacterial invasion. RTN is not often observed in the field. Aquarium corals such as *Acropora* spp., *Pocillopora damicornis*, *Seratopora* spp., and *Euphyllia* spp. also frequently show a characteristic "brown jelly" band of rapidly spreading necrotic tissue, with a ciliated protist, *Helicostoma nonatum*, often found in large numbers actively feeding on coral tissue. A similar parasite has been observed in necrotic tissue of *Acropora grandis* and *Acropora muricata* on the Great Barrier Reef from colonies showing a characteristic brown band of necrotic tissue (*Figure 11.2*). Again, the protist seems to be a secondary invader after initial destruction of the tissue by bacterial toxins and enzymes.

The role of viruses in coral diseases is unclear

It seems very likely that viruses are involved in some diseases of corals and other cnidarians, but only a very small number of studies have been carried out. There is some evidence showing that zooxanthellae may contain latent viruses that are

---

ⓘ THE DIFFICULTIES OF DEFINING THE CAUSE OF CORAL DISEASE

The recent application of microarray technology to the study of white plague type II (WPD-II) disease in the Caribbean illustrates how descriptions of coral diseases based on appearance alone may conceal a wider range of diseases, with different etiological agents. Sunagawa et al. (2009) used a high-density 16S rRNA gene microarray to analyze healthy and diseased samples of the coral *Montastrea faveolata*, off Puerto Rico. The diseased coral exhibited all the signs associated with WPD-II, but the suspected pathogen was never found in any samples, indicating that other unidentified pathogens can cause an identical syndrome. The microarray detected far more bacterial diversity than conventional clone library techniques and showed that the community changed with development of the disease. Lesser et al. (2007) have criticized the experimental basis for the claims of specific bacterial etiology, concluding that "most common coral 'diseases' are a result of opportunistic, nonspecific bacteria that exploit the compromised health of the coral after exposure to environmental stressors." However, Work et al. (2008) argue that our current knowledge of the interactions between coral host, the environment, and infectious agents is insufficient either to support or refute this claim. The legacy of the work by Koch, Pasteur, and the other founders of pathogenic microbiology 150 years ago has perhaps conditioned us always to seek a single microbial cause of a specific disease. The reality is that many diseases are multifactorial and depend on shifts in the normal microbial community and complex changes in the host.

Figure 11.2 Necrotic brown band disease on *Acropora grandis* from the Great Barrier Reef, Australia. (A) Infected branch, showing brown band migrating into healthy tissue, with bare skeleton behind. (Image courtesy of Bette Willis, James Cook University, Australia.) (B) Light micrograph of a ciliated protozoan isolated from the necrotic tissue, identified by Bourne et al. (2008b) as a novel ciliate belonging to the class Oligohymenophorea, subclass Scuticociliatia.

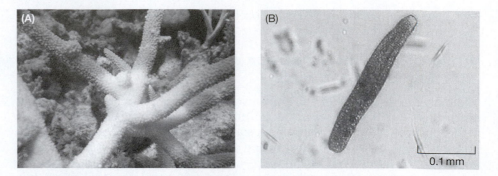

triggered into a lytic infection by environmental changes associated with bleaching, such as elevated temperature, ultraviolet irradiation (see *Figure 2.3*), or pollution. The few studies that have been carried out have mostly relied on the use of transmission electron microscopy to demonstrate virus-like particles (VLPs) in coral tissue, although some recent studies have used metagenomic methods to demonstrate induction of herpes-like viruses in corals subjected to stress. However, at present there is no clear evidence of viral etiology for any specific coral disease. Indeed, there have even been recent suggestions that some viral interactions with corals could be beneficial.

Sponge disease is a poorly investigated global phenomenon

Sponges are highly important members of tropical, temperate, and Antarctic marine ecosystems. As discussed in Chapter 10, sponges are host to a highly diverse microbial population comprising heterotrophic bacteria, archaea, cyanobacteria, protists, and fungi, the cells of which can comprise almost half of the tissue volume in some species. This makes it especially difficult to determine the involvement of pathogenic microbes in disease. Disease outbreaks have been reported worldwide for more than 100 years, with most diseases appearing first as white spots on the sponge surface, followed by necrosis and damage to the internal structures due to bacterial digestion of internal fibers, so that the sponge body crumbles as a result of water pressure. One of the best-documented major epizootics occurred in the Caribbean in 1938, resulting in loss of 70–95% of sponges in some parts. The commercial collection of sponges in the Mediterranean and Ligurian Seas suffered massive losses in the 1980s and some aquaculture systems for sponge cultivation have also suffered disease outbreaks. The microbiological investigation of sponge disease remains limited and the etiological agents and environmental factors have not been identified. Systematic surveys of the health status of sponges in reef ecosystems, accompanied by detailed microbiological investigation, are being conducted to complement our growing knowledge of coral diseases.

Vibrios are a major cause of important diseases of cultured mollusks

Vibrios (members of the family *Vibrionaceae*) are common members of the coastal environment and form part of the normal microbiota of bivalve and gastropod mollusks. Until the recent introduction of advanced molecular methods, the taxonomy of this group has been very confused (see p. 104) and numerous *Vibrio* spp. have been described as disease agents. The persistence of the bacteria in molluskan tissues depends partly on their resistance to the bactericidal activity of the hemolymph. Bacterial toxins and surface factors and the effect of environmental factors (especially temperature) and stress on the neuroendocrine systems of the host are very important in determining whether these interactions are benign or lead to disease. Disease resulting from vibrio infections does not usually occur in adult mollusks, but important exceptions in French

? COULD SNORKELERS' SUNSCREENS INDUCE CORAL BLEACHING?

Sunscreens contain a variety of potentially toxic compounds that could affect marine life. Danovaro et al. (2008) calculated that 10% of the world's coral reefs are at risk from the concentrated release of hundreds of tons of ultraviolet filters and other components of sunscreens from the bodies of swimmers and snorkelers in localized areas. In laboratory studies, they showed that the zooxanthellae from various species of corals showed signs of damage leading to complete bleaching following exposure to sunscreens, especially at higher temperatures. Bleaching was accompanied by the appearance of virus-like particles (VLPs) in the tissue. An obvious explanation for the damage to the zooxanthellae is that it is caused by the induction of a latent, lytic virus as has been previously demonstrated in isolated zooxanthellae by Lohr et al. (2007) or intact corals by Davy et al. (2006) exposed to ultraviolet irradiation or heat shock. It is also possible that the sunscreens induce latent infections in bacteriophages associated with bacteria in the corals, leading to disturbance of the holobiont community. Isolation and characterization of the viruses are needed to resolve this.

marine aquaculture (mariculture) include regular epizootics in the Portugese oyster (*Crassostrea gigas*) caused by *Vibrio splendidus* and related species, and high mortalities of the European abalone (*Haliotis tuberculata*) caused by *V. harveyi*. Recent analysis of the genome sequence of a strain of *V. splendidus* is helping to reveal mechanisms of pathogenicity; a metalloprotease enzyme appears to play a major role. *Vibrio harveyi* also infects cultured pearl oysters (*Pinctada maxima*) in northwestern Australia. *Vibrio tapetis* causes brown ring disease in cultured Manila clams (*Ruditapes philippinarum*); the pathogen attaches to the clam tissue, causing abnormal thickening and a characteristic brown ring along the edge of the shell. Mass mortalities due to *V. tapetis* have caused severe economic losses in France since the 1980s.

Vibrios are also the major cause of disease in hatcheries for cultured bivalve mollusks. Three species have been particularly implicated, namely *Vibrio alginolyticus*, *Vibrio tubiashii*, and *Vibrio anguillarum*. These organisms are all found as members of the normal microbiota of seawater and in association with marine surfaces. Growth is encouraged by accumulation of organic matter, so careful monitoring of water quality and temperature is essential to prevent disease outbreaks in hatcheries. Larvae of different bivalve species seem quite variable in their sensitivity to vibrios and there are also marked differences in the virulence of bacterial isolates. Extracellular toxins (hemolysins, proteases, and a ciliostatic factor) are responsible for larval necrosis.

A wide range of other bacteria can cause infections in bivalve mollusks

Juvenile oyster disease (JOD) results in seasonal mortalities of hatchery-produced juvenile oysters (*Crassostrea virginica*) raised in the northeastern USA. The disease first appeared in the late 1980s, and losses exceeding 90% of total production have since occurred in the states of Maine, New York, and Massachusetts. There are several similarities to brown ring disease, indicating a bacterial etiology. Growth rate is reduced and the shell becomes fragile and uneven, with proteinaceous deposits (known as conchiolin) on the inner shell surfaces (*Figure 11.3A*), followed by sudden high mortality. Infected animals are heavily colonized by a previously undescribed species of the *Roseobacter* group of the *Alphaproteobacteria*, recently named as *Roseovarius crassostreae*. Attachment of the bacteria to the oyster tissue via polar fimbriae (*Figure 11.3B*) is thought to be a key factor in pathogenesis, but this has not been proven. The disease is now known as *Roseovarius* oyster disease (ROD) to emphasize that it is a specific condition (see p. 232).

Gliding bacteria of the *Cytophaga–Flavobacterium–Bacteroides* (CFB) group (see p. 116) can infect the hinge-ligament of the shell, leading to liquefaction via the production of extracellular enzymes, and interference with respiration and feeding. CFB appear to be weak opportunist pathogens, and infection is precipitated by poor nutrition of the animals, rising water temperatures, or other environmental stresses.

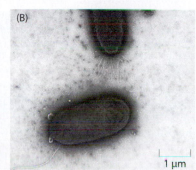

Figure 11.3 *Roseovarius* oyster disease. (A) Shell of infected oyster showing extensive deposition of conchiolin on the inner surface of both valves. (B) Electron micrograph of a negatively stained cell of *Roseovarius crassostreae* showing one bacterium with a flagellum, and the other (partial) with a tuft of polar fimbriae. (Images courtesy of Katherine Boettcher, University of Maine.)

BOX 11.1   RESEARCH FOCUS

Bacteria, coral disease, and global warming
Coral biologists disagree about the role of bacteria in coral bleaching

Development of the bacterial bleaching hypothesis

As discussed in Chapter 10, bleaching is a highly damaging process affecting coral reefs. It is generally perceived to be due to physiological disturbance resulting in disruption of the symbiosis between the host and the photosynthetic zooxanthellae, most often as the result of increased sea water temperatures. The concept that some types of bleaching might be due to bacterial infection stems from the comprehensive studies of the research group led by Eugene Rosenberg of Tel Aviv University, Israel. In 1996, they showed that *Vibrio shiloi* is the causative agent of summer bleaching in the Red Sea coral *Oculina patagonica*. In aquarium experiments, Kushmaro et al. (1998) inoculated *O. patagonica* with *V. shiloi* at different temperatures. At 29°C, bleaching occurred rapidly, whereas at 20 and 25°C the rate of bleaching was slower and less complete. No bleaching at all was observed at 16°C or in uninoculated controls at any temperature, and addition of antibiotics prevented bleaching. Toren et al. (1998) showed that when *V. shiloi* were grown at 25°C they adhered rapidly to coral, whereas bacteria grown at 16°C did not adhere, regardless of whether the corals were grown at 16 or 25°C. The adhesion was strongly inhibited by D-galactose (or a synthetic analog of this sugar), suggesting that the coral surface contains a galactoside receptor for the bacterial adhesin. Banin et al. (2001a) showed that the receptor is indeed in the mucus, and that zooxanthellae (*Symbiodinium*) must be photosynthesizing in order for it to be produced. (Banin et al., 2000). The final piece of evidence linking *V. shiloi* infection to bleaching is the observation that the bacterium secretes a small peptide (12 amino acids) that rapidly inhibits photosynthesis and lyses zooxanthellae in the presence of ammonia (Banin et al., 2001b; Ben-Haim et al., 1999). Again, this factor is only produced at the higher temperatures implicated in bleaching. Other aspects of this interaction, including the possible intracellular replication of the bacterium and the role of a fireworm vector, are reviewed by Rosenberg and Falkowitz (2004). Extrapolating from the *O. patagonica–V. shiloi* model to a general hypothesis that bacteria cause bleaching provoked considerable controversy, because the temperature shifts involved are extreme (16–29°C), whereas mass bleaching events in the Caribbean, Pacific, and Indian oceans typically involve rises of just 1–2°C above normal sea surface temperatures.

Why does *Vibrio shiloi* no longer cause bleaching in *Oculina patagonica*?

The *V. shiloi–O. patagonica* became a model system, and successive doctoral students in Rosenberg's laboratory had worked on it since 1994. Therefore, it must have been an awkward surprise for Leah Reshef at the start of her project when she was unable to isolate *V. shiloi* from bleached corals. Her subsequent investigations revealed that *O. patagonica* has become resistant to *V. shiloi*. Until 2002, all strains of pathogenic *V. shiloi* held in the laboratory caused bleaching in aquarium experiments, but the same strains are now incapable of infecting *O. patagonica*. The "coral probiotic hypothesis" was proposed as an explanation for the phenomenon, stating that "the coral animal lives in a symbiotic relationship with a diverse metabolically active population of microorganisms (mostly bacteria). When environmental conditions change ... the relative abundance of microbial species changes in a manner that allows the coral holobiont to adapt to the new condition" (Reshef et al., 2006). This was further developed to a theory that all the genetic components associated with the partners in the holobiont should be considered as a unit of selection, so microbial symbionts play a key role in adaptation and evolution of higher organisms (Rosenberg et al., 2007). It is impossible to go back in time to investigate what changes may have occurred in *O. patagonica*, but one clue comes from the fact that several bacteria isolated from the coral in 2008 show strong antibacterial activity against *V. shiloi* (Nissimov et al., 2009). Alternatively, the pathogen may have changed; genomic analysis of strains of *V. shiloi* isolated between 1995 and 2005 indicates a shift in population structure, with modern strains lacking genomic islands that might encode virulence factors (Reshef et al., 2008). Finally, changes in the host's natural defenses cannot be ruled out as an alternative or contributory mechanism; we have very limited knowledge of immune responses in corals.

Vibrio spp. cause bleaching and tissue necrosis in a wide range of corals

The discovery by Ben-Haim and Rosenberg (2002) of *Vibrio coralliilyticus* as the cause of temperature-dependent bleaching and necrosis in *Pocillopora damicornis* gave more support to the bacterial bleaching hypothesis, because this coral is widely distributed and susceptible to bleaching throughout the world. Laboratory experiments show that infection and tissue lysis proceeds rapidly at 29°C, but no tissue damage occurs at 25°C or lower. *Vibrio coralliilyticus* is very widely distributed, and numerous strains have been isolated in association with diseases of corals in the Red Sea, Mediterranean Sea, and Indian and Pacific Oceans. Detailed studies by David Bourne and colleagues at the Australian Institute of Marine Science provide revealing insights into the role of vibrios in coral disease (*Plate 11.1E*). Sussman et al. (2008) used a combination of molecular and culture-based techniques to isolate putative pathogens from various species of corals during epizootics of white syndrome (WS) on the Great Barrier Reef and other areas of the Indo-Pacific. Using experimental infections to fulfil Koch's postulates, they showed that two novel strains

BOX 11.1   **RESEARCH FOCUS**

of *V. coralliilyticus* and four closely related *Vibrio* spp. are causative agents of WS. Virulence is highly dependent on temperature, with the most important factor being a powerful zinc-metalloprotease. Sussman et al. (2009) tested the effects of the protease on photosynthetic activity in zooxanthellae isolated from different corals which had varying degrees of sensitivity to WS, and showed that it inactivated photosystem II of zooxanthellae isolated from hosts susceptible to WS, followed by spreading lesions leading to death of the coral tissue. Despite the strong link between the protease and the development of disease signs, Sussman et al. (2009) are careful to point out that WS is a multifactorial disease. In culture, protease production is highly dependent on bacterial cell density, and accumulation of sufficient levels of protease to cause cell damage in the field would require similar high densities; these might occur at the band that is the progressing interface between dead and living tissue, characteristic of WS (see *Plate 11.1E*). Quorum sensing might be involved in regulation of protease production (see p. 80).

Boroujerdi et al. (2009) used nuclear magnetic resonance (NMR) to show that the levels of low-molecular-weight metabolites involved in regulation of energy production and osmotic pressure varied significantly between 24 and 27°C and might be involved in changes in virulence. Motility and chemotaxis to coral mucus are also important in the initiation of infection (Meron et al., 2009).

The structure of coral bacterial communities shifts during bleaching

By tagging specific coral colonies of *Acropora millepora* on the Great Barrier Reef and returning to collect samples, Bourne et al. (2008a) used clone library and denaturing gradient gel electrophoresis (DGGE) analysis to monitor changes in the bacterial communities over a 2.5-year period, which included a severe bleaching event. As the temperature rose, the DGGE "fingerprints" and composition of the clone libraries changed; they became dominated by *Vibrio*-affiliated sequences shortly before the visual signs of bleaching became apparent. The bacterial communities shifted again over a period of a few months as the coral colonies recovered from bleaching and regained their zooxanthellae. The dominant sequences recovered from clone libraries before bleaching and after the recovery period affiliated with *Spongiobacter*, raising the possibility that these bacteria are important in providing antimicrobial activity. This is in agreement with earlier studies on Caribbean corals showing a higher abundance of *Vibrio* spp. during bleaching. The study by Kim Ritchie of the Mote Marine Laboratory, Florida, is of particular relevance. She showed that both *Acropora palmata* mucus itself and the 20% of resident culturable bacteria isolated from the

mucus had antibiotic properties that regulated the structure of the coral microbial community (Ritchie, 2006). Perhaps early physiological effects of rising temperature, such as a change in the composition of the mucus, result in changes in the levels of antibiotics produced by members of the coral holobiont. Ritchie suggested that related vibrios might cooperate to secure a niche by sharing gene products, thus outcompeting the normal residents.

Recently, Littman et al. (2010) compared the effect of heat stress on coral-associated bacterial communities among juveniles of the coral, *Acropora tenuis*, hosting different clade types of *Symbiodinium* (see p. 204), which were heat stressed by placing them at 32°C instead of the control temperature of 28°C. Corals hosting type D *Symbiodinium* showed a dramatic shift in community composition toward vibrios, including *V. coralliilyticus*, when exposed to heat stress; this correlated with a 44% decline in efficiency of the photochemical system. By contrast, major bacterial shifts in *A. tenuis* hosting type C1 *Symbiodinium* were not observed and photochemical efficiency was only slightly affected. Interestingly, type D corals that had previously been exposed to 30°C in the field showed a decrease in vibrios when transferred to the control temperature in the experiment, and bacterial community profiles shifted to resemble those of host harboring the C1 *Symbiodinium*. The effects of thermal stress on coral health are clearly dependent on complex interactions between members of the coral holobiont.

The current status of the bacterial bleaching hypothesis

Despite growing evidence for the role of bacteria in different types of coral bleaching, many coral biologists remain very skeptical about their direct effects, arguing that bleaching is the result of purely physiological effects on the zooxanthellae due to stress, especially increased temperature. Rosenberg's hologenome hypothesis was subject to unusually strong criticism by Leggat et al. (2007), who said that the ideas were based on "flawed interpretation of the literature" and that the authors "dismissed without comment well-established facts about endosymbiosis and the host." In their review of the current thinking on this topic, Rosenberg et al. (2009) remind us that pathogens only cause disease in a host under special conditions that favor the microbe. Increased seawater temperature could cause coral bleaching either by enhancement of microbial virulence or by making the host more susceptible. They conclude that "the concept that bleaching is the result of high temperature acting on the zooxanthellae or on the bacterial community of the coral holobiont are neither mutually exclusive nor inclusive." It is clear that multiple factors are involved and some may act synergistically.

Intracellular rickettsia-like and chlamydia-like pathogens are widespread and have been reported in at least 25 species of bivalves. These infections frequently produce little evidence of tissue damage, and mortality in adult animals is usually low, except in conditions of environmental stress such as sudden temperature change. The larvae are usually very susceptible, leading to problems in aquaculture hatcheries of scallops (*Argopecten irradians*) and other bivalves. The morphology of the bacteria within the cells of the digestive gland and gills is very similar to that of known rickettsias and chlamydia, but detailed identification and taxonomic studies are limited because these are obligate intracellular pathogens that can only be grown in suitable cell cultures.

Virus infections are a major problem in oyster culture

France is one of the leading producers of cultivated oysters and has been particularly susceptible to disease caused by viruses, as well as the vibrio infections previously described. A major epizootic of gill necrosis erupted in France in the 1960s in Portuguese oysters (*Crassostrea angulata*), which demonstrated gill necrosis accompanied by the presence of large inclusion bodies containing icosahedral iridovirus particles in the tissues. This disease spread rapidly along the Atlantic coast of Europe, virtually destroying the European oyster fishery. Pacific oysters (*Crassostrea gigas*) were introduced to replace the lost stock, but these also suffered high mortalities. A serious outbreak of viral disease killed in excess of eight billion oysters in France in 2008. This was shown to be due to the ostreid herpesvirus (OsHV1), which had previously been associated with epizootics in *C. gigas* in Japan, the USA, and Australia since the development of high-density culture methods in the 1940s. It is thought that the 2008 outbreak followed a mild, wet winter with high nutrient levels, which encouraged the young oysters to invest more energy in the development of sex organs, weakening their natural defense mechanisms against viral attack. Susceptibility to the disease has a genetic basis and some resistant lines have been selected.

Another group of viruses known as the marine birnaviruses (MABVs) are widely distributed as disease agents in marine invertebrates (as well as in fish; see infectious pancreatic necrosis virus, IPNV, below). In Japan, infection by MABVs causes considerable losses in the culture of pearl oysters (*Pinctada fucata*). PCR screening shows that 60% of bivalves and 35% of gastropods from a range of wild species in Japanese waters are positive for MABVs and that zooplankton may act as a vector.

Papova-like viruses have been described in the gonads of a number of bivalve species, especially oysters, leading to hypertrophy of gametocytes. Viruses commonly cause secondary infections and may be associated with infection by parasites. Many other VLPs have been observed in marine bivalves and are often associated with disease signs, although proof of etiology and study of pathogenesis is usually lacking, except in cultured species. In most cases, identification of VLPs in electron micrographs has been on the basis of size, morphology, and intracellular location; molecular characterization and propagation are needed.

Bacterial and viral diseases are major problems in aquaculture of crustaceans

A number of crustacean diseases are important in both wild and cultured populations of crustaceans (decapods) such as lobsters, crabs, prawns, and shrimp. Intensive aquaculture of prawn and shrimp now accounts for about half of the total value of aquaculture production. The global demand for prawns and shrimps seems insatiable, and much of the rapid expansion of intensive culture development has occurred in Asia (especially China, Thailand, Vietnam, and Indonesia) and Central America (especially Ecuador). Initial lack of government regulation led to severe problems of habitat destruction and eutrophication of coastal waters. With a better understanding of these problems and improved

---

ⓘ FULFILLING KOCH'S POSTULATES

In many diseases of marine organisms, a specific microbe is strongly associated with disease, but definitive proof of causation is lacking. The "gold standard" evidence for proving disease etiology is fulfillment of Koch's postulates. In the 1880s, Robert Koch proved that a specific bacterium, *Bacillus anthracis*, is the causative agent of anthrax. To obtain such proof, the following conditions must be met: (i) the microbe suspected of causing the specific disease must be associated with all cases; (ii) it must be grown in pure culture outside of the host; (iii) the disease must be reproduced by inoculating healthy individuals with the pure culture; and (iv) the specific microbe must be reisolated from the experimentally infected hosts. The investigation of juvenile oyster disease (JOD) by Maloy et al. (2007) provides an excellent example of a rigorous approach to prove Koch's postulates for a marine disease. They showed that *Roseovarius crassostreae* is consistently associated with diseased animals and forms over 90% of the total recoverable bacterial community, but is never found in healthy oysters. They developed a laboratory model to reproduce the disease and showed that exposure to *R. crassostreae* caused JOD-like mortality including the characteristic deposition of conchiolin in the shell. The timing of colonization by *R. crassostreae* paralleled with the development of disease signs in a natural setting. The bacteria were detected simultaneously to the first microscopic disease signs in oysters that were still healthy and actively growing, providing strong evidence that that they are primary pathogens rather than opportunists.

regulation of location and operation of prawn and shrimp farms, infectious diseases are now considered one of the major limiting factors of further development of the industry, alongside understanding the nutritional needs of animals and improving sustainable supply of feeds. For most crustacean species, culture still depends on the use of larvae produced in hatcheries, mainly from sexually mature females caught in the wild. Thus, genetic selection of disease-resistant animals is not possible, although it is likely that closure of the life cycle for some species will be achieved soon. This will decrease the pressure on natural stocks and enable selection of disease-resistant brood stock. Some crustacean species are cannibalistic or are indiscriminate feeders that will consume feces and molted exoskeletons, leading to rapid disease transmission through ponds and holding tanks. Until recently, knowledge of the ecology of pathogens and application of diagnostic methods has been poor. The development of gene probes and real-time PCR techniques (see p. 44) is assisting in control methods such as selection of larvae and restrictions on movement of stocks unless certified as disease-free. Viruses also remain infectious in frozen seafood products, and molecular methods are used for screening imports and exports to prevent disease transmission. However, many of these techniques are expensive and require highly trained personnel, and it is necessary to develop cheaper and easier biosecurity methods for use by rural farmers in developing countries.

Expansion of intensive prawn culture has been accompanied by a dramatic spread in viral diseases

The expansion of intensive culture of shrimps and prawns, in both volume and geographic distribution, has resulted in the number of well-characterized viruses increasing from six in 1988 to nearly twenty today. One of the most devastating diseases in Asia has been white spot syndrome of penaeid prawns. This was first recognized in eastern Asia and has spread worldwide, with the exception of Australia, with current losses estimated at over US$1 billion. High mortalities (80–100%) result within 2–3 days for juveniles and 7–10 days for adults. The agent (WSSV) is a large, enveloped, rod-shaped, double-stranded DNA virus, recently named as *Whispovirus* and assigned as the sole member of the group *Nimaviridae*. Analysis of the genome sequence of the virus indicates that it differs from all known viruses, although some genes are weakly homologous to herpesvirus genes. *Whispovirus* has a broad host range among crustaceans and infects many tissues, multiplying in the nucleus of the target cell. It is not known how the virus enters the shrimp and spreads within the body, but recent studies suggest that a viral envelope protein may interact with a host chitin binding protein. There have been many studies on the immune responses of the host to infection. *Whispovirus* is a lytic virus and causes destruction of host tissues. Proteomic analysis is being used to elucidate the structural and functional relationships of the various proteins in the virus particle and their role in virus replication and virulence. Disease transmission has been controlled in some regions by use of integrated ELISA- and PCR-based screening of larvae to ensure that they are free of the disease before transfer to ponds; however, there is considerable sequence variation, which may limit the effectiveness of screening techniques. Animals acquire the infection from the water, and show white spots on the inner surface of the shell (*Figure 11.4A*). Interestingly, the white spots appear to be caused by a chitinase, and this enzyme may have been transferred to the viral genome by bacteriophage conversion.

Crustaceans may also be infected by parvoviruses, including infectious hypodermal and hematopoietic necrosis virus (IHHNV), hepatopancreatic parvovirus (HPV), and spawner mortality virus (SMV). SMV emerged in the 1990s in Queensland, Australia, and has been responsible for mortalities of 25–50% in black tiger prawns (*Penaeus monodon*). The virus originates in wild brood stock, as 25% of female spawners carry SMV. Again, integrated PCR and ELISA tests are being developed for high-throughput screening in an attempt to control the

---

(i) DEEP-SEA MUSSELS HAVE MICROBIAL INFECTIONS TOO

Symbiotic interactions between bacteria and bivalve mollusks found at hydrothermal vents and cold seeps have been studied intensively (see Chapter 10), but very little is known about parasitic and pathogenic interactions. Virus-like particles were observed in the mantle tissue of vesicomyid clams showing tissue destruction following heavy mortalities at a methane-hydrate seep (Mills et al., 2005). Viruses also seem to cause tissue destruction in the hydrothermal vent mussels *Bathymodiolus heckerae* and *B. puteoserpentis* (M. E. Ward et al., 2004). Because other parasites were also identified, the role of viral infection in pathology and its importance in population dynamics of these special ecosystems is unknown. Using specific gene probes, Zielinski et al. (2009) described a parasitic bacterium (which they named *Candidatus* Endonucleobacter bathymodioli) that invades the nuclei of six species of deep-sea bathymodiolin mussels from hydrothermal vents and cold seeps. Using FISH and electron microscopy, they showed that the infection of a nucleus begins with a single rod-shaped bacterium, which grows to an unseptated filament of up to 20 μm in length, and then divides repeatedly until the nucleus becomes greatly swollen with up to 80000 bacteria. The nuclear membrane then bursts and the host cell is destroyed. The authors noted that the gill bacteriocytes never become infected, suggesting that the symbiotic bacteria in these cells protect their host nuclei against the parasite. Further investigation of diseases in these deep-sea animals is obviously hampered by the technical challenges of maintaining live specimens for infection challenges.

Figure 11.4 Viral infections of crustaceans. (A) Head of prawn infected with *Whispovirus*, showing characteristic white spots on the exoskeleton. (Image courtesy of Donald Lightner, University of Arizona.) (B,C) Transmission electron micrographs of epithelial cell nucleus of *Cancer pagurus* infected with bacilliform virus (*Cp*BV) showing clusters of characteristic bow-shaped *Cp*BV virions (*arrows*) within an amorphous viroplasm. (Reprinted from Stentiford [2008] by permission of Oxford University Press.)

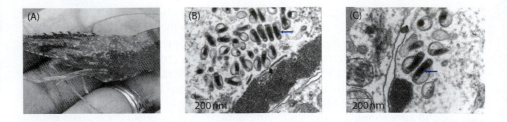

disease. Sequence analysis shows that there is considerable homology between SMV and some insect viruses. The four shrimp parvoviruses fall into two different clades that group with different insect parvoviruses.

Taura syndrome virus (TSV) is a small, nonenveloped, icosahedral, single-stranded RNA virus belonging to the family *Dicistroviridae*; it has caused heavy mortalities in central America and the USA since the 1990s and several variants are known. Other viruses causing significant diseases in shrimp and other crustaceans include baculovirus penaei virus (BPV), monodon baculovirus (MBV) and midgut gland necrosis virus (MGNV).

Animals acquire these viral infections from water, although it has been suggested that seabirds could act as a vector of some shrimp viruses by transmitting infectious virions in their feces. Larvae of aquatic insects may also be reservoirs of infection.

Little work seems to have been done on the importance of viruses in wild crustaceans such as crabs and lobsters, despite their importance in natural populations and fisheries. Baculoviruses, a parvo-like virus, and reoviruses have been reported in the hepatopancreas of *Carcinus* spp., and early reports of viruses resembling herpesviruses, picornaviruses, rhabdoviruses, and bunyaviruses in other crabs have also largely not been pursued since their original descriptions in the 1970s. More recent application of microscopic and molecular methods have demonstrated the importance of viruses in the ecology of wild decapod crustaceans (*Figure 11.4B*).

Bacteria can cause epizootics with high mortalities in crustaceans

As seen with mollusks, various species of *Vibrio* emerge as the major causes of devastating losses in hatcheries and grow-out stages of tropical shrimp and prawn culture, the most important of these being *V. harveyi*. Because of its bioluminescence, a large *V. harveyi* outbreak in prawns can result in a spectacular greenish light in infected ponds, leading to the name "luminous vibriosis." In hatcheries, vibrios attach to the feeding appendages and oral cavity of the larvae, which become weakened and swim erratically. The virulence of different strains of *V. harveyi* varies greatly. Some have a minimum lethal dose (100% mortality) as few as 100 CFU ml^{-1}, whereas other strains are nonvirulent at 10^6 CFU ml^{-1}. *Vibrio harveyi* strains infecting fish produce a siderophore that is essential for virulence, but this does not seem to be the case with prawn isolates, perhaps because invertebrates do not have the efficient iron-sequestering system found in vertebrates (see p. 238). Virulent strains produce two lethal protein exotoxins, which probably act in the larval intestinal tract on the gut epithelial cells by facilitating passage across the gut and colonization of other tissues. The closely related *Vibrio nigripulchritudo* and *Vibrio penaeicida* have recently emerged as major disease agents of prawn and shrimp culture in New Caledonia, Japan, and southwest India. There is some evidence to suggest that acquisition of a plasmid may be responsible for the emergence of highly virulent *V. nigripulchritudo*.

In lobsters and crabs, the exoskeleton is covered by a thin protective lipoprotein layer, and damage to this layer enables various bacteria (probably vibrios) to

---

? DO VIRUSES INFECT CRUSTACEAN ZOOPLANKTON?

The small crustaceans known as copepods have great significance in global marine ecology and nutrient cycling. They are the most abundant members of the zooplankton, and are the main food source for larger zooplankton, many fish, whales, and seabirds. Because of their abundance and importance, a question of special significance is whether their population density could be controlled by viral infection, as occurs with phytoplankton such as coccolithophorids, diatoms, and dinoflagellates (see Chapter 7). Drake and Dobbs (2005) investigated this in a model system by exposing *Arcartia tonsa* to natural concentrations of viruses in seawater. In this experiment, the authors concluded that exposure to viruses had no effect on copepod fecundity, larval survival, or adult survival. They suggest that the effect of viruses on higher trophic levels may be much lower than that on smaller hosts such as protists and that the effects of viral infection in crustaceans may be below the limits of detection. In view of the importance of copepods in ocean processes, it would be interesting to investigate the effect of viruses associated with natural mortality events, perhaps using mesocosm systems.

attach to the chitin exoskeleton. It is possible that biofilm growth of lipolytic bacteria may facilitate penetration resulting in "bacterial shell disease," in which pits in the exoskeleton are produced by bacterial chitinase activity. These can erode to form deep lesions and the bacteria may penetrate into the tissues. The disease has a major impact on some populations of the American lobster (*Homarus americanus*), with consequent serious effects on the lobster fishery of the northeastern coast of the USA and Canada. Another very damaging pathogen of lobsters is gaffkemia, caused by the Gram-positive *Aerococcus viridans* var. *homari* (previously known as *Gaffkya homari*). The disease is usually found in holding ponds and is of considerable concern to lobster fishermen because of the high value of their catch. Schemes to restock depleted fisheries have been badly affected by this disease, and there is considerable evidence that disease spread is linked to the commercial movement of infected animals caught from infected wild stocks and transported over large distances. It is highly contagious and is probably acquired from a reservoir in wild animals, which show an infection rate of 5–10%. The bacterium gains entry to the lobster hemolymph via abrasions in the shell and multiplies very rapidly at higher water temperatures (>10°C). This explains differences in severity of the disease observed during summer and winter impoundment. Virulence seems to be strongly connected with a capsule that prevents agglutination in the hemolymph. Antibiotics such as oxytetracycline are sometimes used as a preventive measure in holding ponds, but there is concern about antibiotic residues in treated lobsters; a withdrawal period of at least 30 days should be observed.

The intracellular rickettsias and mycoplasmas can cause epizootics with high mortalities in crustaceans such as crabs, lobsters, and penaeid prawns, owing usually to infection of the hepatopancreas. Such epizootics can have a marked effect on marine ecology, for example the outbreaks caused by currently unidentified rickettsias affecting Dungeness crabs (*Cancer* and *Carcinus* spp.) in Europe and the USA. Diagnosis of disease is difficult, being based largely on histopathology, but the recent introduction of molecular diagnostic techniques is leading to improved understanding of their ecology.

Parasitic dinoflagellates are major pathogens of crustaceans

Dinoflagellates in the genus *Hematodinium* and related genera can be very significant pathogens of commercially important crustaceans in many parts of the world, with devastating effects on fisheries for crabs and lobsters in North America and langoustines in Europe. Infection causes pathological alterations to the organs, tissues, and hemolymph of the animal. Juveniles and females often show high prevalences of infection, resulting in sudden crashes in populations. At lower levels of infection, biochemical changes in the tissues result in the production of bitter-tasting compounds, making the crabmeat unmarketable.

DISEASES OF VERTEBRATES

Microbial diseases of fish cause major losses in aquaculture, but effects on natural populations are harder to determine

The first scientific description of microbial diseases in fish can be traced to the description of "red pest" in eels in Italy by Canestrini in 1718. This disease, which we now know as vibriosis caused by the bacterium *V. anguillarum*, led to mass mortalities in migrating eels during the eighteenth and nineteenth centuries. Such large-scale fish kills in the wild occur occasionally, particularly in estuaries and on coral reefs, and may be caused by a wide range of bacterial, viral, or protistan infections, or by harmful algal bloom (HAB) toxins. Viral hemorrhagic septicemia (VHS) virus is thought to be responsible for large periodic fluctuations

---

? HOW DO PHAGES SWITCH ON A PRAWN PATHOGEN'S VIRULENCE?

Leigh Owens and colleagues at James Cook University in Australia have provided a fascinating insight into why only some strains of *Vibrio harveyi* are virulent for prawns. They isolated a phage (VHML) from a stable toxin-producing lysogenic strain of *V. harveyi* and showed it to contain a large number of genes with no known function (Oakey et al., 2002). Does this phage convert nonvirulent *V. harveyi* to a virulent form? To answer this question, Munro et al. (2003) first checked that the nonvirulent strain did not contain prophage by treatment with mitomycin C. They then infected the nonvirulent strain with VHML and showed a rise in virulence, associated with the production of several new extracellular proteins. At present, it is not clear how the integration of the phage genome confers virulence. One possibility is that an ADP-ribosylating toxin is present; another idea is that the phage has a gene for a DNA adenine methyltransferase, an enzyme known to alter the expression of bacterial genes. Thus, infection by a phage can change a bacterium from nonvirulent to virulent, either by the transfer of a gene for a new virulence factor or by modifying existing bacterial genes. Phage conversion also results in a significant change in other phenotypic properties (Vidgen et al., 2006), which partly explains why traditional phenotypic identification and typing methods are not always able to resolve *V. harveyi* from closely related species (Cano-Gomez et al., 2009).

in the populations of wild shoaling fish in the North Sea, and major epizootics in Australian pilchards have occurred because of a herpesvirus. Apart from these mass mortality effects, it is hard to estimate the normal impact of disease on fish populations in the wild, since sick fish do not last long in the natural environment and are quickly removed by predators. One of the few pieces of evidence that infection plays a significant role in controlling natural fish populations comes from observations made in the 1970s that hatchery-reared salmon immunized against vibriosis before transfer to the ocean had a 50% greater survival than nonimmunized fish, as shown by the return rate of tagged fish. Fish such as salmon and eels may be particularly susceptible to acute infections owing to the pronounced physiological changes that occur during their migration from fresh to salt water, or *vice versa*. Some pathogens are isolated from a high proportion of wild fish; these tend to be those that cause slow-developing chronic diseases and may affect populations by impairing growth and reproduction.

The development of intensive mariculture in the 1970s led to a rapid growth in the science of fish pathology. Early attempts to farm salmonid fish in intensive offshore pens and cages were frustrated by large-scale mortality and heavy economic losses. This experience has recurred with many different fish species in all parts of the world and at times has threatened the survival of the industry in some countries. The impact of disease should come as no surprise, since it is common in all forms of intensive culture in which single species of animals are reared at high-population densities. Economic factors demand highly intensive systems and this can lead to stress of the cultured fish, which then succumb to disease transmitted rapidly through the dense populations.

Today, marine fish culture is a worldwide industry, with production approaching that of extensive fish capture. Atlantic salmon (*Salmo salar*) in Norway, Chile, Scotland, Canada, and Ireland is the most economically important species, with fish such as gilthead sea bream (*Sparus aurata*) and sea bass (*Dicentrarchus labrax*) dominating culture in the Mediterranean Sea. Japan has a long history of intensive mariculture, dominated by yellowtail (*Seriola quinqueradiata*), ayu (*Plecoglossus altivelis*), flounder (*Paralichthys olivaceus*), and sea bream (*Pagrus major*). A wide range of species are cultured in China and southeast Asia, often in small-scale operations for local consumption. In recent years, there has been expansion in the farming or ranching of high-value fish such as Atlantic cod (*Gadus morhua*), mainly in Norway; or tuna (*Thunnus thynnus, T. orientalis,* and *T. maccoyi*) in the Mediterranean, Japan, and Australia. Whilst there has been some criticism of the sustainability and environmental impact of these ventures, including disease problems, the benefits of closed life-cycle aquaculture of such commercially important fish species relieves the pressure on natural populations, whose stocks are in rapid decline.

The importance of fish diseases in aquaculture has led to the development of specialized branches of veterinary science and diagnostic microbiology

The need to implement effective control measures following an outbreak puts pressure on investigators to determine the causes of mortality quickly. Experience, careful observation of the stock by the fish farmer, and good record keeping play a large part in identifying diseases. Infestation with eukaryotic parasites, such as sea lice and protists (e.g. *Icthyobodo, Cryptocaryon,* and *Tricodina*) typically manifests as an extended course of mortalities, especially in confined fish. Nutritional deficiency is usually indicated by poor growth rates, skeletal abnormalities, and steadily increasing mortalities. Intoxications usually have a very rapid onset, causing mass mortalities within hours. Microbial infection is usually characterized by a rising level of mortality accompanied by characteristic disease signs. These are very varied, but may be broken down into three broad categories: (i) systemic bacteremia or viremia, in which there is rapid growth of the pathogen, often with few external disease signs other than hemorrhaging; (ii) skin, muscle, and gill lesions; and (iii) chronic proliferative lesions. Some highly lethal systemic infections such as those caused by *Aeromonas*, *Piscirickettsia*, and infectious salmon anemia virus (ISAV) may cause high mortality rates even when no external signs are present. In contrast, other agents causing diseases with relatively low mortality rates—for example *Tenacibaculum maritimum*, *Moritella viscosa*, and lymphocystis virus—can result in external lesions such as ulcers, necrosis, and tumor-like growths that make fish unmarketable.

Post-mortem changes are very rapid as a result of overgrowth by the normal microbiota of fish, so it is important to examine fish showing signs of infection before they succumb completely. External examination will often reveal the presence of gill and tissue erosion, eye damage, hemorrhages, abscesses, ulcers, or a distended abdomen. Internal inspection may reveal organ damage and fluids in the body cavity. To the experienced eye, these signs will often indicate a particular disease agent, but the diagnosis must be confirmed by identification of the pathogen.

Identification of bacterial pathogens is usually achieved by plating tissue samples onto various selective media and performing biochemical tests using diagnostic keys. Not all bacterial pathogens are amenable to this approach, as some grow very slowly in culture (notably *Renibacterium* and some mycobacteria). An alternative approach is the use of methods such as ELISA or fluorescent antibody techniques (FAT) to detect bacterial antigens in the blood or tissues, or to detect a high titer of host antibodies against the pathogen (*Figure 11.5*). The diagnosis of viral infections is more difficult and time-consuming because it relies on propagation of the virus in a suitable cell culture. FAT and ELISA are therefore the main methods used for rapid identification of viral infection. There has been some success with "dipstick"-type kits for rapid diagnosis, based on modifications of the ELISA technique. For many bacterial and viral pathogens, there are now accurate molecular diagnostic tests based on PCR amplification and gene probes. Monoclonal antibodies have been produced against specific pathogens, allowing rapid

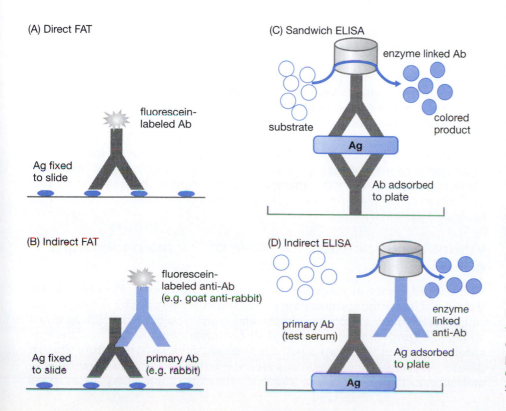

Figure 11.5 Antibody techniques for the detection of pathogens in animal diseases. (A) In the direct fluorescent antibody technique (FAT) a labeled antibody (Ab) reacts with an antigen (Ag), enabling specific microbes to be seen in the microscope, for example from tissue samples or on the surface of skin. (B) In the indirect technique, a stock anti-antibody is used to avoid the need for preparing separate fluorescently labeled primary antibodies. In the enzyme-linked immunosorbent assay (ELISA), the antibody is conjugated to an enzyme (e.g. peroxidase), which catalyzes the conversion of a chromogen to a colored substrate. The sandwich technique (C) can be used for detecting the presence of microbial antigens in animal serum or tissue to detect infection. The indirect ELISA (D) can be used to detect the presence of antibodies to a specific microbe as a way of monitoring immune responses to infection or vaccines. ELISA techniques are usually carried out in 96-well microtiter plates and the intensity of the enzymic reaction is measured in a special spectrophotometer.

identification of disease; they can also be used in fish health programs for screening brood stock for previous exposure to pathogens. Genomic fingerprinting methods (see p. 44) are often used for strain characterization, which is especially important in studies of epizootics.

Bacteria produce infections in fish using a range of pathogenic mechanisms

All bacterial infections involve a number of stages, with different bacterial products important as virulence factors. The possession and expression of genes encoding these factors varies greatly among the broad range of bacteria in various taxonomic groups that have been associated with disease in marine fish.

The initial infection and colonization of the host often depends on the pathogen sensing the presence of exudates and excretion products from the host. Chemotaxis using flagellar or gliding motility may result in association with specific surfaces, such as the gut, skin, or gills. Bacteria often produce pili or some other surface structures that enable firm attachment to epithelial surfaces.

The most critical stage for bacteria during colonization is the ability to overcome the innate host defense mechanisms and adaptive immune system. Fish mucus provides a protective inhibitory function and infection of undamaged skin is rare. Bacteria that penetrate the tissues are subject to the antibacterial collectins and complement proteins found in serum, and resistance to these is often a key determinant of virulence; this is often associated with modifications of the cell envelope. Virulent bacteria may also possess mechanisms to resist phagocytic cells of the host, usually by possession of hydrophobic surface layers or by producing enzymes that cause lysis of phagocytic cells. Some bacteria are able to survive within cells of the host after being taken up within a phagocytic vacuole; these possess a range of mechanisms to overcome the hostile low pH and antibacterial compounds found in the intracellular environment.

As noted on p. 71, nearly all bacteria require iron for the activity of essential cellular functions. Vertebrate animals possess highly efficient systems for the transport and sequestration of iron. The serum protein transferrin binds iron with extreme avidity, reducing the concentration of free iron in tissues to 10^{-18} M, about 10^8 times lower than the concentration that bacteria require for growth. During infection, an even more efficient iron-binding protein (lactoferrin) is released and sequestered iron is removed to storage in the liver and other organs. Thus, successful pathogens must compete with the host's iron-binding system to obtain sufficient iron for growth, and many pathogens achieve this via the production of siderophores and an iron-uptake system.

The final stage in disease is the production of damage. Often this occurs as a result of the host response "overreacting" to the presence of the pathogen, producing damaging cytokines and inflammatory products. Many pathogens produce toxins that can cause death by affecting major organs or extracellular enzymes that cause destruction of tissues.

Vibrios are responsible for some of the main infections of marine fish

As described earlier in this chapter, members of the family *Vibrionaceae* are widespread in marine and estuarine waters and in accompanying sediments and cause disease in many marine animals (and humans; see Chapter 12). As noted above, *V. anguillarum* (also classified as *Listonella anguillarum*) was the first species to be identified as a fish pathogen causing vibriosis in eels (from which it derives its species name), producing ulcers, external and internal hemorrhages, and anemia (*Figure 11.6*). This "classical" vibriosis causes heavy losses in eel

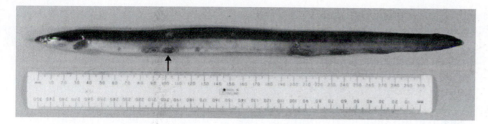

Figure 11.6 Vibriosis in an eel, *Anguilla anguilla*, showing external hemorrhagic lesions. (Image courtesy of Henrik Chart and Colin Munn, University of Plymouth.)

culture, especially in Japan, and in a very wide range of marine species; it has caused particular problems in the culture of salmon in North America, sea bass and sea bream in the Mediterranean, and yellowtail and ayu in Japan. In all species, the disease is characterized by a very rapid generalized septicemia usually accompanied by external hemorrhages. A number of serotypes of the pathogen exist, distinguished by differences in the O-antigen (lipopolysaccharide of the outer membrane), but only the O1 and O2 serotypes are widespread. There are some genetic differences between strains isolated from different geographic regions and fish species; such knowledge is important in the development of vaccines for particular applications (see Chapter 14). Vibriosis was one of the first fish diseases for which vaccines were developed and they have generally been very effective.

One of the major factors in virulence, which explains the extremely rapid growth *in vivo*, is the possession of a 65-Mb plasmid (pJM1) that confers ability to acquire iron in the tissues of the host. Proof of the key role of the plasmid in virulence of *V. anguillarum* was obtained by inducing loss of the plasmid by curing and reintroduction by conjugation and by directed mutagenesis of key genes. Mutation of any of the genes can result in a reduction of virulence by as much as 10^5 times. In experimental infections in which iron uptake by transferrin is swamped by injection of excess iron, possession of the plasmid does not confer an advantage. The plasmid pJM1 contains genes for synthesis of the siderophore anguibactin, an unusual hydroxamate–catechol compound, and outer-membrane transport proteins and regulators of transcription involved in iron uptake (*Figure 11.7*). Expression of the several components occurs only under low-iron conditions, owing to negative control at the transcriptional level by the chromosomally

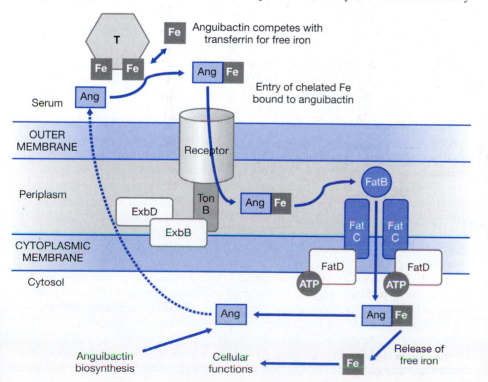

Figure 11.7 Iron uptake via the siderophore anguibactin in *Vibrio anguillarum*. Anguibactin (Ang) is secreted to the external environment, where it competes with the host protein transferrin (T) for circulating iron. Chelated iron (Fe) is reabsorbed into the cell via an outer membrane receptor and membrane transport proteins (TonB, ExbB, ExbD) before transport across the cell membrane via the ATP-dependent fat system. Biosynthesis of anguibactin, the outer membrane receptor and transport proteins are subject to complex regulation at the transcriptional level (not shown). (Based on a drawing by Jorge Crosa, Oregon Health and Science University.)

encoded Fur protein and at least three plasmid-encoded regulators, the best studied of which is AngR. Outer-membrane proteins only expressed under iron-restricted conditions are known as IROMPs. Recently, additional siderophores and iron-uptake mechanisms have been found.

Quorum sensing (QS; see p. 81) via acyl homoserine lactone (AHL) signaling molecules is an important factor in the growth and survival of *V. anguillarum* in its host and in the aquatic environment. There are three separate pathways involved in gene regulation of biofilm formation and production of protease, pigment, and exopolysaccharide. The hierarchical QS system consists of regulatory elements homologous to those found in both *V. fischeri* (the LuxRI homologs VanRI; see *Figure 3.8*) and *V. harveyi* (the LuxMN homologs, VanMN; see *Figure 3.9*). It is possible that some of the AHLs have effects on eukaryotic host cells, including an immunomodulatory function.

Other virulence factors are also involved in *V. anguillarum* pathogenicity, including a metalloprotease and a powerful hemolysin, causing lysis of erythrocytes which may contribute to the acquisition of iron by release of hemoglobin and probably accounts for the pale gills and anemia typical of infection. The structure of the lipopolysaccharide (LPS) confers resistance to the bactericidal effects of complement in normal serum. The flagellum is essential for motility across the fish integument as directed mutations in the *flaA* gene result in reduced virulence. LPS antigens on the sheath of the flagellum (genes *virA* and *virB*) are also essential for virulence—these probably contribute to biofilm formation *in vivo*.

Vibrio ordalii was originally thought to be a biovar of *V. anguillarum*, but was designated as a separate species based on a number of biochemical differences and DNA–DNA hybridization experiments. It has caused major losses in the cage culture of salmon in Pacific coastal waters off Oregon, Washington State, and British Columbia. Infections with *V. ordalii* tend to be localized in muscle tissue rather than the generalized infections seen with *V. anguillarum*. Virulence has not been so well studied as that of *V. anguillarum*, but complement resistance and a leucocytolytic toxin have been described.

Aliivibrio salmonicida (formerly *Vibrio salmonicida*) causes cold-water vibriosis (Hitra disease) in salmon and Atlantic cod (*Gadhus morhua*) farmed in northern regions such as Canada, Norway, and Scotland during the winter, when water temperatures drop below 8°C. Disease signs are broadly similar to those of *V. anguillarum*, but the two bacteria are serologically and genetically distinct. The pathogen is excreted in the feces of infected fish and seems to have good powers of survival in marine sediments, thus causing reinfection, even if farm sites are "fallowed" for a season. Epidemiological studies suggest that there is interchange of the bacterium between populations of cod and salmon. Various strains can be distinguished by different plasmid profiles, but there does not seem to be a close association between plasmid possession and virulence. The bacterium produces various extracellular enzymes, a hydroxamate siderophore and a *fur*-regulated iron uptake system, all of which are implicated in virulence. Interestingly, significant amounts of the siderophore and the IROMPs required for transport are only expressed at temperatures below 10°C, which may explain the increased incidence of disease in the winter. This fact has been important in vaccine manufacture, since temperature-regulated surface proteins are important antigens. Another species that has caused salmon infections in cold northern waters was originally designated as *Vibrio viscosus*, but was reclassified as *Moritella viscosa*. This causes a condition known as winter ulcer disease, in which ulcerous lesions progress from the skin to the underlying muscle. Mortality rates are relatively low, but the appearance of the fish lowers its economic value considerably.

Vibrio vulnificus causes infection of eels in Japan. The fish isolate is closely related to clinical isolates associated with human disease (p. 263), but has been

GENOMIC EVIDENCE SHOWS HOW FISH PATHOGENS ADAPT TO SPECIFIC HOSTS

Analysis of the genomes of bacterial pathogens and comparison with related species provides evidence for their evolution. The genome sequence of a strain of *Aeromonas salmonicida* subsp. *salmonicida* reveals that acquisition of mobile genetic elements, genome rearrangements, and gene loss appear to be responsible for adaptation of the bacterium to its specific host, salmonid fish. Reith et al. (2008) concluded that, since separating from its last common ancestor *A. hydrophila*, the salmon pathogen seems to have acquired multiple plasmids, prophages, genes, and operons, presumably through horizontal gene transfer. Many rearrangements of the genome have occurred, affecting the regulation of gene expression. Another feature of the *Aeromonas salmonicida* genome is the accumulation of a large number of pseudogenes—these are defunct relatives of known genes that no longer encode proteins. The genome sequence of *Aliivibrio salmonicida* also reveals evidence of extensive gene rearrangements through incorporation of insertion sequences and the degeneration of a large number of genes (Hjerde et al., 2008). Of particular interest is the loss of genes involved in the utilization of chitin, which are highly important in enabling vibrios to colonize surfaces in the marine environment (see p. 79). Gene loss or decay is typical of recently evolved genomes, often reflecting adaptation to a specific host (Pallen and Wren, 2007).

designated a separate taxon (biogroup 2) based on phenotypic, cultural, and serological properties. Iron acquisition and production of a capsule, hemolysin, protease, and other toxins have all been implicated in virulence.

Pasteurellosis is a major disease in warm-water marine fish

The expansion in the 1990s of sea bass and sea bream aquaculture in the Mediterranean and yellowtail culture in Japan was accompanied by outbreaks of a new disease that has caused heavy losses. The pathogen responsible was originally identified as *Pasteurella piscicida*, and the disease is still known as pasteurellosis, although the causative agent has been reclassified as *Photobacterium damselae* subsp. *piscicida*. When water temperatures exceed 20°C, acute mortalities up to 70% can occur, especially in the larval and juvenile stages. A more chronic condition occurs in older fish. Interest in virulence mechanisms has focused largely on extracellular proteases, adhesive mechanisms, and the presence of a polysaccharide capsule. Iron concentration seems to be important in the regulation of expression of superoxide dismutase and catalase, which protect bacteria against reactive oxygen species (p. 87) and may be important in intracellular survival. An iron-uptake system is associated with virulence, but as well as uptake by siderophores the bacterium seems able to acquire iron by direct interaction between hemin molecules and outer-membrane proteins. Introduction of rapid diagnostic methods based on PCR technology and vaccines are proving effective in controlling the disease.

Aeromonas salmonicida has a broad geographic range affecting fish in fresh and marine waters

Aeromonas salmonicida was first described as a pathogen of trout in Europe in 1890, but is now known to have a broad host and geographic range. The taxonomy of *A. salmonicida* has been the subject of much debate over the years and it is now generally recognized that acute "typical" furunculosis in salmonids is caused by *A. salmonicida* subsp. *salmonicida*. In mariculture, this usually presents as a severe septicemia with acute mortalities. At the peak of the furunculosis outbreaks in Scotland and Norway in the 1980s, total industry losses neared 50% of stock. Externally, the fish show darkening and hemorrhages around the fins and mouth and internally there is extensive hemorrhaging and destruction of the organs. Other subspecies (*achromogenes*, *masoucida*, *pectinolytica*, and *smithia*) can be distinguished by differences in standard biochemical tests, pigment production, and molecular techniques such as gene probes and DNA–DNA hybridization. These subspecies are generally associated with dermal ulcerations in other marine species such as turbot (*Psetta maxima*) and halibut (*Hippoglossus hippoglossus*). The name furunculosis derives from a boil-like necrotic lesion seen in a chronic form of the infection in older or more resistant fish. Most strains can be easily isolated on laboratory media, although some are fastidious and have a specific requirement for heme.

The virulence mechanisms of *A. salmonicida* are extremely complex, and this pathogen is an excellent example of multifactorial virulence that has attracted the attention of numerous investigators for over half a century. Different components of the factors responsible for entry, colonization, growth *in vivo*, and production of damage interact, as illustrated in *Figure 11.8A*. Genes for many of the virulence factors have been cloned and their properties studied in detail, and the full genome sequence of one strain is available. A key factor in virulence is the "A-protein," composed of a regular structured array of a tetrahedral 49-kDa protein (*Figure 11.8C*). It belongs to the family of surface structures known as S-layers, which occur in a range of bacterial and archaeal species. Virulent and avirulent strains are usually distinguished by the presence or absence of this layer, the hydrophobic nature of which confers autoagglutinating properties in

Figure 11.8 Virulence mechanisms in *Aeromonas salmonicida*. (A) Diagrammatic representation of the interaction of various surface factors, toxins and signaling systems in virulence. The inner line represents the outer membrane (OM) of the cell (cytoplasmic membrane is not shown); the darker outer line represents the S-layer (A-protein). Only part of the *in vivo* capsule and IROMPs (iron-restricted outer membrane proteins) are shown. GCAT, glycerophospholipid:cholesterol acyltransferase; LPS, lipopolysaccharide. (B) Growth of A$^+$ (left) and A$^-$ (right) cultures in broth: A$^+$ cultures autoagglutinate owing to the hydrophobic S-layer (A-protein). (C) Electron micrograph (negatively stained) of an A$^+$ strain showing regular structure of the surface A-protein. (D) Electron micrograph (section) of an A$^+$ strain showing the A-protein as an additional layer (arrowed) external to the outer membrane. Note outer membrane vesicles (V). (Images courtesy of Nigel Parker and Colin Munn, University of Plymouth.)

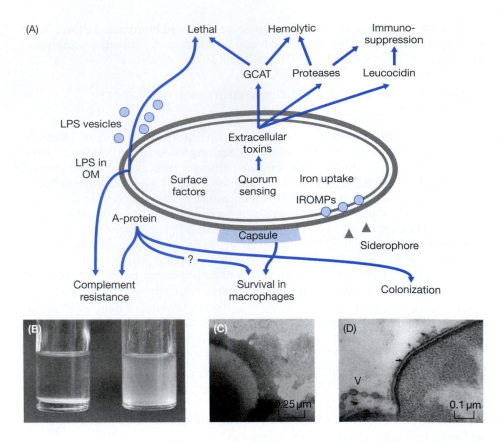

culture (*Figure 11.8B*). Isolates possessing the A-protein (A$^+$) can also be distinguished from A$^-$ strains by their growth on agar media containing Coomassie Blue or Congo Red dyes, which the A-protein absorbs. Electron microscopy shows the A-protein to be present as a layer external to the typical Gram-negative outer membrane (*Figure 11.8C, D*). It is linked to the cell surface via the O polysaccharide side-chain of LPS. The main function of the A-protein is as a protective layer, which contributes to the bacterium's resistance to the bactericidal effects of complement in the serum of the host fish. Because of its hydrophobic nature, it also plays a role in adhesion to host tissue and survival within macrophages. *Aeromonas salmonicida* also produces a range of toxins, and many studies have been carried out on the activity of purified components, although it is now clear that these have synergistic interactions. The enzyme glycerophospholipid:cholesterol acyltransferase (GCAT) and a serine protease are particularly implicated as key virulence factors. GCAT forms a complex with LPS, which is hemolytic, leukocytolytic, cytotoxic, and lethal. Serine protease expression is regulated by an AHL-mediated QS mechanism, and this enzyme (in synergy with GCAT) is responsible for hemolysis. As in *V. anguillarum*, growth *in vivo* is dependent on the production of a siderophore (2,3 diphenol catechol) and the uptake of sequestered iron via IROMPs. Elucidation of the various components in virulence and their immunological properties was a critical step in the formulation of modern vaccines, which have been largely successful in control of the disease in salmon mariculture since the mid-1990s.

The ecology of *A. salmonicida* has been the subject of much controversy, with some investigators suggesting that it is an obligate fish pathogen and others suggesting that it survives in the environment. Historical evidence suggests that it spread throughout Europe more than 100 years ago and spread into the wild salmon population, where it caused major losses in the 1930s. It is very likely that some wild populations are latently infected with *A. salmonicida*. The bacterium may enter a dormant or VBNC state (p. 73), during which the cells undergo various morphological changes. One certainty is that the development of mariculture in enclosed bodies of water such as lochs in Scotland and fjords in

Norway or Chile has led to a shift in the normal microbiota. Whereas *A. salmonicida* seems to have been primarily a freshwater organism, it has now adapted to life in seawater, and recent isolates may have a sodium requirement lacking in earlier strains. Farmed fish can transmit the disease to wild fish around sea cages and these can spread the pathogen to other sites. Wrasse introduced to net pens to remove parasites (sea lice) from salmon can become infected and thus constitute a reservoir of infection. Atlantic salmon have been shown to harbour latent *A. salmonicida* infections when they return from the ocean to spawn, so spread of the disease may be contributing to decline of wild populations.

Marine flexibacteriosis is caused by an opportunist pathogen of low virulence

Members of the CFB group (p. 116), many of which are pigmented and show gliding motility, are responsible for infections in a wide range of fish. Mostly, these bacteria are rather weak pathogens that are best described as opportunists that colonize damaged tissue in fish weakened by stress, especially due to increased water temperature and nutritional deficiency. Most diseases caused by this group occur in fresh water, but *Tenacibacter maritimus* (previously classified as *Flexibacter marinus*) causes a disease called marine flexibacteriosis, sometimes referred to as "marine columnaris" or "black patch necrosis," which is widely distributed in numerous species of wild and farmed fish in Europe, Japan, North America, and Australia. It is characterized by excess mucus production, damage to the gills, tissue necrosis around the mouth and fins, skin lesions, and eventual death. It is most severe in juvenile fish at temperatures above 15°C. The clinical signs and the observation of long Gram-negative gliding bacteria in microscopic preparations are very distinctive. The bacteria are not easy to grow in the laboratory, as they require special low-nutrient media. There appear to be several O-antigen serotypes, which may be related to specific hosts. The disease responds to antibiotics administered by bath and some vaccines have been developed, with variable degrees of success.

Piscirickettsia and *Francisella* are intracellular proteobacteria causing economically important diseases in salmon and cod

Piscirickettsia salmonis was identified in 1989 as the etiological agent of a septicemic disease causing severe economic losses in farmed salmon in Chile. Here, it causes predictable annual epizootics, but it has also caused sporadic outbreaks of disease in Norway, Iceland, and Canada. Infected fish stop feeding, become lethargic, and may show small white lesions or ulcers on the skin, but the disease progresses rapidly and death often occurs with few or no external symptoms. Internally, the most distinctive sign is off-white to yellow nodules in the liver. The disease seems to be transmitted both horizontally by blood-sucking ectoparasites and vertically via the eggs, making the testing and elimination of infected brood stock with ELISA and PCR assays an important element of disease control. *Piscirickettsia* was originally considered to be an obligate intracellular parasite and was formerly propagated only in suitable cell lines, but it can now be grown in broth or agar cultures using appropriate media. A range of serological tests has been developed, and genetic heterogeneity among strains can be studied by analysis of the 23S rRNA operon. Antibiotic treatment is ineffective, since there are very few agents capable of attacking intracellular bacteria without causing damage to host cells; so much effort has been applied to the development of a vaccine. A recombinant vaccine (see p. 323) based on an outer-membrane lipoprotein, OspA, shows promise for disease control.

Epizootics caused by *Francisella* have emerged as a particular problem in culture of the Atlantic cod in Norway. This is a granulomatous inflammatory disease, resulting in extensive internal lesions, with nodules in the spleen, heart, kidney, and liver. The *Francisella* isolated from cod has recently been characterized as closely related to *Francisella philomiragia*, which is widely distributed in aquatic environments and may cause disease in humans with impaired immunity. Other *Francisella* spp. also cause infections in a range of other fish.

Intracellular Gram-positive bacteria cause chronic infections of fish

Renibacterium salmoninarum causes bacterial kidney disease (BKD) and is widely distributed in both wild and cultured salmon in many countries including Canada, the USA, Chile, Japan, Scotland, and Iceland. The expansion of salmon culture through international movement of eggs has assisted the spread of BKD and it causes significant losses in both Pacific and Atlantic salmon. BKD pathology is characterized by chronic, systemic tissue infiltration, causing granular lesions in the internal organs, especially the kidney. External signs include darkening of the skin, distended abdomen, exophthalmia, and skin ulcers. Significant changes in blood parameters are consistent with damage to the hematopoietic and lymphopoietic tissues of the kidney, liver, and spleen. The pathological signs are the result of the interactions between the host's cellular immune response and the pathogen's virulence mechanisms. Tissue destruction forms a focus of necrosis, owing to release of hydrolytic and catabolic enzymes and liberation of lytic agents from the bacteria.

Our understanding of the mechanisms of virulence of *R. salmoninarum* is hampered by the fact that the bacterium takes several weeks to grow on culture plates and does not form discrete colonies. Reproducible infection is also difficult to achieve in aquarium experiments. One important virulence factor is a 57-kDa surface protein, whose hydrophobic properties facilitate attachment to host cells. The key feature of *R. salmoninarum* is its ability to enter, survive, and multiply within host phagocyte cells. Binding of the complement component C3 to the bacterial surface enhances internalization of the bacterium because phagocytic cells possess a receptor for C3. Why does *R. salmoninarum* encourage uptake by cells that normally kill invading pathogens? The answer seems to lie in the pathogen's ability to survive (at least in part) the intracellular killing mechanisms of the macrophage and to replicate (albeit slowly) within the cells. As well as being resistant to reactive oxygen species, *R. salmoninarum* lyses the phagosome membrane in order to escape its strongly antibacterial environment (*Figure 11.9*). In

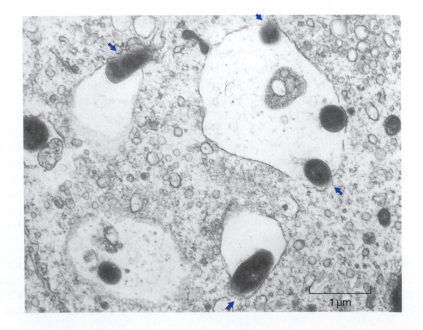

Figure 11.9 Intracellular growth of *Renibacterium salmoninarum*. Transmission electron micrograph showing lysis of the phagosome membrane (*arrows*), prior to entrance of the bacterium into the cytoplasm. (Image courtesy of S. K. Gutenberger and J. R. Duimstra, Oregon State University.)

1 µm

the past, BKD has been difficult to diagnose, but disease management and control are helped by serological techniques (ELISA and FAT), together with recently developed gene probes and techniques for accurate differentiation of clinical isolates (based on PCR amplification of length polymorphisms in the tRNA intergenic spacer regions). These methods are used for certification of brood stock and eggs as disease-free and for implementing quarantine procedures to contain disease outbreaks. There are no effective antibiotic treatments and there have, therefore, been many attempts to develop a vaccine. These are thwarted by the slow growth of the pathogen, and recent work has focused on the use of recombinant DNA technology to produce fusion proteins and DNA vaccines (see Chapter 14), as well as the use of a live preparation of a nonpathogenic *Arthrobacter* sp., which has some antigenic similarity to *R. salmoninarum*. Genes for several virulence factors, including the p57 protein and two hemolysins, have been cloned and expressed in *Escherichia coli* and the nature of the immune response investigated.

Species of the genera *Mycobacterium* and *Nocardia* also cause chronic, persistent infections with long periods of intracellular survival. Many different species of mycobacteria occur; these are widely distributed in seawater and sediments and colonize many species of fish. *Mycobacterium marinum* and *Mycobacterium fortuitum* are the best-known agents of disease. Disease develops slowly and usually affects mature fish; for this reason, mycobacteriosis is a particular problem in marine aquaria, but it has also emerged as a serious problem in the culture of sea bass and turbot. Most species of fish develop few external disease signs other than emaciation, although histopathological investigation reveals extensive granulomatous lesions with caseous necrotic centers. For this reason, the disease is often called fish tuberculosis, since there is some resemblance to the basic pathological mechanisms seen in the human condition caused by *Mycobacterium tuberculosis*. The delayed hypersensitivity reactions and involvement of cell-mediated immunity certainly show some parallels in fish and humans. *Nocardia* spp. cause similar chronic granulomatous conditions to mycobacteria, but the organisms can be distinguished in the laboratory. Again, because of their intracellular nature, there are few antimicrobial agents effective against these pathogens. Valuable aquarium fish are sometimes treated with isoniazid and rifampicin, but this is unwise given the danger of encouraging resistance to these drugs, valuable for the treatment of human tuberculosis. It should also be noted that *M. marinum* (and possibly other species) can cause a zoonotic infection in humans known as fish tank or aquarist's granuloma.

Several Gram-positive cocci cause diseases affecting the central nervous system of fish

The Gram-positive bacteria *Lactococcus garvieae* (formerly *Enterococcus seriolicida*) and *Streptococcus iniae* have been especially problematic as the cause of disease, often termed "streptococcosis" in both fresh and marine warm-water culture, especially in Japan. Several related species have also been described. They cause central nervous damage characterized by abnormal swimming behavior, often with exophthalmia and meningitis. Extracellular cytolytic toxins and an antiphagocytic capsule have been implicated in pathogenicity. *Streptococcus iniae* has caused epizootics in barramundi (*Lates calcarifer*) culture in Australia and was also implicated in an extensive fish kill in the Caribbean Sea in 1999; it can also cause zoonotic infection in humans—infection of wounds in workers handling infected fish leads to severe cellulolytic infection.

Viruses cause numerous diseases of marine fish

Viruses have long been suspected as causative agents of disease in fish, but it was not until the successful development of fish cell culture methods in the 1960s that progress was made. As with bacterial diseases, intensive aquaculture has exacerbated viral disease problems in fish.

The mechanisms by which viruses produce disease in their fish hosts are less well defined than those of bacteria. Viruses must possess a mechanism to enter a susceptible host cell and grow within it; this is usually very specific for a particular type of cell and is determined initially by interactions between surface molecules on the virion and receptors on the host cell. Viruses possess many mechanisms to resist innate host defense mechanisms such as enzymes and antiviral substances in fish mucus. Following local replication at the point of entry, viruses are spread to adjacent cells or carried to other target tissues via the blood, lymph, or nervous systems. The outcome of infection depends on the balance between viral replication and its control by innate defences, interferons, antibodies, and cell-mediated immunity. Host damage is caused by various mechanisms, including killing infected cells because of production of toxic factors during viral replication, cytokines, inflammatory modulators, and necrosis. It is often the intensity of the host response to viral infection that produces the most damage.

Infectious salmon anemia virus is one of the most important pathogens in salmon culture

Infectious salmon anemia (ISA) emerged as a new disease in farmed Atlantic salmon in Norway in the 1980s and has been a continuing problem since that time. Subsequent outbreaks causing major economic losses occurred in Canada, Scotland, the Faroe Islands, the USA, and Chile. European legislation requires that infected stock is killed and the farm site left fallow for at least 6 months, causing great economic hardship to the producers. In the first ISA outbreak in Scotland in 1998, more than 4000 tonnes of fish were slaughtered, with losses approaching 10% of the total industry turnover. This policy was effective in controlling the epizootic, and a second outbreak in 2009 seems to have been quickly contained. However, slaughter policies adopted in Norway and eastern Canada seem to have been ineffective in preventing the disease from becoming endemic. ISA emerged in Chile in 2007, resulting in a devastating economic effect on the industry with more than 100 farms affected and 20% of production lost to the disease. There is significant variation in the severity of ISA, with mortalities in sea pens depending on the virus strain and susceptibility of fish stock. The virus infects the erythrocytes, leading to their lysis. ISA-infected fish are lethargic, with anemia (shown by very pale gills), exophthalmia, and hemorrhages in the eye chamber, skin, and muscle. ISAV is an enveloped negative strand RNA virus of the *Orthomyxoviridae* family, which includes influenza virus. Influenza virus is known for its property of antigenic variation by mutation ("drift") and recombination ("shift") events in the segmented genome, leading to the emergence of new strains. This may also be occurring in ISAV, as isolates from Norway and North America now seem to be genetically distinct. Inactivated virus vaccines have been used in Chile and Norway, but are not permitted under European Union legislation. In the last few years, recombinant DNA vaccines (see p. 323) have been developed and show promise for control of the disease.

Viral hemorrhagic septicemia virus infects many species of wild fish

Viral hemorrhagic septicemia virus (VHSV) is a rhabdovirus most commonly known as a causative agent of disease in the freshwater stage of trout and salmon culture, but which can also cause disease in marine net pens. The VHSV has now been isolated from nearly 50 species of marine fish and is responsible for severe epizootics in wild populations, especially of shoaling species such as herring (*Clupea harengus*), mackerel (*Scomber* spp.) and sprat (*Sprattus sprattus*). As well as the impact of overfishing, fluctuations in natural populations due to viral infection may be driving some commercially important species close to extinction. Detailed phylogenetic studies based on nucleoprotein gene sequences of marine VHSV show that there are distinct populations of the virus and that aquaculture isolates probably originated in wild marine fish. The related infectious

? HOW DID THE ISA VIRUS ORIGINATE?

The sudden emergence of infectious salmon anemia virus (ISAV) in Norway and its rapid spread to other regions raised important questions about its origins. By analyzing the sequence of the hemagglutinin gene, encoding a surface protein critical in the infection process, it is possible to calculate the rates of mutation in highly polymorphic regions. Nylund et al. (2003) concludes that the virus originated in wild salmonids in lakes of Norway, and that transmission probably evolved in the freshwater phase of trout. The development of salmon farming is thought to have changed the balance between ISAV and wild fish, leading to the rapid evolution of a highly pathogenic type. Salmon farmed in sea cages often have heavy infestations of salmon lice and these are thought to be vectors of the virus. Wild herring (*Clupea harengus*) were found to be possible asymptomatic carriers of ISAV (Nylund et al., 2002), but pollock (*Pollachius virens*; a wild species commonly found in and around salmon cages) were considered unlikely to be a reservoir. McClure et al. (2004) and Plarre et al. (2005) were also unable to evaluate definitively the importance of wild and farmed fish as reservoirs of infection for ISAV. Atlantic salmon were introduced to Chile for farming, and the absence of natural hosts for ISAV excludes the possibility of natural reservoirs in this country. Contemporary ISAV strains from farmed Atlantic salmon in Chile and Norway are genetically very similar, and Vike et al. (2009) concluded that the sudden emergence of ISA in Chile is due to vertical transmission via the transport of infected embryos from Norway.

hematopoietic necrosis virus (IHNV) survives for long periods in salt water, but has not been isolated from marine organisms.

Lymphocystis virus causes a highly contagious chronic skin infection of fish

Infection with lymphocystis virus (a member of the iridovirus group) results in hypertrophy of skin cells, causing characteristic papilloma-like lesions, usually on the skin, fins, and tail. The virus seems to have worldwide distribution, causing particular problems in cultured and wild fish in the Mediterranean Sea and in many species of tropical coral reef fish. It is one of the most common infections of aquarium fish. Its role in the ecology of natural populations is unclear; although the disease is not usually lethal, it affects growth rates and probably makes infected fish more susceptible to predation. A large number of other iridoviruses have been reported from more than 140 different species of fish worldwide; there is considerable heterogeneity in genome sequence in this group.

Birnaviruses appear to be widespread in marine fish and invertebrates

Several isolates of birnavirus have been identified in various wild fish in waters around the British Isles, Japan, and the northwest Atlantic. These were originally considered to be closely related to the infectious pancreatic necrosis virus (IPNV), which is best known as a cause of heavy mortalities in small fry in aquaculture hatcheries. Phylogenetic analysis of the marine birnaviruses from wild fish, based on amino-acid sequence of the polyprotein gene, indicates that they form a group distinct from IPNV. Little is known about entry of these viruses and replication in the host, and its impact on wild populations is unclear.

Viral nervous necrosis is an emerging disease with major impact

Viral nervous necrosis is characterized by erratic behavior in infected fish, due to spongiform lesions in the brain (encephalopathy). The causative agent, *Betanodavirus* (a member of the *Nodaviridae*) has been reported in cultured marine fish worldwide and has emerged in the 1990s to have a major impact on the culture of several marine species. PCR analysis shows that wild and cultured fish are subclinically infected with betanodaviruses and constitute a reservoir of infection. The spread of these infections seems to be a direct consequence of the intensity of aquaculture and the movement of stock both within aquaculture areas and between distant geographic regions.

Protists can cause disease in fish via infections, toxins, and direct physical effects

A large number of protists can infect wild, farmed, and aquarium fish. Often these are free-living or benign parasites and the environmental and nutritional conditions that promote disease are largely unknown. For example, *Paramoeba perurans* causes sporadic, severe outbreaks in salmon culture in Tasmania. Diplomonad flagellates are commonly found in the gut of fish, in which they seem to have little effect; but under some circumstances they can cause systemic infection with high mortalities. Myxosporeans such as *Kudoa* sp. commonly cause muscle infections, causing cysts and softening of the tissue that impairs marketability of the fish. *Loma salmonae* is an obligate intracellular microsporidean parasite infecting the gills of many economically important fish, including both wild and cultured salmon and cod. Protistan infections are usually controlled by external dip treatments such as formalin, malachite green, and chloramines T, but there is concern about the use of these agents on fish intended for human consumption because of fears of possible toxicity and carcinogenicity.

Excessive growth of phytoplankton leading to HABs in coastal waters can be responsible for mortalities in fish. Some of these blooms are caused by toxin-

---

? DID MOVEMENT OF TUNA FEED LEAD TO THE WORLD'S BIGGEST FISH KILL?

Two major epizootics occurred in the 1990s in the waters of South and Western Australia, resulting in the biggest ever single-cause fish kills. Mass mortalities of pilchards (*Sardinops sagax*) occurred, spreading at over 30 km a day and covering 6000 km of coastline. There was a reduction in the spawning biomass of more than 75% (T. M. Ward et al., 2001), resulting in severe ecological consequences for other marine life and seabird predators, as well as major economic losses to an important commercial fishery (Murray and O'Callaghan, 2005). The causative agent was identified as a previously unknown herpesvirus. These epizootics coincided with the development of tuna ranching in South Australia; tuna were kept in large offshore sea cages and fed with frozen pilchards imported from California, Peru, Chile, and Japan to fatten them for the sushi and sashimi market. Although conclusive proof is hard to obtain, it seems likely that inadvertent introduction of the virus with feed from a different geographic region may have resulted in a disease outbreak because the Australian pilchard population had not previously encountered this variant of the virus and therefore had no immunity. Alternative explanations might be transmission by ballast water or by seabirds. Since 1998, there have been no more mass mortalities and the virus may now be endemic in Australia; because of the development of host "herd" immunity, it may now cause only low levels of mortality or exist as a latent infection (Whittington et al., 2008).

producing species that can also affect marine mammals (see below) or humans (see Chapter 12). The most dramatic example of a possible toxic effect on fish is that of *Pfiesteria piscicida*, which was identified as a new organism nicknamed "the cell from hell" in the 1990s because of its association with mass fish kills in the Pamlico-Albemarle estuary on the east coast of the USA. Much of the work describing the life cycle of *P. piscicida* and its involvement in fish mortality and illness in humans has caused considerable controversy, which is discussed in *Research Focus Box 12.2*. Other toxic genera, such as *Alexandrium*, *Gymnodinium*, *Karenia*, and *Pseudo-nitzschia* have all been responsible for mass mortality in many parts of the world, both in wild and farmed fish. The increased use of coastal waters for fish and shellfish farming has undoubtedly contributed to stimulation of blooms by input of excess nutrients from the fish excreta and waste food. Caged fish cannot escape the effects.

Nontoxic microalgae can also affect fish. Large blooms reduce light penetration and decrease the growth of seagrass beds, which are often important nursery grounds for the young stages of commercially important fish. Clumping, sinking, and decay of phytoplankton can generate anoxic conditions. Nontoxic algae can also kill fish directly. For example, the diatom *Chaetoceros convolutus* produces long barbs, which clog fish gill tissue causing excess mucus production, leading to death from reduction in oxygen exchange. Incidents involving the loss of more than 250000 farmed salmon at a time have occurred. Blooms of the flagellate *Heterosigma carterae* can cause mortalities in Pacific coast salmon farms costing the industry several million US dollars a year.

Dinoflagellate and diatom toxins can affect marine mammals

Marine mammals are susceptible to a variety of disease conditions, as shown in *Figure 11.10A*, including the effects of toxic microalgae occurring in HABs. The significance of HAB toxins has been extensively investigated because of their effects on human health, and the main discussion of the origin of blooms and the nature of the toxins will be considered in Chapter 13. Most HAB toxins have effects on the nervous system. Ingestion of HAB toxins by humans results in symptoms such as nausea, vomiting, diarrhea, temperature reversal effects, and paralysis, so it is reasonable to conclude that similar effects will also occur in marine mammals. If so, this would clearly affect feeding, buoyancy, heat conservation, breathing, and swimming. Neuropathological symptoms and mortality in marine mammals are being increasingly associated with HABs. One explanation for the stranding or beaching of whales and dolphins (*Figure 11.10B*) is that they become disoriented because of accumulation of toxins after feeding on contaminated fish or shellfish. In some studies, high levels of saxitoxins have been found in tissues of stranded killer whales (*Orcinus orca*) known to have been feeding on mussels in areas with a toxic *Alexandrium* bloom. Although the levels of toxin recovered from most tissues at postmortem do not usually seem high enough to account for symptoms, it is possible that the toxins become concentrated in the brain. Brevetoxin shows a very high affinity for nerve tissue from the manatee (*Trichechus manatus*), which are often killed off the Florida coast during blooms of *Karenia brevis*, both by ingestion and inhalation. Saxitoxins also bind strongly to nerve tissue from several species of whale. Since the blood flow diverts to the brain during diving, this could deliver toxins absorbed from the gut to the brain, where they could accumulate in high enough concentrations to cause disorientation and other neurological effects, leading to stranding.

The toxin domoic acid, produced by the diatom *Pseudo-nitzschia* spp. is responsible for outbreaks of neuropathological illness in marine mammals and birds. The toxin was first associated with amnesic shellfish poisoning in humans (see p. 272) and subsequently linked to mass mortalities of sea lions, pelicans, and cormorants which occur periodically along the coast of northern California.

(A)

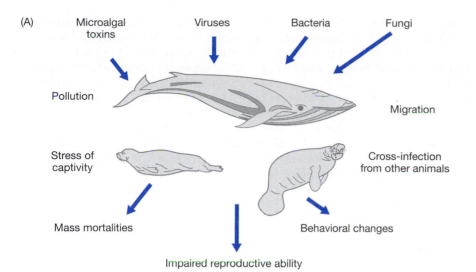

Microalgal toxins   Viruses   Bacteria   Fungi

Pollution

Migration

Stress of captivity

Cross-infection from other animals

Mass mortalities   Behavioral changes

Impaired reproductive ability

(B)

Figure 11.10 Microbes as agents of diseases in marine mammals. (A) Various factors influence the susceptibility of the host and transmission of diseases. (B) Mass stranding of pilot whales, possibly caused by toxic dinoflagellate bloom. (Image courtesy of US National Fisheries and Marine Service.)

Animals are poisoned by feeding on fish that have ingested large amounts of toxic *Pseudo-nitzschia*, which bloom as a result of climatic conditions and upwelling of nutrients. A pulse of domoic acid enters the food chain and the affected animals assimilate the toxin, which accumulates in the brain. Recent studies have found that krill—which is a principal component of the diet of squid, baleen whales, and seabirds—can accumulate domoic acid, which is then passed to higher trophic levels of the food chain. It appears that mortalities of marine animals linked to domoic acid are increasing.

Mass mortalities in the late twentieth century prompted the study of viral diseases of marine mammals

In 1988, there were mass mortalities of more than 18000 common or harbor seals (*Phoca vitulina*) in Northern Europe, with over 60% population loss in some areas. At about the same time, an epizootic with similar mortality rates occurred in bottlenose dolphins (*Tursiops truncatus*) on the east coast of the USA. In 1992, there were mass mortalities of porpoises (*Phocoena phocoena*) in the Irish Sea and of striped dolphins (*Stenella coeruleoalba*) in the Mediterranean Sea, which reduced the population to 30% of its previous level. These disease outbreaks attracted considerable public concern; under pressure from environmentalist lobby groups, a number of government-sponsored research projects were set up

? HOW DID A MICROBIAL TOXIN INSPIRE A FAMOUS HOLLYWOOD MOVIE?

Occasional mass mortality and erratic behavior of seabirds has been reported along the coast of northern California for many years. One such incident is thought to have been the inspiration—with some imaginative additions!—for the classic 1963 Alfred Hitchcock thriller movie *The Birds*, in which residents of a small town are terrified by the erratic behavior of birds. Whilst on vacation in 1961, Hitchcock heard of an incident in the town of Capitola, California, in which seabirds crashed into cars and houses and died in large numbers, and this inspired the idea for the film, adapted from a story set in Cornwall, England, by Daphne du Maurier. In 1991, another incident in Monterey Bay provided the first evidence to suggest that the birds had eaten fish containing the toxin domoic acid, originating from a diatom bloom. Extensive investigation involving scientists from many disciplines of another incident on this coast in 1998 provided conclusive evidence of domoic acid poisoning in birds and sea mammals and explained why this area is a "hot spot" for such events (Scholin et al., 2000).

to investigate the problem. As a result of this work and subsequent studies, we now know that virus diseases are a significant cause of mortality in cetaceans (whales, dolphins, and porpoises) and pinnipeds (seals and sea lions). In the wild, it is difficult to study diseases of marine mammals except when mass mortalities occur, when animals are stranded, or if they become caught in fishing nets. Networks of marine conservation volunteers have helped in the collection of data, and it is now common practice to test stranded cetaceans and pinnipeds for viral diseases using immunological and molecular-based diagnostic tests, so that a worldwide picture of their importance is beginning to emerge. Obtaining blood or tissue samples from live animals is difficult and postmortem deterioration happens very quickly, so data are often sparse. Some knowledge about these diseases has also come from the study of captive animals in zoos and marine parks; the study of diseases of marine mammals has become a specialized branch of veterinary medicine.

Viruses from nine different families have been linked to diseases of marine mammals

The most important viruses infecting cetaceans and pinnipeds are the morbilliviruses; these are RNA viruses of the paramyxovirus group causing a number of mammalian diseases, of which the best-known examples are measles in humans and distemper in dogs. The first clues to the identity of the virus infecting seals in the 1988 epizootic came when serum was found to neutralize canine distemper virus (CDV). At first, it was thought that the disease might be linked to an outbreak of distemper in sled dogs in Greenland. However, the seal virus (now called phocine distemper virus; PDV) was shown to be a new species using sequencing of the viral capsid proteins, although it shares some antigenic cross-reactions with CDV. One possible explanation for the sudden epizootic is transfer from another species (probably the harp seal, *Phoca groenlandica*), which migrated from a different geographic region in which the disease is enzootic. Serological studies have shown that PDV-like viruses are present in several species of marine mammals. By contrast, the virus responsible for outbreaks in seals in the Caspian Sea and Lake Baikal (*Phoca caspica* and *Phoca siberica* respectively) seems to be identical to CDV, and it almost certainly did come from dogs. Infection causes respiratory, gastrointestinal, and neurological symptoms, often with secondary bacterial infections such as pneumonia.

The morbilliviruses isolated from diseased porpoises, dolphins, and several species of whale appear to be closely related antigenically and genetically; they are now recognized as different strains of cetacean morbillivirus (CMV). Phylogenetic studies show that CMV is close to the ancestor of the morbillivirus group, so these viruses probably have a terrestrial origin and may have infected cetaceans for the several millions of years since they have populated the oceans. Infected animals show pneumonia-like symptoms and disturbance of diving, swimming, and navigation ability. In the enzootic state, the virus probably has long-term effects on population dynamics, causing mortalities mainly in young animals in which no immunity exists. Morbilliviruses typically lead to either rapid death or recovery with life-long immunity, with no persistent carrier state. They therefore require large populations to sustain themselves through the input of new susceptible hosts. As with seal distemper, epizootics in cetaceans probably occur as a result of cross-infection from different species or animals from other geographic regions once a sufficiently large population of susceptible individuals has built up to allow spread. The nutritional status of the host population and environmental factors are also important in determining the onset of epizootics. After the 1988 outbreak of phocine distemper in northern Europe, the harbor seal population slowly recovered. In 2002, a new epizootic of phocine distemper in harbor seals emerged, owing to the buildup of a threshold number of nonimmune animals.

Poxviruses have been implicated as the cause of tattoo-like skin lesions in several species of small cetaceans, although they do not usually cause significant mortalities. Papillomaviruses and herpesviruses cause genital warts (proliferation of the squamous epithelium) in porpoises, dolphins, and whales. As with their human equivalent, they are sexually transmitted and may affect population dynamics by disturbing reproduction and social behavior, since sexual "play" is an important part of cetacean social group interactions and genital warts can sometimes be large enough to interfere with copulation.

Various types of caliciviruses occur in a wide range of marine mammals, having been first described as San Miguel sea lion virus. From time to time, influenza A virus has been associated with mortalities in cetaceans and pinnipeds. As occurs in humans, animals infected with influenza become very weak and often succumb to secondary bacterial infections. The isolates are closely related to avian influenza in birds and are highly virulent. Seabirds are often associated with marine mammals during feeding at the surface and this favors transmission.

Several species of bacteria and fungi infect marine mammals

One of the most important globally distributed diseases of cetaceans and pinnipeds is brucellosis, caused by the bacterium *Brucella*, a highly contagious intracellular pathogen that also occurs in cattle, sheep, goats, and pigs. It infects the reproductive tract, especially the placenta and amniotic sac, leading to abortion. Frequent abortions have been observed to occur in closely monitored dolphin pods, and the bacterium could therefore have a significant effect on fertility and population dynamics. Several strains have been isolated from many different species, and serological studies show that up to 30% of small marine mammals surveyed have evidence of exposure. Phylogenetic analyses of the marine isolates of *Brucella* show that these may constitute two distinct new species named *Brucella ceti* and *Brucella pinnipedialis* with cetaceans and seals, respectively, as the preferred hosts.

Leptospirosis, caused by several species of *Leptospira*, occurs in many populations of seals and sea lions. Populations of Californian sea lions suffer a cyclical pattern of infection every 4–5 years. As in the human equivalent (Weil's disease, transmitted by rats), the main symptom is renal failure.

Tuberculosis is a chronic multiorgan disease caused by *M. tuberculosis* and *Mycobacterium bovis*. Other mycobacteria such as *M. marinum* and *M. fortuitum* can cause lesions in the skin and lungs. In seals, sea lions, and cetaceans, there have been several cases in zoos and marine parks. *Burkholderia pseudomallei* can cause septicemia and has been described as the cause of death of 25 dolphins, whales, and sea lions over an 11-year period in an oceanarium in Hong Kong, and sporadically elsewhere. Other bacterial pathogens known to cause sudden mortality in captive cetaceans include *Bordetella bronchiseptica* and species of *Erysipelothrix*, *Streptococcus*, *Salmonella*, and *Pasteurella*. The significance of these pathogens in wild animals is not known.

In addition to primary pathogens, marine mammals can succumb to a host of opportunistic skin and respiratory infections caused by bacteria and fungi, such as pneumonia caused by *Aspergillus fumigatus* and *Aspergillus terrus*, which is a particular problem in stressed captive animals. Direct acquisition of microbes of human origin, especially opportunist bacterial infections of wounds, may also occur when mammals swim in sewage-contaminated waters.

Sea turtles are affected by a virus promoting growth of tumors

Some novel herpesviruses are associated with infection of sea turtles (*Figure 11.11*). The chelonid fibropapilloma-associated herpesvirus has emerged as a major cause of debilitating tumors of the skin and internal organs in green (*Chelonia mydas*), loggerhead (*Caretta caretta*), and olive ridley (*Lepidochelys*

Figure 11.11 Green turtle with a large tumor caused by the fibropapilloma-associated turtle herpesvirus. (Image courtesy of Thierry M. Work, US Geological Survey, Honolulu Field Station.)

olivacea) turtles. The large tumors interfere with feeding, behavior, and reproduction, so there is concern that this disease may cause additional problems for the survival of these endangered species. The disease has been reported in Florida, Hawaii, and Australia and appears to have increased dramatically since its first description in 1938. Genomic analysis shows that viruses belong to the alphaherpesvirus group, but those isolated from different species of turtle from different geographic regions are genetically distinct. The increase in disease seems to be due to unknown environmental factors affecting host susceptibility, rather than emergence of a more virulent form of the virus. A leech, *Ozobranchus* sp., acts as a mechanical vector for the virus. Other herpesviruses and papillomaviruses have also been implicated as the cause of disease in turtles reared in mariculture.

DISEASES OF SEAWEEDS AND SEAGRASSES

Fungi, bacteria, and protists cause ecologically and economically important diseases of seaweeds and seagrasses

Descriptions of diseases of seaweeds (macroalgae) are surprisingly rare. In part this is due to this being a much-neglected area of research, but it is also due to the fact that seaweeds possess effective mechanisms against microbial colonization and infection. Indeed, many of these antimicrobial properties have been investigated for their biotechnological potential (see Chapter 14). Intensive aquaculture of red, brown, and green algae as a source of foods, nutritional supplements, pharmaceuticals, and products such as carageenins and agar is a major industry in many countries, especially China, Japan, Indonesia, Korea, the Philippines, and Chile. The red alga *Porphyra*, harvested as the Japanese food crop nori, is susceptible to a wasting disease resulting in coalescing lesions caused by the fungus *Pythium porphyrae*. Ascomycete fungi cause infections of *Laminaria*, *Sargassum*, and other kelps. Bacterial infections include "ice-ice disease" of the red seaweed *Kappachus alvarezii*, which is one of the most important cultured species for the production of carageenins. As we have seen in many other situations, *Vibrio* spp. are the most likely causative agents, colonizing the algal fronds if they are stressed by low salinity, high temperature, or poor light.

Reports of disease in natural ecosystems include those affecting the coralline algae associated with reefs, originally observed in the 1990s in the Cook Islands and now reported throughout the Indo-Pacific. Bacteria belonging to the genera *Planococcus*, *Bacillus*, and *Pseudomonas* have been identified in association with coralline lethal orange disease (CLOD), an infection that affects encrusting coralline algae. CLOD is identified by the appearance of bright orange coloration spreading to dead areas, although details of the pathogenic mechanisms are scant. Another disease is associated with a black fungal infection. As with other diseases, increased sea-surface temperatures seem to encourage the spread of infection. Other diseases lethal to coralline algae, with as yet unknown etiology, have been reported from the Caribbean. Because coralline algae produce very hard deposits of calcium carbonate, they play an important role in cementing the structure of reefs and protecting them from wave damage, so the emergence of these diseases is a cause for concern.

Diseases also affect the productivity of seagrasses such as eelgrass (*Zostera marina*) and turtlegrass (*Thalassia testudinum*) in shallow coastal habitats (*Figure 11.12*). Seagrass meadows are major habitats for marine life such as juvenile fish, turtles, dugongs, and manatees, and also acts as feeding grounds for ducks and geese. Disease is a further threat to survival of seagrass beds, together with destruction accompanying coastal development, resulting in considerable ecological impact through loss of habitats and economic effects through impairment of recruitment to fishery stocks. Since the 1920s, there have been many reports of "wasting disease" affecting large areas of seagrass meadows in Europe and North

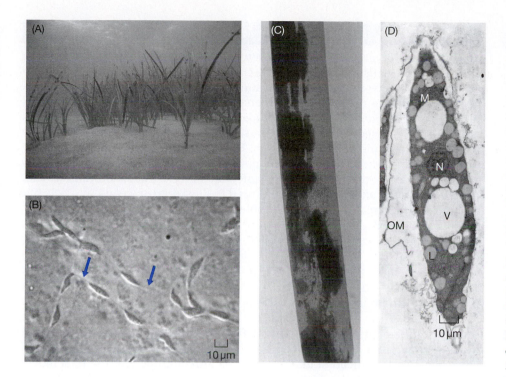

Figure 11.12 Wasting diseases in seagrasses caused by *Labryinthula* spp. (A) Eelgrass (*Zostera marina*) meadow in Peconic Estuary, Long Island, NY, showing signs of infection with *L. zosterae*. (B) Light micrograph of *Labyrinthula* sp. on the surface of agar. Cells can be seen moving through the outer matrix membrane of the trackway (*arrows*). (C) Close-up of infected blades of eelgrass showing characteristic black lesions. (D) Electron micrograph of a single *Labyrinthula* sp. cell within infected tissue of the seagrass *Thalassia testudinum*. L, lipid vesicle; M, mitochondria; N, nucleus; OM, outer matrix membrane; V, vacuole. (Images A, C courtesy of Kimberly Petersen Manzo and Chris Pickerell, Cornell Cooperative Extension Eelgrass Program, www.SeagrassLI.org; images B, D courtesy of Tim Sherman, University of South Alabama.)

America, resulting in the total loss of *Z. marina* from some areas. The causative agent was isolated in the 1980s and originally identified as a fungus or slime mold. Subsequent phylogenetic studies using rRNA gene sequencing revealed the primary pathogen to be a new species, *Labyrinthula zosterae*, a member of a monophyletic group within the heterokont protists. Members of this group (which includes the thraustochytrids; see p. 147) have an unusual morphology; the cells secrete an ectoplasmic membrane leading to a slimy net-like structure of actin-rich filaments, through which the nuclei are transported. The pathogen moves rapidly through the plant tissue through enzymatic degradation of cell walls, leading to blackening and eventually complete destruction of the tissue. However, the photosynthetic ability of tissue further away from the focus of infection is severely inhibited, so the ecological impact of reduced productivity in eelgrass showing even minor diseases signs could be significant.

Many species of algae contain virus-like particles

More than 50 species in most of the algal orders have been shown to contain viruses or VLPs, and about 20 algal viruses have been isolated and characterized. Most studies have been concerned with the role of viruses in infection of planktonic algae, especially prymnesiophytes, diatoms, and dinoflagellates (see Chapter 7). There are surprisingly few reports of viruses in seaweeds. VLPs have been observed in the reproductive organs of several species of filamentous brown algae in different geographic areas. Production of infectious virions is variable and host viability may not always be affected. The *Ectocarpus siliculosus* virus (EsV-1) has been sequenced, and study of the factors involved in viral replication is yielding valuable information about reproductive development in these algae, because virion release is synchronized with the release of spores or gametes, enabling interaction of viruses with their susceptible host cells. EsV-1 is designated as the founding member of the phaeoviruses, a group of the *Phycodnaviridae*. Host-specific phaeoviruses have been identified in eight species of brown algae and are widely distributed in all temperate coastal waters. Since seaweeds are being increasingly grown in intensive culture as a supply of foods, pharmaceuticals, and cosmetic products, it is likely that more evidence of viral disease in these aquaculture species will emerge.

Conclusions

Marine organisms are infected by a variety of bacteria, but members of the vibrio group are especially important in infections of fish and invertebrates because of their abundance in coastal waters and reef systems and their possession of powerful virulence mechanisms that enable them to colonize their hosts and produce disease, either as primary pathogens or opportunists. Bacterial diseases of corals and sponges have resulted in significant ecological changes in some reef systems and threaten many others. Bacteria are major causes of mortality and economic losses in the culture of mollusks, crustaceans, fish, and algae and some produce ecologically important infections in the wild. A number of viral diseases seriously affect intensively cultured mollusks, crustaceans, and fish, but the role of viral diseases in natural populations of corals, mollusks, crustaceans, and algae is poorly documented. By contrast, viruses cause major epizootics in wild fish, marine mammals, and reptiles, some of which threaten the survival of endangered species.

This chapter has provided several examples of evolution of new strains, changes in virulence, and transfer of pathogens between hosts and ecosystems—marine to fresh water, fresh water to marine, and terrestrial to marine. These shifts are often precipitated by movement of organisms between geographic areas, which may occur naturally through migration. However, we have also seen examples of ways in which the spread of intensive aquaculture and the international movement of fish, shellfish, feedstuffs, and seafood products have led to major epizootics and the development of new reservoirs of infection. Another area of particular concern is the effect of climate change and pollution on the incidence of disease in marine ecosystems. Adverse environmental factors may compromise the immune status or general physiology of the host and affect the virulence of pathogens.

Although interest in diseases of marine organisms has increased greatly in the past few years, the number of scientists directly involved in the detailed study of etiology and pathology of microbial infections of marine life remains very small and lags well behind those involved with study of disease in humans or in other natural ecosystems. Since climate change and other anthropological impacts are causing increased incidence and emergence of new diseases in some groups, we need much better baseline knowledge before we can implement strategies to prevent or limit the impact of diseases in the marine environment. This can only be achieved by increasing the number of researchers studying marine diseases and employing the most effective technologies available to obtain data and pooling expertise through increased international cooperation.

References

Acevedo-Whitehouse, K., Rocha-Gosselin, A. and Gendron, D. (2009) A novel non-invasive tool for disease surveillance of free-ranging whales and its relevance to conservation programs. *Anim. Conserv.* 13: 217–225.

Banin, E., Israely, T., Kushmaro, A., Loya, Y., Orr, E. and Rosenberg, E. (2000) Penetration of the coral-bleaching bacterium *Vibrio shiloi* into *Oculina patagonica*. *Appl. Environ. Microbiol.* 66: 3031–3036.

Banin, E., Israely, T., Fine, M., Loya, Y. and Rosenberg, E. (2001a) Role of endosymbiotic zooxanthellae and coral mucus in the adhesion of the coral-bleaching pathogen *Vibrio shiloi* to its host. *FEMS Microbiol. Lett.* 199: 33–37.

Banin, E., Khare, S. K., Naider, F. and Rosenberg, E. (2001b) Proline-rich peptide from the coral pathogen *Vibrio shiloi* that inhibits photosynthesis of zooxanthellae. *Appl. Environ. Microbiol.* 67: 1536–1541.

Ben-Haim, Y. and Rosenberg, E. (2002) A novel *Vibrio* sp. pathogen of the coral *Pocillopora damicornis*. *Mar. Biol.* 141: 47–55.

Ben-Haim, Y., Banim, E., Kushmaro, A., Loya, Y. and Rosenberg, E. (1999) Inhibition of photosynthesis and bleaching of zooxanthellae by the coral pathogen *Vibrio shiloi*. *Environ. Microbiol.* 1: 223–229.

Ben-Haim, Y., Zicherman-Keren, M. and Rosenberg, E. (2003) Temperature-regulated bleaching and lysis of the coral *Pocillopora damicornis* by the novel pathogen *Vibrio coralliilyticus*. *Appl. Environ. Microbiol.* 69: 4236–4242.

Boroujerdi, A. F. B., Vizcaino, M. I., Meyers, A., et al. (2009) NMR-based microbial metabolomics and the temperature-dependent coral pathogen *Vibrio coralliilyticus*. *Environ. Sci. Technol.* 43: 7658–7664.

Bourne, D., Iida, Y., Uthicke, S. and Smith-Keune, C. (2008a) Changes in coral-associated microbial communities during a bleaching event. *ISME J.* 2: 350–363.

Bourne, D. G., Boyett, H. V., Henderson, M. E., Muirhead, A. and Willis, B. L. (2008b) Identification of a ciliate (Oligohymenophorea: Scuticociliatia) associated with brown band disease on corals of the Great Barrier Reef. *Appl. Environ. Microbiol.* 74: 883–888.

Cano-Gomez, A., Bourne, D. G., Hall, M. R., Owens, L. and Høj, L. (2009) Molecular identification, typing and tracking of *Vibrio harveyi* in aquaculture systems: current methods and future prospects. *Aquaculture* 287: 1–10.

Danovaro, R., Bongiorni, L., Corinaldesi, C., et al. (2008) Sunscreens cause coral bleaching by promoting viral infections. *Environ. Health Perspect.* 116: 441–447.

Davy, S. K., Burchett, S. G., Dale, A. L., et al. (2006) Viruses: agents of coral disease? *Dis. Aquat. Organ.* 69: 101–110.

Doucette, G. J., Cembella, A. D., Martin, J. L., Michaud, J., Cole, T. V. N. and Rolland, R. M. (2006) Paralytic shellfish poisoning (PSP) toxins in North Atlantic right whales *Eubalaena glacialis* and their zooplankton prey in the Bay of Fundy, Canada. *Mar. Ecol. Prog. Ser.* 306: 303–313.

Drake, L. A. and Dobbs, F. C. (2005) Do viruses affect fecundity and survival of the copepod *Acartia tonsa* Dana? *J. Plankton Res.* 27: 167–174.

Hjerde, E., Lorentzen, M. S., Holden, M. T. G., et al. (2008) The genome sequence of the fish pathogen *Aliivibrio salmonicida* strain LFI1238 shows extensive evidence of gene decay. *BMC Genomics* 9: 616.

Jepson, P. D., Bennett, P. M., Allchin, C. R., Law, R. J., Kuiken, T., Baker, J. R., Rogan, E. and Kirkwood, J. K. (1999) Investigating potential associations between chronic exposure to polychlorinated biphenyls and infectious disease mortality in harbour porpoises from England and Wales. *Sci. Total Environ.* 4: 339–48.

Kushmaro, A., Rosenberg, E., Fine, M., Ben Haim, Y. and Loya, Y. (1998) Effect of temperature on bleaching of the coral *Oculina patagonica* by *Vibrio* AK-1. *Mar. Ecol. Prog. Ser.* 171: 131–137.

Leggat, W., Ainsworth, T., Bythell, J., Dove, S., Gates, R., Hoegh-Guldberg, O., Iglesias-Prieto, R. and Yellowlees, D. (2007) The hologenome theory disregards the coral holobiont. *Nat. Rev. Microbiol.* 5: doi:10.1038/nrmicro1635C1.

Lesser, M. P., Bythell, J. C., Gates, R. D., Johnstone, R. W. and Hoegh-Guldberg, O. (2007) Are infectious diseases really killing corals? Alternative interpretations of the experimental and ecological data. *J. Exp. Mar. Biol. Ecol.* 346: 36–44.

Littman, R., Bourne, D. G. and Willis, B. L. (2010) Symbiodinium clade determines stability of bacterial associations on temperature stressed juvenile corals. *Mol. Ecol.* 19: 1978–1990.

Lohr, J., Munn, C. B. and Wilson, W. H. (2007) Characterization of a latent virus-like infection of symbiotic zooxanthellae. *Appl. Environ. Microbiol.* 73: 2976–2981.

Maloy, A. P., Ford, S. E., Karney, R. C. and Boettcher, K. J. (2007) *Roseovarius crassostreae*, the etiological agent of Juvenile Oyster Disease (now to be known as *Roseovarius* Oyster Disease) in *Crassostrea virginica*. *Aquaculture* 269: 71–83.

McClure, C. A., Hammell, K. L., Dohoo, I. R. and Gagne, N. (2004) Lack of evidence of infectious salmon anemia virus in pollock *Pollachius virens* cohabitating with infected farmed Atlantic salmon *Salmo salar*. *Dis. Aquat. Organ.* 61: 149–152.

Meron, D., Efrony, R., Johnson, W. R., et al. (2009) Role of flagella in virulence of the coral pathogen *Vibrio coralliilyticus*. *Appl. Environ. Microbiol.* 75: 5704–5707.

Mills, A. M., Ward, M. E., Heyl, T. P. and Van Dover, C. L. (2005) Parasitism as a potential contributor to massive clam mortality at the Blake Ridge Diapir methane-hydrate seep. *J. Mar. Biol. Assoc. UK* 85: 1489–1497.

Mos, L., Morsey, B., Jeffries, S. J., Yunker, M. B., Raverty, S., De Guise, S. and Ross, P. S. (2006) Chemical and biological pollution contribute to the immunological profiles of free-ranging harbor seals. *Environ. Toxicol. Chem.* 25: 3110–3117.

Munro, J., Oakey, J., Bromage, E. and Owens, L. (2003) Experimental bacteriophage-mediated virulence in strains of *Vibrio harveyi*. *Dis. Aquat. Organ.* 54: 187–194.

Murray, A. G. and O'Callaghan, M. (2005) The numerical and programming methods used to implement models of the spread and impact of a major epidemic disease: pilchard herpesvirus, Australia 1995 and 1998/1999. *Environ. Model. Software* 20: 575–585.

Nissimov, J., Rosenberg, E. and Munn, C. B. (2009) Antimicrobial properties of resident coral mucus bacteria of *Oculina patagonica*. *FEMS Microbiol. Lett.* 292: 210–215.

Nylund, A., Devoid, M., Mullins, J. and Plarre, H. (2002) Herring (*Clupea harengus*): a host for infectious salmon anemia virus (ISAV). *Bull. Eur. Assoc. Fish Pathol.* 22: 311–318.

Nylund, A., Devold, M., Plarre, H., Aarseth, M. and Isdal, E. (2003) Emergence and maintenance of infectious salmon anaemia virus (ISAV) in Europe: a new hypothesis. *Dis. Aquat. Organ.* 56: 11–24.

Oakey, H. J., Cullen, B. R. and Owens, L. (2002) The complete nucleotide sequence of the *Vibrio harveyi* bacteriophage VHML. *J. Appl. Microbiol.* 93: 1089–1098.

Pallen, M. J. and Wren, B. W. (2007) Bacterial pathogenomics. *Nature* 449: 835–842.

Plarre, H., Devold, M., Nylund, A. and Snow, M. (2005) Prevalence of infectious salmon anaemia virus (ISAV) in wild salmonids in western Norway. *Dis. Aquat. Organ.* 66: 71–79.

Reith, M. E., Singh, R. K., Curtis, B., et al. (2008) The genome of *Aeromonas salmonicida* subsp. *salmonicida* A449: insights into the evolution of a fish pathogen. *BMC Genomics* 9: 427.

Reshef, L., Koren, O., Loya, Y., Zilber-Rosenberg, I. and Rosenberg, E. (2006) The coral probiotic hypothesis. *Environ. Microbiol.* 8: 2068–2073.

Reshef, L., Ron, E. and Rosenberg, E. (2008) Genome analysis of the coral bleaching pathogen *Vibrio shiloi*. *Arch. Microbiol.* 190: 185–194.

Ritchie, K. B. (2006) Regulation of microbial populations by coral surface mucus and mucus-associated bacteria. *Mar. Ecol. Prog. Ser.* 322: 1–14.

Rolland, R. M., Hamilton, P. K., Kraus, S. D., Davenport, B., Gillett, R. M. and Wasser, S. K. (2006) Faecal sampling using detection dogs to study reproduction and health in North Atlantic right whales (*Euhalaena glacialis*). *J. Cetacean Res. Manage.* 8: 121–125.

Rosenberg, E. and Falkovitz, L. (2004) The *Vibrio shiloi/Oculina patagonica* model system of coral bleaching. *Ann. Rev. Microbiol.* 58: 143–159.

Rosenberg, E., Koren, O., Reshef, L., Efrony, R. and Zilber-Rosenberg, I. (2007) The role of microorganisms in coral health, disease and evolution. *Nat. Rev. Microbiol.* 5: 355–362.

Rosenberg, E., Kushmaro, A., Kramarsky-Winter, E., Banin, E. and Yossi, L. (2009) The role of microorganisms in coral bleaching. *ISME J.* 3: 139–146.

Ross, P. S. (2002) The role of immunotoxic environmental contaminants in facilitating the emergence of infectious diseases in marine mammals. *Hum. Ecol. Risk Assess.* 8: 277–292.

Ross, P. S., de Swart, R. L., Timmerman, H. H., Reijnders, P. J. H., Vos, J. G., Van Loveren, H. and Osterhaus, A. D. M. E. (1996) Suppression of natural killer cell activity in harbour seals (*Phoca vitulina*) fed Baltic Sea herring. *Aquat. Toxicol.* 34: 71–84.

Rypien, K. L. (2008) African dust is an unlikely source of *Aspergillus sydowii*, the causative agent of sea fan disease. *Mar. Ecol. Prog. Ser.* 367: 125–131.

Rypien, K. L., Andras, J. P. and Harvell, C. D. (2008) Globally panmictic population structure in the opportunistic fungal pathogen *Aspergillus sydowii*. *Molec. Ecol.* 17: 4068–4078.

Scholin, C. A., Gulland, F., Doucette, G. J., et al. (2000) Mortality of sea lions along the central California coast linked to a toxic diatom bloom. *Nature* 403: 80–84.

Stentiford, G. D. (2008) Diseases of the European edible crab (*Cancer pagurus*): a review. *ICES J. Mar. Sci.* 65: 1578–1592.

Sunagawa, S., DeSantis, T. Z., Piceno, Y. M., et al. (2009) Bacterial diversity and White Plague Disease-associated community changes in the Caribbean coral *Montastraea faveolata*. *ISME J.* 3: 512–521.

Sussman, M., Willis, B. L., Victor, S. and Bourne, D. G. (2008) Coral pathogens identified for white syndrome (WS) epizootics in the Indo-Pacific. *PLoS ONE* 3: e2393.

Sussman, M., Mieog, J. C., Doyle, J., Victor, S., Willis, B. L. and Bourne, D. G. (2009) Vibrio zinc-metalloprotease causes photoinactivation of coral endosymbionts and coral tissue lesions. *PLoS ONE* 4: e4511.

Toren, A., Landau, L., Kushmaro, A., Loya, Y. and Rosenberg, E. (1998) Effect of temperature on adhesion of *Vibrio* strain AK-1 to *Oculina patagonica* and on coral bleaching. *Appl. Environ. Microbiol.* 64: 1379–1384.

Vidgen, M., Carson, J., Higgins, M. and Owens, L. (2006) Changes to the phenotypic profile of *Vibrio harveyi* when infected with the *Vibrio harveyi* myovirus-like (VHML) bacteriophage. *J. Appl. Microbiol.* 100: 481–487.

Vike, S., Nylund, S. and Nylund, A. (2009) ISA virus in Chile: evidence of vertical transmission. *Arch. Virol.* 154: 1–8.

Ward, J. R. and Lafferty, K. D. (2004) The elusive baseline of marine disease: are diseases in ocean ecosystems increasing? *PLOS Biology* 2: 542–547.

Ward, M. E., Van Dover, C. L. and Shields, J. D. (2004) Parasitism in species of *Bathymodiolus* (Bivalvia: Mytilidae) mussels from deep-sea seep and hydrothermal vents. *Dis. Aquat. Organ.* 62: 1–16.

Ward, T. M., Hoedt, F., McLeay, L., et al. (2001) Effects of the 1995 and 1998 mass mortality events on the spawning biomass of sardine, *Sardinops sagax*, in South Australian waters. *ICES J. Mar. Sci.* 58: 865–875.

Whittington, R. J., Crockford, M., Jordan, D. and Jones, B. (2008) Herpesvirus that caused epizootic mortality in 1995 and 1998 in pilchard, *Sardinops sagax* neopilchardus (Steindachner), in Australia is now endemic. *J. Fish Dis.* 31: 97–105.

Work, T. M., Richardson, L. L., Reynolds, T. L. and Willis, B. L. (2008) Biomedical and veterinary science can increase our understanding of coral disease. *J. Exp. Mar. Biol. Ecol.* 362: 63–70.

Zielinski, F. U., Pernthaler, A., Duperron, S., Raggi, L., Giere, O., Borowski, C. and Dubilier, N. (2009) Widespread occurrence of an intranuclear bacterial parasite in vent and seep bathymodiolin mussels. *Environ. Microbiol.* 11: 1150–1167.

Further reading

General articles

Lafferty, K. D. (2009) The ecology of climate change and infectious diseases. *Ecology* 90: 888–900.

Lafferty, K. D., Porter, J. W. and Ford, S. E. (2004) Are diseases increasing in the ocean? *Annu. Rev. Ecol. Evol. Syst.* 35: 31–54.

McCallum, H. I., Kuris, A., Harvell, C. D., Lafferty, K. D., Smith, G. W. and Porter, J. (2004) Does terrestrial epidemiology apply to marine systems? *Trends Ecol. Evol.* 19: 585–591.

Munn, C. B. (2006) Viruses as pathogens of marine organisms—from bacteria to whales. *J. Mar. Biol. Assoc. UK* 86: 1–15.

Ritchie, K. B., Polson, S. W. and Smith, G. W. (2001) Microbial disease causation in marine invertebrates: problems, practices, and future prospects. *Hydrobiology* 460: 131–139.

Diseases of corals and sponges

Bourne, D. G., Garren, M., Work, T. M., Rosenberg, E., Smith, G. W. and Harvell, C. D. (2009) Microbial disease and the coral holobiont. *Trends Microbiol.* 17: 554–562.

Marhaver, K. L., Edwards, R. and Rohwer, F. (2008) Viral communities associated with healthy and bleaching corals. *Environ. Microbiol.* 10: 2277–2286.

NOAA (2006) Coral health and monitoring program. National Oceanic and Atmospheric Administration, accessed April 1, 2010, http://www.coral.noaa.gov/coral_disease/.

Thurber, R. V., Willner-Hall, D., Rodriguez-Mueller, B., et al. (2009) Metagenomic analysis of stressed coral holobionts. *Environ. Microbiol.* 11: 2148–2163.

Webster, N. S. (2007) Sponge disease: a global threat? *Environ. Microbiol.* 9: 1363–1375.

Webster, N. S., Xavier, J. R., Freckelton, M., Motti, C. A. and Cobb, R. (2008) Shifts in microbial and chemical patterns within the marine sponge *Aplysina aerophoba* during a disease outbreak. *Environ. Microbiol.* 10: 3366–3376.

Diseases of mollusks

Le Roux, F., Zouine, M., Chakroun, N., et al. (2009) Genome sequence of *Vibrio splendidus*: an abundant planktonic marine species with a large genotypic diversity. *Environ. Microbiol.* 11: 1959–1970.

Paillard, C., Le Roux, F. and Borreg, J. J. (2004) Bacterial disease in marine bivalves, a review of recent studies: trends and evolution. *Aquat. Living Res.* 17: 477–498.

Diseases of crustaceans

Chen, K. Y., Hsu, T. C., Huang, P. Y., Kang, S. T., Lo, C. F., Huang, W. P. and Chen, L. L. (2009) Penaeus monodon chitin-binding protein (PmCBP) is involved in white spot syndrome virus (WSSV) infection. *Fish Shellfish Immunol.* 27: 460–465.

Flegel, T. W. (2009) Current status of viral diseases in Asian shrimp aquaculture. *Isr. J. Aquacult.* 61: 229–239.

Flegel, T. W. (2009) Hypothesis for heritable, anti-viral immunity in crustaceans and insects. *Biol. Direct* 4: 32.

Leu, J. H., Yang, F., Zhang, X., Xu, X., Kou, G. H. and Lo, C. F. (2009) Whispovirus. Lesser known large DsDNA viruses. *Curr. Topics Microbiol. Immunol.* 328: 197–227.

Stentiford, G. D. and Shields, J. D. (2005) A review of the parasitic dinoflagellates *Hematodinium* species and *Hematodinium*-like infections in marine crustaceans. *Dis. Aquat. Organ.* 66: 47–70.

Stentiford, G. D., Bonami, J. R. and day-Sanz, V. (2009) A critical review of susceptibility of crustaceans to Taura syndrome, yellowhead disease and white spot disease and implications of inclusion of these diseases in European legislation. *Aquaculture* 291: 1–17.

Vogan, C. L., Powell, A. and Rowley, A. F. (2008) Shell disease in crustaceans—just chitin recycling gone wrong? *Environ. Microbiol.* 10: 826–835.

Diseases of fish

Austin, B. and Austin, D. A. (2007) *Bacterial Fish Pathogens. Disease of Farmed and Wild Fish*, 4th Edn. Springer Praxis Books, Chichester, UK.

Buchmann, K., Bresciani, J., Ariel, E., Pedersen, K. and Dalsgaard, I. (2009) *Fish Diseases: An Introduction*. Biofolia Publishing, Copenhagen.

Yin, L. K. (ed.) (2004) *Current Trends in the Study of Bacterial and Viral Fish and Shrimp Diseases*. World Scientific Publishing, Singapore.

Diseases of marine mammals

Bargu, S., Powell, C. L., Coale, S. L., Busman, M., Doucette, G. J. and Silver, M. W. (2002) Krill: a potential vector for domoic acid in marine food webs. *Mar. Ecol. Prog. Ser.* 237: 209–216.

Bossart, G. D. (2006) Marine mammals as sentinel species for oceans and human health. *Oceanography* 19: 134–137.

Bossart, G. D. (2007) Emerging diseases in marine mammals: from dolphins to manatees. *Microbe* 2: 544–549.

Gulland, F. M. D. and Hall, A. J. (2007) Is marine mammal health deteriorating? Trends in the global reporting of marine mammal disease. *EcoHealth* 4: 135–150.

Trainer, V. L. (2001) Marine mammals as sentinels of environmental biotoxins. In: *Handbook of Neurotoxicology*, Vol. 1 (ed. E. J. Masaro). Humana Press, Totowa, NJ, pp. 349–361.

Van Bressem, M. F., Van Waerebeek, K. and Raga, J. A. (1999) A review of virus infections of cetaceans and the potential impact of morbilliviruses, poxviruses and papillomaviruses on host population dynamics. *Dis. Aquat.Organ.* 38: 53–65

Diseases of sea turtles

Chaloupka, M., Balazs, G. H. and Work, T. M. (2009) Rise and fall over 26 years of a marine epizootic in Hawaiian green sea turtles. *J. Wildl. Dis.* 45: 1138–1142.

Greenblatt, R. J., Work, T. M., Balazs, G. H., Sutton, C. A., Casey, R. N. and Casey, J. W. (2004) The Ozobranchus leech is a candidate mechanical vector for the fibropapilloma-associated turtle herpesvirus found latently infecting skin tumors on Hawaiian green turtles (*Chelonia mydas*). *Virology* 321: 101–110.

Stacy, B. A., Wellehan, J. F. X., Foley, A. M., et al. (2008) Two herpesviruses associated with disease in wild Atlantic loggerhead sea turtles (*Caretta caretta*). *Vet. Microbiol.* 126: 63–73.

Diseases of algae and seagrasses

Largo, D. B., Fukami, K. and Nishijima, T. (1995) Occasional pathogenic bacteria promoting ice-ice disease in the carrageenan-producing red algae *Kappaphycus alvarezii* and *Eucheuma denticulatum* (Solieriaceae, Gigartinales, Rhodophyta). *J. Appl. Phycol.* 7: 545–554.

Kapp, M., Knippers, R. and Müller, D. G. (2006) New members of a group of DNA viruses infecting brown algae. *Phycol. Res.* 45: 85–90.

McKone, K. L. and Tanner, C. E. (2009) Role of salinity in the susceptibility of eelgrass *Zostera marina* to the wasting disease pathogen *Labyrinthula zosterae*. *Mar. Ecol. Prog. Ser.* 377: 123–130.

Ralph, P. J. and Short, F. T. (2002) Impact of the wasting disease pathogen, *Labyrinthula zosterae*, on the photobiology of eelgrass *Zostera marina*. *Mar. Ecol. Prog. Ser.* 226: 265–271.

Chapter 12

Marine Microbes as Agents of Human Disease

The overwhelming majority of marine microbes that are naturally present in the sea (indigenous or autochthonous organisms) are nonpathogenic for humans. However, there are several important diseases caused by infection with bacteria naturally associated with marine animals as part of their normal microbiota. Humans become infected when the bacteria colonize the intestinal tract through ingestion or enter the body via wounds, multiply within the body, and produce disease via various pathogenic mechanisms. Such diseases are not normally transmitted from person to person (although cholera provides a very important exception). Humans may also become ill following exposure to toxins produced by bacteria, dinoflagellates, and diatoms, usually via ingestion of fish or shellfish contaminated by the toxins. Zoonoses are infectious diseases of fish and marine mammals that can be transmitted to humans. Infectious diseases are also caused by viruses and bacteria introduced into the sea via sewage pollution; these are discussed in Chapter 13.

Key Concepts

- Several species of *Vibrio* are major human pathogens, whose normal habitat is seawater, plankton, and sediments in coastal and marine environments.

- Interactions with phages, extensive exchange of genetic information, and complex regulatory mechanisms are responsible for the virulence of vibrios in human infection.

- Acquisition of genes encoding virulence determinants via horizontal gene transfer has enabled *Vibrio cholerae*, a member of the normal microbiota of estuarine waters, to become a human pathogen.

- Several types of "food poisoning" are caused by consumption of fish or shellfish containing certain bacteria, their toxins, or metabolic products.

- Zoonotic infections can be acquired by handling infected fish or marine mammals.

- Toxins produced by cyanobacteria, dinoflagellates, or diatoms may cause illness, usually after ingestion of fish or shellfish that have accumulated the toxins.

- The incidence, distribution, and effects of toxic harmful algal blooms are increasing and detection and management of outbreaks is a major activity of public health and fisheries authorities.

Pathogenic vibrios are common in marine and estuarine environments

Most members of the genus *Vibrio* can tolerate a range of salinities and they are therefore widely distributed in marine and estuarine environments throughout the world, especially in warmer waters above 17°C. They are associated with plankton and the surfaces of marine animals and occur in high densities in filter-feeding shellfish. As discussed in Chapter 11, several species are important pathogens of marine fish and invertebrates, but there are also about 12 species that have been associated with human diseases. Of these, the most important are *Vibrio cholerae*, *Vibrio vulnificus*, and *Vibrio parahaemolyticus*, discussed in detail in this chapter. Other species (see *Table 4.1*) are occasionally associated with diarrhea or wound infections, usually following consumption or handling of seafood.

Cholera is a major human disease with a reservoir in coastal environments

The disease cholera (caused by *V. cholerae*) was originally confined to parts of India until the early nineteenth century, when six major pandemics of the classical form of the disease spread to Europe and North America. Throughout the twentieth century, cholera continued to spread to become a major cause of mortality throughout the world, aided by increased intercontinental transport, wars, and natural disasters. The 1960s saw the emergence of the El Tor biotype from the Celebes Islands of Indonesia, which has spread to become the seventh global pandemic. The El Tor biotype has reduced virulence compared with the previous strains, but is better adapted for transmission.

Until quite recently, it is unlikely that cholera would be included in a book about marine microbiology. The classical view of cholera is that it is a disease spread from person to person through fecally contaminated drinking water; this concept developed from John Snow's famous deduction in 1854 that a particular point source (the Broad Street water pump) was responsible for clusters of disease outbreaks in London. Cholera is characterized by profuse diarrhea caused by fluid and electrolyte loss from the small intestine and is therefore easily spread from person to person via contamination of water and food. However, the history of cholera through the ages reveals a striking association with the sea, and epidemiologists have long suspected that there might be an aquatic environmental reservoir of *V. cholerae*. Beginning in the 1970s, work by Rita Colwell and colleagues produced significant findings that changed our thinking about the ecology of *V. cholerae*.

There are 200 serotypes of the bacterium, distinguished by their lipopolysaccharide O-antigens on their surface, although they are biochemically similar. Until 1993, all epidemic cases of cholera were caused by *V. cholerae* possessing antigen O1 (including the El Tor biotype), but in 1993 a new serotype, O139, emerged as the cause of epidemics that began in the Bay of Bengal. Non-O1 and non-O139 strains are commonly isolated from estuaries and coastal waters, even in areas where cholera is not endemic. The survival of the pathogen in water is greatly affected by environmental conditions, particularly salinity, temperature, and nutrient concentration. *Vibrio cholerae* can survive for long periods in seawater, but the numbers that can be isolated from water are often very low and insufficient to initiate infection. Comparison of direct epifluorescence and viable counts shows that a large proportion of the cells enter a type of dormant state termed "viable but nonculturable" (VBNC). Use of molecular methods has confirmed the VBNC phenomenon, and it has subsequently been shown to occur in many other bacteria, although there is controversy about use of the term VBNC and how the state should be defined (see p. 73). During transition to the VBNC state, the cells initiate an active program of change, resulting in reduced size, alteration of mRNA, and changes to the cell surface. This adaptation allows bacteria to survive adverse changes in nutrient concentration, salinity, pH, and temperature. *Vibrio cholerae* associates with a range of phytoplankton and zooplankton, especially copepods, and to egg masses of chironomid flies. Like many other vibrios, *V. cholerae* possesses chitin-binding proteins and chitinase, which play key roles in colonization of surfaces in the environment, as well as the regulation of pathogenic processes (see *Research Focus Box 12.1*).

Vibrio cholerae produces disease in humans owing to production of a toxin and other pathogenic factors

Susceptibility of humans to cholera is affected by a range of genetic (e.g. blood group antigens) and nutritional (vitamin A deficiency) factors. During infection, the bacterium encounters dramatic changes in its environment, particularly the shock of temperature rise and low pH as it encounters the acid of the stomach. There is a complex process of gene regulation as the bacterium responds to chemotactic signals, swims toward the surface of the lumen of the small intestine, and attaches to the epithelium by the production of pili on the bacterial surface. The bacteria then produce a toxin that affects membrane permeability, stimulating loss of electrolytes and water from the small intestine, but it does not lead to permanent damage to the cells. Each molecule of the toxin is composed of two different protein subunits. There are five B-subunits that bind to membrane receptors (carbohydrate-containing lipids known as gangliosides) on the surface of the gut cells, and one A-subunit that is inserted into the host cell membrane. The A-subunit is an ADP-ribosylating enzyme that modifies a host regulatory protein. This results in continued activation of the enzyme adenyl cyclase in the gut cells, leading to increased levels of cyclic AMP, responsible for excessive fluid and electrolyte loss by activating a transmembrane regulator chloride channel. Fluids in the gut cells are replaced from interstitial tissue, plasma, and deeper tissues, and cholera victims usually die from extreme dehydration as fluids flow uncontrolled from the body. The dehydration can be limited by oral rehydration therapy or intravenous drips; in these circumstances, a patient can pass up to 20 liters of fluid a day. However, despite the effectiveness of this treatment, it is often very difficult to deliver during epidemics in field hospitals in remote areas.

Control of cholera remains a major world health problem

Transmission of cholera can be controlled within the human population by good sanitation and clean drinking water, but this is still not available in many parts of the world. Cholera results in untold human misery in Asia and Africa with major epidemics occurring periodically, especially during floods, earthquakes, wars, and other disasters. Although cholera is a disease requiring notification to the World Health Organization (WHO) under international health regulations, the majority of cases are not reported because countries fear damage to trade and tourism. In 2005, nearly 132000 cases and 2300 deaths were officially recorded, but these figures probably represent only 5–10% of actual cases worldwide. Outbreaks in Africa are increasing, and epidemics are showing higher case fatality rates and increased duration.

Despite many efforts, attempts to develop effective vaccines have had only partial success. The WHO has coordinated the use of vaccines in areas where cholera is endemic. Various approaches have been used, including the use of killed cells, purified B-subunits of the toxin, and live attenuated vaccines (in which the active A-subunit of the toxin has been deleted by genetic modification). A combination of killed whole cells combined with purified B-subunits seems to be the most promising. The challenge is to produce a vaccine that is inexpensive to produce in the country where it is needed, easy to administer, and effective under field conditions.

? ARE CHOLERA OUTBREAKS AFFECTED BY THE CLIMATE?

The sudden explosive outbreak of cholera that occurred in South America in 1991 was unexpected as epidemics had not occurred there for over a century. In 1996, Colwell hypothesized that a climatic event was responsible, namely the El Niño Southern Oscillation, which resulted in warmer coastal waters and influx of nutrient runoff from the land following heavy rains. Detailed studies of the epidemiology and environmental factors affecting cholera have also been obtained by extensive field studies in Bangladesh, where a bi-annual seasonal cycle occurs. Regular screening of water for *Vibrio cholerae* and monitoring of phytoplankton and zooplankton blooms, together with analysis of sea-surface temperatures and mean wave height, revealed strong seasonal patterns, which help to explain the frequency and severity of cholera epidemics (review, Lipp et al., 2002). These findings are of particular significance given current concerns about climate change. Human and animal diseases caused by other *Vibrio* species, such as *V. parahaemolyticus*, *V. vulnificus*, *V. shiloi*, *V. coralliilyticus*, and *V. harveyi* also show variations with climatic conditions. Elucidation of the complex ecological interactions occurring in these pathogens is vital if we are to predict and manage changes in the pattern of infectious diseases caused by climate change.

Mobile genetic elements play a major role in the virulence of *Vibrio cholerae*

There is evidence that the ancestor of the modern organism has undergone dramatic shifts in its genetic makeup, which explains its transition from an aquatic nonpathogenic microbe to a human pathogen. Genetic information in *V. cholerae* is divided between two distinct chromosomes; the larger is about 2.9 Mb and the smaller about 1.1 Mb. The large chromosome contains most of the "housekeeping" genes and major pathogenicity factors; these are clustered in distinct regions that have a different G:C base pair ratio to other parts of the chromosomes and are therefore known as pathogenicity islands. These seem to have been acquired relatively recently by gene transfer mediated by phages. The genes encoding the two subunits of the toxin (*ctxA* and *ctxB*) are derived from integration into the chromosome of the genome of a phage (CTXΦ). Outside of areas contaminated by excreta from infected persons, most environmental isolates do not possess the *ctx* genes. Other genes, encoding accessory toxins, are involved in the morphogenesis of CTXΦ. As noted above, the second major factor in pathogenicity is the pili that coat the bacterial surface; because the expression of these is closely linked to toxin production, they are known as toxin coregulated pili (TCP). The *tcp* genes encoding the pili are contained in a 40-kb chromosomal pathogenicity island (VPI-1). This region may also be a phage genome (VPIΦ) and some authorities argue that the pili act as a receptor for CTXΦ, but this idea has not been supported by subsequent analysis of numerous TCP-positive strains in which the phages could not be demonstrated. Nevertheless, the genetic structure of this region suggests that it is a mobile element that may have been recently acquired, perhaps by transformation or generalized transduction by a phage (see p. 160). VPI-1 also encodes ToxT, which is a key regulatory element for the transcription of both *ctxAB* and the *tcp* operon; thus, the remarkable situation occurs in which a regulator on one mobile element controls the expression of a toxin on another. ToxR and ToxS are transmembrane proteins that also regulate pili and toxin production in response to environmental signals. Intriguingly, homologs of both proteins have been found to be widely distributed in the family *Vibrionaceae*, in both pathogenic and nonpathogenic species, so it appears that they evolved in the distant past as part of the ancestral genome. Their function is to control the synthesis of outer-membrane porins that regulate transport of ions and small molecules with their relative abundance in the membrane altered in response to osmotic pressure. In *V. cholerae*, these signals have evolved to control toxin and pilus gene expression through a complex signal transduction cascade involving other membrane signaling proteins that are encoded by the CTX and TCP pathogenicity islands.

The evolution of pandemic strains can be explained by these shifts in genetic composition. Two islands in the genome of seventh pandemic (El Tor) strains (VSP-1 and VSP-2) encode genes responsible for altered carbohydrate metabolism and functions thought to be essential for the characteristic environmental persistence and spread of the El Tor biotype. Another shift involved the horizontal gene transfer of genes encoding enzymes involved in synthesis of lipopolysaccharide; as discussed in *Research Focus Box 12.1*, this led to the emergence of strain O139. The key to this extensive genome evolution is the large integron island (123.5 kb) found on the small chromosome; this is a cluster of genes that captures open reading frames from a wide variety of sources, integrating them by site-specific recombination and rearranging them to form functional genes.

Possibly, in the natural aquatic situation, environmental factors induce the lytic cycle in toxigenic *V. cholerae*, resulting in the release of extracellular phages. Under appropriate conditions, these could infect nontoxigenic bacteria, resulting in the emergence of new toxigenic strains. Passage through the human gut results in enrichment of the virulent forms, and hundreds of genes are expressed *in vivo* which are not expressed outside the host. A model summarizing the ecology of *V. cholerae* is given in *Figure 12.1*.

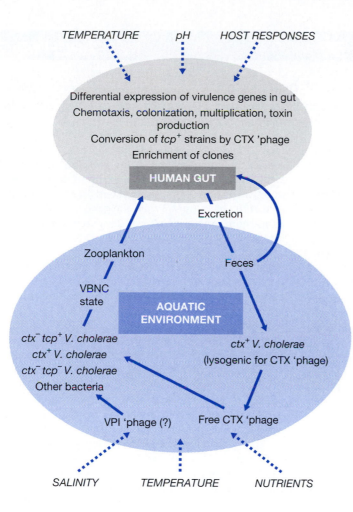

TEMPERATURE   pH   HOST RESPONSES

Differential expression of virulence genes in gut
Chemotaxis, colonization, multiplication, toxin
production
Conversion of *tcp*⁺ strains by CTX 'phage
Enrichment of clones

HUMAN GUT

Excretion

Zooplankton

Feces

VBNC
state

**AQUATIC
ENVIRONMENT**

ctx⁻ tcp⁺ V. cholerae
ctx⁺ V. cholerae
ctx⁻ tcp⁻ V. cholerae
Other bacteria

ctx⁺ V. cholerae
(lysogenic for CTX 'phage)

VPI 'phage (?)   Free CTX 'phage

SALINITY   TEMPERATURE   NUTRIENTS

Figure 12.1 Model for the ecological interactions of *Vibrio cholerae*. The ecosystem comprises the human host, the estuarine or coastal environment, copepods and other plankton, *V. cholerae* and other aquatic bacteria, plus phages mediating gene transfer.

Non-O1 and non-O139 serotypes of *Vibrio cholerae* are widely distributed in coastal and estuarine waters

There are numerous other strains of *V. cholerae* that do not possess either the O1 or O139 antigens characteristic of epidemic cholera strains. These non-O1 and non-O139 serotypes lack the major cholera toxin but they can cause localized mild to moderate gastroenteritis and watery diarrhea, possibly as a result of the production of accessory toxins that damage the membranes or disrupt the cytoskeleton of gut cells. These strains are widely distributed as autochthonous components of coastal and estuarine waters and can survive and multiply in a wide range of seafoods; infections are most commonly associated with eating raw or undercooked seafood, especially oysters. Infections are seasonal and show a peak with increased water temperatures in late summer and early fall, coinciding with the warmest water temperatures. They are also capable of infecting wounds and causing redness and swelling at the site of infection. As with *Vibrio vulnificus* (see below), septicemia due to non-O1 and non-O139 *V. cholerae* can occur in immunocompromised people and in people with underlying liver disease.

Vibrio vulnificus causes serious illness associated with seafood

Vibrio vulnificus is part of the natural microbiota of estuarine and coastal water, sediment, plankton, and shellfish throughout the world. The greatest incidence of infection from consumption of contaminated shellfish or from wounds occurs in the Gulf States of the USA. Increased awareness has recently led to recognition of cases in the Caribbean islands, Japan, and Taiwan, and infections regularly

BOX 12.1   RESEARCH FOCUS

The best of both worlds

How the cholera pathogen uses layers of signaling and regulatory systems to adapt for a life in the aquatic environment and in its human host

The cholera pathogen possesses an amazingly complex set of interconnected signaling and regulatory pathways to maximize its dissemination and survival, both in the host and in the aquatic reservoir (Nelson et al., 2009). Infection results in the copious secretion of water and salts by cells of the small intestine, leading to the characteristic "rice-water stool" containing over 10^7 viable bacteria per milliliter. Merrell et al. (2002) showed that the excreted bacteria are hyperinfectious, with an infectious dose about 100–1000 times lower than laboratory-cultured bacteria. This transient phenotypic change drives the rapid person-to-person transfer during a cholera epidemic, but decays when the bacterium enters the aquatic environment. Through transcriptional analysis of *Vibrio cholerae*, LaRocque et al. (2005) showed that the toxin coregulated pili (TCP) genes were highly expressed early in infection and suggested that the toxin genes are only expressed when the bacterium has passed into the lower part of the small intestine. Further studies by Schild et al. (2007) support the idea that in the late stages of infection, *V. cholerae* expresses genes that prepare it for survival outside of the host. These include genes involved in motility, transport, and regulation of attachment to chitin and its subsequent utilization. Recently, Bourassa and Camilli (2009) showed that storage of glycogen improves persistence in the aquatic environment and transmission to new hosts.

As shown in *Figure 12.1*, lytic and lysogenic bacteriophages play an essential role in the ecology and pathogenicity of *V. cholerae*. The density of lytic bacteriophages increases during epidemics, as large numbers are shed in the rice-water stools (Faruque et al., 2005). When the bacterium enters the aquatic environment, bacteriophage lysis, decline of the hyperinfective state, and induction of the viable but nonculturable (VBNC) state contribute to a decline in transmission.

One of the most intriguing aspects of the aquatic phase of *V. cholerae* is its association with chitin (reviewed by Pruzzo et al., 2008). This polymer is the most abundant polymer in aquatic environments, and many bacteria attach to chitin and secrete enzymes in order to degrade it as a carbon and nitrogen source. *Vibrio cholerae* and other *Vibrio* spp. have numerous adaptations for colonization and utilization of chitin. Meibom et al. (2004) exposed sterile fragments of crab shell to *V. cholerae* and isolated mRNA from the attached and unattached bacteria. The RNAs were labeled with fluorescent markers and applied to a DNA microarray of the *V. cholerae* genome, which showed that when the bacterium attached to chitin, the expression of 104 genes was increased and expression of 93 genes was reduced. The attached bacteria upregulated genes for chitin degradation and protein synthesis, whilst genes for motility and chemo-

taxis were downregulated as the bacteria sensed the chitin surface and attached to it. Li and Roseman (2004) had earlier shown that the regulation of chitin catabolism is due to a two-component sensor system (see p. 76) on the surface of the bacterium. When the gene encoding the sensor kinase component (ChiS) was disrupted, expression profiling enabled the identification of the relevant genes.

Significant advances have recently been made in the laboratory of Bonnie Bassler (Princeton University) and others in understanding how *V. cholerae* uses quorum sensing (QS; see p. 81) to monitor changes in cell population density and to coordinate expression of more than 70 genes involved in biofilm formation and production of virulence factors. Hammer and Bassler (2003) showed that *V. cholerae* mutants, which are "locked" into a regulatory state that mimics high cell density, have impaired formation of biofilms. *Vibrio cholerae* produces two autoinducer molecules, which work in conjunction with a phosphorelay system, and a master regulator protein, HapR. At high cell density, genes associated with virulence are repressed. This is thought to coincide with elimination of bacteria from the host, when toxin and pili are not needed, but chitin association and biofilm formation are. This is a very different "strategy" from the majority of pathogens that use QS to regulate virulence, in which coordinated expression of virulence factors occurs at high cell densities in the host. Higgins et al. (2007) showed that adding a synthetic autoinducer to cultures of *V. cholerae* prevented synthesis of virulence factors, suggesting an opportunity for therapy of cholera infections. Another layer of complexity comes from the finding that QS works in conjunction with a second signaling system involving the small di-cyclic nucleotide molecule di-GMP (Waters et al., 2008).

One aspect of binding to chitin that is of particular importance is that it induces natural competence, that is, the ability to acquire exogenous DNA by transformation. Meibom et al. (2005) showed that a type IV pilus produced on the cell surface can attach to DNA and retract into the cytoplasm. This process is subject to regulation by several systems (including the HapR QS regulator) associated with cell density, declining growth, or stress in biofilm-associated bacteria. By co-cultivating the O1 and O139 serotypes of *V. cholerae* in a biofilm on the surface of chitin, Blokesch and Schoolnik (2007) showed that large segments of DNA may be incorporated into the genome. A cassette of genes that transformed the O1 serotype to the O139 serotype was transferred in a single step and was associated with development of resistance in the transformant to bacteriophages that infected O1 strains. This provides an explanation for the sudden emergence of the epidemic O139 Bengal serotype in 1992, and its rapid spread in India and Bangladesh.

occur in other parts of the world. In healthy individuals, consumption of contaminated raw shellfish (especially oysters) usually results in an unpleasant but not life-threatening gastroenteritis. However, in individuals with chronic underlying diseases (especially alcohol abuse, liver disease, diabetes, or AIDS) a septicemic form of the disease may result, with about 90% of cases requiring hospitalization and a mortality rate of about 50%. Several states in the USA require health warnings to be displayed wherever raw oysters are served. The infective dose is believed to be very low, with estimates ranging from 100 to 300 CFU needed to infect susceptible persons. Males are much more susceptible to infection, possibly owing to hormonal differences. A key feature of this pathogen is its extremely rapid growth in host tissues, and death can occur as little as 24 h after infection. Infected persons show fever, chills, shock, and large skin lesions filled with bloody fluid. These become necrotic and so extensive that surgical removal of tissue or amputation of limbs is often necessary (*Figure 12.2A*).

Vibrio vulnificus is clearly a very aggressive organism able to circumvent many of the body's defense mechanisms, and a number of factors seem to be involved in this extraordinary virulence. An extracellular capsule is especially important in conferring resistance to phagocytosis and the bactericidal effect or complement in the serum. Strains of the pathogen that naturally lack the capsule, or mutants in which the genes encoding the capsule are disrupted, lose their virulence. A second major factor explaining the rapid growth of the pathogen is its ability to scavenge iron as a nutrient from the host. *Vibrio vulnificus* produces powerful cytolytic and proteolytic enzymes that are probably responsible for the extensive tissue damage. In fatal cases, death is primarily due to the effect of endotoxin (LPS) which stimulates overproduction of cytokines with consequent host damage.

Distribution of *Vibrio vulnificus* in the marine environment is affected by temperature and salinity

Vibrio vulnificus can be detected at low levels in seawater from many parts of the world, but it is most common in warm coastal waters, where it can reach densities of up to 10^4 colony forming units per milliliter. Filter-feeding shellfish such as oysters concentrate high levels of the bacterium. There is a strong seasonality to *V. vulnificus* infections, and this correlates with an inability to detect the pathogen in the winter, when the temperature drops below about 15°C. Like *V. cholerae* and many other pathogens, the bacterium enters a VBNC state as a stress response (in this case, reduced temperature rather than starvation). VBNC cells have been shown to be virulent and can be resuscitated and cultured if protected from oxidative stress (see p. 75).

Vibrio vulnificus and other marine vibrios can cause wound infections

As well as causing food-borne disease, *V. vulnificus* can also cause a very serious infection following contamination of wounds. This can occur if existing open wounds are contaminated by seawater containing the pathogen or by wounds such as those from shucking oyster shells, fishing hooks, fish spines, or bones (*Figure 12.2B*). The pathogen produces extensive tissue necrosis without the predisposing conditions described above, and sometimes progresses to septicemia. Many other vibrios are also responsible for infection of "coral cuts," which are a common hazard of scuba diving. Unless treated promptly, these wounds may take many weeks to heal and can sometimes lead to cellulitis, which may be fatal. Organisms such as *Vibrio alginolyticus*, *Vibrio coralliilyticus* and *Vibrio harveyi*, which are commonly associated with corals, multiply rapidly at 37°C and produce powerful proteolytic enzymes and other virulence factors that can cause tissue necrosis.

ⓘ IRON ACQUISITION IS IMPORTANT IN BACTERIAL INFECTIONS

All bacteria require iron for the activity of essential cellular functions. Vertebrate animals possess highly efficient systems for the transport and sequestration of iron. The serum protein transferrin binds iron with extreme avidity, reducing the concentration of free iron in tissues to 10^{-18} M, about 10^8 times lower than the concentration that bacteria require for growth. During infection, an even more efficient iron-binding protein (lactoferrin) is released, and this removes iron to storage in the liver and other organs. Thus, successful pathogens must compete with the host's iron-binding system in order to obtain sufficient iron for growth. *Vibrio vulnificus* is particularly likely to cause septicemic infection in people who have underlying disease involving liver damage because the iron-storage and transport systems are disturbed, leading to excess levels of iron in the serum, allowing the bacteria to grow so quickly that they overwhelm the body's defenses (Jones and Oliver, 2009).

Figure 12.2 Consequences of *Vibrio vulnificus* infection. (A) Gangrene and hemorrhage in a patient with septicemia following infection by ingestion. (B) Infection of a fish bone injury from which septicemia developed. (C) Gram-negative curved bacilli isolated from a blood sample. (Reprinted from Hsueh et al. [2004].)

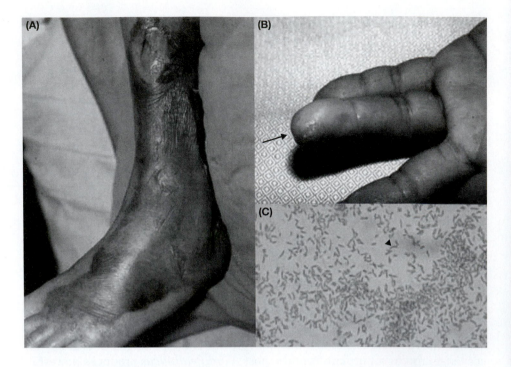

Seafood-borne infection by *Vibrio parahaemolyticus* is common throughout the world

Vibrio parahaemolyticus occurs in marine and coastal waters throughout the world, colonizing the surface of many types of marine animals and plankton. It is a leading cause of gastroenteritis associated with the consumption of seafood and is the commonest cause of food poisoning in Asia (especially Japan), because of the popularity of raw seafood. *Vibrio parahaemolyticus* is becoming more important in the USA, Australasia, and Europe, owing largely to the increasing international trade in seafood products, but also possibly to rising seawater temperatures. After an incubation period of 12–24 h, infection results in watery diarrhea, abdominal cramps, nausea, vomiting, headache, fever, and chills. Most infected persons recover within a few days, but antibiotic treatment may be necessary for more serious infections. Mild cases are usually not reported, so the true incidence of infection is unknown.

The disease is associated with raw or incompletely cooked seafood (particularly shrimp, crabs, and bivalve mollusks). It is thought that a large infective dose is required for sufficient numbers to survive the acidity of the stomach (use of antacid medicines may reduce this), and outbreaks often follow the cross-contamination of seafood that has been kept in a warm kitchen. The doubling time of the bacterium at 37°C can be as low as 10 min, permitting a massive expansion of the population within a few hours.

Although high numbers of *V. parahaemolyticus* can usually be recovered from harvested fish and shellfish, especially those from estuarine and coastal environments, only a small proportion of isolates appear to carry factors that make the bacterium pathogenic. Isolates from environmental samples are usually much less virulent than those isolated from clinical samples, so it is likely that there are either multiple types of the organism, or that it undergoes genetic changes and selective enrichment in the gut as seen with *V. cholerae*. The exact mechanisms by which the bacterium produces disease remain unclear, despite extensive research. Most strains isolated from clinical samples are "Kanagawa positive"—this is a reaction on a specialized type of blood agar—because they possess a thermostable direct hemolysin (TDH) and/or a related toxin (TRH). By contrast, most environmental isolates express neither TDH nor TRH, even if they contain the *tdh* or *trh* genes, which can be identified using gene probes. There are numerous serotypes of *V. parahaemolyticus*, characterized by the O (LPS) and K (capsular) antigens. Since the mid-1990s, a clonal pandemic strain O3:K6 has spread throughout the world and is now the cause of most seafood-associated outbreaks. Genomic analysis of *V. parahaemolyticus* shows that there are several genetic regions characteristic of pathogenicity islands, mainly present in the O3:K6 pandemic strains. It seems that these strains have evolved within a short time frame by acquiring large regions of new DNA, and this may explain their prevalence as human pathogens.

Further insight into the mechanisms of pathogenicity of *V. parahaemolyticus* came with the recognition of genes for two type III secretion systems (TTSS-1 and TTSS-2) in the first sequenced genome. Like other vibrios, *V. parahaemolyticus* has two chromosomes. TTSS-1 is encoded on the large chromosome and is present in environmental and clinical isolates, whereas TTSS-2 is encoded on the small chromosome in a pathogenicity island that also encodes TDH, and is present only in clinical isolates. All isolates of the O3:K6 pandemic serotype contain the genes for TTSS-2 and TDH. The TTSS can be likened to a molecular syringe which pathogens use to inject proteins into host cells; this is known to be a major factor in the pathogenicity of other enteric pathogens such as *Shigella, Salmonella*, and enteropathogenic *Escherichia coli*. The TTSS-1 of *V. parahaemolyticus* injects a number of proteins that induce autophagy in the host cell (a process of self-digestion of cell contents), followed by rounding and release of nutrients for the pathogen on cell death (*Figure 12.3*), whereas TTSS-2 is thought to be responsible for enterotoxicity.

Scombroid fish poisoning is a result of bacterial enzyme activity

This type of food-borne intoxication is associated with eating fish of the family *Scombridae*, which include tuna, mackerel, marlin, and bonito. The tissue of these fish contains high levels of the amino acid histidine. If there is a delay or breakdown at any stage in the refrigeration process between catching the fish and consumption, bacteria from the normal microbiota can multiply and convert histidine to histamine (*Figure 12.4*). Bacteria isolated from fish associated with scombroid fish poisoning include *Raoultella planticola, Morganella morganii, Hafnia alvei*, and *Photobacterium phosphoreum*. The histamine is heat

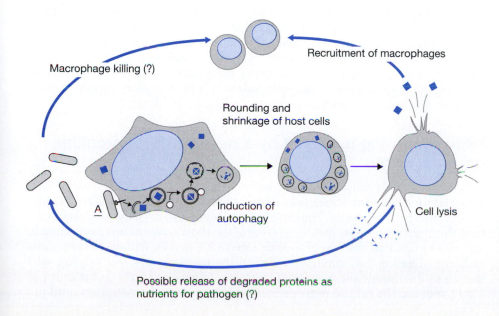

Macrophage killing (?)

Recruitment of macrophages

Rounding and shrinkage of host cells

Induction of autophagy

Cell lysis

Possible release of degraded proteins as nutrients for pathogen (?)

Figure 12.3 *Vibrio parahaemolyticus* induces a series of events that culminates in the efficient death of host cells. (Modified and reproduced from Burdette et al. [2008] by permission of the National Academy of Sciences.)

Figure 12.4 Enzymatic conversion of histidine to toxic histamine. (Reproduced from McLauchlin et al. [2006] by permission of Oxford University Press.)

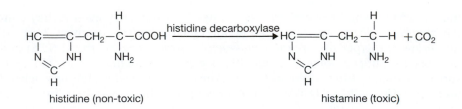

histidine (non-toxic)   histamine (toxic)

stable and will withstand normal processing such as canning. The spoilage may not be enough to alter the taste or smell of the fish, but levels of histamine above 100 mg per 100 g can be sufficient to induce an allergic-type response. Reddening of the face and neck, shortness of breath and, in severe cases, respiratory failure can result within a few minutes of eating contaminated fish. Treatment with antihistamine drugs is beneficial and the person usually recovers fully. In Europe, the incidence of scombroid poisoning is increasing because of the rising popularity of sushi and the import of fresh tuna and swordfish by airfreight over long distances, with opportunities for breaks in the cold chain.

Botulism is a rare lethal intoxication from seafood

Botulism is one of the most feared food-borne diseases. It is caused by *Clostridium botulinum*, a Gram-positive bacterium that forms endospores that can survive high temperatures. The bacterium is a strict anaerobe and under appropriate conditions it produces a powerful neurotoxin in foodstuffs. Botulinum toxin is the most lethal toxin known; if ingested, tiny amounts of the toxin (estimated to be about 200 ng) cause a severe flaccid paralysis. The toxin is a protease that cleaves proteins involved in the docking of neurotransmitter vesicles, blocking release of the neurotransmitters and therefore preventing muscle contraction. *Clostridium botulinum* is normally incapable of growth within the adult gut, so disease is dependent on ingestion of preformed toxin, with symptoms usually occurring after 12–36 h. Unlike other fast-acting neurotoxins, the delay is caused by absorption of the toxin and transport via the central nervous system. (Note that the bacterium can grow in the gut of babies, and infant botulism is typically associated with spore-contaminated honey.) *Clostridium botulinum* is classified into types A to G based on the serological properties of the toxin produced. Type E is most commonly associated with seafood and can be isolated from marine sediments and the intestinal contents of fish and crustaceans. Fortunately, botulism is rare and most cases occur from home-prepared or ethnic foods. Botulism is a serious health problem in Arctic Inuit communities because of the use of traditional methods for preparing fermented salmon eggs or whale and seal meat. Several outbreaks have occurred following consumption of smoked fish in modified atmosphere packaging (see *Figure 13.3*), since smoking kills most but not all the spores; such products should always be refrigerated to inhibit growth of any survivors. Most salting and drying methods of fish preservation are safe. Wound botulism can occur in workers handling fish, causing local paralysis.

Fugu poisoning is caused by a neurotoxin of probable bacterial origin

This intoxication is caused by the ingestion of tetrodotoxin (TTX; *Figure 12.5A*), which is found in the intestines, liver, and gonads of pufferfish, especially *Takifugu rubripes* and related species. The toxin is probably synthesized by bacteria inhabiting the intestine of these fish. TTX is one of the most active neurotoxins known, and acts by blocking the flow of sodium ions in the nerves. Within a few minutes or hours of eating fish contaminated with the toxin, victims feel tingling sensations in the mouth and a sense of lightness, followed quickly by the onset of total paralysis. The person remains conscious but totally immobilized until the

---

RUMMAGING IN THE RUBBISH REVEALS CAUSE OF BOTULISM OUTBREAK

Commercial canning, the most common method of preserving fish, is designed to ensure destruction *Clostridium botulinum* spores. The "12 D process" of cooking at high temperature ensures a 10^{12}-fold reduction in spore numbers, offering a very high safety margin. However, one famous outbreak of botulism in 1978 in Birmingham, England, involved one of the world's best known fish-processing companies. In this case, four people were infected (two died) following consumption of part of a can of Alaskan salmon. Careful investigation by public health officials resulted in the discarded can being recovered from a rubbish bin. It was found to have a minute hole in the seal. Despite being properly heated, the investigators concluded that a small amount of contaminating material from fish viscera was sucked into the can as it cooled on the production line, and this contained enough viable spores to germinate and produce a lethal dose of toxin. The economic consequences for the company were very serious because of the costs of product recall and adverse publicity.

(A)

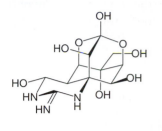

(B)

Figure 12.5 (A) The structure of tetrodotoxin. (B) A typical fugu restaurant in Tokyo, where live puffer fish are displayed in tanks, to be freshly prepared by licensed chefs.

moment of death, which occurs in up to 50% of cases. In Japan, fugu is a prized delicacy for which diners in specialist restaurants (*Figure 12.5B*) are prepared to pay large amounts for the thrill of eating tiny portions of this risky food, with the added frisson of ingesting a miniscule dose of the toxin in order to give a "buzz." Fishing for *T. rubripes* and the sale of fugu are now tightly regulated in Japan and fugu chefs must be specially licensed, so most of the 50 or so cases each year occur among fishermen and amateur cooks who prepare the dish. There are a few licensed fugu restaurants in the USA, but it is banned in Europe. Fish farmers in some parts of Japan have recently been producing nonpoisonous fugu by rearing fish on special diets. The development of safe fugu has met with problems of consumer acceptance and controversy about whether rules prohibiting the sale of fugu liver ("kimo," the most expensive and prized delicacy) should be relaxed. Fugu has very deep cultural and economic significance in Japan!

Some diseases of marine mammals and fish can be transmitted to humans

Contact between humans and animals such as whales and seals has existed for a very long time in hunting communities, and in these cases there have been some well-described zoonoses of microbial origin. Infection by metazoan parasites acquired from consumption of whale and seal meat is also important. Bacterial zoonoses include "seal finger," an extremely painful infection of the hands caused by *Staphylococcus* or *Erysipelothrix* which occurs in seal hunters and research workers, especially in the Arctic regions of Norway and Canada. Seal bites can also cause serious infections, caused by a variety of poorly characterized bacteria, and are notoriously difficult to treat unless antibiotics are used promptly. Large outbreaks of *Salmonella* infection, causing diarrhea and vomiting, have occurred in Inuit communities that have eaten infected whale or seal meat. Research workers and aquarium staff may be at special risk of contracting disease. There have been a few cases of infection by *Mycobacterium* sp. (causing respiratory and skin lesions) and *Brucella* infection (causing severe lethargy, fever, and headaches) in such people. Of the viruses, influenza poses the most important risk and several instances of direct transmission from persons in close contact with infected seals have occurred. Evolution of new strains of influenza virus occurs regularly through antigenic changes, owing to recombination events in the fragmented genome. It is therefore possible that marine mammals could provide an opportunity for transmission of new variants to humans. Protozoan parasites such as *Giardia* may have a reservoir in marine mammals.

? DO SYMBIOTIC BACTERIA MAKE DEADLY TOXIN?

Tetrodotoxin (TTX) is widespread amongst marine animals, including pufferfishes, triggerfishes, sunfishes, xanthid crabs, seastars, and marine snails, in which its main function seems to be in defense against predators. A few species of terrestrial salamanders, newts, and toads also produce the toxin. Animals that contain TTX appear to have single-point mutations in the amino-acid sequence of a sodium-channel protein, which makes them resistant to the toxin's effects. Some marine animals also use TTX as a mechanism of attack. The blue-ringed octopus (*Hapalochlaena maculosa*) of northern Australia contains TTX in its venom, which it uses to paralyze its prey of fish and crustaceans, and in all parts of the animal's body (Yotsu-Yamashita et al., 2007). A tropical flatworm, *Planocaeta multicentaculata*, uses TTX to paralyze its prey of cowrie mollusks (Ritson-Williams et al., 2006). What is the explanation for the wide distribution of such a complex molecule with no critical metabolic role? It seems unlikely that the toxin is synthesized by the animals themselves. The most likely explanation is that it is acquired through the food web. This is supported by the fact that farmed pufferfish that have been fed special diets are nontoxic (Noguchi et al., 2006; Noguchi and Arakawa, 2008). Various isolates of *Vibrio, Pseudomonas, Alteromonas, Shewanella, Flavobacterium*, and other bacteria isolated from TTX-containing animals are reported to produce the toxin in culture, although there is much controversy about these results (Lee et al., 2000; Matsumura, 2001). The functions of TTX for these bacteria and the evolution of this possibly symbiotic relationship are unknown.

There is a definite, albeit small, risk of acquiring diseases from marine mammals, and marine conservation organizations have highlighted the need for caution when engaging in the rescue of stranded animals. Of some concern is the increasing popularity of tourist activities such as "petting zoos" and swimming with dolphins, which could lead to a rise in transmission of zoonoses. Such activities also increase the likelihood of transmission *from* humans *to* marine mammals, although there is limited evidence of this to date.

Some diseases of fish may also be transmitted to humans, usually via skin abrasions or wounds. Infection by *Mycobacterium marinarum* (fish tank granuloma) is an important hazard for aquarium workers. It usually causes lesions on the hands, but can cause more serious infections of the bones and joints in people with impaired immunity. *Streptococcus iniae* and *Edwardsiella tarda* have been responsible for small outbreaks of serious invasive infections in workers handling infected warm-water fish from aquaculture facilities. In addition, there are a very large number of parasitic infections of fish caused by flukes, helminths, and nematodes that cause human disease following ingestion of undercooked fish.

Toxic dinoflagellates and diatoms pose serious threats to human health

The growth of marine plankton is affected by a wide range of factors that influence their spatial and temporal distribution. Seasonal periodic increases in plankton density (blooms) obviously have great ecological importance in ocean food webs. In addition, some planktonic microbes produce toxins that affect human health, and many of these can form exceptional blooms under certain conditions. These are frequently referred to as "red tides," although this colloquial term is something of a misnomer since not all toxic blooms are red in color or may not even reach sufficient densities to discolor the water. A more generally accepted term is "harmful algal blooms" (HABs). This term is also not entirely satisfactory as it includes excessive growth of nontoxic algae and because health effects are not always associated with a distinct "bloom." Among the many thousands of species of phytoplankton, only about 150 are known to produce toxins, and only a few of these actually cause problems for human health. Examples of some of the main toxic dinoflagellates and diatoms are illustrated in *Figure 12.6*. The main health hazard for humans comes from eating fish or shellfish that have accumulated the toxin in their tissues through feeding in water containing high levels of toxin-producing plankton. Some toxins may also produce disease symptoms as a result of direct contact with contaminated water or aerosols. Representative structures of the major toxins are shown in *Figure 12.7*. All of the toxins are nonproteinaceous substances that remain active after cooking. This, together with the usually very rapid onset and neurological symptoms, is an important distinction of HAB intoxications from most bacterial and viral infections associated with consumption of shellfish. Most of the toxins inhibit transmission of the nerve impulse by interfering with the passage of sodium ions through the nerve cells.

Figure 12.6 Scanning electron micrographs of some toxic diatoms and dinoflagellates. (A) *Pseudonitzschia pseudodelactissima*; (B) *Karenia brevis*; (C) *Dinophysis fortii*; (D) *Gambierdiscus toxicus*. (Images courtesy of Paula Scott, Florida Fish and Wildlife Conservation Commission.)

Paralytic shellfish poisoning is caused by saxitoxins produced by dinoflagellates

Paralytic shellfish poisoning (PSP) has a worldwide distribution and is the best known human illness associated with HABs. *Alexandrium* spp. and *Gymnodinium catenatum* have been described for many years in temperate waters of North America, northern Europe, Scandinavia, and southeast Asia, but in the past 25 years these taxa have been increasingly reported from countries in the southern hemisphere, including Australia, New Zealand, and Chile. This is thought to be due to spread via transport in ballast water of large ships. In tropical waters, PSP is largely caused by *Pyrodinium bahamense*. PSP results from eating shellfish, especially clams, oysters, mussels, and certain species of crabs, in which toxins have built up through filter feeding. Crabs and lobsters can also accumulate the toxin by feeding on contaminated bivalve mollusks. There are about 20 related water-soluble toxins responsible for PSP; these are known as saxitoxins.

PSP has a dramatic onset, often within minutes of consuming contaminated shellfish. Tingling around the lips is quickly followed by numbness of the face and neck, nausea, headache, and difficulty in speech. In severe cases, muscular and respiratory paralysis can occur, and mortality can be more than 10% unless there is rapid access to medical services. Treatment consists of stomach pumping and administration of charcoal to absorb the toxins. In severe cases, artificial respiration may be required, but there are usually no long-lasting effects.

Management of paralytic shellfish poisoning depends on assaying toxins in shellfish

Many countries operate PSP-management programs, which include monitoring seawater for the density of toxin-producing dinoflagellates and the assay of shellfish samples for saxitoxins. Careful analysis has shown that there does not seem to be a straightforward relationship between the presence of toxin-producing dinoflagellates and the level of particular toxins in shellfish tissue. Shellfish

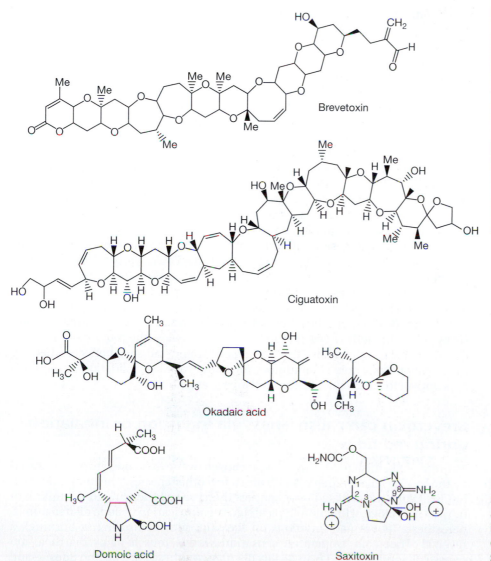

Brevetoxin

Ciguatoxin

Okadaic acid

Domoic acid

Saxitoxin

Figure 12.7 Representative structures of major dinoflagellate and diatom toxins. Note that there is considerable variation of structure of brevetoxins, ciguatoxins, and saxitoxins.

Figure 12.8 Principle of receptor binding and cell culture assays for sodium-channel-blocking toxins. (A) Native toxin (*blue triangles*) in a standard or the sample to be tested will compete with a radioactive analog of the toxin (*triangles with flash*) for binding sites on the receptor. By measuring the amount of radioactivity bound, a standard curve is generated that can be used to measure the amount of toxin. Membrane preparations from rat (for PSP) or frog (for ASP) brain are used as a source of receptor; recently, a cloned glutamate receptor for ASP has been expressed in cultured cells. (B) In the MIST® cytotoxicity assay, cultured mouse neuroblastoma cells are treated with the alkaloids ouabain (inhibitor of Na$^+$/K$^+$ ATPase) and veratridine (Na$^+$ activator). In the presence of Na$^+$ channel-blocking toxins, the action of veratridine is inhibited. The intensity of the color in viable cells is inversely proportional to toxin concentration and is read in a plate colorimeter in high-throughput assays using 96-well microtiter trays.

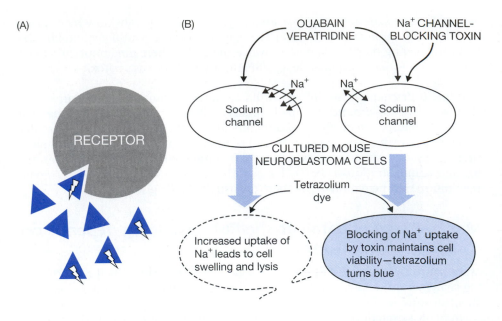

may modify the toxins or excrete them differentially according to climatic and physiological conditions, making management of outbreaks more difficult. Commercial collection of shellfish is prohibited when saxitoxin levels in the tissue exceed a certain threshold concentration (typically, 800 μg kg^{-1}). The traditional, internationally accepted method involves intraperitoneal injection of an extract of shellfish meat into mice. The toxin is quantified as "PSP equivalents" by the time taken for the mice to die, but the assay does not give any information about the precise amounts of individual toxin components or their breakdown products. There have been many attempts to develop alternative methods that do not depend on the use of animals and that give more reliable and informative results. High-pressure liquid chromatography (HPLC) is now recognized as the official method in the European Union, but its use has not been approved in some countries because it is time-consuming and technically complex, with problems arising from the lack of standards. Assays based on receptor binding and measuring the effects of toxins on sodium transport in cultured cells have also been developed (*Figure 12.8*). These are useful in research, but have not found wide application for routine monitoring. A commercially available "dipstick" test based on polyclonal antibodies to detect PSP toxins has been used for several years and offers many advantages as a simple test for screening shellfish for public health protection, as it rarely gives false negative results. However, in validation studies, the test has been reported to show false positive results, and subsequent analysis by other methods shows that toxin levels are below the required limits. The closure of fisheries can have devastating economic and social effects, affecting the livelihood of whole communities for many years, so validation of methods appropriate to a particular fishery and area is essential.

Brevetoxin can cause illness via ingestion or inhalation during red tides

Neurotoxic shellfish poisoning (NSP) is caused by the toxin brevetoxin, produced by the dinoflagellate *Karenia brevis* and a few other species. The lipophilic toxin binds to voltage-gated sodium channels in nerve cells, leading to disruption of normal neurological function. Typical signs of intoxication include dilated pupils, paresthesia (abnormal sensations on the skin, such as tingling or burning), a reversal of hot–cold temperature sensation, vertigo, muscle pains, diarrhea, and nausea. Blooms of *K. brevis* are highly distinctive—these are true "red tides"—and are usually seasonal, starting in late summer and lasting for many months. They have been known in the Gulf of Mexico, especially the Florida coast, for several centuries; however, in recent years blooms have occurred more frequently and have been of longer duration. Since the late 1980s, blooms have also occurred on the eastern US coast and in New Zealand, probably involving other *Karenia* spp. Like PSP, NSP is caused by consumption of shellfish that have accumulated the toxin, although symptoms are usually milder. An additional effect occurs because the blooms can be so extensive that lysis of *K. brevis* cells and disruption by surf action leads to aerosols containing the toxin. Toxic aerosols can be carried by wind up to 100 km inland, causing shortness of breath and eye irritation. In Florida, beach closures are necessary during red tides and health advisory notices are issued warning asthmatics and other susceptible people to stay indoors. The toxin also kills marine mammals, birds, and turtles (see p. 248) and the blooms also cause massive fish kills, leading to further problems as coastal waters become anoxic. Thus, red tides have a major impact on fisheries and tourism, and government agencies are involved in close monitoring using satellite imagery and other techniques, so that precautionary management procedures can be put in place at the first signs of a bloom.

Diarrhetic shellfish poisoning and azaspiracid poisoning result in gastrointestinal symptoms

Diarrhetic shellfish poisoning (DSP) is caused by eating shellfish (usually bivalve mollusks) that contain the toxin okadaic acid and possibly some other toxins produced by dinoflagellates of the genera *Dinophysis* and *Prorocentrum*. Okadaic acid is an inhibitor of protein phosphatases and leads to symptoms of vomiting, diarrhea, and abdominal pain, which usually lasts for a day or two. The symptoms resemble those caused by viral infection from shellfish grown in sewage-polluted waters (see p. 296), but a growing body of evidence since the 1980s has shown that DSP is caused by the toxin okadaic acid and can result from consuming shellfish from waters free of sewage contamination. Many outbreaks have been described in Japan and in European countries, particularly on the Atlantic coast of Spain, France, and the British Isles. A European Union Directive requires the monitoring of shellfish for the toxin (by HPLC assay), and this has resulted in closure of many traditional shellfish harvesting and aquaculture operations in areas where toxins have been detected. Unlike the other shellfish poisonings, a DSP risk can occur even in the absence of an exceptional increase in plankton density; *Dinophysis* densities as low as 200 cells per milliliter can lead to accumulation of unacceptable levels of toxin in shellfish. Okadaic acid has been shown to be carcinogenic in rats, and long-term exposure has been suggested as a possible cause of cancer in the digestive tract, although there seems to be little epidemiological evidence to raise human health concerns.

In 1995, an outbreak of disease associated with mussels harvested from Irish waters was linked to the consumption of mussels (*Mytilus edulis*), and a new group of closely related toxins called the azaspiracids was shown to be responsible. Since then, the incidence of azaspiracid poisoning seems to be rising and it has been reported throughout Europe and in Canada. The symptoms are severe diarrhea and vomiting, nausea, and headache, which may persist for several days. In a mouse bioassay, azaspiracids cause death from serious tissue and organ damage, with neurotoxic effects at high doses. However, their mode of action seems quite different from the other shellfish toxins. Development of improved chromatographic assays means that routine monitoring for azaspiracids is now being introduced, with the result that they are much more widespread than previously thought. The toxins seem to be associated particularly with *M. edulis* and there are no obvious links with bloom events. Rather, it seems as if the toxin accumulates gradually in the tissue of the mussel as a result of long-term exposure to low levels of toxic microalgae (possibly *Protoperidinium* sp.).

Amnesic shellfish poisoning is caused by toxic diatoms

Amnesic shellfish poisoning (ASP) was discovered in 1987, following an unusual outbreak of seafood poisoning in Prince Edward Island, Canada, in which more than 100 people were affected and three died. The symptoms included rapid onset of vomiting, disorientation, and dizziness that did not match those of other known diseases. During the investigation, it emerged that those affected had suffered short-term memory loss, which persisted for many months. This suggested that a new type of neurotoxin was involved, and this was identified as domoic acid produced by the diatom *Pseudo-nitzschia multiseries*. We now know that several other species of *Pseudo-nitzschia* can produce domoic acid and these have been isolated from many parts of the world. Human cases of ASP are rare, but they are almost certainly underreported, and in areas where *Pseudo-nitzschia* blooms occur they may be associated with eating planktivorous fish such as anchovies. As discussed on p. 249, domoic acid from *Pseudo-nitzschia* blooms also causes disease in marine birds and mammals and is increasingly recognized in the marine food chain.

Ciguatera fish poisoning has a major impact on the health of tropical islanders

In terms of public health, ciguatera fish poisoning (CFP) or "ciguatera" is undoubtedly the most important of the diseases caused by marine toxins, especially for inhabitants of tropical islands who rely on fishing. It is also a well-known hazard among sailors and travelers in the tropics (35°N–35°S) and was first documented by explorers of the Caribbean and Pacific in the fifteenth and sixteenth centuries. In 1774, Captain James Cook gave a detailed account in his ship's log of poisoning that affected him and his entire crew after eating fish in the New Hebrides, which was almost certainly a description of CFP. The causative agent of CFP was not discovered until 1977, when the toxin-producing dinoflagellate *Gambierdiscus toxicus* was isolated in the Gambier Islands in French Polynesia. Related species occur worldwide, although isolates vary greatly in the amount and types of toxin produced. Cyanobacteria have also recently been implicated in some outbreaks. The fat-soluble toxins become concentrated as they pass up the food chain, and disease is almost always associated with eating large predatory fish. More than 400 species of fish have been implicated in CFP, with moray eel and barracuda being the most common. CFP is characterized by a distinctive sequence of diarrhea, vomiting, abdominal pain, and neurological effects within a few hours of eating contaminated fish. Unusual symptoms include numbness and weakness in the extremities, aching teeth, and reversal of temperature sensation (cold things feel hot and hot things feel cold). In severe cases, this can progress rapidly to low blood pressure, coma, and death. However, the type and severity of symptoms depends on the particular "cocktail" of toxins consumed, and this varies with species of fish and geographic origin. There seem to be important differences in structure between the main CFP toxins implicated in Pacific and Caribbean cases. Doses that cause severe symptoms in one person may be harmless in another. It is also likely that other ancillary toxins are involved. In some people, neurological problems can persist for many years, and a repeat attack can be triggered by eating any fish (whether or not it contains toxin) or by drinking alcohol. The toxin has also been shown to cross the placenta to affect the fetus, and there are even reports of transmission to another person via sexual intercourse. In countries where the disease is endemic, there are a range of local remedies, but these are of unproven value. In severe cases, it is necessary to administer intravenous mannitol, which reverses the effect on sodium transport.

The incidence of CFP can be quite localized but unpredictable. Local fishermen will often believe that fish from a particular island or reef will be hazardous, whilst others nearby are not. Many travelers have learnt, to their cost, that

---

? DID CIGUATERA CAUSE THE MIGRATION OF PACIFIC ISLANDERS?

Between AD 1000 and 1450, multiple waves of migration occurred between the island of Polynesia and distant lands, such as New Zealand. The reasons why islanders undertook such perilous journeys across large expanses of ocean have always been unclear. By studying past climate conditions and examination of archaeological artifacts, a native of the Cook Islands has suggested that the migrations may have been due to increased chronic incidence of ciguatera affecting the population, who were heavily dependent on ciguatera-susceptible fish at the time (Rongo et al., 2009). Reliance on fishing in the islands declined after this time and the remaining islanders seem to have moved to "safer" small fish species, rather than large carnivorous fish. However, ciguatera still occurs today and modern outbreaks have led to mass emigration and reliance on processed foods rather than fish.

such claims are not always reliable. *Gambierdiscus toxicus* adheres to macroalgae on the surface of dead coral and on the seabed, but little is known about the factors promoting colonization of the macroalgae by the dinoflagellates and its subsequent proliferation and toxicity. Damage to the reef, for example by fishing, diving, boat activities, military action, mining, or construction work, promotes the initial colonization by the macroalgae. Increased frequency of bleaching and diseases of corals may also be leading to higher occurrence of CFP. However, this is only part of the story, as the incidence of toxic fish is very dependent on environmental factors such as rainfall and nutrient runoff from the land. Also, it is likely that there are genetic differences among different strains of *G. toxicus* and factors that affect dinoflagellate growth may have a different effect on toxin production. Recently, reports of ciguatera-like illness associated with dinoflagellates in the genus *Ostreopsis* have been reported in Italy and Puerto Rico. The complex ecological, toxicological, and physiological factors seen with CFP illustrate dramatically how careful we must be when looking for a simple "cause and effect" in the etiology of a disease.

Dinoflagellates and diatoms probably produce toxins as antipredator defense mechanisms

The toxins produced by dinoflagellates and diatoms are usually complex secondary metabolites, and diversion of energy to their synthesis implies some ecological benefit. The extreme sensitivity of humans and other animals to these toxins and the fact that most act on similar mechanisms of nerve action is presumably coincidental. It is usually presumed that toxins have evolved to deter predation by zooplankton. If this is correct, once physical and chemical conditions are favorable for the initial bloom of a toxic species, the production of toxins will deter predation and prolong the maintenance of high densities of the species. It seems that many zooplankton species, such as copepods, can discriminate dinoflagellates with low toxin content and feed selectively, and some flagellates and ciliates appear to be killed by dinoflagellate toxins. However, toxins do not give universal protection against grazing by all species of predator, and it may be that the toxins offer a selective advantage to organisms that produce them under low-nutrient conditions, by directing predation pressure onto competitors. It is notable that the production of toxin, and its cellular concentration, is very dependent on the supply of nutrients (especially phosphorus). This has led to the suggestion that the toxins may act as a reserve for storage when the nutrient supply is unbalanced. Many dinoflagellates have the capacity for both phototrophic and heterotrophic nutrition, and the production of such secondary metabolites may be associated with major switches in metabolic pathways. The possible involvement of bacteria in toxin synthesis has attracted considerable interest and this is considered in *Research Focus Box 12.3*.

The incidence of harmful algal blooms and toxin-associated diseases is increasing owing to the interaction of many complex factors

Fossil studies show that periodic blooms have occurred for millennia and that these have been linked to mass mortalities of marine life in the past. Some phenomena like the Florida *K. brevis* red tide and human diseases like PSP and CFP have been described for many years, but there is little doubt that the reported frequency and distribution of harmful algal bloom (HABs) have increased markedly in the past few years. For example, until the 1970s, PSP-producing blooms were known only in the temperate waters of North America, Japan, and Europe, but since 1990 PSP outbreaks have been reported increasingly in the southern hemisphere. As noted above, apparently new problems like ASP, DSP, azaspiracid poisoning, and *Pfiesteria*-associated illness have emerged in the last few years.

BOX 12.2   RESEARCH FOCUS

The controversial "cell from hell"
The effects of *Pfiesteria* dinoflagellates on fish and human health

The story of *Pfiesteria piscicida* and related organisms forming the "*Pfiesteria* complex" is an exciting example of scientific detective work surrounded by conflicting views, with important political and social repercussions (Kaiser, 2002). The story began in 1988, when water from the Pamlico estuary appeared to cause death of fish in aquaria at the North Carolina State University. After death of the fish, accompanied by hemorrhagic sores, the dinoflagellate seemed to disappear until more fish were added to the aquaria, leading to its description as a toxic "phantom dinoflagellate" linked to massive fish deaths in the estuaries of North Carolina (Burkholder et al., 1992, 1995). Unidentified chemical signals from fish (possibly excreta or secretions) were thought to stimulate the formation of toxin-producing zoospores. The organism was named *Pfiesteria piscicida*—in honor of the phycologist Lois Pfiester, with the species epithet meaning "fish killer"—and reported to have an unusually complex life cycle including at least 24 flagellated, amoeboid, and encysted stages (Steidinger et al., 1996).

In 1997, large fish kills occurred in the Chesapeake Bay estuary, causing severe losses to the Maryland fishery. Fishermen and others who had regular contact with the estuary reported various skin ulcers, mental effects, and other health problems, suggesting that chronic exposure to a toxin from *P. piscicida* might be responsible (Grattan et al., 1998) and leading to a new nickname—"the cell from hell." This caused great public concern, with intense coverage on television, in newspapers, and in a dramatically titled popular book, *And the Waters Turned to Blood* (Barker, 1997). Laboratory workers who handled dilute cultures of *Pfiesteria* had also reported symptoms of changes in personality and cognitive function, lethargy, and breathing difficulties, which were thought to be due to exposure to the putative *P. piscicida* toxin through aerosol formation or skin contact (Glasgow et al., 1995).

Many aspects of the biology and proposed toxicity of *P. piscicida* have led to unparalleled controversy among scientists working in this field (see Kaiser, 2002; Trainer, 2002). Using molecular detection techniques, Litaker et al. (2002a) were unable to find evidence of many of the stages of the proposed life cycle and concluded that the amoeboid stages reported in tanks in which *P. piscicida* is actively feeding on fish were contaminants introduced with the fish. Burkholder and Glasgow (2002) criticized this study, reaffirming that they had evidence of transformation to amoebae and cysts, which Litaker et al. did not refer to, and claiming that they were working with strains of *P. piscicida* that are noninducible and do not produce toxins, which was further rebutted by Litaker et al. (2002b). Many of the difficulties undoubtedly arose because—contrary to normal practice—the *Pfiesteria* strains were not freely available at this time to the scientific community. In another study, Berry et al. (2002) reported that the closely related *Pfiesteria shumwayae* kills fish as rapidly as the allegedly toxigenic strains described by Burkholder, but Berry found no evidence of toxin production. Furthermore, attempts to PCR-amplify genes known to be involved in synthesis of polyketides—to which all known dinoflagellate fish toxins belong—were unsuccessful. The apparently toxic effects seen during fish kills seem to be due to a combined effect of other algae, fungal infection, and predation by *Pfiesteria*. Vogelbein et al. (2002) also found that *P. shumwayae* do not produce toxins and concluded that fish mortality results only when the dinoflagellate is in close contact with the fish, owing to micropredatory feeding via a peduncle that sucks fluids from the fish tissue (see *Figure 6.3A*).

Progress in evaluating the role of a toxin was hampered by difficulties in working safely with *Pfiesteria*, since the discovery of possible human health effects necessitated category 3 biohazard containment with special air filtration. Numerous contradictory claims have been made for the identification and characterization of a toxin, including reports of neurotoxic effects in laboratory animals, but many of these studies have been criticized for lack of standardization of assay procedures and work with impure cultures (reviewed by Place et al., 2008). Moeller et al., (2007) recently reported that *Pfiesteria* produces a highly labile small molecule containing copper and sulfur that can break down to produce free radicals. However, this finding has not been repeated and the study can be criticized because there is no clear mode of action to link the putative toxin with previously described effects in either humans or fish, and because it was not unequivocally proven that the toxin was produced by *Pfiesteria*. Place et al. (2008) propose an alternative scenario—involving another dinoflagellate *Karlodinium*—as an explanation for the Maryland fish kills. The role of a toxin in human health still remains unclear. Following extensive public health investigations of the "estuary-associated syndrome" over 4 years, Morris et al. (2006) found no evidence linking neuropsychological symptoms with the presence or absence of *Pfiesteria* in water samples, measured using a PCR assay. They concluded that routine exposure to estuarine waters in which *Pfiesteria* may be present does not represent a significant risk to human health. However, they note that "There is no question that persons exposed to the Pocomoke River in the summer of 1997 had profound, reversible (and well-documented) neuropsychological deficits," and that this, together with the studies of laboratory-exposed persons and animal experiments, makes it "plausible that *Pfiesteria*, in unique, isolated instances and/or in association with specific, unusually toxic strains, can cause human health effects." After 20 years of study, *Pfiesteria* remains as mysterious as ever.

Blooms occur owing to the occurrence of particular combinations of physical conditions (temperature, sunlight, water stratification, circulation) and chemical conditions (levels of oxygen and specific nutrients). Species dispersal due to large-scale water movements is important, and unusual currents and storms may account for appearance of atypical blooms. For example, in some recent years, the *K. brevis* red tide off Florida has moved much further north because of formation of unusual circulation patterns in the Gulf Stream. The occurrence of some HABs has been closely correlated with the El Niño Southern Oscillation phenomenon. The increased incidence of HABs is often cited as evidence of disturbance to our oceans and atmosphere due to increased carbon dioxide levels.

Nutrient enrichment of coastal waters is often believed to be responsible for the increased incidence of HABs. Upwelling of nutrients into cold oligotrophic waters is responsible for many natural blooms, such as those that occur regularly off the California coast. Such natural phenomena may now be overlaid with anthropogenic sources of excess nutrients, and some studies have provided clear evidence for this. Although there are few long-term studies and it is difficult to compare the results from different areas, it seems likely that a significant increase in plankton growth can result from nutrient input—from sewage and runoff of agricultural fertilizers—into coastal waters and estuaries with limited exchange with the open ocean. For example, regular blooms of nontoxic microalgae such as *Phaeocystis* and *Emiliania* have become a regular occurrence in the North Sea and English Channel (see *Figure 6.5*). Could such eutrophication also lead to an increase in toxic species? One widely held idea is that nutrient inputs from sewage or agricultural land runoff alter the ratio of particular nutrients, as well as the total loading, and that this may change the balance of the different plankton groups. Sewage is rich in nitrogen and phosphorus, but has a low silicon content. Since diatoms specifically require silicon, this could favor the selective growth of dinoflagellates. In the case of the emergence of dinoflagellate blooms (including *Pfiesteria*) as a problem in the North Carolina and Maryland estuaries, there seemed to be a strong association with increased nutrient levels due to an expansion in poultry and pig farming on the surrounding land. However, not all *Pfiesteria* outbreaks occurred in nutrient-enriched waters, and not all areas of the estuaries with increased nutrients showed high populations of *Pfiesteria*. The expansion of mariculture, such as high-density salmon or shrimp culture, has been directly implicated in the increase in frequency of HABs due to nutrient enrichment from uneaten food and excreta. For example, the increased incidence of high levels of PSP toxins in shellfish in Scotland during the 1990s was probably due to poorly sited salmon pens in lochs, bays, and inlets with limited water exchange.

One obvious explanation for the apparent increase in HABs and associated diseases is increased awareness of toxic species and more extensive and effective monitoring. This is exacerbated by the increased use of coastal waters for aquaculture of shellfish and finfish. These activities act as sensitive indicators of potential problems and undoubtedly account for at least part of the apparent spread of PSP and DSP. Mass mortalities in fish farms can be due to the effects of HABs, although these do not always involve toxic species (see p. 247).

Coastal waters must be regularly monitored to assess the development of harmful algal blooms

Regular surveys by microscopic examination of the dynamics of phytoplankton populations within an area may give advance warning of an increase in particular species, which may sometimes precede a toxic bloom. Such surveys are time-consuming and it is often difficult to distinguish toxic species or strains using morphological criteria. Some improvements can be achieved using flow cytometry or epifluorescence microscopy after labeling the target species with specific fluorescent antibodies. Unfortunately, background fluorescence is often

---

ⓘ TOXIC HABs CAN BE SPREAD VIA SHIP'S BALLAST

One explanation for the increased geographic distribution of some HABs is transport of the causative organisms via ship movements. Modern large vessels pick up many thousands of liters of water as ballast, which can be transported to completely different geographic regions. Several studies have shown that the cysts of toxic dinoflagellates and other harmful "alien" organisms can be transported in this way. For example, the introduction of paralytic shellfish poisoning into southern Australia in the 1980s is suspected to be due to transport from Japan and Korea. As a consequence, regulations of the International Maritime Organization require ships to have programs for management of ballast water, such as off-shore exchange of ballast water, disinfection of the water before it is discharged, or transfer to shore-based disinfection facility.

BOX 12.3

Poisonous partners
Do symbiotic bacteria produce toxins in dinoflagellates and diatoms?

Early in the study of the origin of paralytic shellfish poisoning (PSP), microbiologists observed bacteria-like structures in electron micrographs of dinoflagellate cells. Could these be symbionts? And are they involved in toxin synthesis or breakdown? Besides the intrinsic interest in the possible role of symbiosis in providing their eukaryotic host with a possible defensive mechanism, answering these questions is of importance in the monitoring and mitigation of fish and shellfish toxicity. However, despite a number of studies in this area, there are no definitive answers to these questions. It is generally accepted that certain cyanobacteria can synthesize toxins with similar structures to the PSP toxins such as saxitoxins in culture (Aráoz et al., 2009), but synthesis by other bacteria is less certain. Many studies have shown that various dinoflagellate-associated bacteria can produce toxins in pure culture; but different research groups are often fiercely critical of the assay methods used by others and suspect that some suspected toxins are structurally unrelated (Piel, 2004).

Dinoflagellates form tight associations with numerous bacteria; many experiments have been conducted to compare levels of toxin production in dinoflagellate cultures before and after curing of contaminating bacteria, by treatment with antibiotics or enzymes. For example, Green et al. (2004) examined the link between the toxicity of *Gymnodinium catenatum* (the most important cause of PSP) and the diversity of bacteria on the surface of cysts of this dinoflagellate, using culture-based methods and 16S rRNA gene sequencing. They found no evidence for autonomous production of the PSP toxins by the bacteria, but concluded that the microbiota influences the biosynthesis of the toxins, perhaps by supplying cofactors or precursors necessary for toxin production. Azanza et al. (2006) studied toxin production by bacteria associated with *Pyrodinium bahamense* var. *compressum*, one of the major causes of PSP in the tropics. They treated cultures of the dinoflagellate with a cocktail of antibiotics to remove surface bacteria, then sonicated the cells to release endosymbiotic bacteria. Various bacterial isolates, including *Moraxella* sp., *Bacillus* sp. and *Erythrobacter* sp. were identified and cultured. Culture extracts were shown to produce high levels of saxitoxins using chromatographic methods and to cause PSP-like effects when injected into mice. The authors speculated

that association of the dinoflagellate cysts with specific bacteria in the sediments and subsequent resuspension during a bloom event might be critical factors in the development of a toxic event.

Brevetoxin, produced by the red tide dinoflagellate *Karenia brevis*, belongs to a class of secondary metabolites known as polyketides, the carbon framework of which is assembled by a group of polyketide synthase (PKS) enzymes. Snyder et al. (2005) used a fluorescence-activated cell sorter (see p. 34) to separate dinoflagellate cells from associated bacteria. PCR amplification of PKS genes, using different primer sets, combined with fluorescence *in situ* hybridization (FISH) showed that two genes localized exclusively to *K. brevis* cells, while a third localized to both *K. brevis* and associated bacteria. This study demonstrates that the dinoflagellate cells do possess the intrinsic biosynthetic capacity to synthesize the toxins, but leaves open the possibility that bacterial synthesis could also be involved in determining the levels of toxins produced.

Bacterial involvement also seems to be implicated in the production of domoic acid by the diatom *Pseudo-nitzschia multiseries*. Various experiments have shown that axenic cultures have greatly reduced levels of domoic acid, whereas levels are increased when certain species of bacteria isolated from the original culture are reintroduced. By cultivating an axenic strain of the diatom in a cellophane tube surrounded by the original nonaxenic culture, Kobayashi et al. (2009) showed that direct contact with living bacteria is necessary for *P. multiseries* to produce a high level of domoic acid.

In conclusion, although the precise role of bacteria in synthesis of these toxins remains uncertain, there seems little doubt that the presence of bacteria has a considerable influence on the levels and nature of toxins released by the algae. An additional complication is that some of the bacteria naturally associated with dinoflagellates and diatoms can cause lysis of the cells, leading to sudden release of high levels of toxin, whilst other bacteria may metabolize algal products (Kodama et al., 2007). The importance of such algicidal bacteria as a possible method of control of HABs is discussed in Chapter 14.

a problem. Remote sensing can be used by equipping satellites with spectral scanners that detect chlorophyll and other pigments in surface waters. When coupled with physical measurements such as sea surface temperatures and current flows, satellite images are especially useful in tracking the development and movement of blooms. As genetic sequence data for toxic dinoflagellates and diatoms become available, gene probes are being increasingly used. A common method is to amplify microalgal DNA encoding 18S rRNA using eukaryote-specific primers. The resulting PCR products can be cloned and sequenced, leading to a specific oligonucleotide probe that will hybridize with DNA of the organism in water samples. Increased sensitivity and real-time assays can be achieved using molecular beacons. It would be useful to develop probes for specific genes involved in toxin biosynthesis.

Surveys can be used to limit the economic and health impacts of HABs, but whether we can use such information to control their development or spread is questionable, since HABs often occur over huge areas. The effects of weather, ocean currents, and the many physical and biological factors that determine the development and eventual demise of a bloom are unpredictable. Chemical treatments, such as adding agents that promote agglutination and sinking of microalgal cells, can be applied locally to control blooms, but the ecological impact of such intervention on a large scale needs careful assessment. Flocculating clays are used widely in Japan and China to control blooms around fish farms. Algicidal bacteria and viruses capable of initiating lysis of dinoflagellates and diatoms have been isolated from seawater and investigated as possible biological control agents. Also, some microalgae have been shown to contain lysogenic viruses, which can be induced into the lytic cycle by particular conditions. Some success in initiating the collapse of bloom populations has been achieved in microcosm experiments, but much more research and evaluation of the ecological and "scale-up" issues is needed before biological control becomes a practical proposition. Such biotechnological approaches to the control of HABs are considered further in Chapter 14.

Conclusions

The most important human diseases associated with autochthonous marine bacteria are infections caused by bacteria in the genus *Vibrio*. These are adapted to a life in the aquatic environment but possess complex regulatory systems coordinating the expression of virulence factors when they transfer to the human host. *Vibrio cholerae* and *V. parahaemolyticus* both show evidence of extensive gene transfer that explains the emergence of new pathogenic variants. Numerous intoxications are caused by dinoflagellates and diatoms. Factors that affect plankton blooms and the production of toxins harmful to humans are numerous, complex, and poorly understood. A combination of natural and anthropogenic effects is undoubtedly responsible for a worldwide increase in HABs, a wider range of toxic species and the emergence of "new" human diseases. Climate change is likely to increase the threat of disease in the human population. Bacterial pathogens show increased abundance in seawater at higher temperatures and people are more likely to become infected following coastal flooding storms or hurricanes, whilst the increased intensity of HABs and their spread to new regions can be linked to climate change.

> ### ? IS IT POSSIBLE TO DEVELOP RELIABLE "HAB FORECASTS"?
>
> The ability to provide early warning of the development of specific HABs would be of great value to managers of fisheries and conservation programs, and for residents and tourists in coastal zones. Computer modeling based on environmental factors, such as water temperature and salinity, river flow, winds and tides, and biological factors such as abundance and behavior of toxic dinoflagellates have been used successfully to predict red tides in the USA. In the near future, it is likely that automatic sensing devices using gene probe technology could be placed on offshore buoys, and information could be beamed to satellites and integrated with improved remote-sensing signal-detection systems linked to powerful computer models (Paul et al., 2007). This could lead to reliable real-time forecasts.

References

Aráoz, R., Molgó, J. and Tandeau de Marsac, N. (2009) Neurotoxic cyanobacterial toxins. *Toxicon* 56: 813–828.

Azanza, M. P., Azanza, R. V., Vargas, V. M. and Hedreyda, C. T. (2006) Bacterial endosymbionts of *Pyrodinium bahamense* var. compressum. *Microb. Ecol.* 52: 756–764.

Barker, R. (1997) *And the Waters Turned to Blood*. Touchstone Books, New York.

Berry, J. P., Reece, K. S., Rein, K. S., et al. (2002) Are *Pfiesteria* species toxicogenic? Evidence against production of ichthyotoxins by *Pfiesteria shumwayae*. *Proc. Natl Acad. Sci. USA* 99: 10970–10975.

Blokesch, M. and Schoolnik, G. K. (2007) Serogroup conversion of *Vibrio cholerae* in aquatic reservoirs. *PLoS Pathog.* 3: e81, doi:10.1371/journal.ppat.0030081.

Bourassa, L. and Camilli, A. (2009) Glycogen contributes to the environmental persistence and transmission of *Vibrio cholerae*. *Mol. Microbiol.* 72: 124–138.

Burdette, D. L., Yarbrough, M. L., Orvedahl, A., Gilpin, C. J. and Orth, K. (2008) *Vibrio parahaemolyticus* orchestrates a multifaceted host cell infection by induction of autophagy, cell rounding, and then cell lysis. *Proc. Natl Acad. Sci. USA* 105: 12497–12502.

Burkholder, J. M. and Glasgow, H. B. (2002) The life cycle and toxicity of *Pfiesteria piscicida* revisited. *J. Phycol.* 38: 1261–1267.

Burkholder, J. M., Noga, E. J., Hobbs, C. W. and Glasgow, H. B. (1992) New "phantom" dinoflagellate is the causative agent of major estuarine fish kills. *Nature* 358: 407–410.

Burkholder, J. M., Glasgow, H. B. and Hobbs, C. W. (1995) Fish kills linked to a toxic ambush-predator dinoflagellate—distribution and environmental conditions. *Mar. Ecol. Prog. Ser.* 124: 43–61.

Colwell, R. R., Huq, A., Islam, M. S., et al. (2003) Reduction of cholera in Bangladeshi villages by simple filtration. *Proc. Natl Acad. Sci. USA* 100: 1051–1055.

Faruque, S. M., Islam, M. J., Ahmad, Q. S., Faruque, A. S. G., Sack, D. A., Nair, G. B. and Mekalanos, J. J. (2005) Self-limiting nature of seasonal cholera epidemics: role of host-mediated amplification of phage. *Proc. Natl Acad. Sci. USA* 102: 6119–6124.

Glasgow, H. B., Burkholder, J. M., Schmechel, D. E., Tester, P. A. and Rublee, P. A. (1995) Insidious effects of a toxic estuarine dinoflagellate on fish survival and human health. *J. Toxicol. Environ. Health* 46: 501–522.

Grattan, L. M., Oldach, D., Perl, T. M., et al. (1998) Learning and memory difficulties after environmental exposure to waterways containing toxin-producing *Pfiesteria* or *Pfiesteria*-like dinoflagellates. *Lancet* 352: 532–539.

Green, D., Blackburn, S. I., Bolch, C. J. S. and Llewellyn, L. (2004) Phylogenetic and functional diversity of the cultivable bacterial community associated with the paralytic shellfish poisoning dinoflagellate *Gymnodinium catenatum*. *FEMS Microbiol. Ecol.* 47: 345–357.

Hammer, B. K. and Bassler, B. L. (2003) Quorum sensing controls biofilm formation in *Vibrio cholerae*. *Mol. Microbiol.* 50: 101–104.

Higgins, D. A., Pomianek, M. E., Kraml, C. M., Taylor, R. K., Semmelhack, M. F. and Bassler, B. L. (2007) The major *Vibrio cholerae* autoinducer and its role in virulence factor production. *Nature* 450: 883–886.

Hsueh, P. R., Lin, C. Y., Tang, H. J., Lee, H. C., Liu, J. W., Liu, Y. C. and Chuang, Y. C. (2004) *Vibrio vulnificus* in Taiwan. *Emerg. Infect. Dis.* 10: 1363–1368.

Jones, M. K. and Oliver, J. D. (2009) *Vibrio vulnificus*: disease and pathogenesis. *Infect. Immun.* 77: 1723–1733.

Kaiser, J. (2002) The science of *Pfiesteria*: elusive, subtle and toxic. *Science* 298: 346–349.

Kobayashi, K., Takata, Y. and Kodoma, M. (2009) Direct contact between *Pseudo-nitzschia multiseries* and bacteria is necessary for the diatom to produce a high level of domoic acid. *Fisheries Sci.* 75: 771–776.

Kodama, M., Doucette, G. and Green, D. (2006) Relationships between bacteria and harmful algae. In: *Ecology of Harmful Algae* (ed. E. Graneli and J. T. Turner). Springer, Berlin, pp. 243–255.

LaRocque, R. C., Harris, J. B., Dziejman, M., et al. (2005) transcriptional profiling of *Vibrio cholerae* recovered directly from patient specimens during early and late stages of human infection. *Infect. Immun.* 73: 4488–4493.

Lee, M. J., Jeong, D. Y., Kim, W. S., et al. (2000) A tetrodotoxin-producing *Vibrio* strain, LM-1, from the puffer fish *Fugu vermicularis radiatus*. *Appl. Environ. Microbiol.* 66: 1698–1701.

Li, X. and Roseman, S. (2004) The chitinolytic cascade in vibrios is regulated by chitin oligosaccharides and a two-component chitin catabolic sensor/kinase. *Proc. Natl Acad. Sci. USA* 101: 627–631.

Lipp, E. K., Huq, A. and Colwell, R. R. (2002) Effects of global climate on infectious disease: the cholera model. *Clin. Microbiol. Rev.* 15: 757–770.

Litaker, R. W., Vandersea, M. W., Kibler, S. R., Madden, V. J., Noga, E. J. and Tester, P. A. (2002a) Life cycle of the heterotrophic dinoflagellate *Pfiesteria piscicida* (Dinophyceae). *J. Phycol.* 38: 442–463.

Litaker, R. W., Vandersea, M. W., Kibler, S. R., Noga, E. J. and Tester, P. A. (2002b) Reply to comment on the life cycle and toxicity of *Pfiesteria piscicida* revisited. *J. Phycol.* 38: 1268–1272.

Matsumura, K. (2001) Letter to the Editor: No ability to produce tetrodotoxin in bacteria. *Appl. Environ. Microbiol.* 67: 2393–2394.

McLauchlin, J., Little, C. L., Grant, K. A. and Mithani, V. (2006) Scombrotoxic fish poisoning. *J. Pub. Health* 28: 61–62.

Meibom, K. L., Li, X. B., Nielsen, A. T., Wu, C. Y., Roseman, S. and Schoolnik, G. K. (2004) The *Vibrio cholerae* chitin utilization program. *Proc. Natl Acad. Sci. USA* 101: 2524–2529.

Meibom, K. L., Blokesch, M., Dolganov, N. A., Wu, C. Y. and Schoolnik, G. K. (2005) Chitin induces natural competence in *Vibrio cholerae*. *Science* 310: 1824–1827.

Merrell, D. S., Butler, S. M., Qadri, F., et al. (2002) Host-induced epidemic spread of the cholera bacterium. *Nature* 417: 642–645.

Moeller, P. D. R., Beauchesne, K. R., Huncik, K. M., Davis, W. C., Christopher, S. J., Riggs-Gelasco, P. and Gelasco, A. K. (2007) Metal complexes and free radical toxins produced by *Pfiesteria piscicida*. *Environ. Sci. Technol.* 41: 1166–1172.

Morris, J., Grattan, L. M., Wilson, L. A., et al. (2006) Occupational exposure to *Pfiesteria* species in estuarine waters is not a risk factor for illness. *Environ. Health Perspect.* 114: 1038–1043.

Nelson, E. J., Harris, J. B., Morris, J. G., Calderwood, S. B. and Camilli, A. (2009) Cholera transmission: the host, pathogen and bacteriophage dynamic. *Nat. Rev. Microbiol.* 7: 693–702.

Noguchi, T. and Arakawa, O. (2008) Tetrodotoxin—distribution and accumulation in aquatic organisms, and cases of human intoxication. *Mar. Drugs* 6: 220–242.

Noguchi, T., Arakawa, O. and Takatani, T. (2006) Toxicity of pufferfish *Takifugu rubripes* cultured in netcages at sea or aquaria on land. *Comp. Biochem. Physiol. D Genomics Proteomics* 1: 153–157.

Paul, J., Scholin, C. A., van den Engh, G. and Perry, M. J. (2007) *In situ* instrumentation. *Oceanog.* 20: 70–78.

Piel, J. (2004) Metabolites from symbiotic bacteria. *Nat. Prod. Rep.* 21: 519–538.

Place, A. R., Saito, K., Deeds, J. R., Robledo, J. A. F. and Vasta, G. R. (2008) A decade of research on *Pfiesteria* spp. and their toxins: unresolved questions and an alternative hypothesis. In: *Seafood and Freshwater Toxins: Pharmacology, Physiology, and Detection*, 2nd ed. (ed. L. M. Botana). CRC Press, Boca Raton, FL, pp. 717–751.

Pruzzo, C., Vezzulli, L. and Colwell, R. R. (2008) Global impact of *Vibrio cholerae* interactions with chitin. *Environ. Microbiol.* 10: 1400–1410.

Ritson-Williams, R., Yotsi-Yamashita, M. and Paul, V. J. (2006) Ecological functions of tetrodotoxin in a deadly polyclad flatworm. *Proc. Natl Acad. Sci. USA* 103: 3176–3179.

Rongo, T., Bush, M. and van Woesik, R. (2009) Did ciguatera prompt the late Holocene Polynesian voyages of discovery? *J. Biogeog.* 36: 1423–1432.

Schild, S., Tamayo, R., Nelson, E. J., Qadri, F., Calderwood, S. B. and Camilli, A. (2007) Genes induced late in infection increase fitness of *Vibrio cholerae* after release into the environment. *Cell Host Microbe* 2: 264–277.

Snyder, R. V., Guerrero, M. A., Sinigalliano, C. D., Windshell, J., Perez, R., Lopez, J. V. and Rein, K. S. (2005) Localization of polyketide synthase encoding genes to the toxic dinoflagellate *Karenia brevis*. *Phytochem.* 66: 1767–1780.

Steidinger, K. A., Burkholder, J. M., Glasgow, H. B. Jr, Hobbs, C. W., Garrett, J. K., Truby, E. W., Noga, E. J. and Smith, S. A. (1996) *Pfiesteria piscicida* gen. et sp. nov. (*Pfiesteriaceae* fam. nov.), a new toxic dinoflagellate with a complex life cycle and behavior. *J. Phycol.* 32: 157–164.

Trainer, V. L. (2002) Marine biology—unveiling an ocean phantom. *Nature* 418: 925–926.

Vogelbein, W. K., Lovko, V. J., Shields, J. D., Reece, K. S., Mason, P. L., Haas, L. W. and Walker, C. C. (2002) *Pfiesteria shumwayae* kills fish by micropredation not exotoxin secretion. *Nature* 418: 967–970.

Waters, C. M., Lu, W., Rabinowitz, J. D. and Bassler, B. L. (2008) Quorum sensing controls biofilm formation in *Vibrio cholerae* through modulation of cyclic di-GMP levels and repression of vpsT. *J. Bacteriol.* 190: 2527–2536.

Yotsu-Yamashita, M., Mebs, D. and Flachsenberger, W. (2007) Distribution of tetrodotoxin in the body of the blue-ringed octopus (*Hapalochlaena maculosa*). *Toxicon* 49: 410–412.

Further reading

Anderson, D. M. (2009) Approaches to monitoring, control and management of harmful algal blooms (HABs). *Ocean Coast. Manag.* 52: 342–347.

Belkin, S. and Colwell, R. R. (2005) *Oceans and Health. Pathogens in the Marine environment.* Springer, New York.

Graneli, E. and Turner, J. T. (2006) *Ecology of Harmful Algae.* Springer, Berlin.

Grattan, L. M., Oldach, D. and Morris, J. G. (2001) Human health risks of exposure to *Pfiesteria piscicida*. *Bioscience* 51: 853–857.

Llewellyn, L. E. (2001) Ecology of microbial neurotoxins. In: *Handbook of Neurotoxicology*, Vol. 1 (ed. E. J. Masaro). Humana Press, Totowa, NJ, pp. 239–255.

National Research Council (1999) *From Monsoons to Microbes.* National Academy Press, Washington, DC.

Thompson, F. L., Austin, B. and Swings, J. (2006) *The Biology of Vibrios.* ASM Press, Washington, DC.

Walsh, P. J., Smith, S., Gerwick, W., Solo-Gabriele, H. and Fleming, L. (2008) *Oceans and Human Health. Risks and Remedies from the Seas.* Elsevier Science and Technology, Burlington, MA.

Chapter 13

Microbial Aspects of Marine Biofouling, Biodeterioration, and Pollution

The first part of this chapter reviews some of the detrimental activities of marine microbes that have direct consequences for human health, wealth, and welfare. Besides their direct impact as pathogens of humans or as the cause of losses in aquaculture and fisheries discussed in earlier chapters, marine microbes have some other economically important detrimental effects through biofouling of marine surfaces and structures, the biodeterioration of materials, and the spoilage of seafood. Biotechnological approaches to overcome the effects of biofouling are discussed in Chapter 14. The second theme of this chapter concerns marine pollution. The health hazards arising from sewage pollution are discussed, together with methods of monitoring water and shellfish using microbial indicators and new approaches for direct detection of pathogens. In the final section, the role of marine microbes in biodegradation and bioremediation of oil and other chemical pollutants is discussed. Finally, the role that microbes play in mobilizing mercury into marine food webs is considered.

Key Concepts

- Microbial colonization often provides the first stage in biofouling of surfaces in the sea, leading to economic losses through interference with the efficient operation of ships and damage to structures such as piers and aquaculture facilities.

- Corrosion and deterioration of metal and wooden structures is initiated by microbes.

- Microbial activity results in harmful spoilage of fish and shellfish; it can also be used to produce seafood products with altered desirable properties.

- Pathogenic microbes are introduced to the sea via sewage and constitute a health hazard through recreational use of coastal environments and contamination of shellfish.

- Bacterial indicators are used for monitoring coastal pollution but may be unreliable predictors of health risks, which are mainly associated with viruses.

- Many microbes degrade oil naturally and this may be enhanced using bioremediation.

- Microbial activity in sediments and pelagic aggregates leads to production of toxic mercury compounds that can contaminate fish.

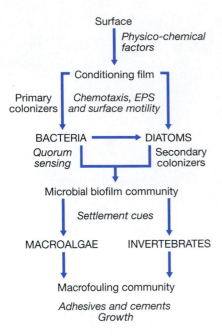

Surface

Physico-chemical factors

↓

Conditioning film

Primary colonizers | *Chemotaxis, EPS and surface motility*

BACTERIA ⟶ DIATOMS

Quorum sensing | Secondary colonizers

↓

Microbial biofilm community

Settlement cues

MACROALGAE   INVERTEBRATES

↓

Macrofouling community

Adhesives and cements Growth

Figure 13.1 Development of biofouling, showing the critical role of microbial activities. Italicized captions indicate processes amenable to disruption for control of biofouling. Invertebrate macrofouling organisms produce soft fouling (e.g. soft corals, sponges, anemones, and tunicates) and hard fouling (e.g. barnacles, mussels, and tubeworms). EPS, exopolymeric substances.

BIOFOULING AND BIODETERIORATION

Microbial biofilms are often the first phase in biofouling

As previously discussed in Chapter 3, the surfaces of inanimate objects and living organisms in the sea are colonized by mixed microbial communities that form biofilms showing complex physical structures and chemical interactions. The process of fouling usually begins with the formation of a molecular conditioning film formed by the deposition of organic matter on the surface within a few minutes of immersing a material into seawater. The conditioning film is composed of many compounds including amino acids, proteins, lipids, nucleic acids, and polysaccharides. This is usually followed over the next few hours by attachment of bacteria and/or diatoms as primary colonizers, leading to a slimy biofilm up to 500 μm thick and consolidated by the production of sticky exopolymers (*Figure 13.1*). This is followed by the settlement of other microbes, planktonic algal spores, and invertebrate larvae, but in some cases colonization by macrofouling organisms such as barnacles can occur very rapidly. The complex dense community that develops leads to the biofouling of all types of marine surfaces, including coastal plants, macroalgae, animals, piers and jetties, oil-drilling rigs, fishing gear, aquaculture cages, engineering materials, and boat hulls (*Figure 13.2A*). Deleterious effects include deterioration of materials, blockage of pipes, reduced efficiency of heating and cooling plants, and interference with the efficient operation of boats and ships. The economic effects of biofouling are immense. Sailing enthusiasts know the cost (in lost weekends!) of scraping the bottom of yachts, whilst biofouling of large ships can cause extreme increases in fuel usage due to frictional drag. Formation of a microbial biofilm alone can cause a 1–2% increase in drag. Subsequent colonization by macroalgae increases this to about 10% and, if unchecked, extensive colonization by hard-shelled invertebrates such as barnacles, tubeworms, bryozoans, or mussels can lead to 30–40% drag. This costs shipping companies and navies throughout the world tens of billions of dollars because of losses in fuel efficiency and the cost of antifouling measures. Increased fuel usage also leads to increased emissions of CO_2 and sulfurous pollutants. Various methods for the prevention of macrofouling of ships' hulls, such as the use of copper sheeting, have been in use for hundreds of years, and self-polishing copolymer paints containing copper or tin compounds have been used extensively in the last 50 years. Tributyl tin (TBT) is particularly effective, but its use has been banned since 2008 by the International Maritime Organization, following the recognition that it has serious environmental consequences, with particular effects on the ecology and reproductive behavior of marine invertebrates and the immunity of marine mammals (see p. 252). There is, therefore, an active search for effective "environment-friendly" alternatives—microbial aspects of antifouling biotechnology are considered in the next chapter.

It is important to note that whilst the negative aspects of these processes are considered here because of their deleterious consequences, microbial biofilms also play a very important, positive role. Microbial products may stimulate the settlement and metamorphosis of algae and invertebrate larvae, which is very important in marine ecology (e.g. in reef formation) and aquaculture (e.g. in settlement of mussels in suspended rope culture).

Microbially induced corrosion occurs as a result of the activities of microorganisms within biofilms on metals, alloys, and composite materials

Sulfate-reducing bacteria (SRB) have long been known to be the main cause of marine corrosion(*Figure 13.2B*), but the importance of mixed consortia with fermentative acid-producing bacteria in biofilms is increasingly recognized. Fermentative bacteria produce acid themselves as well as supplying acetate as a substrate for SRB metabolism. Such corrosion is a particular problem for the offshore petroleum industry because it causes corrosion of pipelines and oil-drilling

Figure 13.2 Biofouling and corrosion of marine structures. (A) Extensive colonization of seabed monitoring structure after immersion for several years in coastal waters off Boston, USA. (Image courtesy of US Geological Survey, Woods Hole Science Center.) (B) Corrosion pits on the surface of a ship's tank due to action of sulfate-reducing bacteria. (Image courtesy of Graham Hill, ECHA Microbiology, Cardiff.)

platforms and also leads to the "souring" of crude oil by the production of hydrogen sulfide. Corrosion of steel structures can be limited by the use of cathodic protection, but this can lead to structural weakness unless the electrical potential is carefully applied and monitored; formation of excess hydrogen can cause metal fatigue. Routine monitoring by measuring levels of culturable acid-producing bacteria together with the application of biocides to fluids injected into the well can help to control the problem. Testing for SRB is more problematic because of the need for anaerobic culture; development of molecular-based test kits could prove useful to the industry. Problems with the use of cathodic protection and environmental concerns about large-scale use of biocides at sea favor a biotechnological solution to the control of corrosive biofilms. Some success has been achieved by addition of nitrite and/or nitrate together with sulfide-oxidizing bacteria such as *Thiobacillus denitrificans* to injection fluids.

Microbes cause biodeterioration of marine wooden structures and timber

Microbial colonization and decomposition of wooden structures (e.g. wharves, jetties, piers, and boats) and of timber transported or stored at sea causes an immense amount of damage costing billions of dollars. The main cause of damage is penetration by wood-boring invertebrates, of which the most important are the shipworms. These are not worms, but small bivalves in the family *Teredinidae*, which tunnel into wooden structures using their serrated shell. As is common with most animals that feed on wood (termites are a well-studied example), shipworms and other marine invertebrates that digest wood rely on an obligate symbiotic association with a consortium of cellulose and wood-degrading microbes. In one species of shipworm, *Bankia setacea*, the gammaproteobacterium *Teredinibacter turnerae* has been identified as a symbiont of a specialized region of the host gill. The bacterial symbiont is transmitted vertically via the larvae, as shown by specific rRNA probes that reveal the bacterium to be present in reproductive tissue and eggs. This ensures a broad distribution of larvae. Wood is also degraded by gribbles; these are small isopods such as *Limnoria lignorum* and related species. The use of various preservative wood treatments delays deterioration, but these are expensive, have an adverse environmental impact, and cannot be used on new timber during transport. Identification of the symbionts and their mode of transmission provide a potential weak link in the process of biodeterioration, as it may be possible to identify compounds that inhibit bacterial colonization or metabolism.

The preservation of archaeological timbers, such as those recovered from shipwrecks, is a highly specialized process. Waterlogged wood is largely protected when buried in anoxic sediments, but is subject to rapid decay by a range of fungi once it is brought to the surface.

Microbial growth and metabolism are the major cause of spoilage of seafood products

Fish and shellfish deteriorate very rapidly as a result of microbial activities, resulting in the rapid production of discoloration, slime, and unpleasant odors and

---

BACTERIAL ACTIVITY CAN PRODUCE DEADLY GAS IN OIL PRODUCTION

The activity of sulfate-reducing bacteria in oil and gas reserves can lead to high levels of hydrogen sulfide gas. High levels of sulfide cause "souring" of the oil that can cause problems in refining and reduces the value of the product. More importantly, toxic levels of hydrogen sulfide can be produced on drilling rigs, and there have been numerous examples of leaks on offshore drilling platforms, for example in the Gulf of Mexico, North Sea, and Java Sea, which have resulted in fatalities to oil-rig workers. Hydrogen sulfide is toxic at very low levels, and health and safety authorities set exposure limits of 5–20 ppm over a working day. Trace amounts of the gas are very easily detected at low levels by its characteristic "rotten eggs" smell, but one of its insidious features is that it paralyses the olfactory nerve at levels above 30 ppm, so a worker may smell the gas, investigate its source, and then be unable to smell higher concentrations. At low levels, hydrogen sulfide causes headaches, blurred vision, and conjunctivitis. It can be instantly fatal at high levels, and even a few minutes exposure to 500–1000 ppm can cause long-term illness. Modern rigs employ a variety of sensors to detect escaping gas and workers receive special training in dealing with leaks, but in some oil-well "blowouts" drilling companies have been forced to set fire to rigs to burn off the escaping hydrogen sulfide.

flavors. The composition of microbial communities in freshly caught seafood varies considerably according to the animal species and the methods of fishing, handling, processing, and storage. Natural transformation of compounds within the fish tissue begins soon after death owing to autolytic processes, making catabolites available for decomposition by members of the bacteria that are normal inhabitants of the gut and skin of fish. Bacteria such as *Pseudomonas*, *Shewanella*, *Alteromonas*, *Moraxella*, *Acinetobacter*, *Cytophaga*, and *Flavobacteria* are particularly prominent as agents of rapid spoilage of unpreserved fish. Psychrophilic strains of *Pseudomonas* and *Shewanella* become the dominant spoilage organisms when fish is chilled on ice. Specific spoilage organisms are most often associated with the production of ammonia, amines, ketones, organic acids, and sulfur compounds that characterize "off" seafood. Trimethylamine is the most important of these metabolites, produced by reduction of trimethylamine oxide, naturally present in the tissue of many fish.

Bacteria grow rapidly to high levels (over 10^8 CFU per gram) in the nutrient-rich environment of fish tissue, which contains high concentrations of readily utilizable substrates such as free amino acids. Metabolic consortia develop in the mixed microbial community; for example, lactic acid bacteria (e.g. *Lactobacillus* and *Carnobacterium*) degrade the amino acid arginine to ornithine, which is further degraded to putrescine by enterobacteria. The production of siderophores is important for the acquisition of iron, since fish tissue contains limiting concentrations (see p. 239). Lactic acid bacteria characteristically produce bacteriocins, which are highly specific membrane-active peptide antibiotics that inhibit certain other bacteria.

Most fish spoilage organisms are easily cultured and identified using standard microbiological methods, although there has been some recent use of molecular methods for characterization and early detection of spoilage organisms (e.g. gene probes for *Shewanella putrefaciens*) and chemical assays for total volatile basic nitrogen after extraction with trichloracetic or perchloric acid. The growth kinetics of specific spoilage organisms can be determined by absorbance measurements in culture and used to construct mathematical models to predict rates of spoilage and shelf-life of fish products.

Processing, packaging, and inhibitors of spoilage are used to extend shelf-life

The food industry has devised various methods of extending shelf-life in "value-added" products by inhibiting or delaying spoilage, but many processes used with other foods are unsuitable because the texture and flavor of fish products is easily destroyed by processing, leading to reduced consumer acceptance. Many processes (e.g. salting, pickling, and smoking) are based on traditional methods, whilst recent innovations include packaging in a CO_2-modified atmosphere (*Figure 13.3*). These methods can shift the ecology of the microbial flora; for example, whilst CO_2 packaging of fresh fish in ice is highly effective, it can suppress the growth of respiratory bacteria so that fermentative *Photobacterium*, lactic acid bacteria, and enterobacteria become dominant. Spore-forming bacteria (*Bacillus* and *Clostridium*) can survive mild heat treatment (pasteurization) of vacuum-

Figure 13.3 Fish preservation, old and new. (A) Fishwives packing herring into barrels for salting, circa 1920. (Image courtesy of North Shields Library Services.) (B) A factory for processing basa fish reared by aquaculture for international shipping as frozen fillets, showing preparation under high levels of hygiene before packaging in a CO_2-modified atmosphere. (Image courtesy of CL-Fish Co. Ltd, Vietnam.)

packed products. As well as spoilage, some fish products are the source of human pathogens. Fish- and shellfish-associated bacterial, viral, and toxic diseases were considered in Chapter 12, but other pathogens (e.g. *Salmonella* spp., *Staphylococcus aureus*, and *Listeria monocytogenes* may be introduced during handling and processing. *Listeria* is of particular concern in lightly preserved ready-to-eat products such as cold-smoked fish and shellfish because of its ability to grow over a very wide temperature range and salinity. Control of these pathogens is an important requirement for commercial processors.

Besides improvements in packaging methods, a number of biotechnological approaches to the control of specific organisms have been investigated. Antibiotics such as tetracyclines were once added to ice on board fishing vessels and in markets to prevent spoilage of fresh fish. This practice is now prohibited because of concerns over antibiotic residues and resistance. However, bacteriocins from lactic acid bacteria are generally regarded as safe and are permitted in foods (e.g. nisin is widely used in dairy products). With fish, addition of purified bacteriocins has produced some promising results, as has the addition of nonspoilage lactic acid bacteria as competitors of pathogens or spoilage organisms. It has been shown that quorum-sensing signaling molecules (see p. 81) are produced during development of the microbial spoilage community, and interference with the quorum-sensing mechanism might offer a potential new method of control.

Some seafood products are produced by deliberate manipulation of microbial activities

The spoilage activities of microbes in fish are generally regarded as detrimental, but in some parts of the world, there are a number of ethnic food products that depend on microbial activities for preservation and flavors. Most of these are encountered in Asian countries, especially Indonesia, Thailand, the Philippines, and Japan. Many processes are conducted according to traditional recipes, but the microbiology of some has been investigated during commercialization of production for export, and pure starter cultures may sometimes now be used. Perhaps the best-known product is *nam-pla* (fish sauce), which is now widely used in the West because of the popularity of Thai cuisine. *Nam-pla* is made by fermentation of fish hydrolyzed by a high salt concentration (15–20%), which encourages growth of the extreme halophile *Halobacterium salinarum*, leading to characteristic flavors and aromas. This appears to be the only food product that relies on activities of a member of the *Archaea*. Other high-salt products include *som-fak*, *burong-isda* and *jeikal*, a traditional Korean food made from fermented shrimp. Other fermented fish products such as *plaa-som* use lower salt concentrations; in these, a microbial flora dominated by lactic acid bacteria and yeasts leads to a characteristic aroma. There must be enough salt present (2–8%) and a final pH of less than 4.5 to inhibit the growth of pathogens; nevertheless, contamination by *Staphylococcus* can be a problem. Garlic is a major ingredient of some recipes; not only does this serve as a carbohydrate source, but it also has antibacterial activity, and it is interesting that some constituents of garlic inhibit quorum sensing. *Ika-shiokara* is made from squid and fish guts pickled in 2–30% salt in a process that depends on growth of the yeast *Rhodotorula*; if this sounds appealing, you will find it as a delicacy in Hokkaido, Japan.

MICROBIAL ASPECTS OF MARINE POLLUTION BY SEWAGE

Coastal pollution by wastewater is a significant source of human disease

A large proportion of the world's population lives near the coast. Many of the largest urban settlements have grown up around river estuaries and natural harbors because of their importance for trade. As these towns and cities developed, it was an easy option to dispose of untreated sewage directly into the rivers and sea; the adage "the solution for pollution is dilution" applied. In many developed countries, awareness of the potential problems arising from disposal in close proximity to the population led to longer and longer pipelines off the coast, but the grounds for this were usually esthetic rather than health-related. In many countries, vast quantities of untreated sewage are still disposed directly to sea. Even in developed countries, where most medium sized communities will have sewage treatment works, human waste containing potential pathogens still finds its way into coastal waters from isolated dwellings, houseboats, and marinas. Mixture of untreated sewage with storm-water runoff is a regular occurrence, even in large cities with well-developed sewage systems.

Since the mid-twentieth century, increased wealth and leisure time (especially in Europe and North America) has led to greater use of the sea for recreational use, and a trip to the seaside became an important feature of many peoples' lives. In the 1950s, awareness began to be focused on the potential hazards of swimming in sewage-polluted water. Public awareness of the problem has become more acute because of the growing popularity of seawater bathing and sports such as surfing, sail boarding, and diving. As discussed in *Research Focus Box 13.1*, there are few well-documented large outbreaks of serious disease associated with recreational use of marine waters, but there are many epidemiological studies showing health risks for swimmers in waters polluted by wastewater. In most adults, these diseases are troublesome but not usually life-threatening; however, young children, old people, and those with an impaired immune system are at risk of acquiring more serious infections. Coastal waters used for recreation contain a mixture of pathogenic and nonpathogenic microorganisms derived from sewage effluents, bathers themselves, seabirds, and runoff from agricultural land contaminated by waste from livestock, as well as indigenous marine bacterial pathogens such as the vibrios discussed in Chapter 12. Most of these pathogens are transmitted via the fecal–oral route and cause disease when swimmers unwittingly ingest seawater. Infection via the ears, eyes, nose, and upper respiratory tract may also occur, and it is possible that aerosols may be a significant route of infection for surfers, causing mild respiratory illness. The number of organisms required to initiate infection will depend on the specific pathogen, the conditions of exposure, and the susceptibility and immune status of the host. Viruses and parasitic protozoa may require only a few viable units (perhaps 100 or less) to initiate infection, whereas most bacteria require large doses. The types and numbers of various pathogens in sewage vary significantly according to the distribution of disease in the population from which sewage is derived, as well as geographic and climatic factors. As a general rule, the more serious infections transmitted via the fecal–oral route—diseases such as cholera, hepatitis, typhoid, and poliomyelitis—only become significant for swimming-associated illness if there is an epidemic in the local population. In this case, direct person-to-person transmission is likely to be far more important (however, recall the special case of cholera transmission discussed in Chapter 12).

Although bacteria are used as indicators of sewage pollution in seawater, as discussed below, they are much less important than viruses as a source of infection. Pathogenic enteric bacteria include *Salmonella*, *Shigella* and pathogenic strains of *Escherichia coli*. Dermatitis ("swimmers' itch") and infection of cuts and grazes can be caused by bacteria such as *Staphylococcus*, *Pseudomonas* and *Aeromonas* and the yeast *Candida*. These organisms can also cause ear and eye infections.

A range of human viruses are present in seawater contaminated by sewage

More than 100 human enteric viruses belonging to three main families are transmitted via the fecal–oral route. The *Adenoviridae* are nonenveloped, icosahedral double-stranded DNA viruses most commonly associated with infections of the

---

ⓘ SEWAGE TREATMENT REDUCES ENTRY OF PATHOGENS TO THE ENVIRONMENT

Each day, every human being produces between 100 and 500 g of feces and 1–1.5 liters of urine, which are disposed via sewers, together with water used for flushing and washing. The sewage treatment involves three stages, which successively reduce the load of biological and chemical contaminants. Primary treatment consists of screening and separating solid material, before a second stage in which microbial processes occur in biofilms or flocs in trickling filters, activated sludge systems, or other types of bioreactors. Before discharge, sewage is usually subject to tertiary treatment such as filtration through sand or activated charcoal, or coagulation with iron or aluminum salts. Tertiary treatment usually results in approximately 10^7-fold reductions in fecal bacteria and 10^5-fold reductions in the number of viruses (Bitton, 2005). Disinfection via ultraviolet light, chlorination, or ozone reduces pathogen levels further still, but these are expensive options requiring complex engineering works. Raw or partially treated sewage is still discharged in huge quantities into coastal waters in many parts of the world.

upper respiratory tract, but there are several adenovirus types that cause gastroenteritis, leading to diarrhea and vomiting, primarily in children. The family *Caliciviridae* includes caliciviruses, astroviruses, and noroviruses. These are single-stranded RNA viruses, with small round or hexagonal virions. *Norovirus*, previously known as the Norwalk agent, is the most significant enteric virus in developed countries, probably causing about half of all cases of gastroenteritis. Millions of cases of infection by noroviruses occur each year through contamination of drinking water and foods by infected persons; the disease is a major problem as a cause of local epidemics in hospitals, hotels, holiday camps, and cruise ships. The infectious dose is thought to be very low (perhaps 1–10 virus particles) as stomach acid does not inactivate the virus and immunity to the virus is often incomplete, so they are a major cause of disease associated with swimming or consumption of seafood. The disease causes severe nausea, vomiting, and diarrhea, often accompanied by painful abdominal cramps, headaches, and chills.

The *Picornaviridae* are single-stranded RNA viruses with small icosahedral virions. This group includes *Poliovirus*, which infects the central nervous system, causing paralysis and death. Infections were at their peak in the 1940s and 1950s and prompted the development of effective vaccines and a World Health Organization global eradication program; there are now only a few thousand cases a year in a few regions of Africa and the Indian subcontinent. Coxsackieviruses also belong to this group and are probably the most abundant type in sewage; besides gastroenteritis, they are associated with meningitis and can sometimes cause paralysis.

Infection with hepatitis A virus is endemic in many developing countries and over 90% of children may be infected and excreting the virus. It is most commonly acquired by travelers to such countries, although there have been occasional outbreaks in Europe and North America associated with consumption of shellfish and other foods. Infection damages the function of the liver, but is not usually serious.

Finally, the *Reoviridae* family includes the rotaviruses, which are the most serious causes of infant mortality in Africa and large parts of Asia, transmitted via contaminated drinking water. Even in developed counties, a large proportion of young children are infected, meaning that sewage contains high numbers of the viruses.

Fecal indicator bacteria have been used for many years to test public health risks in marine water

The concept of an indicator organism stems from the idea that pathogens may be present in such low numbers in water that detection may be too difficult or expensive. An indicator is an easily cultivable organism present in sewage, whose presence indicates the possibility that pathogens may be present. Ideally, there will be a built in "safety margin" because the indicator should be present in greater numbers and survive longer in the environment than the pathogens. This methodology was developed from the system used to assess the safety of drinking water, developed over 100 years ago. As we shall see, the traditional bacterial indicators for monitoring pollution of seawater and shellfish fall far short of this ideal and major efforts are in progress to develop more reliable indicators and to improve methods for direct detection of the pathogens themselves.

The coliform bacteria are facultative anaerobic Gram-negative bacilli characterized by sensitivity to bile salts (or similar detergent-like substances) and the ability to ferment lactose at 35°C. The definition of the coliform group is operational rather than taxonomic, with 80 species in 19 genera; the best known of these are *Escherichia*, *Citrobacter*, *Enterobacter*, *Erwinia*, *Hafnia*, *Klebsiella*, *Serratia*, and *Yersinia*. The fecal coliforms—they may also be termed "thermotolerant

Figure 13.4 Outline of membrane filtration method for enumeration of *Escherichia coli* and enterococci in marine water samples. There is a wide variation in the media and conditions used; those shown here are based on the USEPA method. With modern selective and chromogenic media, confirmatory tests are usually not necessary for routine analysis and results can be obtained within 24 h. Filters for *E. coli* are preincubated at 35°C before transfer to 44°C. mEI, membrane enterococci IBG; mTEC, membrane thermotolerant *E. coli*; IBG, indoxyl-β-D-glucoside; IMG, 5-bromo-6-chloro-3-indolyl-β-D-glucuronide; ONPG, o-nitrophenol-β-D-galactopyranoside.

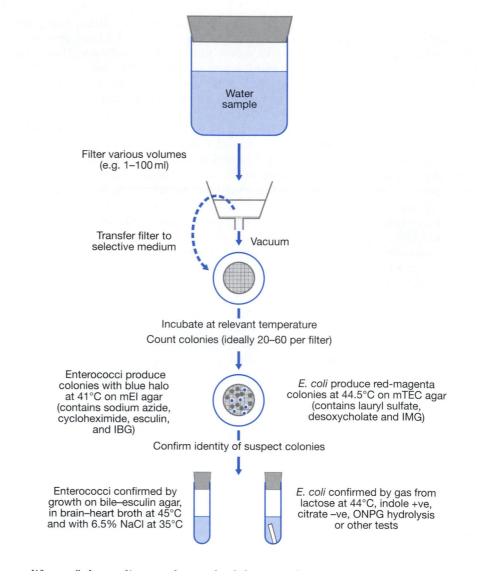

COULD CHILDREN PICK UP DEADLY *E. COLI* INFECTION ON THE BEACH?

We usually think of *Escherichia coli* as a harmless member of the gut microbiota, but several types are pathogenic, of which the most serious is the serotype O157:H7. This causes severe kidney damage and can be fatal, especially in children. Cattle are the main reservoir of *E. coli* O157:H7 and large numbers of the bacteria may be excreted in their feces. Serious epidemics in humans are usually associated with consumption of meat contaminated in the abattoir or direct contact with infected animals, but there have also been links between infection and environmental sources (Muniesa et al., 2006). In southwest England, the death of a child in 1999 and a number of other cases of O157:H7 infections since that date have been linked with pollution of beaches by streams draining from land where cattle are grazing. Further research is needed to track and confirm the sources. Williams et al. (2007) and others have shown that *E. coli* O157:H7 can survive in seawater and beach sand for various time periods, depending on the conditions.

coliforms," depending on the methodology used—are a subset of the coliforms originally distinguished by their ability to grow and ferment lactose at 44°C; the main member of this group is *E. coli*. Growth on selective media and fermentative properties are used in the traditional methods of identification and enumeration in water (*Figures 13.4* and *13.5*). However, traditional tests involve several stages and reactions in biochemical tests (including lactose fermentation) are very variable, so there has been a move over the last decade toward enzyme- and molecular-based tests. A defining feature of the coliforms is the enzyme β-galactosidase, which can be detected using o-nitrophenol-β-galactopyranoside (ONPG) as a substrate, resulting in a yellow color. *Escherichia coli* can be distinguished by the production of β-glucuronidase, which can cleave either methylumbelliferyl-β-glucuronide (resulting in fluorescence under UV light) or proprietary chromogenic compounds (resulting in distinctive colored colonies on agar plates). These reactions form the basis of several commercial testing methods, which have gained widespread acceptance because they produce results more rapidly.

Early microbiologists recognized that *E. coli* and other coliform bacteria are present in large numbers (over 10^8 per gram) in the gut of warm-blooded animals and their feces; therefore they became the standard indicator for fecal pollution of water. In fact, obligately anaerobic bacteria such as *Bacteroides*, *Bifidobacterium*, *Lactobacillus*, and *Clostridium* are about 100 times more abundant in the human gut, but these have not been used so extensively until recently because they are more difficult to cultivate.

Escherichia coli and coliforms are unreliable indicators of human fecal pollution of the sea

It is now generally recognized that, whilst appropriate for testing drinking water, coliform and *E. coli* counts are woefully inadequate indicators of marine water quality. The total coliforms can be derived from a wide variety of sources, including plants and soil, so they often reflect runoff from the land as well as fecal pollution. It is also very difficult to distinguish whether *E. coli* in coastal waters originates from human or animal fecal contamination. Most significant of all, as shown in *Table 13.1*, the survival of indicator coliforms and *E. coli* in the environment does not seem to reflect the survival of viral pathogens. Thus, they fail to meet one of the major requirements of a reliable indicator organism. Many studies have been carried out on the survival of bacteria and viruses in water. The effects of temperature, sunlight, pH, water turbidity, salinity, and presence of organic matter are highly complex interacting variables. Of these, the bactericidal effects of ultraviolet irradiation are the most important. Bacterial indicators have different survival characteristics in marine and fresh waters, while human viruses are inactivated at similar rates in both. Although many different studies have produced conflicting results, depending on the environmental conditions, a general conclusion is that levels of total coliforms are virtually useless as an indicator of health risks associated with marine waters, whilst levels of fecal coliforms (including *E. coli*) do not necessarily indicate good or poor water quality and are unreliable as predictors of health risks (see *Research Focus Box 13.1*). In the USA, *E. coli* has not been used as an indicator for regulatory purposes since 1989 (a review of criteria by the US Environmental Protection Agency is expected in 2010).

The fecal streptococci or enterococci are more reliable indicators for monitoring marine water quality

As early as 1980, it was proposed that the group known as the fecal streptococci might be an alternative indicator for monitoring environmental water quality. These bacteria are consistently associated with the intestines of humans and warm-blooded animals and, as shown in *Table 13.1*, generally have longer survival times in water than coliforms and *E. coli*. It was previously thought that different species were more consistently associated with humans or animals, and a high ratio of fecal coliforms to fecal streptococci has been interpreted as indicative of primarily human fecal contamination, whereas a low ratio was thought to indicate animal sources. These conclusions are now thought to be unreliable

Table 13.1 Typical survival times of indicator organisms and viruses in temperate coastal waters[a]

Type	Conditions[b]		
	Extreme	Moderate	Protected
Coliforms	1 to a few hours	Hours to days	A few days
E. coli	Hours to 1 day	A few days	Days to weeks
Fecal streptococci	1 to a few days	Days to 1 week	Weeks
Human viruses	A few days	Days to weeks	Weeks to months

[a]This is a very simplified outline of survival times as a general guide, since the results of T_{90} or T_{99} calculations (the time taken to reduce the initial population by 90% or 99%) are highly variable.
[b]Conditions that affect microbial survival in water are highly complex and interdependent. Extreme conditions would be typified by clear water, bright sunlight, high temperatures; protected conditions would include turbid water, cloud cover, low temperatures. However, some studies show prolonged survival of coliforms under certain conditions, for example in tropical waters.

because of variable survival times of different species in seawater. Later reclassification led to recognition of a subgroup including the species *Enterococcus avium*, *Enterococcus faecalis*, *Enterococcus faecium*, *Enterococcus durans*, and *Enterococcus gallinarum*. These are differentiated from other streptococci by their ability to grow in 6.5% sodium chloride, at pH 9.6, and in a wide temperature range of 10–45°C; they are also resistant to 60°C for 30 min. Since the most common types found in the environment fulfil these criteria, the terms "fecal streptococci," "intestinal enterococci," and "*Enterococcus*" are used interchangeably. Although the number of enterococci excreted in feces is lower than *E. coli*, they are nevertheless much higher than the numbers of pathogenic bacteria and viruses. Methods of enumeration follow the same principles as those for the coliforms (see *Figure 13.4*) and a variety of specialized media incorporating chromogenic substrates, antibiotics, and tests for specific biochemical reactions can be used. A fluorigenic substrate, 4-methylumbelliferyl-β-D-glucoside tests for the distinctive enzyme β-glucuronidase and is the basis of the Enterolert® testing system, which is now being widely adopted as a standard method. Unfortunately, indigenous marine vibrios and aeromonads can interfere with the reactions and lead to false positive results. Runoff from farmland after heavy rainfall can significantly distort the counts of enterococci, and some studies have attempted to distinguish bacteria of human and animal origin.

Molecular-based methods permit quicker analysis of indicator organisms and microbial source tracking

The conventional culture-based methods used to assess microbial quality of water are widely used because they require relatively simple laboratory equipment, but they suffer from the disadvantage that it usually takes 18–96 h to obtain results. Whilst regular water monitoring will give information about the long-term trends at particular sites, it does not provide information to the public about the state of the water on a particular day. Knowing that the levels of indicator organisms were unacceptably high a couple of days ago and that the beach users should have been advised not to swim does not inspire public confidence and does not adequately protect public health. Conversely, some authorities close beaches for 2–3 days after heavy rains as a precaution because past experience has shown that heavy rains may be associated with increased levels of indicator organisms; the reality may be that any "spike" in the levels of pathogens may have been and gone more quickly. This can be of considerable importance to the economy of coastal communities relying on beach trading as a source of income. For these reasons, there have been many efforts to develop more rapid methods of testing that will produce results in 2–4 h. This would allow management decisions to be made the same day and also permit better tracing of the source of contamination to its origin. Indeed, in the USA the Environment Protection Agency is required by a law passed in 2007 to approve methods that give results within 2 h.

A number of PCR-based studies have been carried out using primers targeting sequences of 16S rRNA genes or functional genes of indicator bacteria. Microarray techniques (see p. 54) have the advantage that they can simultaneously assay for hundreds of different gene sequences. For water testing, the usual slide-based microarray chips are not as successful as suspension systems. Luminex™100® is a commercial system, which uses beads carrying oligonucleotide sequences. DNA is extracted from water and amplified by multiplex PCR using a range of primers. The beads are coated with probes designed to hybridize with these amplicons and are also labeled with spectrally specific fluorescence markers so that quantification of specific targets can be measured in a fluorimeter. This technology is yielding promising results, but improvements in probe design and methods for DNA extraction are required.

Molecular biological-based methods suffer from the requirement for highly trained laboratory personnel and expensive equipment. Generally, they also detect genetic sequences from both dead and viable bacteria, which may overestimate concentrations when compared with culture-based methods. One promising alternative involves the immunomagnetic separation of *E. coli* or enterococci, using magnetic beads coated with antibodies directed against these bacteria. The beads are mixed with a water sample and selectively capture the bacteria; the beads are then collected from the sample using a magnet. The concentration of bacteria present is assessed by removing them from the beads and measuring the amount of ATP present by measuring bioluminescence in a luminometer, using a luciferin-luciferinase assay (see p. 26). This seems to be a reliable and accurate method that is relatively inexpensive and does not require highly trained personnel. However, at present it suffers from the limitation that available antibodies do not detect all strains of *E. coli* or enterococci.

One of the main applications of new methods has been for microbial source tracing. The aim here is to identify populations of bacteria that are specific to particular animals, because recreational waters are often contaminated by non-point sources of fecal pollution of animal origin through runoff of farm waste via streams, or from wildlife including seabirds. Until recently, microbial source tracking was usually carried out by culturing isolates of indicator bacteria from a wide range of sources and building up a library of profiling data, using phenotypic characteristics (such as antibiotic sensitivity or carbohydrate utilization) or genetic fingerprinting methods (such as RFLP, RAPD, AFLP, repPCR, PFGE, or ribotyping; see *Table 2.2*). These methods are very labor-intensive and time-consuming but reveal important information about temporal and geographic changes in communities associated with different animals.

A variety of alternative indicator species have been investigated

Recognition of the limitations of the coliforms, *E. coli*, and (to a lesser degree) enterococci as reliable predictors of health risks in marine waters has led to investigations of a number of alternative indicators. *Clostridium perfringens* has been used as an indicator because it forms resistant endospores that survive for long periods in the environment. This is particularly valuable in monitoring long-term effects of sewage disposal, for example in sediments and offshore sewage sludge dumps. Bacteriophages have been studied extensively, the rationale being that these viruses could be expected to have similar survival characteristics to viruses pathogenic for humans. The most widely investigated bacteriophages are the F+-specific RNA coliphages that infect *E. coli* expressing sex pili on their surface. Coliphages can be measured using plaque assays (plating concentrated water samples on lawns of susceptible host bacteria; see *Figure 7.2*) or with group-specific oligonucleotide probes. Some success has been achieved in distinguishing animal and human sources and in establishing that they may have survival times similar to the human viruses when exposed to ultraviolet light under various conditions. However, some studies have detected enteric viruses in water even when very low levels of coliphages were present.

The order *Bacteroidales* is a group of Gram-negative anaerobic bacteria including *Bacteroides*, *Prevotella* and *Porphyromonas*. *Bacteroides fragilis* is one of the most abundant organisms in human feces and early studies showed that it dies rapidly on transfer to seawater, making it a good indicator of recent human pollution. Detection of *B. fragilis* bacteriophages has also been employed. Despite some promising research results, the requirement for anaerobic cultivation makes these methods unsuitable for many laboratories undertaking routine monitoring. However, detection and quantification of *Bacteroides* and *Prevotella* by quantitative PCR (see p. 44) has proved very useful in tracking sources of pollution. A variety of primer sets have been developed that distinguish between strains from humans and those of other animals such as cattle, horses, pigs, and dogs. Real-time quantitative PCR is used to determine relative concentrations of host-specific strains.

? COULD A PLANT VIRUS INDICATE FECAL POLLUTION?

Recent research by Mya Breitbart's group at South Florida University on metagenomic analyses of viral sequences in the human gut has revealed a surprising result that may lead to the development of new pollution indicators (Breitbart et al., 2003). To obtain baseline values for the distribution and diversity of viruses in the marine environment, Rosario et al. (2009) identified and enumerated a range of viral groups by quantitative PCR. Further surveys revealed that the pepper mild mottle virus (PMMoV; an RNA virus in the genus *Tobamovirus*) is very widespread and abundant in human feces and in treated wastewater and coastal water samples exposed to wastewater. The virus was also found in Atlantic and Pacific coastal waters near discharge sites. However, the virus was not found in the feces of a large number of animals (apart from chicken and seagulls) from the USA. PMMoV could be detected for about 1 week after introduction of sewage, suggesting that it has potential as an indicator of recent sources of human-specific pollution. Why such high numbers of this plant virus are found in the human gut is unknown, although PMMoV RNA sequences are present in many spicy food products containing peppers, suggesting that this might be a source of viruses through ingestion (Colson et al., 2010). However, Colson and colleagues provided preliminary evidence indicating that it might be associated with certain pathogenic conditions; if so, this would be the first example of a plant virus infecting humans.

Table 13.2 Microbiological criteria for quality of coastal and transitional waters in the European Union[a]

Indicator	Excellent quality[b]	Good quality[b]	Sufficient quality[c]
Escherichia coli	100	200	185
Intestinal enterococci	250	500	500

[a]Bathing Water Directive 2006/7/EC. [b]Based on the 95th percentile. [c]Based on the 90th percentile.

Methanogenic bacteria have also been investigated as potential indicator species. Methanogens are widely distributed in marine sediments and activated sludge in sewage plants, but most members of the genus *Methanobrevibacter* seem to be restricted to the mouth and intestines of animals. The species *Methanobrevibacter smithii* appears to be a human-specific strain and is present in human feces at densities between 10^7 and 10^{10} per gram. PCR assays targeting specific sequences of the *nifH* gene have been developed.

Different countries use different quality standards for marine waters

There have been many studies attempting to prove an epidemiological link between the levels of microbial indicators and adverse health risks from recreational use of marine waters. The methodological or statistical basis of such studies has been the subject of much controversy. Public health authorities in many countries have developed monitoring programs and quality standards based on microbiological criteria, and in 2003 the World Health Organization used an evidence-based approach in the development of international guidelines (see *Research Focus Box 13.1*). Nevertheless, there are numerous different classification systems used by different authorities. The nature of the indicators, the frequency of sampling, the methods of quantification and the threshold values for compliance with standards all vary. Within the USA, for example, different states use different criteria. Some variations are probably appropriate depending on local conditions; for example, *Clostridium perfringens* may be a more reliable indicator in tropical waters. Because there can be considerable day-to-day variation in the numbers of indicator organisms detected, many systems are based on percentage compliance levels, such as the requirement in the European Union that 90 or 95% of measurements taken during the sampling period meet the prescribed standard (*Table 13.2*). Other authorities such as the US Environmental Protection Agency use geometric mean values of data to even out unusually high or low results. The latter method is statistically more stable but has the disadvantage of overlooking the occasional high values at the top of the statistical distribution, which might be of public health concern. The 90/95% compliance system is statistically less robust, and different results can be obtained depending on the method of calculation. However, it has the advantage of reflecting the occasional high levels of concern and is better understood by the public.

Sewage pollution of water in which shellfish are harvested for human consumption poses a serious health hazard

Molluskan bivalve shellfish such as oysters, mussels, clams, and cockles concentrate pathogens from their environment by filter feeding. For example, a single oyster filters about 5 liters of water a day and each square meter of a dense bed of mussels filters more than 100000 liters a day. Thus, even if there are low concentrations of pathogenic microbes in the water, there is the potential for very large numbers to accumulate in the bodies of the animals. The rate of accumulation of microbes and their subsequent survival within the tissue depends on the species involved and numerous environmental factors, especially temperature.

ⓘ HIGH PRESSURE PROCESSING MAKES SHELLFISH SAFER

The only sure way of eliminating pathogens—such as viruses from sewage contamination or indigenous marine *Vibrio* bacteria—from shellfish is by thorough cooking. However, many consumers reject this as it alters the desired taste and texture of shellfish, especially premium products such as oysters. Very high hydrostatic pressure (typically 275–400 MPa) can be used as an alternative to heat for the processing of sensitive foodstuffs and this method has proved effective in reducing bacterial viability and prolonging shelf-life of shellfish (Murchie et al., 2005). This process is now used widely in commercial processing of oysters and some other shellfish; it also has the advantage of making oysters easier to open and "plumping up" the volume of the meat. Importantly, high-pressure treatment has been shown to inactivate *Vibrio vulnificus, V. parahaemolyticus,* and *V. cholerae,* and pathogenic viruses such as caliciviruses and hepatitis A virus (Calci et al., 2005). Unfortunately, there is little evidence that this method leads to the inactivation of algal toxins, another significant health hazard associated with shellfish.

Shellfish are often subjected to a process of disinfection known as depuration before they are sent to market for human consumption. Depuration involves transferring the animals for 24–48 h to free-flowing or recirculating clean seawater, which is disinfected by ultraviolet light, chlorination, or ozone treatment, so that the natural filtering activity of the bivalve mollusks results in elimination of their intestinal contents. The processes of depuration were developed in the early twentieth century to eliminate the risk of typhoid fever caused by the bacterium *Salmonella typhi*, which caused serious epidemics at that time, associated with eating shellfish grown in polluted water. Depuration is effective in removing *S. typhi* and other fecal bacteria including indicator organisms (see below) from shellfish. However, it is much less effective in the removal of sewage-associated viruses, naturally occurring bacterial pathogens such as *Vibrio parahaemolyticus* and *V. vulnificus* (see p. 263), dinoflagellate and diatom toxins (see p. 270), or chemical pollutants. Hepatitis (of the A or "non-A, non-B" serotypes) acquired from shellfish is of particular concern; although rarely fatal, it can cause a long, severe debilitating illness and some large outbreaks involving hundreds of cases have occurred; for example, following the serving of contaminated shellfish at banquets or receptions. Norovirus is the most common cause of seafood-associated gastroenteritis but usually only comes to the attention of health authorities when outbreaks involving large groups of people occur. It is likely that thousands of undocumented cases of seafood-associated gastroenteritis occur each year, since most victims do not visit their doctor and, in any case, accurate diagnosis is difficult. As occurs with illnesses caused by the indigenous marine bacterial pathogens *V. parahaemolyticus* and *V. vulnificus*, the problem is exacerbated by the fact that shellfish are often eaten raw or only partially cooked. Many of the enteric viruses are resistant to moderate heating, low pH, freezing, and drying, so normal food-processing methods have little effect.

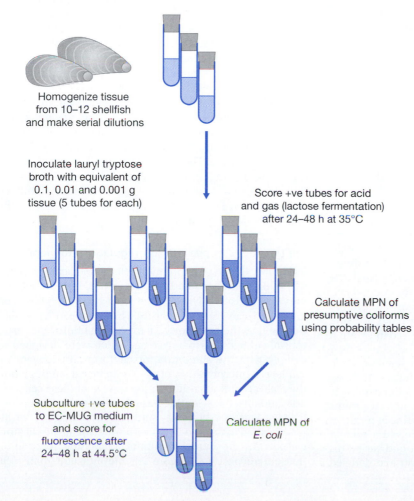

Homogenize tissue from 10–12 shellfish and make serial dilutions

Inoculate lauryl tryptose broth with equivalent of 0.1, 0.01 and 0.001 g tissue (5 tubes for each)

Score +ve tubes for acid and gas (lactose fermentation) after 24–48 h at 35°C

Calculate MPN of presumptive coliforms using probability tables

Subculture +ve tubes to EC-MUG medium and score for fluorescence after 24–48 h at 44.5°C

Calculate MPN of *E. coli*

Figure 13.5 Outline of most probable number (MPN) technique for determining coliforms and *Escherichia coli* in shellfish. There is a wide variation in the media and conditions used; those shown here are based on the USEPA method. The test is done in two stages to ensure recovery of damaged bacteria and reduce interference by substances in shellfish tissue. EC-MUG medium contains 4-methyl-umbelliferyl-β-D-glucuronide, a fluorogenic substrate for glucuronidase (distinctive of *E. coli*). Calculation of MPN is based on the statistical probability that some tubes will be inoculated with a single organism from the dilutions used.

BOX 13.1   RESEARCH FOCUS

Is it safe to swim?
The difficulties of linking water quality and health risks

Establishing a correlation between the health effects of bathing and the levels of indicator organisms

The first attempts to investigate this area were carried out by the US authorities in the 1950s in freshwater bathing sites. These showed that there was a significant increase in gastrointestinal, ear, nose, and throat symptoms in swimmers exposed to waters where the coliform count exceeded 2700 per 100 ml. This was used by the US authorities to set the first standards for recreational waters. The epidemiological basis for defining these standards was questioned by many authors, most notably Cabelli et al. (1982), who undertook more detailed studies at US beaches. Rigorous attempts to ensure standardization of microbiological methods, reporting of symptoms by participants, and the use of suitable control groups were included. Cabelli's group found that levels of fecal coliforms were not a reliable indicator of health risks, but that there was a statistically significant correlation between the levels of enterococci and an increased incidence of gastrointestinal symptoms associated with swimming. As a result, the US Environment Protection Agency (USEPA) dropped the use of fecal coliforms as indicators in marine waters and set a new standard based on a geometric mean of 35 fecal streptococci per 100 ml in five samples over a 30-day period. Many similar surveys have been undertaken, and meta-analyses of the data from numerous epidemiological studies have been conducted (Prüss, 1998; Wade et al., 2003; USEPA, 2007). There is a consistent correlation between increased risk of gastrointestinal symptoms and the counts of fecal indicator bacteria, especially enterococci. Some studies have reported increased rates of skin symptoms among swimmers, but controlled studies have not produced convincing evidence of links with fecal contamination.

The development of controlled studies

One of the major limitations of the types of study conducted by Cabelli and others is that they rely on *perception* of illness by those taking part, who are usually asked to respond to a questionnaire. It is also very difficult to distinguish the true effect of swimming from other effects; for example, illness might result from food consumed during a trip to the beach. In the 1990s, growing public concern and pressure from groups such as Surfers against Sewage and the Marine Conservation Society led the UK Government to sponsor new epidemiological research designed to overcome some of the criticisms of earlier studies. Carefully controlled beach surveys were conducted, in which volunteers were given a medical examination before randomized groups undertook supervised swims with water testing at frequent intervals. One week later, participants received a second medical examination plus testing of throat and ear swabs and fecal specimens. Careful randomization and detailed

analysis of questionnaires allowed other risk factors such as food intake to be taken into account. The UK study (Fleisher et al., 1998; Kay et al., 1994) confirmed that there was close correlation between levels of enterococci (measured at chest depth) in bathing water and the increased risk of gastrointestinal illness. A significant increase in gastrointestinal disease occurred when the number exceeded 30–40 per 100 ml and a convincing dose–response curve was observed.

World Health Organization guidelines

Based on a systematic review of these and other studies linking health risks with bathing, the World Health Organization (2003) published new guideline values for the microbiological quality of recreational waters. These use a statistical approach in which 95% of the counts must lie below a threshold value. Using this system, the estimated additional risk of bathing (compared with a control group) in waters with less than 40 colony-forming units (CFU) of enterococci per 100 ml is less than 1% (grade A water). Between 41 and 200 CFU per 100 ml (grade B water), this rises to a 1–5% increased risk, and between 201 and 500 CFU per 100 ml (grade C water) the increased risk is 5–10%. In grade D water, containing more than 500 enterococci per 100 ml, there is a significant risk of contracting gastrointestinal illness (>10% increased risk). The risk of acquiring acute febrile respiratory infections is significant in grade C and D waters, when it rises to 3.9%. It should be noted that these risk factors apply to healthy adults, as children are usually excluded from this type of epidemiological study for ethical reasons. The risk of infection is probably much higher for lower age groups. It should also be noted that the data on which the standards are based are mainly obtained from Northern Europe and North America, so may not be applicable in all regions.

Are the new European Union regulations too lenient?

In Europe, the first definition of microbiological standards was published in 1976 as the European Community Bathing Water Directive. This defined testing protocols based on total and fecal coliform counts. The European legislation was used as the basis for very expensive improvements to coastal sewage disposal systems which were designed to comply with the standards for coliform and *Escherichia coli* levels. Interpretation of epidemiological studies led to the realization that, although the motive of these schemes was laudable, the scientific basis for evaluating their effectiveness was questionable, since an increased risk of illness seemed to be associated with water-quality indicators well within the European limits. Increasing recognition that enterococci are a more reliable indicator than fecal

BOX 13.1   RESEARCH FOCUS

coliforms led to the inclusion of a guide level for enterococci in amendments to the European legislation in 1998. After a long period of discussion and argument, revised microbiological conditions were adopted in 2006, which is expected to shift the emphasis from treatment and disposal of sewage to expanding the "Blue Flag" classification system for beaches and improvements in public information (Mansilha et al., 2009). Health authorities will have to make decisions on what constitutes an "acceptable risk," but public information on the interpretation of the water quality results posted at many seaside beaches is woefully inadequate to enable people to make an informed personal choice about the risks of bathing or undertaking water sports. However, in view of the importance of international tourism, public confidence in monitoring procedures is essential; and the current diversity of approaches is confusing.

Surprisingly, the current European level of 100 enterococci per 100 ml (indicating "excellent" water quality) is considerably greater than the level predicted by the epidemiological studies as carrying a significant increased probability of illness and is much higher than that in many other developed countries. For example, the USA and Canada use a threshold value of 35 enterococci per 100 ml (based on a geometric mean of 5 samples over a 30-day period). In Australia and New Zealand, the standards for primary contact activities such as swimming and surfing are 150 fecal coliforms and 35 enterococci per 100 ml (with maximum permissible levels of 600 and 100 respectively); the authorities also define less stringent standards for secondary contact activities such as boating and fishing. It is surprising that the rank of "excellent quality" water under European standards is only

the equivalent of grade B in the WHO guidelines. Efstratiou and Tsirtis (2009) tested seawater samples from coastal sites in Greece and found significant levels of the pathogens *Salmonella* and *Candida albicans* in waters that scored "excellent" or "good" results in the 2006 EU standards, and concluded that the lower levels of the single indicator required by US standards is a more reliable proxy for the risk of contact with pathogens.

Are the risks higher for other recreational users?

Participants in water sports such as body boarding, surfing and scuba diving may have prolonged exposure and increased risks of ingestion of seawater or inhalation of aerosols. Despite many anecdotal reports of illnesses among participants, there have been very few studies of health risks from microbial infections. In a survey of surfers who spent a mean of 77 days a year surfing in Oregon, USA, Stone et al. (2008) concluded that surfers had a mean exposure of 11–86 CFU enterococci per day, with an epidemiological model predicting that they have a 23% probability of exceeding the exposure equivalent to the USEPA maximum acceptable gastrointestinal illness rates. In a web-based survey of surfers, Turbow et al. (2008) found a strong correlation between water-quality impairment and the number of complaints of gastrointestinal and upper respiratory symptoms. Water sports activities may also take place outside of the official "bathing season," meaning that the authorities do not undertake checks of water quality. For example, in southwest England the best surfing conditions often occur in the winter, when potential pathogens may survive for longer or be resuspended from sediments by wave action (Bradley and Hancock, 2003).

Many countries have microbiological standards for the classification of waters in which shellfish are cultivated

The bacteriological quality of shellfish is usually determined by a most probable number method to quantify *E. coli* as an indicator of fecal pollution (*Figure 13.5*). For example, in the USA and Australia, waters used for harvesting shellfish for direct human consumption must not contain more than 70 total or 14 fecal coliforms per 100 ml. In Europe, testing is carried out on an "end-product" basis by testing bacterial counts in replicate samples of shellfish flesh, the results of which are used to classify harvesting waters, as shown in *Table 13.3*.

Again, the use of *E. coli* as an indicator is a very unreliable predictor of health risk, and the standards have no epidemiological basis. Many outbreaks (sometimes large and severe) of viral illness have been associated with shellfish that appear to be of high standard as judged by the levels of bacterial indicators. Shellfish for human consumption are also tested for the presence of microbial toxins, for which there are separate standards (see p. 271).

Direct testing for pathogens in shellfish is possible using molecular techniques

The above discussion suggests that the indicator concept has major shortcomings when applied to marine waters and shellfish. Would it not be better to undertake direct testing for the pathogens themselves? The problem here is that

Table 13.3 Classification of shellfish (bivalve mollusks) harvesting areas in the European Union according to microbiological quality[a]

Category	E. coli per 100 g	Fecal coliforms per 100 g	Comments
A	<230	<300	May be harvested for direct human consumption
B	<4600	<6000	Must be depurated in approved unit, heat treated, relayed in an approved category A area, or heat treated by an approved method before sale for human consumption
C	<46000	<60000	Must be relayed for a minimum of 2 months to meet category A or B requirements, or be heat treated
Prohibited	–	>60000	Unsuitable for shellfish production

[a]Shellfish Hygiene Directive 1991.

most viruses, the pathogens that cause most concern, are difficult or impossible to culture. The usual method is to inoculate cell cultures of suitable human cells (such as enterocytes) with water samples (after concentration) and monitor for cell lysis or other cytopathic effects. Special safety precautions must be taken in laboratories using human cell culture, and the effects, if any, may take several weeks to appear. All attempts to cultivate norovirus—the most significant shellfish-associated pathogen—have failed. Attention has therefore turned to molecular biological techniques, in which the most promising approach has been detection of viral RNA sequences by reverse transcriptase PCR method. Suitable primers that target the RNA virus group under study are added to samples of water or shellfish tissue extracts in the presence of the enzyme reverse transcriptase to make a cDNA copy, which is then amplified by Taq polymerase, usually using a nested PCR (see p. 44). Difficulties with the technique include selection of suitable primers that are representative of virus strains circulating in a particular region, and interference with the PCR amplification by inhibitors, which can be a particular problem in shellfish tissue. The methods are still too time-consuming and expensive to be used in routine testing. However, they are valuable in confirming the sources of disease outbreaks and special studies such as environmental impact assessments of new sewage disposal schemes and classification of shellfish harvesting areas.

OIL AND OTHER CHEMICAL POLLUTION

Oil pollution of the marine environment is a major problem

The world economy is totally dependent on petroleum products as a source of energy and raw materials for a vast range of products. Over 3 billion tonnes of crude oil are extracted annually, and about 0.1% of this finds its way into the sea during the extraction, transportation, and consumption of crude oil and petroleum products (*Figure 13.6*). Although it is not the largest source, pollution by oil tankers attracts the greatest public concern. Tankers take on water for ballast and, when this is discharged, considerable local contamination of the sea occurs. Occasionally, tankers collide or run aground, releasing large quantities of oil into the sea. Fortunately, in recent years, tighter regulations on ballast and greater use of double-hulled tankers has reduced the amount of oil spilt in this way, but major incidents still occur. The immediate damaging effect of this on marine wildlife and the economic impact on fisheries and tourism means that environmental agencies are under intense pressure to alleviate the problem quickly and effectively. Apart from immediate effects, there are concerns about toxic residues and long-term disruption of ecosystem communities. There are various strategies for dealing with spilled oil, including trapping with booms and skimming the surface to remove the oil, using absorbent materials to soak up oil, applying

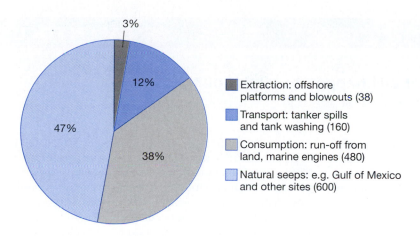

Figure 13.6 Sources of oil inputs to the sea (based on data from Oil in the Sea Report, Ocean Studies Board and Marine Board of the National Academy of Sciences, 2003). Figures in parentheses show approximate input in kilotonnes.

dispersant chemicals, or setting fire to oil slicks. The efficacy of these processes is very dependent on the location of the spill and weather conditions, and the use of dispersants or burning can have highly damaging effects on marine ecology.

A range of microbes are responsible for biodegradation of oil at sea

Fortunately, naturally occurring microbes break down most of the components of petroleum, which is a complex mixture of thousands of compounds, predominantly hydrocarbons. Large quantities of crude oil have been seeping naturally into the oceans for millions of years, and many different microbes in more than 160 genera have evolved mechanisms to degrade the different hydrocarbon components—these have been isolated from waters all over the world. Bacteria seem to be the most important, but yeasts and filamentous fungi have also been described. Most hydrocarbon-degrading isolates are heterotrophs belonging to the *Gammaproteobacteria* (e.g. members of the orders *Oceanospirillales*, *Alcanivorax*, *Cycloclasticus*, *Pseudomonas*, and *Acinetobacter*). *Cyanobacteria* (e.g. *Agmenellum*, *Microcoleus*, and *Phormidium*) and the alga *Ochromonas* have also been linked with hydrocarbon degradation; some species appear to accumulate hydrocarbons within vesicles, but not degrade them. It is possible that these photoautotrophs form consortia with heterotrophic bacteria, leading to breakdown. In anoxic environments, SRB and archaeal methanogens can degrade hydrocarbons as a sole source of carbon and energy. The breakdown of oil depends on the activities of a consortium of microorganisms, each responsible for transformation of a particular fraction. Normally, these constitute only a small proportion (<1%) of the microbiota of seawater, but in a polluted environment they multiply rapidly to as much as 10% of the population. Immobilization of mixed microbial communities on biofilms may be particularly important in efficient biodegradation. *Research Focus Box 13.2* describes how the use of dispersants at the wellhead enabled rapid degradation of oil in the *Deepwater Horizon* blowout in 2010, one of the worst ever pollution incidents.

The fate of oil depends on a combination of physical and biological processes

When oil is released into the seas as a result of a spillage, it floats on the surface and the low-molecular-weight fractions evaporate quickly. Some components are water-soluble and photochemical oxidation occurs through the action of sunlight. The fate of an oil slick is very dependent on wave action and weather conditions. Sometimes, droplets of oil will become emulsified in the water and disperse quickly; in other cases a water-in-oil emulsion will form to produce a thick viscous mousse, which takes a very long time to disperse and gives rise to the familiar lumps of beach tar. Different crude oils vary, but usually over 70% of the hydrocarbons are biodegradable. Only the asphaltenes and resin components are recalcitrant to breakdown.

BOX 13.2   RESEARCH FOCUS

What will happen to all the oil?
Bacteria clean up deep-sea oil after the *Deepwater Horizon* disaster

In April 2010, the *Deepwater Horizon* platform was drilling an exploratory well for oil at a depth of 1500 m, 200 miles off the coast in the Gulf of Mexico. A blowout led to a methane explosion and fire that killed 11 workers. The fire raged for 36 h, before the platform sank, leaving the wellhead seriously damaged and gushing huge quantities of crude oil into the sea. A massive operation involving hundreds of ships and shore-based operations was put in place to remove surface oil or trap it in containment booms in an attempt to protect the thousands of miles of beaches, estuaries, and wetlands on the Gulf Coast. Throughout the summer of 2010, nearly five million barrels of oil (780 thousand cubic meters) of crude oil flowed unchecked into the sea, until the well was sealed with a concrete cap on July 15. By the first week of August, oil could no longer be detected in the deep water column or on the ocean surface, and on September 19, the US Coast Guard declared the well dead.

From a microbiological perspective, one of the most interesting consequences of the spill was the fate of the massive plumes of oil released in the deep water. This is a cold, high-pressure environment with very low levels of carbon, so it was not known how the microbial community would respond to such a sudden influx of hydrocarbons. We know that there have been natural oil seeps from the seabed over millions of years, so it is likely that the microbial community has adapted to utilize oil efficiently. Because of the scale of the release, more than four million liters of chemical dispersant (Corexit 9500) were used to treat the oil as it escaped from the wellhead. It was reasoned that dispersing the oil would facilitate its breakdown by heterotrophic microbes, but the main reason for using it at the well head was for safety reasons—the light crude oil was highly volatile and would be highly flammable if it came to the surface. The injection of dispersant successfully forced the oil to come to the surface at a distance from the wellhead as a result of creating small droplets that were less buoyant and moved more easily by ambient currents.

In June 2010, scientists from the Woods Hole Oceanographic Institution showed the presence of a plume of highly dispersed droplets of hydrocarbons more than 200 m high and 2 km wide, which extended over 35 km from the source (Camilli et al., 2010). They used underwater sampling equipment to assess oxygen levels and some of the more easily measured hydrocarbons. They reported that there seemed to be no significant reduction in oxygen concentration, suggesting that the rate of microbial degradation of the oil might be low and that it would take many months to disappear. However, in a second major study of the plume published just 1 week later, a team from the Lawrence Berkeley National Laboratory found evidence of rapid degradation of the oil (Hazen et al, (2010). The apparent contradiction in the findings was perplexing, but Hazen's team analyzed the microbial composition of the plume, rather than simply measuring oxygen consumption. The analysis was performed using the PhyloChip microarray that was developed in the Berkeley laboratory as a rapid method for the accurate detection of up to 50000 different species of bacteria and archaea in an environmental sample. They found a significant stimulation of psychrophilic bacteria that were closely related to organisms known to degrade oil rapidly, with the dominant microbe in the oil plume being a new species assigned to the order *Oceanospirillales*, closely related to *Oleispirea antarctica* and *Oceaniserpentilla haliotis*. This bacterium was present at about 90% of the population within the plume, but only 5% in the surrounding water.

Analysis of microbial genes in the dispersed oil plume correlated with changes in the concentration of various components of the oil and rapid biodegradation rates were demonstrated in laboratory studies, with the half-life of alkanes ranging from 1.2 to 6.1 days. Hazen et al. (2010) concluded that the degradation proceeded so rapidly because the oil contained a high component of readily degradable volatile components, and because the oil was dispersed into small droplets, providing a large surface area amenable to attack by the bacteria. However, there does seem to be a constraint on the rate of degradation and that is reassuring in view of the fears about massive oxygen depletion creating a "dead zone." Hazen and colleagues concluded that the rates of degradation were checked because the enzymes involved depend on iron as a cofactor and iron levels were very low in the water, and because the average concentration of oil in the plume was only 1 ppm. The oxygen saturation inside the plume was only a few percent lower than outside the plume.

At the time of writing (September 2010) there is still some uncertainty about the fate of the deep-sea oil, and some scientists are very skeptical that it has disappeared completely. Certainly, there will be some recalcitrant components that will take much longer to degrade, and there are probably deposits of undispersed oil on the seabed. Also, the dispersants used are known to be toxic to marine life and—as they had never been used in the deep sea previously—their long-term effects are not known. Whatever happens 1500 m down, oil that has contaminated the surface water and the shores, creeks, mangroves, and salt marshes will behave very differently. The legacy of the *Deepwater Horizon* disaster will undoubtedly affect the ecology and economy of the Gulf region for many decades to come and some sensitive areas may never recover.

The biochemical processes involved in the biodegradation of oil have been studied extensively. Aerobic processes are responsible for the efficient biodegradation of oil. The initial step involves incorporation of molecular oxygen into *n*-alkanes by oxygenases, resulting in primary alcohols, which are then further oxidized to aldehydes and fatty acids. The fatty acids are then metabolized to acetyl-coenzyme A, which enters primary metabolism via the tricarboxylic acid (TCA) cycle. In this way, the *n*-alkanes are completely converted to carbon dioxide and water. There are many different routes for the degradation of the aromatic hydrocarbons, the key step being the cleavage of the aromatic ring; oxygen is again essential for this process. Polyaromatic hydrocarbons (PAHs) such as naphthalene, phenanthrene, and pyrene contain four or more aromatic rings and are only degraded slowly by a few types of bacteria. Consequently, PAHs that are not degraded in the water column accumulate in sediments and can persist for very long periods. Many harbors and estuaries that are used extensively by shipping have sediments that are chronically polluted with PAHs. In such anoxic sediments, anaerobic degradation occurs with the oxidative processes linked to the reduction of nitrate or sulfate, or the production of methane. Degradation probably begins with carboxylation reactions, followed by ring reduction and cleavage. The activity of burrowing benthic animals can enhance aerobic PAH degradation by microbes, owing to the mixing of sediments and introduction of oxygen through the elaborate ventilation systems that they construct; isolation of bacteria from such burrows may prove a rich source of new species.

Biodegradation is enhanced by addition of emulsifiers

Biodegradation proceeds most quickly when the oil is emulsified into small droplets and hydrocarbon uptake by microorganisms is stimulated by the production of biosurfactants, surface-active agents containing both hydrophilic and hydrophobic regions which reduce surface tension. Many microbes have hydrophobic surfaces and adhere to small droplets of oil, and many also produce extracellular compounds that disperse the oil. There has been great interest in developing natural biosurfactants as an alternative to chemical dispersants. For example, the bacterium *Acinetobacter calcoaceticus* is particularly effective in biodegradation of oil because it both adheres to hydrocarbons and produces an extracellular glycolipid biosurfactant called emulsan. Emulsan-deficient mutants grow very poorly on hydrocarbons. It may be possible to select naturally occurring strains or use genetic engineering to produce large amounts of biosurfactants, although optimizing the industrial-scale production of these compounds may be difficult. Emulsan is also used to reduce viscosity to aid in the extraction of crude oil.

Addition of nutrients is necessary to increase the rate of oil biodegradation

The hydrocarbons in oil provide a carbon source for bacteria, but oil is deficient in other nutrients (especially nitrogen and phosphorus) and the supply of these is the main factor limiting the rate of degradation. Therefore, the most successful approach to bioremediation is the addition of inorganic or organic nutrients as fertilizers to speed up natural processes. The process of seeding oil spills with exogenous microorganisms shown to have high degradative activity in the laboratory (bioaugmentation) has been less successful because they are rapidly outcompeted by the enrichment of naturally occurring microbes. Furthermore, the idea that genetic modification could be used to create a "superbug" to digest different components of oil in the marine environment has proved to be misguided. Bioremediation of petroleum products is applied extensively to clean up contaminated soil (e.g. to reclaim land polluted by spillage from oil tanks), and many commercial products have been developed. However, the scientific rigor with which these have been tested for marine bioremediation is questionable. A major problem in testing the effectiveness of bioremediation is monitoring the extent of degradation. Because breakdown proceeds in a progressive fashion, disappearance of compounds such as the alkanes and small aromatic compounds can be measured easily, but monitoring removal of the more recalcitrant compounds is more problematic. One internal standard method is to measure the disappearance of biodegradable components in comparison with the concentration of hopanes, which are highly recalcitrant to breakdown.

Bioremediation has been used to lessen the impact of oil spills on vulnerable coasts

Bioremediation is defined as a biological process to enhance the rate or extent of naturally occurring biodegradation of pollutants, although in a broader sense it can be used for deliberate use of any biological process that reverses environmental damage. Laboratory studies provide little information about how well bioremediation treatments will work in the field. Some mesocosm and controlled release experiments in the field have been done, but these obviously have to be limited in size and scope, as few authorities are willing to allow the deliberate pollution of coastal waters. Therefore, most of our knowledge about the efficacy of bioremediation comes from studies of opportunity following large-scale spills from tanker accidents, in which investigators have little control over the prevailing conditions. Of such incidents, the cleanup after the *Exxon Valdez* spill in Alaska has been the most extensive study of bioremediation. Deliberate experimental contamination of shorelines in Norway and Canada had shown previously that, even under Arctic conditions, oil applied to shorelines would be degraded naturally within a few years, but that addition of agricultural fertilizers like ammonium phosphate, ammonium nitrate, or urea increased the initial rate of biodegradation up to 10 times. The best approach to fertilization seems to be to use oleophilic compounds, which stick to the oil and/or release nutrients slowly. One such slow-release fertilizer is a proprietary compound called Inipol™ EAP22; this is a microemulsion of urea in brine, encapsulated in an external phase of oleic acid and lauryl phosphate, co-solubilized by butoxyethanol. Bioremediation has also been used in other major oil spills, such as the *Prestige* that affected the coast of Galicia in Spain in 2002 and in oil released in Kuwait during the 1991 Gulf War. Bioremediation of sediment contamination is more problematic. In terrestrial situations, biodegradation is enhanced by tilling, which introduces O_2, but this is obviously impractical in marine situations. It may be possible to introduce chemical oxidants or alternate electron acceptors, but this requires careful evaluation of ecosystem effects. There have been some successes with the introduction of O_2 by aeration pumps. Bioremediation in sensitive habitats like mangroves and salt marshes is particularly difficult.

In summary, bioremediation by the application of oleophilic or slow-release fertilizers is now generally accepted to have proved its worth as one component in the response to an oil spill. However, before it is used, careful attention must be paid to the nature of the substratum and the degree of penetration of oil into the sediments. Addition of exogenous organisms has not been successful, but further development of surfactant-producing strains and their formulation into products that allow them to compete with the stimulation of indigenous microorganisms may hold promise for the future. More research on changes in microbial community composition in response to introduction of oil containing different mixtures of hydrocarbons may also yield valuable information about the best ways to enhance natural processes of degradation.

Microbes are important in the distribution of persistent organic pollutants

Persistent organic pollutants include the organochlorine insecticides (e.g. DDT, aldrin, and chlordane), industrial chemicals (e.g. polychlorinated biphenyls; PCBs) and by-products (e.g. dioxins and furans). They reach the sea via terrestrial runoff and atmospheric deposition. In numerous countries the manufacture and use of many of these chemicals has been prohibited, but they are highly persistent in sediments and dump sites and produced during incineration of waste. PCBs occur in many electronic products, and the disposal of unwanted equipment is a significant source of these chemicals. These chemicals are very resistant to photochemical, biological, and chemical degradation. Most persistent compounds are halogenated and highly soluble in lipids and they accumulate particularly in fatty tissues. Their semivolatile nature allows them to vaporize or to be adsorbed onto atmospheric particles and they are therefore transported over great distances. For example, animals and humans living in polar regions have high levels of organic pollutants in their tissues despite these compounds not being used there. Extensive evidence links these compounds to reproductive failure, impairment of the immune system, deformities, and other malfunctions in a wide range of marine life (see p. 252), and they are highly toxic to humans through ingestion of fish and other routes.

Microbial cycling of organic pollutants acquired by plankton in the upper parts of the ocean plays an important role in their distribution through ocean food webs. The bacterioplankton presents a large surface area for the adsorption of organic pollutants, and microbial loop processes release compounds during settlement of plankton debris and organic particles through the water column, but some particles with adsorbed pollutants will be buried in sediments. Disturbance of sediments by tides, currents, dredging, and the activity of benthic animals can release large quantities of the chemicals into the water. PCBs are highly recalcitrant to degradation and there is an active search for microbes capable of breaking down the chlorine bonds, thus offering a potential use in bioremediation. Many aerobic bacteria can degrade the biphenyl ring in PCBs, but not the heavily chlorinated congeners. Anaerobic degradation of PCBs is known to occur, but isolation and identification of the organisms responsible has been elusive; different bacteria with distinct dehalogenases and congener specificities occur. Denaturing gradient gel electrophoresis (DGGE) analysis of rRNA genes from communities, coupled with selective enrichment, has recently been used to isolate new species of PCB degraders. Syntrophic consortia of SRB and dechlorinating bacteria may be responsible.

Bacteria are effective in the removal of heavy metals from contaminated sediments

Some SRB, *Bacillus* spp. and *Pseudomonas* spp. are particularly effective at immobilizing metals into a nonbioavailable form or respeciating them into less toxic forms. Proteins localized on the surface of dormant spores of *Bacillus* spp. isolated from marine sediments can catalyze the oxidation of metals such as manganese. Bacterially mediated oxidation of soluble Mn (II) to insoluble Mn (III, IV) oxides and oxyhydroxides is particularly important because the products oxidize a variety of organic and inorganic compounds and serve as electron acceptors for anaerobic respiration, leading to immobilization of heavy metals by adsorption. This bacterial precipitation of metal oxides is also important because it leads, over millennia, to the formation of manganese nodules on the seabed. Besides manganese, these small balls contain a high percentage of other valuable metals, especially nickel and cobalt. They occur in large fields, but harvesting is difficult because these sources are mostly in water over 5000 m deep, at long distances from major landmasses. If the problems of harvesting could be solved, communities in South Pacific islands could obtain considerable economic returns. Oxidation of Fe (II) by anaerobic bacteria, including anoxygenic phototrophs and nitrate-reducing organisms, also occurs in sediments.

Microbial systems can be used for monitoring the environment for toxic chemicals

The Microtox® system is an established bioassay for the rapid toxicity testing of water, sediment, and soil samples. It depends on inhibition of bioluminescence of *Vibrio fischeri* (see p. 104), which is supplied as a standardized freeze-dried

WHERE DOES ALL THE PLASTIC GO?

Every year, society produces 260 million tonnes of plastic (Thompson et al., 2009), and a very large proportion of this finds its way into the marine environment. Beaches throughout the world are littered with plastic debris and it is estimated that more than 100000 tonnes of plastic waste circulates in two huge "garbage patches" in the North Pacific gyre. Plastic waste is extremely harmful to marine mammals, birds, and fishes. Ropes, discarded fishing nets, and plastic bags cause entanglement and suffocation, whilst syringes, cigarette lighters, nurdles (pellets used in plastic manufacture), and other debris is mistaken for food by seabirds, fish, and turtles (Gregory, 2009). Plastics also adsorb, release, and transport toxic chemicals. After breaking up by wave action, smaller debris is ingested by marine invertebrates, but little is known of the consequences of this, nor about the role of microorganisms. Recent work combining microscopy and molecular analysis has shown that mixed biofilms of bacteria, picoeukaryotes and archaea rapidly colonize plastic microparticles, which may be different from those in bulk (Harrison et al., 2010). This offers some hope that removal of some of the toxic components might occur by microbial action, but the hydrocarbon portions of plastics are virtually indestructible. So-called biodegradable plastics (e.g. shopping bags incorporating corn starch) break up very slowly in cold marine waters and do not truly degrade, but release many small particles of plastic.

? HOW DOES MERCURY FIND ITS WAY INTO PELAGIC FISH?

High levels of methyl mercury (MeHg) in coastal sediments can be easily explained by deposition of mercury from industrial sources. However, MeHg levels are also rising rapidly in open ocean waters of the North Pacific, even though they are very distant from sources of mercury. It is especially important, since one of the major sources of mercury intake by humans is via consumption of Pacific tuna. Sunderland et al. (2009) measured MeHg at various locations in the North Pacific and found that levels were particularly elevated at mesopelagic depths of 200–700 m. They concluded that mercury emitted from Asian coal-fired power plants is transported over long distances by ocean currents and that most of the mercury adheres to algae, which die and sink into deeper waters. Anerobic bacteria must be methylating the mercury during breakdown of organic matter within oxygen-depleted niches within marine snow aggregates. Similar findings have been reported in the Mediterranean Sea by Cossa et al. (2009). Additional evidence comes from the results of a study by Senn et al. (2010), who measured distribution of carbon, nitrogen, and mercury stable isotope signatures to show that high MeHg levels in tuna could only be explained if methylation of mercury is occurring in open ocean waters. Senn et al. (2010) and Choy et al. (2009) showed that the mesopelagic habitat is a major entry point for mercury into marine food webs and that the depth at which different fish species feed affects the tissue levels of MeHg. Lowenstein et al. (2010) used a DNA barcoding technique to show that sushi made from different species of tuna can contain very different levels of MeHg; this could be an important factor in the formulation of health advice for fish consumption.

culture. Light emission is measured using a photometer and is very sensitive to the presence of toxic chemicals at sublethal concentrations. The more recently introduced QwikLite™ bioassay uses the same principles, but is based on inhibitory effects on bioluminescence of the dinoflagellate *Pyrocystis lunula* (see p. 139). These methods are reported to be many times more sensitive and much easier to carry out than conventional bioassays using fish or amphipods.

The study of natural marine microbial communities is an important aspect of monitoring the effects of pollution, temperature shifts, and other environmental disturbance. The DGGE technique for community analysis (see p. 48) has found wide application in such studies. Pollution-degrading organisms can be genetically modified to link the expression of degradative genes to reporter systems such as *lux* genes from *V. fischeri* or GFP genes from jellyfish (see p. 33). These reporter systems have wide applications in cell biology as well as environmental investigations.

Mobilization of mercury by bacterial metabolism leads to accumulation of toxic methylmercury

An indirect consequence of the activity of bacteria in soil and water is the transformation by microbial methylation reactions of trace elements such as mercury, arsenic, cadmium, and lead. These methylated elements are mobilized and can enter the food chain, causing deleterious health effects. The accumulation of mercury in marine fish is a particular health concern. Although it is a naturally occurring element, the main source of mercury in the environment is from coal-fired power stations and waste incinerators; it is also used in some industrial processes. Mercury is also present in some pesticides and is an important component of discarded electronic equipment and batteries. About 2000 tonnes of mercury are deposited in the oceans each year, and levels are rising because of the increased burning of coal in rapidly developing industrialized countries such as China and India. Most mercury enters the sea in the form of the Hg^{2+} ion and adsorbs readily to particles in which it can be transformed by several types of bacteria, especially anaerobic SRB, to methylmercury (CH_3Hg^+) in a reaction involving the compound methylcobalamin as a methyl donor. A second methyl group can also be attached to give dimethylmercury ($(CH_3)_2Hg$). Methylmercury is a potent neurotoxin; it causes liver, kidney, and cardiovascular damage in adult humans and severely affects development of the fetal nervous system if consumed by pregnant women. Because methylmercury is lipid soluble, it is concentrated in the food chain, especially in fish. One of the world's most serious environmental disasters occurred in Japan in the 1970s, when thousands of people who had eaten fish from the heavily polluted Minamata Bay were seriously affected by methylmercury poisoning. In this case, hundreds of deaths and long-term health effects occurred. Because of its sequential concentration at each step of the food chain, high levels of methylmercury occur in top predators such as tuna, shark, and marine mammals. Very high levels of mercury have been found in the tissues of Eskimo and Inuit communities who eat large amounts of fish and whale meat. The general health advice that fish is a good source of nutrition with significant health benefits is complicated by the possibility of consuming unsafe levels of methylmercury. Health authorities in many countries advise consumers—particularly pregnant women, women of child-bearing age, nursing mothers, and young children—to avoid fish with high levels of methylmercury, limit their intake of fish with moderate levels of methylmercury, and consume fish with low levels of methylmercury no more than twice a week.

Conclusions

This chapter has shown that microbes have some serious deleterious effects due to colonization of marine surfaces, biodeterioration and corrosion of structures and materials, and spoilage of seafood products. These activities cause significant economic losses to maritime industries. Some control methods have been discussed here, whilst other biotechnological solutions are considered in the next chapter. We have seen that microbes play an important role in pollution of coastal waters by sewage, both as human pathogens introduced with fecal waste and also as indicators of pollution for monitoring effects on public health. There are severe limitations with conventional techniques, but considerable improvements are now being achieved through the use of new methodologies. By contrast, microbes play a highly beneficial role in the degradation of oil, most of which finds its way into the ocean from natural seepage. Attempts to augment or enhance these natural processes of degradation for the bioremediation of coastal pollution from oil spills have been used with varying degrees of success. It remains to be seen whether microbes will be effective in the removal of other industrial pollutants of the modern world, especially persistent organic pollutants and plastics. One of the most worrying developments discussed is the realization that ocean microbes can convert mercury to a toxic form that enters the food chain, as part of the normal food web processes. This threatens the safety of pelagic fish as a food source and can only be controlled by limiting emissions from power plants in newly industrialized countries—another problem facing our oceans, to be considered alongside rising carbon dioxide levels and the development of oxygen-minimum zones.

References

Atlas, R. and Bragg, J. (2009) Bioremediation of marine oil spills: when and when not—the Exxon Valdez experience. *Microb. Biotechnol.* 2: 213–221.

Bitton, G. (2005) *Wastewater Microbiology*, 3rd Edn. John Wiley & Sons Inc., Hoboken, NJ.

Bradley, G. and Hancock, C. (2003) Increased risk of non-seasonal and body immersion recreational marine bathers contacting indicator microorganisms of sewage pollution. *Mar. Poll. Bull.* 46: 792–794.

Breitbart, M., Hewson, I., Felts, B., et al. (2003) Metagenomic analyses of an uncultured viral community from human faeces. *J. Bacteriol.* 185: 6220–6223.

Cabelli, V. J,. Dufour, A. P., McCabe, L. J. and Levin, M. A. (1982) Swimming-associated gastroenteritis and water quality. *Am. J. Epidemiol.* 115: 606–616.

Calci, K. R., Meade, G. K., Tezloff, R. C. and Kingsley, D. H. (2005) High-pressure inactivation of hepatitis A virus within oysters. *Appl. Environ. Microbiol.* 71: 339–343.

Camilli, R., Reddy, C. M., Yoerger, D. R., et al. (2010) Tracking hydrocarbon plume transport and biodegradation at Deepwater Horizon. *Sci. Express*, doi: 10.1126/science.1195223.

Choy, C. A., Popp, B. N., Kaneko, J. J. and Drazen, J. C. (2009) The influence of depth on mercury levels in pelagic fishes and their prey. *Proc. Natl Acad. Sci. USA* 106: 13865–13869.

Colson, P., Richet, H., Desnues, C., et al. (2010) Pepper mild mottle virus, a plant virus associated with specific immune responses, fever, abdominal pains, and pruritus in humans. *PLoS ONE* 5: e10041, doi:10.1371/journal.pone.0010041.

Cossa, D., Averty, B. and Pirrone, N. (2009) The origin of methylmercury in open Mediterranean waters. *Limnol. Oceanogr.* 54: 837–844.

Efstratiou, M. A. and Tsirtsis, G. (2009) Do 2006/7/EC European Union Bathing Water Standards exclude the risk of contact with *Salmonella* or *Candida albicans*? *Mar. Poll. Bull.* 58: 1039–1044.

Fleisher, J. M., Kay, D., Wyer, M. D. and Godfree, A. F. (1998) Estimates of the severity of illnesses associated with bathing in marine recreational waters contaminated with domestic sewage. *Int. J. Epidemiol.* 27: 722–726.

Gregory, M. R. (2009) Environmental implications of plastic debris in marine settings—entanglement, ingestion, smothering, hangers-on, hitch-hiking and alien invasions. *Philos. Trans. Roy. Soc. B Biol. Sci.* 364: 2013–2025.

Harrison, J., Ojeda, J., Sapp, M., Schratberger, M. and Osborn, M. A. (2010) The formation and structure of microbial biofilms associated with synthetic microplastics in coastal sediments. Proceedings of the Society of General Microbiology Spring 2010 Meeting, March 29–April 1, 2010, Edinburgh, UK.

Hazen, T. C., Dubinsky, E. A., DeSantis, T. Z., et al. (2010) Deep-sea oil plume enriches indigenous oil-degrading bacteria. *Sci. Express*, doi: 10.1126/science.1195979.

Kay, D., Fleisher, J. M., Salmon, R. L., Wyer, M. D., Godfree, A. F., Zelenauch-Jacquotte, Z. and Shore, R. (1994) Predicting likelihood of gastroenteritis from sea bathing; results from randomized exposure. *Lancet* 344: 905–909.

Lowenstein, J. H., Burger, J., Jeitner, C. W., Amato, G., Kolokotronis, S. O. and Gochfeld, M. (2010) DNA barcodes reveal species-specific mercury levels in tuna sushi that pose a health risk to consumers. *Biol. Lett.* 6: 692–695.

Mansilha, C. R., Coelho, C. A., Heitor, A. M., Amado, J., Martins, J. P. and Gameiro, P. (2009) Bathing waters: new directive, new standards, new quality approach. *Mar. Poll. Bull.* 58: 1562–1565.

Muniesa, M., Jofre, J., Garcia-Aljaro, C. and Blanch, A. R. (2006) Occurrence of *Escherichia coli* O157: H7 and other enterohemorrhagic *Escherichia coli* in the environment. *Environ. Sci. Technol.* 40: 7141–7149.

Murchie, L. W., Cruz-Romero, M., Kerry, J. P., Linton, M., Patterson, M. F., Smiddy, M. and Kelly, A. L. (2005) High pressure processing of shellfish: a review of microbiological and other quality aspects. *Innov. Food Sci. Emerg. Technol.* 6: 257–270.

Pacquit, A., Lau, K. T., McLaughlin, H., Frisby, J., Quilty, B. and Diamond, D. (2006) Development of a volatile amine sensor for the monitoring of fish spoilage. *Talanta* 69: 515–520.

Peterson, J., Garges, S., Giovanni, M., et al. (2009) The NIH Human Microbiome Project. *Genome Res.* 19: 2317–2323.

Prüss, A. (1998) A review of epidemiological studies from exposure to recreational water. *Int. J. Epidemiol.* 27: 1–9.

Rosario, K., Symonds, E. M., Sinigalliano, C., Stewart, J. and Breitbart, M. (2009) Pepper mild mottle virus as an indicator of fecal pollution. *Appl. Environ. Microbiol.* 75: 7261–7267.

Senn, D. B., Chesney, E. J., Blum, J. D., Bank, M. S., Maage, A. and Shine, J. P. (2010) Stable Isotope (N, C, Hg) study of methylmercury sources and trophic transfer in the northern Gulf of Mexico. *Environ. Sci. Technol.* 44: 1630–1637.

Stone, D. L., Harding, A. K., Hope, B. K. and Slaughter-Mason, S. (2008) Exposure assessment and risk of gastrointestinal illness among surfers. *J. Toxicol. Environ. Health* 71: 1603–1615.

Sunderland, E. M., Krabbenhoft, D. P., Moreau, J. W., Strode, S. A. and Landing, W. M. (2009) Mercury sources, distribution, and bioavailability in the North Pacific Ocean: insights from data and models. *Global Biogeochem. Cycles* 23: GB2010, doi:10.1029/2008GB003425.

Thompson, R. C., Swan, S. H., Moore, C. J. and vom Saal, F. S. (2009) Our plastic age. *Philos. Trans. Roy. Soc. B Biol. Sci.* 364: 1973–1976.

Turbow, D. J., Kent, E. E. and Jiang, S. C. (2008) Web-based investigation of water associated illness in marine bathers. *Environ. Res.* 106: 101–109.

USEPA (2007) *Improved Enumeration Methods for the Recreational Water Quality Indicators: Enterococci and Escherichia coli.* US Environmental Protection Agency, Office of Science and Technology, Washington, DC.

Wade, T. J., Pai, N., Eisenberg, J. N. S. and Colford, J. M. Jr (2003) Do US Environmental Protection Agency water quality guidelines for recreational waters prevent gastrointestinal illness? A systematic review and meta-analysis. *Environ. Health Perspect.* 111: 1102–1109.

Williams, A. P., Avery, L. M., Killham, K. and Jones, D. L. (2007) Persistence, dissipation, and activity of *Escherichia coli* O157: H7 within sand and seawater environments. *FEMS Microbiol. Ecol.* 60: 24–32.

World Health Organization (2003) *Guidelines for Safe Recreational Water Environments*, Vol. 1: *Coastal and Fresh Waters*. WHO, Geneva.

Yang, J. C., Madupu, R., Durkin, A. S., et al. (2009) The complete genome of *Teredinibacter turnerae* T7901: an intracellular endosymbiont of marine wood-boring bivalves (shipworms). *PLoS ONE* 4: e6085, doi:10.1371/journal.pone.0006085.

Further reading

Biofouling and biodeterioration

Callow, M. E. and Callow, J. A. (2002) Marine biofouling: a sticky problem. *Biologist* 49: 1–5.

Durr, S. and Thomason, J. (2009) *Biofouling.* Wiley-Blackwell, Chichester, UK.

Harder, T. and Lee, L. H. (2009) Bacterial adhesion and marine fouling. In: *Advances in Marine Antifouling Coatings and Technologies* (ed. C. Hellio and D. Yebra). Woodhead Publishing and CRC Press, Boca Raton, FL, pp. 113–131.

Lau, S. C. K., Dahms, H.-U., Dobretsov, S. and Harder, T. (2007) Marine biofilms as mediators of colonization by marine macroorganisms: implications for antifouling and aquaculture. *Mar. Biotechnol.* 9: 339–410.

Little, B. J., Lee, J. S. and Ray, R. I. (2008) The influence of marine biofilms on corrosion: a concise review. *Electrochim. Acta* 54: 2–7.

Nemati, M., Jenneman, G. E. and Voordouw, G. (2008) Impact of nitrate-mediated microbial control of souring in oil reservoirs on the extent of corrosion. *Biotechnol. Prog.* 17: 852–859.

Videla, H. A. (2000) An overview by which sulphate-reducing bacteria influence corrosion of steel in marine environments. *Biofouling* 15: 37–47.

Spoilage and safety of seafood

Gram, L. and Dalgaard, P. (2002) Fish spoilage bacteria—problems and solutions. *Curr. Opin. Biotechnol.* 13: 262–266.

Jothikumar, N., Lowther, J. A., Henshilwood, K., Lees, D. N., Hill, V. R. and Vinje, J. (2005) Rapid and sensitive detection of noroviruses by using TaqMan-based one-step reverse transcription-PCR assays and application to naturally contaminated shellfish samples. *Appl. Environ. Microbiol.* 71: 1870–1875.

Lee, R., Lovatelli, A. and Ababouch, L. (2010) *Bivalve Depuration: Fundamental and Practical Aspects.* Food and Agriculture Organization of the United Nations, Rome.

Schultz, A. C., Saadbye, P., Hoorfar, J. and Nørrung, B. (2007) Comparison of methods for detection of norovirus in oysters. *Int. J. Food Microbiol.* 114: 352–356.

Teplitski, M., Wright, A. C. and Lorca, G. (2009) Biological approaches for controlling shellfish-associated pathogens. *Curr. Opin. Biotechnol.* 20: 185–190.

Water quality and microbial source tracking

Baums, I. B., Goodwin, K. D., Kiesling, T., Wanless, D., Diaz, M. R. and Fell, J. W. (2007) Luminex detection of fecal indicators in river samples, marine recreational water, and beach sand. *Mar. Poll. Bull.* 54: 521–536.

Bushon, R. N., Brady, A. M., Likirdopulos, C. A. and Cireddu, J. V. (2009) Rapid detection of *Escherichia coli* and enterococci in recreational water using an immunomagnetic separation/adenosine triphosphate technique. *J. Appl. Microbiol.* 106: 432–441.

Converse, R. R., Blackwood, A. D., Kirs, M., Griffith, J. F. and Noble, R. T. (2009) Rapid QPCR-based assay for fecal *Bacteroides* spp. as a tool for assessing fecal contamination in recreational waters. *Water Res.* 43: 4828–4837.

Griffin, D. W., Donaldson, K. A., Paul, J. H. and Rose, J. B. (2003) Pathogenic human viruses in coastal waters. *Clin. Microbiol. Rev.* 16: 129–143.

Griffith, J. F., Cao, Y., McGee, C. D. and Weisberg, S. B. (2009) Evaluation of rapid methods and novel indicators for assessing microbiological beach water quality. *Water Res.* 43: 4900–4907.

Kay, D. and Fawell, J. (2007) *Standards for Recreational Water Quality.* Foundation for Water Research, Marlow, UK.

Noble, R. T. and Weisberg, S. B. (2005) A review of technologies for rapid detection of bacteria in recreational waters. *J. Water Health* 3: 381–392.

Ufnar, J. A., Wang, S. Y., Christiansen, J. M., Yampara-Iquise, H., Carson, C. A. and Ellender, R. D. (2006) Detection of the *nifH* gene of *Methanobrevibacter smithii*: a potential tool to identify sewage pollution in recreational waters. *J. Appl. Microbiol.* 101: 44–52.

Zielinski, O., Busch, J. A., Cembella, A. D., Daly, K. L., Engelbrektsson, J., Hannides, A. K. and Schmidt, H. (2009) Detecting marine hazardous substances and organisms: sensors for pollutants, toxins, and pathogens. *Ocean Sci.* 5: 329–349.

Oil biodegradation and bioremediation

Head, I. M., Jones, D. M. and Röling, W. F. M. (2006) Marine microorganisms make a meal of oil. *Nat. Rev. Microbiol.* 4: 173–182.

Mckew, B. A., Coulon, F., Yakimov, M. M., et al. (2007) Efficacy of intervention strategies for bioremediation of crude oil in marine systems and effects on indigenous hydrocarbonoclastic bacteria. *Environ. Microbiol.* 9: 1562–1571.

Chapter 14

Marine Microbes and Biotechnology

Biotechnology is defined broadly as the application of scientific and engineering principles to provide goods and services through mediation of biological agents. The first part of this chapter considers examples of products from marine microorganisms, including enzymes, pharmaceuticals, antifouling agents, polymers, and biofuels. This is followed by discussion of biotechnological processes for the diagnosis and prevention of disease in marine systems, especially aquaculture. Biotechnological approaches are also used for monitoring and remediating pollution—these were considered in Chapter 13. Major economic benefits derive from the industrial and biomedical exploitation of microbes isolated from the sea—their enormous natural diversity and range of metabolic activities provide great opportunities for future exploitation. As our appreciation of the great variety of marine microbes grows with the study of different habitats, so the collection of microbes, their genes, and their products with useful properties continues to expand. A summary of some beneficial applications of marine microbes is given in *Table 14.1*.

Key Concepts

- Marine microbes are the source of important enzymes and polymers used in many branches of industry.

- Microalgae may provide a significant new source of biofuels.

- Many pharmaceuticals and other health products have been obtained from marine microbes, but their potential has not yet been fully realized commercially.

- Study of bacterial colonization of marine surfaces has led to new approaches to antifouling and prevention of infection through interference with signaling mechanisms.

- Structural components of marine microbes are being exploited in nanotechnology, bioelectronics, and the development of new materials.

- Marine microbes and their products are increasingly used in vaccines, probiotics, and immunostimulants for aquaculture of finfish and shellfish, in order to overcome the problems of resistance to antimicrobial compounds.

- Phages have potential uses in the biological control of bacterial diseases of fish, shellfish, and corals.

Table 14.1 Some beneficial effects of activities or products from marine microbes

Application	Examples
Aquaculture	Disease diagnostics Nutritional supplements Pigments Probiotics Vaccines
Cosmetics	Liposomes Polymers Sunscreens
Environmental protection	Bioremediation of pollution Disease diagnostics Nontoxic antifouling agents Toxicology bioassays Waste processing
Food processing	Enzymes Flavors Preservatives Texture modifiers
Manufacturing industry	Bioelectronics Polymers Structural components
Minerals and fuels	Desulfurization of oil and coal Manganese nodules Oil extraction Production of biofuels
Nutraceuticals	Antioxidative compounds Dietary supplements "Health foods"
Pharmaceuticals	Antibacterial, antifungal, and antiviral agents Antitumor and immunosuppressive agents Drug delivery Enzymes Neuroactive agents Self-cleaning implants
Textiles and papers	Enzymes Surfactants

Enzymes from marine microbes have many applications

Enzymes are widely used by industry and the global market in 2010 is estimated at about US$6 billion, growing by 5–10% a year. Many research institutes and commercial organizations have developed culture collections of marine bacteria—and to a lesser extent fungi—with initial attention being focused on culturable organisms that are easily collected from near-shore habitats. The full potential of the thousands of marine microbes already in culture collections has not yet been fully investigated. The most commonly exploited enzymes are those that degrade polymers, especially proteins and carbohydrates. Examples of early successes include the production by *Vibrio* spp. of various types of extracellular proteases, some of which are tolerant of moderate salt concentrations and detergents. Another is the extraction of glucanases and other carbohydrate-degrading enzymes from *Bacillus* spp. isolated from mud samples.

The most successful developments have occurred with the isolation of enzymes from extremophilic microorganisms; some of the products that have resulted are shown in *Table 14.2*. Enzymes from thermophiles and hyperthermophiles have particular attractions in industrial processes, which often require high temperatures. Even at milder temperatures, thermophilic enzymes are beneficial because of their much greater stability. The structural features of thermophilic enzymes that confer stability and function at high temperatures (sometimes >100°C) are discussed in Chapter 3.

The use of proteases and lipases as stain removers in detergents is a particularly important application, since properties that confer high thermostability and activity are often combined with resistance to bleaching chemicals and surfactants used in these products. Some washing powders incorporate thermophilic cellulases and hemicellulases, which digest loose fibers and help to prevent "bobbles" on clothes after washing; these enzymes are also important in the manufacture of "stone-washed" denim. Enzyme production for "biological" detergents is

Table 14.2 Some biotechnological applications of extremophilic microbes

Product	Applications
Thermophiles and hyperthermophiles	
Amylases, pullulanases, lipases, proteases	Baking, brewing, food processing
DNA polymerases	PCR amplification of DNA
Lipases, pullulanases, proteases	Detergents
S-layers	Ultrafiltration, electronics, polymers
Xylanases	Paper bleaching
Halophiles	
Bacteriorhodopsin	Bioelectronic devices, optical switches, photocurrent generators
Compatible solutes	Protein, DNA and cell protectants
Lipids	Liposomes (drug delivery, cosmetics)
S-layers	Ultrafiltration, electronics, polymers
γ-linoleic acid, β-carotene, cell extracts	Health foods, dietary supplements, food colors, aquaculture feeds
Psychrophiles	
Ice nucleating proteins	Artificial snow, frozen food processing
Polyunsaturated fatty acids	Food additives, dietary supplements
Proteases, lipases, cellulases, amylases	Detergents
Alkaliphiles and acidophiles	
Acidophilic bacteria	Fine papers, waste treatment
Elastases, keratinases	Leather processing
Proteases, cellulases, lipases	Detergents
Sulfur-oxidizing acidophiles	Recovery of metals, sulfur removal from coal and oil

a very large market accounting for approximately 30% of the total global production of enzymes. The first source of these enzymes was soil bacteria of the genus *Bacillus*, but enzymes from marine thermophiles have higher temperature optima and superior stability.

Modern food processing uses a wide range of enzymes. Almost all processed foods now rely on some form of modified starch product for the improvement of texture, control of moisture, and prolonged shelf life. Amylases hydrolyze α-1,4-glycosidic linkages in starch to produce a mixture of glucose, malto-oligosaccharides, and dextrins. All the remaining α-1,4-glycosidic branches in the products are hydrolyzed by pullulanase. When starch is treated with amylase and pullulanase simultaneously at high temperatures, it produces higher yields of desired end products. Pullulanases and other enzymes derived from the hyperthermophilic bacterium *Thermotoga maritima* have recently been introduced into food processing. A range of other carbohydrate-modifying enzymes have been isolated from marine thermophiles.

Agar-degrading enzymes can be isolated from many species of bacteria, including *Vibrio*, *Pseudomonas*, *Pseudoalteromonas*, *Alteromonas*, *Thallassomonas*, *Cytophaga*, and *Agarivorans*. Zones or craters of agar degradation are often seen surrounding colonies of bacteria isolated from seawater, sediments, algae, and marine invertebrates when grown on agar plates. Agar, obtained from red seaweeds such as *Gracilaria*, is added to many foods such as ice cream, glazes, and processed cheese to improve texture. Industrially, agar-degrading enzymes (agarases) are used to produce diverse oligosaccharides with a variety of properties and numerous applications in health foods and cosmetics. Agarases are also widely used in the molecular biology laboratory, as a method of purifying DNA after separation on agarose gels. The enzymes are also useful for degradation of the cell wall of algae for extraction of labile substances with biological activities such as unsaturated fatty acids, vitamins, and carotenoids.

As with many enzymes and other proteins isolated from bacteria, it is often more convenient for manufacturing to clone the genes responsible and express them in *Escherichia coli* or another recombinant host. However, there can be problems with expression and correct folding of thermostable proteins in mesophilic hosts, and if post-translational modifications are required for enzyme function, the recombinant enzyme may not function as required.

Enzymes from psychrophilic microbes have many uses and isolates from the deep sea and polar regions have been exploited for use in food-processing applications in which low temperatures are required to prevent spoilage, destruction of key ingredients (e.g. vitamins), or loss of texture. For example, galactosidases from cold-adapted bacteria have been used for the removal of lactose from milk to improve digestibility, and xylanases are used in baking to improve crumb texture in bread. Proteases from psychrophiles can be used for tenderizing meat at low temperatures. Significant savings in energy costs result from the use of cold-water laundry detergents that incorporate proteases and lipases active at temperatures of 10°C or less.

DNA polymerases from hydrothermal vent organisms are widely used in molecular biology

The discovery of thermostable DNA polymerases used in the polymerase chain reaction (PCR) has arguably been one of the most spectacular scientific advances since the discovery of DNA itself. The use of the PCR in research, disease diagnostics, and forensic investigations has led to a huge market; the market value of PCR enzymes, kits, and equipment is growing by 7–10% each year and is forecast to reach US$38 billion by 2015. The original PCR enzyme, *Taq* polymerase,

SUNKEN WHALE CARCASSES YIELD RICH SOURCE OF NOVEL ENZYMES

As described on p. 212, when the carcass of a dead whale reaches the seafloor, it supports a diverse community of organisms for many years. After initial removal of soft tissue by fish, crabs, and other scavengers, the skeleton is colonized by bacteria, which degrade the bones. Whalebones contain up to 60% lipids, and microbial sulfate reduction during degradation of the lipids and other organic compounds leads to the production of hydrogen sulfide, which is then used by microbial mats of chemosynthetic bacteria such as *Beggiatoa* (Treude et al., 2009). Cloning of genes encoding lipases has yielded enzymes with potential uses in a variety of biotransformations, such as the processing of vegetable oils for biofuels. Often, the skeleton is colonized by *Osedax* worms (see *Plate 10.4*), which contain symbiotic bacteria that break down collagen, cholesterol, and lipids (Goffredi et al., 2005). Sequencing of the genome of this symbiont is in progress and is likely to yield more novel enzymes with potential industrial applications.

Table 14.3 Some thermostable DNA polymerases and their sources

DNA polymerase	Organism	Source	Half-life at 95°C (min)	Proofreading
Taq	*Thermus aquaticus*	T (N or R)	40	–
Amplitaq®	*Thermus aquaticus*	T (R)	40	–
Vent™	*Thermococcus litoralis*	M (R)	400	+
Deep Vent™	*Pyrococcus* GB-D	M (R)	1380	+
Tth	*Thermus thermophilus*	T (R)	20	–
Pfu	*Pyrococcus furiosus*	M (N)	120	+
ULTm™	*Thermotoga maritima*	M (R)	50	+

M, marine hydrothermal vent; (N), natural; (R), recombinant; T, terrestrial hot spring.

was isolated from *Thermus aquaticus* from a terrestrial hot spring. Although *Taq* is still the least expensive and most widely used enzyme for both research and diagnostic uses of the PCR, a number of alternative enzymes are now available that are superior in certain applications, and some of these were originally found in marine bacteria (*Table 14.3*). The choice of enzyme depends on the specific activity and sensitivity required with different amounts and lengths of template, nucleotide specificity, and various other factors, including cost. Enzymes from the hyperthermophilic vent bacteria have greater thermostability and activity at higher temperatures, particularly useful when amplifying GC-rich sequences. They may also have a higher fidelity of replication (due to integral proofreading ability), although careful optimization of the reaction is required; enzymes such as Vent™ or *Pfu* are often used in a mixture with *Taq*.

Metagenomics and bioinformatics lead to new biotechnological developments

Molecular biology methods have led to exciting new approaches in the search for new products from marine microbes. In an analogy with the gold rush, "bio-prospectors" can use oligonucleotide hybridization probes to "pan" for genetic sequences of interest in the hope of reaping rich rewards. This enables the screening of communities without the need for culture. For example, comparison of sequence data for genes encoding proteins with a particular function can allow the identification of consensus sequences. These can be used to construct complementary oligonucleotide probes, which are then used to search for genes in DNA amplified from a community of interest. Bioinformatics data-mining tools can be used to identify gene homologs, and genes can then be cloned and expressed in suitable hosts. The disadvantage of this method is that it may fail to detect truly novel proteins with sequences unlike those previously described. This limitation is borne out by the results from metagenomic analyses, such as the Global Ocean Survey (see *Research Focus Box 4.1*) and genome sequencing of cultured microbes. In both cases, large numbers of open reading frames (ORFs) in the genomes have no match with existing genes in databases. To overcome this, direct functional screening of metagenomic libraries can be employed to identify candidate genes associated with observed enzymatic activities. PCR primers for amplification of sequences from libraries can be designed using sequence features characteristic of particular enzyme genes. All of these approaches are leading to new enzymes such as esterases, lipases, chitinases, amidases, cellulases, alkane hydrolases, and proteases. Advances in sequencing technology and bioinformatics mean that we will undoubtedly be able to develop numerous new enzymes carrying out reactions that are completely unknown to us at present.

ⓘ **LIFE FROM SCRATCH**

For many years, the J. Craig Venter Institute has been attempting to build "life from scratch," and the technical obstacles were finally overcome when Gibson et al. (2010) announced that they had created an artificial bacterium by replacing the entire genome of a species of *Mycoplasma* with DNA designed on a computer and assembled by chemical synthesis. This heralds the start of a new era of synthetic biology. The huge diversity of marine microbes means that their genes will play a major role in the future of biotechnology, leading to many exciting new products and processes.

Polymers from marine bacteria are finding increasing applications

Microbial polymers are used in bioremediation, industrial processes, manufacturing industry, and food processing. The best-exploited compounds are extracellular polysaccharides that form the glycocalyx associated with biofilm formation and protection from phagocytosis. Applications in bioremediation were mentioned in Chapter 13. Other potential applications include underwater surface coatings, bioadhesives, drag-reducing coatings for ship hulls, dyes, and sunscreens. It has been suggested that the high production (up to 80% cell dry weight under appropriate conditions) of poly-hydroxy-β-hydroxyalkanoates by some marine bacteria as food reserves could be exploited for the production of biodegradable plastics. Oil-derived plastics will continue to dominate the market for some time, but bacterially produced plastics are becoming more competitive because of increasing oil prices and the polluting effects of nondegradable plastics.

Microalgae are promising new sources of biofuels

The global economy is dependent on oil for energy and as a source of chemicals. With dwindling supplies of fossil fuels and concerns about carbon dioxide emissions and political instability in oil-producing countries, there has been considerable recent interest in using renewable biological materials for the production of biofuels. Crops such as corn, sugar, soybeans, sunflowers, or palms can be fermented to alcohol for addition to gasoline, or used to produce biodiesel. However, the costs of production, competition with food production for land use, and the ecological consequences of deforestation associated with such schemes make the use of land crops unsustainable. Therefore, there has been much interest in growing microalgae as a source of biodiesel. Some of the most promising species of microalgae that may be suitable for large-scale culture include marine cyanobacteria (e.g. *Synechococcus*), diatoms (e.g. *Phaeodactylum tricornutum* and *Chaetoceros muelleri*), prymenesiophytes (e.g. *Emiliania huxleyi*), and chlorophytes (e.g. *Dunaliella tertiolecta* and *Botryococcus braunii*). Microalgae are easy to cultivate and can grow almost anywhere, requiring sunlight and some simple nutrients. Growth is very efficient because, unlike vascular plants, all cells in a microalgal culture are photosynthetic and carbon dioxide and nutrients are available directly to the cells. In addition, the concentration of carbon dioxide in an algal suspension is much higher than that in the atmosphere above a land plant. Algae therefore have the potential to produce about 30 times more energy per unit area than arable crops such as soybeans.

The starting points for oil production from microalgae are the lipids and fatty acids found in their membranes and stored in intracellular compartments as an energy reserve. Algal strains vary greatly in the types and quantity of lipids produced; selection of appropriate strains for biodiesel production depends on many factors. Careful control of growth conditions can lead to the cells containing over 75% of their mass as lipids. These lipids can be converted into a variety of compounds with similar properties to petroleum products, including diesel. Algal culture has many other applications, especially for the production of food products, food additives, and for aquaculture feeds. Whilst there have been considerable successes in small-scale production of biofuels, developing this process into a sustainable and cost-effective industry will require much ingenuity. Supplies of carbon dioxide and sunlight are clearly not a problem, but organisms require other major nutrients, especially nitrogen and phosphorus. Microbial nitrogen fixation could perhaps be harnessed for supply to algal culture systems, but the only source of phosphorus is from the mining of natural deposits and supplies are running out quickly, which will threaten all forms of agriculture. The challenge will be, therefore, to develop sustainable systems with almost perfect systems for recycling nutrients after harvesting hydrocarbons.

? CAN MICROALGAE BE "MILKED" FOR OIL?

One of the drawbacks of conventional methods for culture of microalgae for production of biodiesel is that the batch culture method requires harvesting of the algae, followed by lysis of the cells and subsequent extraction of the oil, which can be difficult. The company OriginOil Inc. (www.originoil.com) uses a new process combining ultrasound and an electromagnetic pulse to break the algal cell walls. Then the algal suspension is force-fed with carbon dioxide, which lowers its pH and separates the biomass from the oil. Other strategies under development include the use of viruses to lyse the algae. An alternative suggested method is to genetically modify the algae to secrete oil droplets that can be "milked" from the culture. Ramachandra et al. (2009) propose a system for growing diatoms in continuous culture in specially designed solar panels to obtain a high growth rate and high oil content. Secreted oil would rise to the top of the culture and be easily collected. Many university and industrial research groups are actively pursuing the development of biofuels from microalgae (review, Huang et al., 2010), with significant funding from major oil and aviation companies. One of the companies most active in this field is Synthetic Genomics Inc. (www.syntheticgenomics.com) who are engineering strains of microalgae to secrete large amounts of high chain-length hydrocarbons in a continuous process.

Marine microbes are a rich source of biomedical products

Natural products provide many compounds used in medicine and health promotion. Throughout much of the twentieth century, the pharmaceutical industry was engaged in a continual search for new compounds with biological activities that can be exploited in therapy. Many of our most successful drugs are secondary metabolites obtained from bacteria and fungi isolated in the terrestrial environment, especially soil. However, in the last twenty years, the effort necessary to isolate valuable new compounds has become increasingly disproportionate to the returns. Companies moved away from natural product-based drug discovery and came to rely more on the use of structure–function analysis to modify chemically existing compounds. Now, there is a resurgence of interest in natural products from marine habitats because of the realization that their great biodiversity is likely to yield many new compounds.

Several hundred bioactive compounds from marine bacteria have been investigated, with most agents produced by cyanobacteria (especially the genera *Lyngbya* and *Symploca*) and actinomycetes (especially the genera *Streptomyces* and *Salinospora*) (*Table 14.4*). Most cyanobacterial compounds are peptides or complex polyketides, but there is enormous diversity in structure, so that each strain produces a variety of compounds with differences in pharmacological properties.

Bioactive compounds from marine invertebrates may be produced by microbial symbionts

For many years, marine invertebrates (especially sponges and bryozoans) have been investigated as a source of bioactive compounds, and many antibiotics and antitumor agents have been isolated. Many of these soft-bodied animals have evolved chemical defenses against predators or to prevent colonization or disease. Different species of these animals are host to a wide range of symbionts (see Chapter 10), and it now seems likely that many of the compounds discovered are actually produced by the microbes inhabiting the animals, rather than by the host's own metabolism. However, although compounds isolated from sponges

Table 14.4 Examples of pharmaceutical compounds from marine bacteria

Agent	Producing organism	Activity
Abyssomicin	*Verrucispora* sp.	Inhibits para-amino-benzoic acid synthesis; broad-spectrum antibiotic
Apratoxins	*Lyngbya* spp.	Interfere with signaling and transcription in tumor cells
Bryostatins	*Candidatus* Endobugula sertula	Inhibit protein kinase C; antitumor; treatment of leukemia and esophageal cancer; may be useful in treatment of Alzheimer's disease
Coibamide A	*Leptolyngbya* sp.	Induces apoptosis in tumor cells.
Cryptophycins	*Nostoc* spp.	Inhibit polymerization of tubulin in tumor cells
Curacin	*Lyngbya majuscula*	Blocks mitosis in tumor cells.
Largazole	*Symploca* sp.	Inhibits histidine decarboxylase; antitumor; antiepileptic; prevents mood swings
Psymberin	Uncultivated sponge symbiont	Cytotoxic
Salinosporamide A	*Salinospora* sp.	Inhibits protein breakdown in the proteasome; effective against a range of cancers; antimalarial; may relieve some symptoms of Alzheimer's disease

Figure 14.1 The structure of bryostatin-1. This compound is produced by an endosymbiotic bacterium colonizing the bryozoan *Bugula neritina*.

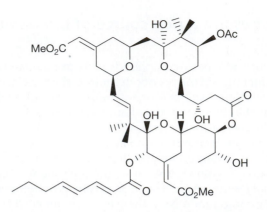

and bryozoans often share structural similarity with known microbial metabolites, it is not easy to prove that they are indeed of microbial origin in the symbiosis. One notable example is the family of compounds known as bryostatins, isolated from the bryozoan *Bugula neritina*. Bryostatins are cytotoxic macrolides (*Figure 14.1*), which are very likely synthesized by an as-yet uncultured gammaproteobacterium named *Candidatus* Endobugula sertula. Bryostatin-1 shows great promise for the treatment of certain types of leukemia and esophageal cancer in clinical trials and has also been found to have promising potential in the treatment of Alzheimer's disease. If the phenomenon of production of useful metabolites by symbionts is found to be widespread, it overcomes the major problem of the need to harvest scarce marine animal life and the consequent disruption of ecosystems. Exploitation of microbial symbionts for pharmaceutical production could involve isolation or culture of the microbes, but since many cannot be cultured, it is likely to be more productive to identify and clone genes for recombinant expression.

Sponges are likely to be a particularly good source for new microbial compounds. As noted in Chapter 10, sponges contain highly diverse microbial communities comprising up to half of the sponge volume, and there appear to be sponge-specific microbiota that have coevolved for ~600 million years. Cultivation of sponge-associated microorganisms that produce bioactive compounds is being actively pursued by several research groups using a range of culture media and conditions; if successful, this approach has obvious advantages for large-scale inexpensive production of compounds. Several species of sponge-associated actinomycetes, vibrios, and fungi have been successfully isolated and shown to produce active compounds with antitumor and antibacterial activities. However, despite employing various strategies to improve isolation, such as the addition of sponge tissue extracts to media, only a small proportion of the microbes present in sponges have been cultured. It is likely that many sponge microbes are obligate symbionts and will prove very difficult or impossible to culture as they may require specific metabolic intermediates from the host sponge.

It is likely that metagenomic approaches to identify and clone appropriate genes will be increasingly used. Many of the most interesting secondary metabolites with biotechnological potential are produced by the action of polyketide synthases. These are giant modular enzymes composed of multiple protein units that contain a coordinated group of active sites. The complex polyketide molecules are synthesized in a stepwise manner from simple 2-, 3-, or 4-carbon building blocks. Bacterial polyketide synthases contain many catalytic domains organized into repeated sets of modules—each module incorporates one carbon unit into the growing polyketide chain. Several of these enzyme complexes have now been cloned, including that from *Candidatus* E. sertula, responsible for bryostatin synthesis, and a gene cluster responsible for synthesis of psymberin (a highly potent antitumor agent) from an uncultivated symbiont of the sponge *Psammocinia* aff. *bulbosa* (*Figure 14.2*).

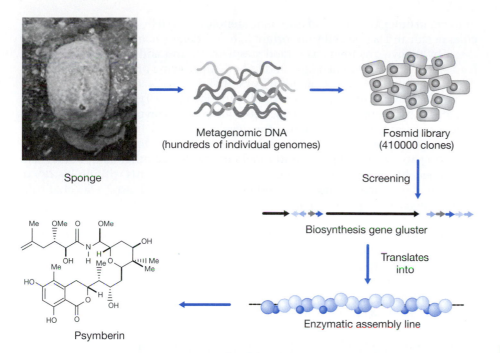

Sponge

Metagenomic DNA
(hundreds of individual genomes)

Fosmid library
(410000 clones)

Screening

Biosynthesis gene gluster

Translates
into

Enzymatic assembly line

Psymberin

Figure 14.2 Example of the cloning of a biosynthesis gene cluster from the metagenome of a complex microbial consortium in a sponge. This strategy was developed by Fisch et al. (2009) to target polyketide synthase genes from metagenomes of sponge–bacterial associations. (Reprinted from Hertweck [2009] by permission of Macmillan Publishers Ltd, © Nature Chemical Biology.)

New antimicrobial compounds may be discovered through study of complex microbial communities

Most of the antimicrobial antibiotics used in the treatment of microbial infections are derived from terrestrial fungi (e.g. penicillins) and bacteria, especially actinomycetes (e.g. tetracyclines, aminoglycosides, and macrolides). Cephalosporin was obtained from the fungus *Cephalosporium*, isolated near a coastal sewage outfall. Drug-discovery companies have begun a renewed search for novel microbes (especially actinomycetes) in the marine environment. Large-scale rapid-throughput screening programs are underway in several parts of the world, investigating marine sediments, sponges, corals, and other habitats. It is likely that the intense competitive pressure found in dense mixed microbial communities, such as those on algal surfaces and in corals, will select for microbes producing antimicrobials. Biofilms are likely to be a rich source, and attention should be given to using new types of bioassays to detect metabolites that interfere with cell signaling or chemotaxis, as well as those that cause outright growth inhibition in the standard detection method (see p. 80 for examples of research in this area).

Marine microbes are the source of a range of health-promoting products

The term "nutraceuticals" is used to describe the wide range of substances used for health promotion and includes functional foods, probiotics, and nutritional supplements. This is an expanding market, valued at present in excess of US$180 billion worldwide. New marine microbial products, especially those from microalgae, are likely to contribute to this area. For example, the polyunsaturated

? DRUGS FROM THE SEA—WITH SO MUCH POTENTIAL, WHY ARE THERE SO FEW SUCCESSES?

There are numerous examples of potential drugs from marine sources that have shown great promise in initial trials but have not been fully developed by pharmaceutical companies for widespread use in medicine. After demonstration of *in vitro* activity, agents must be rigorously tested for toxicity and potential adverse effects in cell cultures and laboratory animals. Probably only 1 in 1000 agents reaches the next stage of phase 1 clinical trials in human volunteers. The drug must then go through progressively larger phase 2 and phase 3 clinical trials involving thousands of patients before it is licensed by government authorities for use as a drug. The process is very expensive and can take 10–15 years, so at each stage of development the company has to evaluate the costs of continuing the process against the potential market returns—this cost–benefit analysis is especially important as patent protection on new drugs may last less than 20 years. The estimated cost of bringing a drug to market is estimated at US$ 200–800 million, so it is not surprising that fewer than 1 in 5000 potential agents become licensed for human use.

long-chain omega-3 fatty acid docosahexaenoic acid (DHA) promotes cardiovascular health and is especially important in fetal brain development; it is therefore taken by many consumers as a food supplement and added to infant formula feed and omega-3-enriched eggs and milk. With growing pressure on fish stocks and the possibility of contamination by methylmercury (see p. 304), the production of DHA by the marine heterotrophic dinoflagellate *Crypthecodinium cohnii* is finding favor with manufacturers. This source is also favored by strict vegetarians. A spin-off from this work was the development of DHA-enriched diets for larval stages in aquaculture. Newly isolated deep-sea psychrophilic bacteria may also be a good source of polyunsaturated fatty acids and other compounds with health benefits. It is necessary to prove that microorganisms with no history of use in food products are nontoxic and nonpathogenic and also to develop suitable methods for large-scale cultivation.

A major part of the health-food industry is concerned with antioxidative effects. Research with bacteria from coral reefs exposed to very high levels of visible light and ultraviolet irradiation indicates that they may have novel mechanisms of reversing the resultant oxidative damage. This could lead to products that overcome some aspects of the aging process. The production of protective pigments by marine bacteria is leading to the development of new sunscreen preparations.

New approaches to antifouling have been discovered through study of microbial colonization of surfaces

As described in Chapter 13, microbes play a critical role in the biofouling of marine surfaces, with serious economic consequences. As a result of environmental problems caused by the use of tributyl tin (TBT) paints and their subsequent prohibition by the International Maritime Organization, there has been great interest in developing natural products as antifouling agents. Ideally, products isolated from marine organisms should have low toxicity, be effective at low concentrations and break down quickly if released into the environment. Hundreds of molecules with antifouling properties have been investigated, with the richest sources being sessile marine organisms, especially sponges, corals, seaweeds, and ascidians as such organisms often possess mechanisms to avoid overgrowth by other organisms. As shown in *Figure 13.1*, there are various critical points at which the process of biofouling may be inhibited. From a microbiological perspective, our interest is in the prevention of the initial formation of a biofilm. One example of a successful development in this field is the discovery of brominated furanone compounds produced by the seaweed *Delisea pulchra*, which prevent biofilm formation by interfering with quorum sensing (see *Research Focus Box 14.1*).

The surfaces of some species of algae are colonized by multispecies bacterial biofilms that seem to inhibit fouling by interfering with cues for settlement of invertebrate larvae or spores of other algae. Bacteria belonging to the genera *Pseudoalteromonas* and *Phaeobacter* produce a range of inhibitory compounds that may have potential applications in antifouling treatments. A number of antifouling coatings for plastic surfaces have also been developed by mixing extracts of marine isolates of *Bacillus*, *Pseudomonas*, and *Streptomyces* with water-based resins. These "living paints" have shown promise in experimental trials, but problems with formulation and delivery of active compounds means that laboratory assays are often not confirmed in field trials.

Marine microbes are a rich source for biomimetics, nanotechnology, and bioelectronics

In materials science and technology, biomimetics is the term given to the process of "taking good designs from nature." Nanotechnology involves the construction of materials and functional objects assembled from the basic molecular building blocks, which offers the potential of new products ranging from new computer technology to microscopic machines. Marine microbes are proving to be a rich source for these new technologies. Bacterial and archaeal S-layers are finding applications in nanotechnology because of the ordered alignment of functional groups on the surface and in the pores, which allow chemical modifications and the binding of molecules in a very precise fashion. Isolated S-layer subunits can self-assemble as monolayers on solid supports (e.g. lipid films, metals, polymers, and silicon wafers), which has a range of applications in colloid and polymer science and the electronics industry. The uniform size and alignment of pores in S-layers also makes them suitable for use as ultrafiltration membranes. Fusion proteins incorporating a range of specific properties such as antibodies or enzymes can be developed to allow highly precise construction and alignment of functional membranes for use in biosensors.

As described in Chapter 6, different species of diatoms construct their silica shells in a variety of beautiful forms. Understanding the molecular basis by which diatoms achieve the construction of their frustules may lead to advances in nano-assembly of materials into desired structures. The microscopic rotary motor of bacterial flagella has attracted the interest of engineers for some years, with suggestions that isolated basal bodies could form the basis of self-propelled micromachines (e.g. for targeted drug delivery systems). The recent discovery of chemotaxis and ultrafast swimming in marine bacteria (p. 76) could lead to advances in this field, as further research leads to an understanding of the molecular basis of these processes. Magnetotactic bacteria (p. 108) produce magnetic crystals with a uniform structure that is difficult to achieve in industrial processes. These could have important applications in the electronics industry and for biomedical applications (e.g. in the production of magnetic antibodies and in magnetic resonance imaging). Understanding the mechanisms by which bacteria construct the magnetosomes and introduction of changes via genetic modification could lead to the ability to display particular proteins.

Biomolecular electronics relies on the use of native or genetically modified biological molecules such as proteins, chromophores, and DNA. One of the best examples to date is the use of bacteriorhodopsin isolated from *Halobacterium salinarum* (p. 126). Bacteriorhodopsin changes its structure every few milliseconds to convert photons into energy. A chromophore embedded in the protein matrix absorbs light and induces a series of changes that change the optical and electrical properties of the protein. Bacteriorhodopsin can store many gigabytes of information in three-dimensional films (holographic memories), and genetic modification produces proteins with various desirable properties. Bacteriorhodopsin has also been used to construct artificial retinas. The newly discovered proteorhodopsins (p. 64) may also have biotechnological applications.

Microbial biotechnology has many applications in aquaculture

As discussed in Chapter 13, the health of marine animals depends on interactions between the host, the pathogen and the environment. In aquaculture, the most important practical measures to prevent or limit diseases are those that reduce stress and maintain good hygiene and the overall health of the stock (*Table 14.5*). These factors are largely a matter of good husbandry and management practice. However, microbial biotechnology has an important role to play, especially in the development of fast and effective diagnostic procedures, control of disease with antibiotics and vaccines, and the development of probiotics for stimulating growth and feed conversion.

DIATOM CELL WALLS LEAD TO NEW COMPUTER CHIPS AND CLOTHES THAT CHANGE COLOR

The cell walls (frustules) of diatoms are constructed by laying down submicron-sized lines of silicon dioxide (silica) embedded with small amounts of organic material. The nano- and micrometer sized patterns are laid down with extraordinary precision, with unique patterns characterizing more than 100000 species. The genome of a strain of the diatom *Thalassiosira pseudonana* was sequenced in 2004, and subsequent analysis has revealed a set of 75 genes specifically involved in bioprocessing of silica. Using whole genome expression profiling and microarrays (see p. 54), Mock et al. (2008) identified 30 genes that were upregulated when the diatoms were grown in low levels of silicic acid, of which 25 had no homology to known genes. Identification of such genes enables the possibility of genetic control or modification of this process, leading to new methods of manufacturing semiconductor computer chips with faster processing speeds than conventional fabrication (photolithography) allows. The complex network of tiny holes in diatom frustules also produces interference patterns when light is shone from different angles. Mass culture of diatoms under controlled laboratory conditions has been developed, raising the possibility of color-changing and iridescent hologram effects when the diatom shells are mixed with paints, polymers, cosmetics, and clothing (Parker and Townley, 2007).

BOX 14.1

Interrupting bacterial conversations
Study of the inhibition of quorum sensing leads to important spin-offs

How the seaweed *Delisea pulchra* protects itself from colonization by other organisms

The overgrowth of one organism by another is widespread in the marine environment. Such natural biofouling is especially important for sessile organisms like corals and seaweeds because colonization by epibionts can limit availability of nutrients and light or interfere with growth and reproduction. Understanding the mechanism by which marine organisms cope with such challenges provides an exciting opportunity for biotechnological exploitation (de Nys and Steinberg, 2002). Many algae are thought to inhibit colonization by the production of secondary metabolites on or near their surfaces, which prevent settlement of fouling organisms. Peter Steinberg, a marine ecologist at the University of New South Wales, investigated the surface of the red seaweed *Delisea pulchra* after noticing that it usually had a very clean surface uncontaminated by biofilms and epibionts. He discovered that *D. pulchra* produces a group of compounds known as brominated furanones, and analysis of the molecular structure showed that they are very similar to homoserine lactones. As previously discussed (see p. 81) acyl homoserine lactones (AHLs) are the key molecules in LuxI/LuxR-based quorum sensing. Steinberg, Kjelleberg, and colleagues subsequently showed that the algal compounds prevented the formation of bacterial biofilms on the surface of the alga by interfering with bacterial communication.

How furanones block quorum sensing

One method to investigate quorum sensing is the use of reporter strains of bacteria that have an easily detectable property that is regulated by the signal molecules. For example, the bacterium *Chromobacterium violaceaum* produces an intense violet pigment, the synthesis of which is regulated by the LuxI/LuxR system. If a piece of *D. pulchra* tissue is placed on a plate of *C. violaceum*, the furanones diffuse out and compete with the AHLs, blocking pigment production. Another detection method is to measure the intensity of bioluminescence using reporter strains of *Escherichia coli* that have been engineered to express AHL-regulated bioluminescence genes. Manefield et al. (1999) used this system to show that furanones bind to the same site on the LuxR regulatory protein as do the AHLs. Once bound, furanones accelerate turnover of LuxR by protease action, thus further limiting the expression of regulated genes (Manefield et al., 2002). Rasmussen et al. (2000) used a model system with the bacterium *Serratia marcescens*, in which it is known that AHLs regulate the expression of at least 28 genes involved in the coordinated multicellular behavior (swarming motility) connected with biofilm formation in this bacterium. Furanones were shown to compete with AHLs and interfere with bacterial communication in mixed-species biofilms by preventing the synthesis of a surface-active agent (serrawettin) that is essential for the swarming process. Dworjanyn et al. (2006) proved that this was an ecologically relevant method of protection for the alga by showing that *D. pulchra* produces furanones in vesicles in specialized cells near the algal surface at ~100 ng per cm^2—a concentration sufficient to inhibit bacterial attachment and biofilm formation. Furthermore, when furanone production drops, which can occur in shallow waters during the summer, fouling by other algae can occur. Dobretsov et al. (2007) used TRFLP community analysis (see p. 48) to show that biofilm communities developing on plastic experimental surfaces had a much lower diversity in the presence of furanones and that this in turn affected the settlement of bryozoan and polychaete larvae.

Quorum sensing inhibitors have diverse applications

The furanone molecules have been synthesized and patented, and developed for a variety of applications by the company Biosignaling Pty. Ltd. Specific products for which proof of concept data have been generated include oil and gas, marine antifouling paints, water-treatment applications, cleaners, and deodorants. However, costs of synthesis and marketing considerations mean that these very promising applications have not yet been fully exploited. Some of the most successful applications are likely to be in high-value products, such as integration into plastics to prevent biofilm formation on surfaces such as contact lenses, medical implants, and catheters.

Perhaps one of the most important and unexpected spin-offs from marine microbiology is the use of furanones as drugs to control human infections in which biofilm formation is a key factor. For example, the bacterium *Pseudomonas aeruginosa* is normally an opportunist pathogen of low virulence, but it is the leading cause of death in over 90% of children with the genetic disease cystic fibrosis. Biofilm formation and production of a wide range of toxins and enzymes is regulated by a network of quorum-sensing systems. Hentzer and Givskov (2002) showed that furanones promoted clearance of bacteria in experimental *P. aeruginosa* infections of mice, and additional studies have demonstrated the potential use of furanones as anti-infective agents (Kim et al., 2008). Although this approach has yet to be proven safe and effective in humans, interfering with bacterial signaling systems is seen as an attractive new approach to infection control because it is thought that bacteria are less likely to develop resistance, as occurs readily with agents that kill the bacteria (Dong et al., 2007).

Most bacterial pathogens can be killed or inhibited by antimicrobial agents

Although many antimicrobials are effective against bacterial pathogens in the laboratory, the number of effective treatments for fish or shellfish disease is quite limited for a number of reasons. First, antimicrobials must be proven to be active against the pathogen, but should produce minimal side effects in the host. The best chemotherapeutic antimicrobials work by targeting a process present in bacteria that is absent or different in their eukaryotic host. For example, penicillins (e.g. amoxicillin) target peptidoglycan synthesis whilst oxytetracycline targets the 30S ribosomal subunit; both of these are unique to bacteria, Second, the agent must reach the site of infection in adequate concentrations

Table 14.5 Important factors for optimizing the health of fish and shellfish in aquaculture

Health factor	Practices
Design and operation of culture systems	Separation of hatchery and growing-on facilities Good management practices and record keeping
Hygiene	Disinfection of nets Protective clothing and equipment Prompt removal of moribund and dead fish Bioremediation for improvement of water quality
Nutrition	Careful monitoring of optimal growth rates at all stages of the life cycle Immune stimulants as feed additives Probiotics for growth promotion and disease prevention
Minimizing stress	Avoid netting, grading, overcrowding Maintain good water quality Avoid feeding before handling Use anesthesia during handling Breed "domesticated" lines of fish
Breaking the pathogen's life cycle	Disinfection of tanks and equipment Separate fish of different ages Fallowing sites for 6 months to 1 year
Diagnosis of disease	Development of rapid methods (e.g. antibody tests and PCR assays)
Eliminating vertical transmission	Production of specific pathogen-free brood stock, testing eggs and sperm for pathogen (e.g. antibody tests and PCR assays)
Preventing geographic spread	Licensing system for egg and larval suppliers Notifiable disease legislation Movement restrictions from infected sites
Eradication	Slaughter policy for notifiable diseases Government compensation
Antimicrobial treatment	Antibiotics and synthetic antimicrobial agents Sensitivity testing Limit use to prevent evolution of resistance
Vaccination	Immersion, oral, and injectable vaccines Ensure strains used for vaccines are appropriate for local disease experience Well-designed tests for evaluation of efficacy in appropriate species Assess the need for reimmunization (boosters) DNA vaccines
Genetic improvement of stock	Select for disease resistance traits Transgenics—incorporation of genes for disease resistance and growth promotion

Items in blue type show practices dependent on microbial biotechnology.

to kill the pathogen or, more usually, inhibit its growth sufficiently to allow the host's immune system to eliminate the pathogen. The rate of uptake, absorption, transport to the tissues, and excretion rates vary greatly among different fish species. Very few agents are capable of inhibiting intracellular pathogens such as *Renibacterium salmoninarum* or *Piscirickettsia salmonis*. Furthermore, because fish are poikilothermic, these processes are very dependent on temperature. Therefore, the efficacy of a particular compound should ideally be evaluated for each host–pathogen interaction under various environmental conditions. A third factor is the need to evaluate the rate of elimination of the drug to ensure that there are no unacceptable residues in the flesh of fish intended for human consumption. Again, because fish are poikilothermic, the rate of excretion and degradation depends on temperature, so it is necessary to calculate a "degree-day" withdrawal period between the last administration of the antimicrobial and the slaughter of the fish for the market. For example, the withdrawal period for oxytetracycline is 400 degree-days (e.g. 40 days at 10°C or 20 days at 20°C). With the growth of aquaculture, government agencies and large retailers in many countries now test farmed fish for antimicrobial residues, in the same way that they test meat or milk. Government regulatory authorities require a considerable amount of testing before licensing a drug for use, and the high costs of testing deter pharmaceutical companies from introducing new agents. The range of treatments is therefore very limited. In Europe, the only antimicrobial agents formally licensed for use in aquaculture are flumequine, oxytetracycline, sulfadiazine-trimethoprim, oxolinic acid, florfenicol, sarafloxacin, and amoxicillin (although licensing varies in different countries of the European Union). Regulatory control is similarly strict in Canada, the USA, and Norway, although different agents are approved. By contrast, more than 30 compounds are licensed in Japan. In some parts of the world, there are no effective controls at all on antimicrobial usage. The routine use of antibiotics for disease prevention or growth promotion has been banned in many countries because of the problems of resistance and residues discussed below.

Antimicrobial agents are most commonly administered in medicated feed. Unfortunately, reduction in feeding is often one of the first signs of disease, so infected fish may not receive the appropriate dose of antimicrobial agents from medicated feed. Antimicrobials may sometimes be given by immersion of infected fish in a bath containing the agent, especially for gill and skin infections. Problems arise with both of these routes of administration because of wastage and contamination of the environment. Injection is rarely used, except for brood stock and aquarium fish.

Resistance to antimicrobial agents is a major problem in aquaculture

Bacteria possess three main strategies for resistance, as shown in *Table 14.6*. Individual bacterial isolates often possess more than one resistance mechanism, and

Table 14.6 Examples of the biochemical basis of acquired bacterial resistance to antimicrobials used in aquaculture

Strategy	Example	Mechanism
Modification of the target binding site	Penicillins	Altered penicillin-binding membrane proteins
	Quinolones	Altered DNA gyrase
Enzymic degradation	Penicillins	β-lactamase production
Reduced uptake or accumulation	Tetracyclines	Altered membrane transport proteins → active efflux
Metabolic bypass	Sulfonamides	Hyperproduction of substrate (*p*-aminobenzoic acid)

individual antimicrobials may be affected by different resistance mechanisms in different bacteria. Bacteria possess *intrinsic* resistance to certain agents because of inherent structural or metabolic features of the bacterial species—this is almost always expressed by chromosomal genes. This type of resistance is relatively easy to deal with, but *acquired* resistance causes major problems in all branches of veterinary and human medicine. The use of almost every antimicrobial sooner or later leads to the selection of resistant strains from previously sensitive bacterial populations. This occurs via spontaneous mutations in chromosomal genes, which occur with a frequency of about 10^{-7}, or by the acquisition of plasmids or transposons. Resistance genes carried on conjugative plasmids (R-factors) may spread rapidly within a bacterial population and may transfer to other species. Plasmid-borne resistance to various antimicrobials is frequently encountered in bacteria pathogenic to fish and shellfish. Emergence of a resistant strain at a fish site renders particular antimicrobials useless, and the resistance can easily spread until it is the norm for that species. Antimicrobials do not cause the genetic and biochemical changes that make a bacterium resistant, but they do select for strains carrying the genetic information that confers resistance. The more an antimicrobial is used, the greater the selection pressure for resistance to evolve. If a particular antimicrobial agent is withdrawn from use, the incidence of resistant strains often declines, because the resistant bacteria now have no advantage and the additional burden of extra genetic information makes them less competitive. However, this is not always the case, as plasmids may confer resistance to several antibiotics. Resistance causes considerable problems in aquaculture. For example, at some times, between 20 and 30% of cases of furunculosis in Scottish salmon culture have been due to strains of *Aeromonas salmonicida* that are resistant to three or more antimicrobials.

Besides the obvious economic losses caused by inefficiency in disease control, there is great concern about the risks of antimicrobial usage in aquaculture to human health and environmental quality. Several studies have shown a buildup of resistant strains in sediments underneath sea cages in sites with poor water exchange, due largely to the accumulation of uneaten food. Transfer of resistance genes to marine bacteria is known to occur, and antimicrobial-resistant bacteria have been isolated from fish that have escaped from facilities where these agents are used excessively. Experimental studies have shown transfer to human pathogens and commensals such as *Salmonella* and *E. coli*, and this raises concerns about risks of transfer of resistance genes into the gut flora of consumers or fish-farm workers. Antibiotic residues could also prove toxic or cause allergies in some people who eat the fish. These concerns about aquaculture are part of a general awareness of the folly of indiscriminate use of antimicrobial agents in medicine, agriculture, and everyday products.

Vaccination of finfish is widely used in aquaculture

Teleost fish possess an efficient immune response and respond to the administration of microbial antigens by the production of antibodies (B-cell response) and cell-mediated immunity (T-cell response). The most common method of administering vaccines to small fish (up to about 15 g) is via brief immersion in a dilute suspension. Particulate antigens (such as bacteria) probably stimulate immunity after passage across the gills. Intraperitoneal injection is necessary for reliable protection with some vaccines, especially viral vaccines and bacterial vaccines that contain soluble components (e.g. most furunculosis vaccines). Injection vaccines are usually administered with an oil adjuvant, which ensures slow release of the antigen and a heightened immune response. Injection vaccines have high efficacy, but have a number of drawbacks. Despite devices to convey fish from the water onto an injection table and the use of repeater syringes (*Figure 14.3*), injection vaccines incur high labor costs and they cannot be used on fish weighing less than about 15 g. Unless carefully managed, the stress associated with crowding, removal from the water, and injection causes mortalities and

? WHAT ARE THE HAZARDS OF ANTIBIOTIC USAGE IN AQUACULTURE?

Allen et al. (2010) note that the presence and spread of antibiotic resistance in nonclinical and nonagricultural environments such as aquaculture are poorly understood and that further research on the environmental reservoirs of resistance determinants is needed. In many parts of the world, antimicrobial usage is unregulated, and large quantities are used as prophylactic treatments with little regard for testing for sensitivity or proper withdrawal periods. This is especially problematic in prawn and shrimp aquaculture in parts of Asia and South America. In a review of the topic, Cabello (2006) argues that aquaculture in China and Chile is the major selective pressure for evolution of resistance to quinolones (an important antibiotic for human diseases). On the other hand, some scientists think that the risk is overstated; for example, Macmillan (2000) states that "there is very limited actual evidence to indicate that the use of antibiotics in aquaculture is of any risk to public health." Nevertheless, for all of the reasons mentioned above, disease prevention is better than cure, and the continued success of marine aquaculture is only possible with the advent of effective vaccines. Indeed, where vaccines have been successfully introduced, antimicrobial usage has dropped sharply. For example, the use of antimicrobials in Norway has fallen from 50000 kg to about 750 kg per annum in the last 15 years.

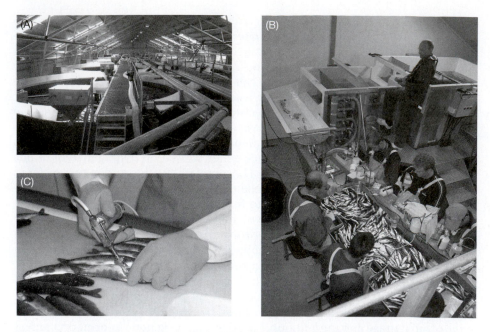

Figure 14.3 Large-scale vaccination of juvenile Atlantic salmon. Fish are pumped from the holding tanks by pipes (A) to an anesthetic bath (B) before intraperitoneal injection by staff (C). (Image courtesy of Ingunn Sommerset, Intervet Bio AS, Norway. Reprinted from Sommerset et al. [2005] by permission of Expert Reviews Ltd.)

may even precipitate infection. The most desirable form of vaccine is one that can be administered orally. The main difficulty with this approach is that microbial antigens are degraded in the fish's stomach and fore-gut before reaching the gut-associated lymphoid tissue in the hind-gut, where the immune response occurs. This is overcome by microencapsulation of the vaccine in biodegradable polymers such as poly DL-lactide-co-glycolide. Many commercial vaccines using this, or similar, approaches are now available.

The simplest type of bacterial vaccine is a bacterin, which consists of a dense culture of bacterial cells killed by formalin treatment. Although technically simple, careful attention must be given to the quality control of media composition and incubation conditions, to ensure that the bacterin contains the appropriate protective antigens. One of the earliest successes in vaccine development was the vibriosis vaccine, which is effective against *Vibrio anguillarum* infection in salmon. The protective antigen in this case seems to be lipopolysaccharide in the bacterial outer membrane. Commercial vaccines usually incorporate two or more serotypes to allow for antigenic variation and they often work well in a range of situations. Development of effective bacterins for *Vibrio* [now *Aliivibrio*] *salmonicida* and *Vibrio ordalii* also proved relatively straightforward, but the early success with the vibrios was not repeated with other diseases. For example, the breakthrough in development of an effective, long-lasting furunculous vaccine was only achieved after recognition of the crucial role of extracellular proteases and iron-regulated outer membrane proteins (see p. 242), and manipulation of the culture and formulation conditions to ensure the correct blend of particulate and soluble antigens.

Recombinant DNA technology is used to produce vaccines for diseases caused by viruses and some bacteria

Viruses can only be propagated in cell culture and some bacterial pathogens, notably *Renibacterium salmoninarum* and *Piscirickettsia salmonis*, are slow growing and difficult to culture. The cost of production of inactivated vaccines using cell culture is prohibitively expensive. Therefore, attention has turned to the use of recombinant DNA technology. Genes important in virulence—such as viral capsid proteins, bacterial toxins, or surface antigens—can be cloned and expressed in a recombinant host to produce a subunit vaccine. The most common method of achieving this is to fuse the gene of interest to a gene for β-galactosidase or maltose-binding protein, leading to production of a fusion or hybrid protein in a high-expression system such as *E. coli*. The fusion protein is produced in large amounts, sometimes up to 30% of total protein. If desired, the carrier protein can be removed and the cloned microbial antigen can be purified using chromatography, but in practice this is often not necessary as crude cell lysates make efficient vaccines, thus reducing processing costs. Sometimes, yeast or insect cell cultures (using a baculovirus vector) are used, especially if glycosylation of the protein is required. Additional genes for T-cell epitopes (e.g. from tetanus toxin or measles virus) may also be introduced to enhance the immunostimulatory potential of a vaccine by stimulating function of the T-cells in the immune system. This approach has been used to generate vaccines against several bacterial diseases, including furunculosis (based on the *A. salmonicida* serine protease), piscirickettsiosis (*P. salmonis* OspA membrane protein), and bacterial kidney disease (*R. salmoninarum* hemolysin and metalloprotease). Subunit vaccines for the viral diseases infectious salmon anemia (ISA) and infectious pancreatic necrosis (IPN) are based on capsid proteins.

DNA vaccination or genetic immunization depends on expression of a sequence encoding the protective antigen

Unlike conventional vaccination, which depends on the administration of an antigen, DNA vaccination, also known as genetic immunization, is based on the delivery of naked DNA containing the sequence for a region of the protective antigen (epitope). The DNA containing this sequence is incorporated into a plasmid with appropriate promoters; if this is done correctly, the gene will be expressed in the fish host cells. Thus, the fish makes foreign antigens internally and then mounts an immune response to them. An alternative strategy is based on the expression of an antibody fragment inside the cell that can bind to and inactivate the pathogen. Fish cells efficiently express foreign proteins encoded by eukaryotic expression vectors. The first experiments using this technology were carried out using glycoprotein genes of the rhabdoviruses causing viral hemorrhagic septicemia (VHS) and infectious hematopoietic necrosis (IHN). These were highly successful, and large-scale trials have confirmed their efficacy. However, DNA vaccines for other viral diseases (such as IPN, ISA, and halibut nodavirus) have produced mixed results, and refinement of the vector plasmids and delivery system are required. In experimental trials, many of these vaccines have been effective after intramuscular injection. An alternative method of gene delivery into the skin or muscle is to use a "gene gun," which fires tiny gold particles coated with the DNA into the tissue. Issues of longevity of protection, stimulation of the correct immune responses, and safety need further investigation.

Unfortunately for the proponents of this technology, public opinion—especially in Europe, but much less so in North America and Asia—is against such genetic engineering in the food chain. Attempts by the industry to introduce transgenic fish that incorporate genes for faster growth have so far not been acceptable. DNA vaccination is actually not transgenic modification, because the DNA is not introduced into the germ line, but this is a subtle distinction for those who object to this new technology. DNA vaccines are thought by most experts to have very low risks and with better understanding and acceptance they hold considerable potential in aquaculture, especially for control of diseases caused by viruses,

? ARE LIVE VACCINES SAFE?

Live attenuated vaccines, in which the pathogen is rendered avirulent, are very effective because the bacterium or virus replicates within the host and delivers antigens over a prolonged period. Live vaccines are also better at stimulating mucosal and cell-mediated immunity, and they are more suitable for oral delivery. Many human viral vaccines are based on this principle, although live bacterial vaccines have been less favored because their more complex genomes lead to the possibility of incomplete attenuation or subsequent reversion to virulence by recombination. Recombinant DNA technology allows the deletion and replacement of specific genes necessary for virulence and survival *in vivo* in a more controlled and targeted approach to attenuation. Such an approach was used to construct a live *A. salmonicida* vaccine by deletion of the *aroA* gene, which encodes an essential amino acid biosynthesis pathway, not present in animals. Allelic replacement of this gene and subsequent further attenuation guards against the possibility of reversion. In trials, this vaccine was highly effective, and the vaccine strain could be engineered to deliver other antigens (Marsden et al., 1996). A similar approach was used for *P. damselae* subsp. *piscicida* by removing a siderophore gene and for *S. iniae* by removing the gene for a surface virulence protein (Locke et al., 2008). Live vaccines have also been developed for the disease VHS, which normally occurs in salmon and trout in the freshwater stage, but is also known in marine turbot. Live vaccines for IPN have been successful in freshwater salmonid culture, but are not effective against marine birnaviruses (Sommerset, 2005). Even with good evidence of protection and no evidence of reversion, live vaccines have not found favor with licensing authorities for use in fish, largely prompted by concerns about deliberate release into the environment.

BOX 14.2   RESEARCH FOCUS

"The enemy of my enemy is my friend"
The revival of an old idea to use viruses for the treatment of disease and biological control of harmful marine organisms

A brief history of phage therapy

Soon after their discovery in 1915, it was proposed that phages (see p. 158) might be useful for treating bacterial infections. Over many years, research institutes in Tbilisi, Georgia, and Wroclaw, Poland, developed methods for isolating and propagating phages and using them to treat a variety of human infections. One of the earliest use of phages was the treatment of cholera in India; between 1925 and 1934, more than one million doses of vibriophages were administered at the start of cholera outbreaks and released into community drinking water for prophylaxis, resulting in significant reduction in death rates (reviewed by Nelson et al., 2009), and there is now interest in reviving this approach because vaccines have had such limited success (Brüssow, 2007). In the 1930s, pharmaceutical companies in the USA began to produce phages commercially, but there were problems with standardization and poor design of clinical trials. Furthermore, the discovery and rapid success of antibiotics meant that the idea of phage therapy was quickly forgotten in Western countries in the 1940s, although it continued to be used successfully in Georgia and Eastern Europe. In the 1980s, the widespread development of antibiotic resistance prompted scientists to reassess the potential of phage therapy and carefully designed trials have proved that it can be highly successful in treatment of bacterial infections of mammals. Housby and Mann (2009) note that there are now about 20 companies developing phage products for commercial use in plant, animal, and human diseases, especially those caused by antibiotic-resistant bacteria. Various other promising applications include control of plant diseases and treatment of food to prevent *Listeria* infections (Brüssow, 2007; O'Flaherty et al., 2009).

Advantages and disadvantages of phages as therapeutic and biocontrol agents

Phages have a narrow host range that can be directed against specific pathogens, and they should therefore not affect the normal host microbiota. They are self-perpetuating in the presence of susceptible bacteria, so repeated administration is not necessary. Phages coevolve with their host bacteria, so even if a bacterial strain acquires resistance, it should be possible to isolate new phages that are infective. However, this variability means that if the exact strain of bacterium causing an infection is unknown, it is necessary to inoculate with a "cocktail" of phages. Additional problems can occur because animals or humans may mount an immune response to the phage proteins and phage particles may not reach all parts of the body. Another potential problem may arise owing to the phenomenon of lysogeny, gene transfer and conversion of phenotypic prop-

erties by phage infection. For example, infection of *Vibrio harveyi* by VHML phage causes enhanced virulence (see p. 235); and Flegel et al. (2005) argue that the use of phages for control of pathogenic *Vibrio* spp. should include careful consideration of the potential for transfer of virulence or antibiotic resistance genes.

Use of phages in aquaculture of fish

In Japan, Nakai et al. (1999) were seeking to control a serious opportunistic disease problem caused by *Lactococcus garvieae* affecting *Seriola quinqueradiata* (yellowtail). They found that over 90% of *L. garvieae* strains were sensitive to two phages. In an experimental infection, groups of fish were injected intraperitoneally with either the bacteria or phage, or with a combination. If phage was administered together with the bacterial inoculum, 100% of fish survived, whereas the survival rate was only 10% in fish injected with the bacteria alone. If phage was administered to fish 1 h after infection, the time of onset of disease was delayed and partial protection occurred; if 24 h elapsed between bacterial infection and injection of the phage, the protection afforded was only about 50%. When phages were incorporated into the fish feed, mortality was reduced from 65% to 10%. Nakai and Park (2002) also found that the freshwater fish *Plecoglossus altivelis* (ayu) was protected against oral challenge with *Pseudomonas plecoglossicida*. In another freshwater infection trial in aquaria, addition of phages to the water delayed the onset of furunculosis caused by *Aeromonas salmonicida* in brook trout *Salvelinus fontinalis* (Imbeault et al., 2006). However, Verner-Jeffreys et al. (2007) showed that Atlantic salmon injected with phage immediately after injection of *A. salmonicida* died more slowly, but eventual mortality was the same as the control treatment and there was some evidence of evolution of resistance to specific phage strains. These authors concluded that phage therapy may not be suitable for such a highly virulent pathogen. Whilst initially promising, phage therapy seems at present to have limited scope for application in finfish culture—prevention of disease by vaccines is more efficient and cost-effective.

Phage therapy in aquaculture of crustaceans

Prospects for the use of phage therapy in control of bacterial diseases of crustaceans are more promising. Here, the need for emergency intervention to eliminate catastrophic infections by *Vibrio* spp. could make the effort of maintaining collections of different phage "cocktails" commercially worthwhile. Although there are a number of anecdotal reports of trials of phage therapy in prawn and shrimp culture in Asia, and of lobster culture in North America, there are few published reports in scientific journals. In India, Vinod et al. (2006) isolated several lytic phages of

BOX 14.2   RESEARCH FOCUS

V. harveyi from farm or hatchery water used to rear the shrimp *Penaeus monodon.* In laboratory microcosm experiments, addition of phages to water produced a reduction in the bacterial counts from 10^6 to 10^3 CFU per milliliter and 80% of larvae survived, compared with 25% in the control.

In the same laboratory, Karunasagar et al. (2007) showed that there was a dose-dependent effect on the formation of *V. harveyi* biofilms and suggested that phage could be useful for preventing the buildup of *V. harveyi* in tanks. Treatment of four tanks with *V. harveyi* infection showed that phage was more effective in controlling infection than antibiotic treatment, although the treatments were not replicated. More recently, Crothers-Stomps et al. (2010), working at the James Cook University and Australian Institute of Marine Science, isolated a collection of phages from two virus families with lytic activity against *V. harveyi* and are developing phage "cocktails" to control vibriosis in culture of larvae from the tropical rock lobster *Panulirus ornatus*.

Could phages be used to control coral disease?

The threat to coral reefs from infectious diseases, especially those caused by bacteria, was discussed in Chapter 11. It is difficult to envisage methods for prevention or treatment of disease in natural reef ecosytems. Large-scale use of antibiotics would not be practicable and the lack of an antibody response in corals means that vaccination is not possible—although Teplitski and Ritchie (2009) suggest that it may be possible to prime the immune defenses of certain corals. Eugene Rosenberg and coworkers at the University of Tel Aviv, Israel, have proposed that phage therapy can be used to control bacterial coral diseases. In aquarium experiments, Efrony et al. (2007) isolated phages of the pathogen *Thallassomonas loyaena* from the Red Sea and showed that addition of one type of phage at a density of 10^4–10^6 plaque-forming units per milliliter protected the coral *Favia favus* against infection caused by incubation in water taken from an aquarium containing diseased coral. The phages increased in number, and levels remained high even after water in the aquarium was changed—indicating that the phage replicates and remains associated with coral tissue. The *T. loyaena* phage was characterized and shown to prevent transmission of the disease to healthy corals (Efrony et al., 2009). Similarly, the coral *Pocillopora damicornis* was protected against infection by *Vibrio coralliilyticus* using phages specific for this bacterium; furthermore, the coral was protected against reinfection by the bacterium (Efrony et al., 2007). As before (see *Research Focus Box 11.1*), Rosenberg's ideas have not been well received by some coral biologists, who argue that his proposal to treat reefs in this way is impractical and environmentally unsound. However, in a situation in which some reefs are in critical danger of being lost forever because of disease,

it is important to give this proposal serious consideration. Recently, permission was obtained to conduct a small-scale experiment on a threatened reef in the Red Sea, and it gave promising results—none of the phage-treated coral colonies showed transmission of the diseases, whereas 40% of the untreated colonies did. Would it really be technically feasible to treat a reef? Rosenberg calculates that to treat a reef 10 km × 100 m wide × 10 m deep (a water volume of 10^7 m^3) would require only 10 liters of phage lysate to produce a concentration of 10^6 phage per liter—this is an amount easily obtainable from a microbiological fermenter. In the USA, Kellogg (2007) has suggested that prospects are good for using phages to treat diseased corals in the seriously impacted reefs of the Florida Keys. Whilst generally supportive of the concept, Teplitski and Ritchie (2009) argue that a better understanding of the role of vibrios as commensal bacteria in corals is needed before therapy is used in the field, especially the effect of phage therapy on other members of the coral holobiont.

Could viruses be used to control harmful algal blooms and other problems?

Viruses are known to cause lysis and natural population crashes of phytoplankton (see p. 161), so it is not surprising that they have been proposed as biological control agents. Keizo Nagasaki and coworkers at the Fisheries Research Center, Hiroshima, Japan, have isolated numerous viruses of toxic and nuisance diatoms and dinoflagellates (Nagasaki, 2008). In particular, the virus HaV has been investigated as an agent to control red tides caused by the raphidophyte *Heterosigma akashiwo*. The virus has a high rate of replication and appears nontoxic to other marine organisms and can be produced in large volumes at relatively low cost. However, the complexities of host–virus specificities in natural populations mean that single viral isolates would have little effect and "cocktails" of viral isolates of different host specificities will be needed (Tomaru et al., 2008).

Increasing knowledge of viruses of other protists could lead to novel approaches to disease control in ecologically important species. For example, the discovery of a virus in a thaustochytrid (Takao et al., 2005) suggests that it would be highly profitable to screen for viruses of the closely related labyrinthulids that cause wasting disease in seagrass (see p. 253).

In conclusion, although the idea of viruses as biological control agents is appealing, the diversity and complexities of marine life need careful consideration, and considerable further research with well-validated experiments is needed—we should heed some of the lessons learned from agricultural biocontrol measures on land (Secord, 2003).

eukaryotic parasites and hard-to-culture bacteria like *R. salmoninarum* and *P. salmonis*. Interestingly, a DNA vaccine approach may also confer protection against viral infection in crustaceans.

Probiotics, prebiotics and immunostimulants are widely used in marine aquaculture

Interest in this area has occurred largely as a result of the need to reduce antibiotic usage because of concerns over resistance and residues. Probiotics have been very successful in other branches of agriculture such as poultry and pig farming and commercial production is now of considerable economic importance to the animal feed industry, with a global market of US$7 billion in 2010, growing by over 13% a year. Numerous studies in various species of finfish (especially salmonids) have shown that addition of certain bacteria to feeds leads to disease resistance or other benefits. In general, probiotic strains have been isolated from naturally occurring microbiota of fish. They are distinguished from vaccines because they do not require participation of the host immune system and may act directly by inhibition of pathogens via production of antimicrobial metabolites or bacteriocins, or by competition for nutrients or sites for colonization of the gut mucosa. Besides direct effects on pathogens or enhancement of nutrition, some probiotic products (in the broadest sense of the term) have immunostimulatory effects. A range of preparations incorporating lipopolysaccharides, glucans, and other microbial components have been shown to enhance host defenses. Whilst many studies lack scientific rigor and proof of the mechanisms of disease prevention by specific challenge, evidence of their efficacy is provided by major reductions in antibiotic usage. Some statistically rigorous studies have shown that probiotics improve the survival and growth of larvae and lessen the survival of pathogenic bacteria in the gut.

Many probiotic agents show antagonistic activity against pathogens *in vitro*, but this is no guarantee that they have this effect *in vivo*. Most antagonistic compounds are secondary metabolites produced during the stationary phase in culture, and it is unknown whether they are produced in significant amounts *in vivo*. Electron microscopy has been used to show how probiotic bacteria may compete with pathogens for colonization sites. Some probiotics may produce beneficial dietary compounds such as vitamins, antioxidants, and lipids or may provide their host with digestive enzymes, enabling the degradation of complex nutrients such as chitin, starch, or cellulose. These factors are becoming very important in the development of sustainable aquaculture, as there is a need to use alternative nutrient sources to replace the industry's reliance on fishmeal.

The gastrointestinal tract of fish and shellfish larvae becomes colonized by bacteria present in the water; therefore, control of water quality in the early larval stages is very important to prevent colonization by pathogens. Agents may be added to the water, incorporated into pelleted feeds, or used to enrich live feed such as rotifers or *Artemia*. A variety of commercial products incorporating dried endospores of *Bacillus* spp. (e.g. *B. subtilis*, *B. licheniformis*, and *B. megaterium*) have proved effective in limiting disease. The spores are cheap to manufacture, stable, and easy to administer. Other agents, such as lactic acid bacteria, *Roseobacter* spp., *Pseudoalteromonas*, *Pseudomonas*, nonpathogenic *Vibrio* spp., yeasts, and microalgae have also been effective, but these are more difficult to distribute commercially as live cultures.

There does seem to be a specific indigenous microbiota of the gut of marine fish, composed mainly of Gram-negative, aerobic, anaerobic, and facultatively anaerobic bacteria. The most common genera isolated by culture from the gut of marine fish are *Vibrio*, *Photobacterium*, *Pseudomonas*, and *Acinetobacter*. Recent application of molecular methods such as PCR-DGGE analysis (see p. 48) shows that various other organisms are important and that the microbiota is very variable dependent on genetic, nutritional, and environmental factors. Unlike terrestrial animals, the gut microbiota of aquatic animals probably consists of relatively small populations of attached indigenous bacteria, complemented by large numbers of allochthonous bacteria maintained by their constant ingestion from the surrounding water and feed. Metagenomic studies of the gut of various fish species would be highly desirable for future studies.

There is a highly developed industry supplying a great range of products for use in shrimp, prawn, and lobster culture. These products are promoted as immunostimulants and sometimes as vaccines, although this is usually considered an incorrect use of the term since it should only be applied to an agent that induces protection through long-lasting immune memory that is dependent on primary challenge with an antigen and the stimulation of clones of specific lymphocytes. Invertebrates lack this system, although crustaceans do possess active cells such as hemocytes that are responsible for inflammatory-type reactions, phagocytosis, and the killing of microbes via oxidative burst or microbicidal proteins. Immune stimulation can occur by enhancing nonspecific complement activation, phagocytosis, or cytokine production. A range of treatments, including live or killed bacteria, glucans, lipopolysaccharide, extracts of yeast, and algal cell walls, have been used, and promising effects have been reported.

Conclusions

The untapped hidden treasure in the sea is immense. Microbes and their genes and proteins are so diverse that we have exploited only a tiny fraction of the opportunities for biotechnology. Advances in molecular biology, especially high-throughput sequencing, powerful bioinformatics tools, and functional screening methods mean that many new products and processes for industry and medicine will be discovered in the coming years and exploited through the developing field of synthetic genomics. Marine microbial biotechnology will play a major role in ensuring new sources of energy and sustainable aquaculture to feed the world's growing population, and it may help to mitigate some of the environmental damage caused by human activities. We will face important ethical and environmental issues to ensure that the exploitation of these discoveries can be properly harnessed for the benefit of mankind and our planet.

? CAN CRUSTACEANS BE VACCINATED?

Smith et al. (2003) question the validity of many of the studies that have investigated treatments that claim immune protection for crustaceans and argue for more rigorous testing to show that they are cost-effective methods of control. By contrast, Johnson et al. (2008) review the compelling evidence that shrimp can be successfully vaccinated against white spot syndrome virus with viral envelope protein subunits, with double-stranded RNA or with plasmids expressing the viral proteins. Recently, a number of studies have challenged the view that invertebrate immune systems lack specificity and memory and suggest that this may be explained by enhanced phagocytosis (Rowley and Powell, 2007). The case of antiviral immunity in crustaceans is particularly interesting and the subject of much current debate. Flegel (2009) proposes a hypothesis to explain the process of viral accommodation, by which crustaceans and insects can persistently carry viral pathogens at low levels without signs of disease and transmit them to their offspring or to naive individuals. It is thought that specific responses may be based on viral sequences inserted into the host genome, and Flegel proposes how this might be exploited for development of vaccines.

References

Allen, H. K., Donato, J., Wang, H. H., Cloud-Hansen, K. A., Davies, J. and Handelsman, J. (2010) Call of the wild: antibiotic resistance genes in natural environments. *Nat. Rev. Microbiol.* 8: 251–259.

Brüssow, H. (2007) Phage therapy: the Western perspective. In: *Bacteriophage: Genetics and Molecular Biology* (eds S. McGrath and D. van Sinderen). Caister Academic Press, Norfolk, UK, pp. 159–192.

Cabello, F. C. (2006) Heavy use of prophylactic antibiotics in aquaculture: a growing problem for human and animal health and for the environment. *Environ. Microbiol.* 8: 1137–1144.

Crothers-Stomps, C., Høj, L., Bourne, D. G., Hall, M. R. and Owens, L. (2010) Isolation of lytic bacteriophage against *Vibrio harveyi*. *J. Appl. Microbiol.* 108: 1744–1750.

de Nys, R. and Steinberg, P. D. (2002) Linking marine biology and biotechnology. *Curr. Opin. Biotechnol.* 13: 244–248.

Dobretsov, S., Dahms, H. U., Huang, Y. L., Wahl, M. and Qian, P. Y. (2007) The effect of quorum-sensing blockers on the formation of marine microbial communities and larval attachment. *FEMS Microbiol. Ecol.* 60: 177–188.

Dong, Y. H., Wang, L. H. and Zhang, L. H. (2007) Quorum-quenching microbial infections: mechanisms and implications. *Philos. Trans. Roy. Soc. B Biol. Sci.* 362: 1201–1211.

Dworjanyn, S. A., de Nys, R. and Steinberg, P. D. (2006) Chemically mediated antifouling in the red alga *Delisea pulchra*. *Mar. Ecol. Prog. Ser.* 318: 153–163.

Efrony, R., Loya, Y., Bacharach, E. and Rosenberg, E. (2007) Phage therapy of coral disease. *Coral Reefs* 26: 7–13.

Efrony, R., Atad, I. and Rosenberg, E. (2009) Phage therapy of coral white plague disease: properties of phage BΛ3. *Curr. Microbiol.* 58: 139–145.

Fisch, K. M., Gurgui, C., Heycke, N., et al. (2009) Polyketide assembly lines of uncultivated sponge symbionts from structure-based gene targeting. *Nat. Chem. Biol.* 5: 494–501.

Flegel, T. (2009) Hypothesis for heritable, anti-viral immunity in crustaceans and insects. *Biol. Direct* 4: 32.

Flegel, T. W., Pasharawipas, T., Owens, L. and Oakey, H. J. (2005) Phage induced virulence in the shrimp pathogen *Vibrio harveyi*. In: *Diseases in Asian Aquaculture* V. *Fish Health Section* (eds

P. Walker, R. Lester and M. B. Bondad-Reantaso). Asian Fisheries Society, Manila, pp. 329–337.

Gibson, D. G., Glass, J. I., Lartigue, C., et al. (2010) Creation of a bacterial cell controlled by a chemically synthesized genome. *Science* 329: 52–58.

Goffredi, S. K., Orphan, V. J., Rouse, G. W., et al. (2005) Evolutionary innovation: a bone-eating marine symbiosis. *Environ. Microbiol.* 7: 1369–1378.

Hentzer, M. and Givskov, M. (2003) Pharmacological inhibition of quorum sensing for the treatment of chronic bacterial infections. *J. Clin. Invest.* 112: 1300–1307.

Hentzer, M., Riedel, K. and Rasmussen, T. B. (2002) Inhibition of quorum sensing in *Pseudomonas aeruginosa* biofilm bacteria by a halogenated furanone compound. *Microbiolgy* 148: 87–107.

Hertweck, C. (2009) Hidden biosynthetic treasures brought to light. *Nat. Chem. Biol.* 5: 450–452.

Housby, J. N. and Mann, N. H. (2009) Phage therapy. *Drug Discov. Today* 14: 536–540.

Huang, G., Chen, F., Wei, D., Zhang, X. and Chen, G. (2010) Biodiesel production by microalgal biotechnology. *Appl. Energy* 87: 38–46.

Imbeault, S., Parent, S., Lagacé, M., Uhland, C. F. and Blais, J. F. (2006) Using bacteriophages to prevent furunculosis caused by *Aeromonas salmonicida* in farmed brook trout. *J. Aquat. Anim. Health* 18: 203–214.

Johnson, K. N., Van Hulten, M. C. W. and Barnes, A. C. (2008) "Vaccination" of shrimp against viral pathogens: phenomenology and underlying mechanisms. *Vaccine* 26: 4885–4892.

Karunasagar, I., Shivu, M. M., Girisha, S. K., Krohne, G. and Karunasagar, I. (2007) Biocontrol of pathogens in shrimp hatcheries using bacteriophages. *Aquaculture* 268: 288–292.

Kellogg, C. A. (2007) *Phage Therapy for Florida Corals?* US Geological Survey Fact Sheet 2007-3065.

Kim, C., Kim, J., Park, H. Y., Park, H. J., Lee, J., Kim, C. and Yoon, J. (2008) Furanone derivatives as quorum-sensing antagonists of *Pseudomonas aeruginosa*. *Appl. Microbiol. Biotechnol.* 80: 37–47.

Lee, H. S., Kwon, K. K., Kang, S. G., Cha, S.-S., Kim, S.-J. and Lee, J.-H. (2010) Approaches for novel enzyme discovery from marine environments. *Curr. Opin. Biotechnol.* 21: 351–357.

Locke, J. B., Aziz, R. K., Vicknair, M. R., Nizet, V. and Buchanan, J. T. (2008) *Streptococcus iniae* M-like protein contributes to virulence in fish and is a target for live attenuated vaccine development. *PLoS ONE* 3: e2824.

Macmillan, J. R. (2000) Antibiotics in aquaculture: questionable risk to public health. *Alliance for the Prudent Use of Antibiotics Newsletter* 18: 1–5.

Manefield, M., de Nys, R., Kumar, N., Read, R., Givskov, M., Steinberg, P. and Kjelleberg, S. A. (1999) Evidence that halogenated furanones from *Delisea pulchra* inhibit acylated homoserine lactone (AHL)-mediated gene expression by displacing the AHL signal from its receptor protein. *Microbiology* 145: 283–291.

Manefield, M., Rasmussen, T. B., Henzter, M., Andersen, J. B., Steinberg, P., Kjelleberg, S. and Givskov, M. (2002) Halogenated furanones inhibit quorum sensing through accelerated LuxR turnover. *Microbiology* 148: 1119–1127.

Marsden, M. J., Vaughan, L. M., Foster, T. J. and Secombes, C. J. (1996) A live (Delta aroA) *Aeromonas salmonicida* vaccine for furunculosis preferentially stimulate T-cell responses relative to B-cell responses in rainbow trout (*Oncorhynchus mykiss*). *Infect. Immun.* 64: 3863–3869.

Merrifield, D. L., Dimitroglou, A., Foey, A., et al. (2010) The current status and future focus of probiotic and prebiotic applications for salmonids. *Aquaculture* 302: 1–18.

Mock, T., Samanta, M. P., Iverson, V., et al. (2008) Whole-genome expression profiling of the marine diatom *Thalassiosira pseudonana* identifies genes involved in silicon bioprocesses. *Proc. Natl Acad. Sci. USA* 105: 1579–1584.

Nagasaki, K. (2008) Dinoflagellates, diatoms, and their viruses. *J. Microbiol.* 46: 235–243.

Nakai, T. and Park, S. C. (2002) Bacteriophage therapy of infectious diseases in aquaculture. *Res. Microbiol.* 153: 13–18.

Nakai, T., Sugimoto, R., Park, K. H., Matsuoka, S., Mori, K., Nishioka, T. and Maruyama, K. (1999) Protective effects of bacteriophage on experimental *Lactococcus garvieae* infection in yellowtail. *Dis. Aquat. Organ.* 37: 33–41.

Nelson, E. J., Harris, J. B., Morris, J. G., Calderwood, S. B. and Camilli, A. (2009) Cholera transmission: the host, pathogen and bacteriophage dynamic. *Nat. Rev. Microbiol.* 7: 693–702.

O'Flaherty, S., Ross, R. P. and Coffey, A. (2009) Bacteriophage and their lysins for elimination of infectious bacteria. *FEMS Microbiol. Rev.* 33: 801–819.

Parker, A. R. and Townley, H. E. (2007) Biomimetics of photonic nanostructures. *Nat. Nanotechnol.* 2: 347–353.

Ramachandra, T. V., Mahapatra, D. M., Karthic, B. and Gordon, R. (2009) Milking diatoms for sustainable energy: biochemical engineering versus gasoline-secreting diatom solar panels. *Ind. Eng. Chem. Res.* 48: 8769–8788.

Rasmussen, T. B., Manefield, M., Andersen, J. B., et al. (2000) How *Delisea pulchra* furanones affect quorum sensing and swarming motility in *Serratia liquefaciens* MG1. *Microbiology* 146: 3237–3244.

Ringø, E., Olsen, R. E., Gifstad, T. Ï., Dalmo, R. A., Amlund, H., Hemre, G. I., and Bakke, A. M. (2010) Prebiotics in aquaculture: a review. *Aquacult. Nutr.* 16: 117–136.

Rowley, A. F. and Powell, A. (2007) Invertebrate immune systems: specific, quasi-specific, or nonspecific? *J. Immunol.* 179: 7209–7214.

Secord, D. (2003) Biological control of marine invasive species: cautionary tales and land-based lessons. *Biol. Invasions* 5: 117–131.

Smith, V. J., Brown, J. H. and Hauton, C. (2003) Immunostimulation in crustaceans: does it really protect against infection? *Fish Shellfish Immunol.* 15: 71–90.

Sommerset, I., Krossøy, B., Biering, E. and Frost, P. (2005) Vaccines for fish in aquaculture. *Expert Rev. Vaccines* 4: 89–101.

Takao, Y., Honda, D., Nagasaki, K., Mise, K. and Okuno, T. (2005) Isolation and characterization of a novel single-stranded RNA virus infectious to a marine fungoid protist, *Schizochytrium* sp. (Thraustochytriaceae, Labyrinthulea). *Appl. Environ. Microbiol.* 71: 4516–4522.

Teplitski, M. and Ritchie, K. (2009) How feasible is the biological control of coral diseases? *Trends Ecol. Evol.* 24: 378–385.

Tomaru, Y., Shirai, Y., and Nagasaki, K. (2008) Ecology, physiology and genetics of a phycodnavirus infecting the noxious bloom-forming raphidophyte *Heterosigma akashiwo*. *Fisheries Sci.* 74: 701–711.

Treude, T., Smith, C. R., Wenzhofer, F., et al. (2009) Biogeochemistry of a deep-sea whale fall: sulfate reduction, sulfide efflux and methanogenesis. *Mar. Ecol. Prog. Ser.* 382: 1–21.

Verner-Jeffreys, D. W., Algoet, M., Pond, M. J., Virdee, H. K., Bagwell, N. J. and Roberts, E. G. (2007) Furunculosis in Atlantic salmon (*Salmo salar* L.) is not readily controllable by bacteriophage therapy. *Aquaculture* 270: 475–484.

Vinod, M. G., Shivu, M. M., Umesha, K. R., Rajeeva, B. C., Krohne, G., Karunasagar, I. and Karunasagar, I. (2006) Isolation of *Vibrio harveyi* bacteriophage with a potential for biocontrol of luminous vibriosis in hatchery environments. *Aquaculture* 255: 117–124.

Further reading

Antifouling biotechnology

Dobretsov, S., Teplitski, M. and Paul, V. (2009) Mini-review: quorum sensing in the marine environment and its relationship to biofouling. *Biofouling* 25: 413–427.

Holmstrom, C., Egan, S., Franks, A., McCloy, S. and Kjelleberg, S. (2002) Antifouling activities expressed by marine surface associated *Pseudoalteromonas* species. *FEMS Microbiol. Ecol.* 41: 47–58.

Enzymes and pharmaceuticals

Antranikian, G. (2007) Industrial relevance of thermophiles and their enzymes. In: *Thermophiles: Biology and Technology at High Temperatures* (ed. F. Robb and D. Grogan). CRC Press, Boca Raton, FL, pp. 113–186.

Bull, A. T. (2004) *Microbial Diversity and Bioprospecting*. American Society for Microbiology, ASM Press, Washington, DC.

Gerwick, W. H., Coates, R. C., Engene, N., Gerwick, L., Grindberg, R. V., Jones, A. C. and Sorrels, C. M. (2008) Giant marine cyanobacteria produce exciting potential pharmaceuticals. *Microbe* 3: 277–284.

Gulder, T. A. M. and Moore, B. S. (2009) Chasing the treasures of the sea—bacterial marine natural products. *Curr. Opin. Microbiol.* 12: 252–260.

Haygood, M. G., Schmidt, E. W., Davidson, S. K. and Faulkner, D. J. (1999) Microbial symbionts of marine invertebrates: opportunities for microbial biotechnology. *J. Molec. Microbiol. Biotechnol.* 1: 33–43.

Kennedy, J., Marchesi, J. R. and Dobson, A. D. W. (2008) Marine metagenomics: strategies for the discovery of novel enzymes with biotechnological applications from marine environments. *Microb. Cell Factories* 7: 27.

Marx, J. C., Collins, T., D'Amico, S., Feller, G. and Gerday, C. (2007) Cold-adapted enzymes from marine Antarctic microorganisms. *Mar. Biotechnol.* 9: 293–304.

Newman, D. J. and Hill, R. T. (2006) New drugs from marine microbes: the tide is turning. *J. Ind. Microbiol. Biotechnol.* 33: 539–544.

Penesyan, A., Kjelleberg, S. and Egan, S. (2010) Development of novel drugs from marine surface associated microorganisms. *Mar. Drugs* 8: 438–459.

Taylor, M. W., Radax, R., Steger, D. and Wagner, M. (2007) Sponge-associated microorganisms: Evolution, ecology, and biotechnological potential. *Microbiol. Mol. Biol. Rev.* 71: 295–347.

Thomas, T. R. A., Kavlekar, D. P. and LokaBharathi, P. A. (2010) Marine drugs from sponge-microbe association—a review. *Mar. Drugs* 8: 1417–1468.

Walsh, P. J., Smith, S., Gerwick, W. H., Solo-Gabrielle, H. and Fleming, L. (eds) (2008) *Oceans and Human Health: Risks and Remedies from the Seas.* Academic Press, London.

Wijffels, R. H. (2008) Potential of sponges and microalgae for marine biotechnology. *Trends Biotechnol.* 26: 26–31.

Williams, P. G. (2009) Panning for chemical gold: marine bacteria as a source of new therapeutics. *Trends Biotechnol.* 27: 45–52.

Biofuels

Chisti, Y. (2010) Fuels from microalgae. *Biofuels* 1: 233–235.

Mata, T. M., Martins, A. N. A. and Caetano, N. (2010) Microalgae for biodiesel production and other applications: a review. *Renew. Sust. Energy Rev.* 14: 217–232.

Biotechnology in aquaculture

Gillund, F. Y., Dalmo, R., Tonheim, T. C., Seternes, T. and Myhr, A. I. (2008) DNA vaccination in aquaculture—expert judgments of impacts on environment and fish health. *Aquaculture* 284: 25–34.

Ninawe, A. S. and Selvin, J. (2009) Probiotics in shrimp aquaculture: avenues and challenges. *Crit. Rev. Microbiol.* 35: 43–66.

Raghukumar, S. (2008) Thraustochytrid marine protists: production of PUFAs and other emerging technologies. *Mar. Biotechnol.* 10: 631–640.

Ringø, E., Løvmo, L., Kristiansen, M., et al. (2010) Lactic acid bacteria vs. pathogens in the gastrointestinal tract of fish: a review. *Aquacult. Res.* 41: 451–457.

Vine, N. G., Leukes, W. D. and Kaiser, H. (2006) Probiotics in marine larviculture. *FEMS Microbiol. Rev.* 30: 404–427.

Wang, Y. B., Li, J. R. and Lin, J. D. (2008) Probiotics in aquaculture: challenges and outlook. *Aquaculture* 281: 1–4.

Nanotechnology

Lebeau, T. and Robert, J. M. (2003) Diatom cultivation and biotechnologically relevant products. Part II: Current and putative products. *Appl. Microbiol. Biotechnol.* 60: 624–632.

Schuler, D. and Frankel, R. B. (1999) Bacterial magnetosomes: microbiology, biomineralization and biotechnological applications. *Appl. Microbiol. Biotechnol.* 52: 464–473.

Sleytr, U. B., Huber, C., Ilk, N., Pum, D., Schuster, B. and Egelseer, E. M. (2010) S-layers as a tool kit for nanobiotechnological applications. *FEMS Microbiol. Lett.* 267: 131–144.

Chapter 15
Concluding Remarks

This final chapter attempts to review some of the major themes that link the various concepts to emerge throughout the book. What have we learned about the importance of marine microbes and what future developments might occur? Marine microbiology has shown dramatic advances in the past two decades. This book has described some of the most significant advances in marine microbial ecology, most of which have been made possible by the development and application of new methods. Undoubtedly, the next few years will see the increasing application of metagenomics, metatranscriptomics, and metaproteomics, with an exponential growth in information fuelled by ever-decreasing sequencing costs. One of the major challenges facing microbial ecologists is how to interpret the huge datasets of information that are being generated. New bioinformatics systems and collective interpretation of data will become necessary if we are to understand the role of specific microbes in ecological and biogeochemical processes. The ability to identify and sequence individual microbial cells and groups of interacting organisms and study their activities *in situ* is now becoming possible through the application of new techniques such as single-cell genome amplification, atomic force microscopy, and nanoSIMS. These approaches are revealing new insights into biogeochemical processes and the interconnectedness of microbes in the sea. Despite our great advances, we are still far from a good description of those organisms that catalyze essential parts of the biogeochemical cycles. A common theme that has emerged throughout the book is the fact that individual microbes interact with each other and with their immediate environment, so that microscale or nanoscale processes have ecosystem, or even global, effects. For example, who could have predicted that bacterial motility and intercellular communication affect carbon flux in the oceans, or that viral lysis of algae can affect the global climate? Interdisciplinary research linking microbiology and molecular biology with chemical and physical oceanography will become increasingly important to understand such processes. Integration of molecular tools with remote-sensing technology means that we will soon have reliable real-time *in situ* monitoring of microbial processes. Concerns about global climate change, ocean acidification, and the effects of anthropogenic pollutants have been the major stimulus—and the rationale for funding by government agencies and private benefactors, such as the Gordon and Betty Moore Foundation—for much of the recent research in marine microbial ecology and oceanography. Continued support for major long-term international projects is essential if we are to understand the effects of human activities on the oceans in the coming century.

This book contains numerous examples of the deterioration in the health of marine ecosystems, with profound implications for biodiversity and for human health and welfare. Striking parallels between the effect of climatic conditions on the epidemiology of diseases in humans and marine animals are emerging. If we are to avoid or mitigate the worst effects of these changes, we must use scientific understanding and human ingenuity to limit further degradation of the environment and to devise methods to monitor and control disease and ecosystem disruption. We need to have a better understanding of how complex systems emerge from the interactions of biological entities at all levels. The fact that most activities of marine microbes are "unseen" by everyday human experience has been emphasized in this book. It is important that future generations have a sound understanding of marine microbial processes and interactions, which will help human society to make the difficult decisions that lie ahead.

I hope that the book has revealed how important marine microbiology has been in revealing fundamental scientific information. For example, we have seen that detailed analysis of the interactions between symbiotic microbes and their animal hosts provides fascinating insights into evolutionary processes and animal development. The study of marine viruses has led to exciting new theories about the early evolution of life and revealed how important viruses are as drivers of genetic diversity and biological processes.

As well as improving our scientific understanding of life processes and Earth systems, many economic benefits result from the applications of marine microbiology. We are now beginning to see significant commercial benefits from the exploitation of marine microbes and their metabolic activities and unforeseen "spin-offs" have resulted. For example, who could have foreseen that studying bioluminescent bacteria or the microbial colonization of seaweeds could lead to a completely new approach to the control of human disease, through inhibition of quorum sensing? Could we have predicted that studying how halophiles survive extreme salt concentrations could lead to an artificial eye or that studying the microbial communities that develop on whale carcasses on the sea floor could lead to improved laundry detergents? Who would have thought that studying the ultrastructure of diatoms could lead to clothes that change color? The term "blue sky research" is used to describe research undertaken with no obvious foreseeable application—"blue sea research" would be a much more appropriate description. The huge diversity of genes and proteins in marine microbes will provide a rich repository for the production of many useful products such as enzymes, pharmaceuticals, and new materials as we enter the era of synthetic biology. Most importantly, these sources may provide new sources of energy to replace our unsustainable reliance on fossil fuels. Hopefully, we may even be able to exploit these new discoveries to mitigate some of the damage to our environment.

The vastness and complexity of the marine environment is still largely unexplored. At the beginning of the twenty-first century, we have discovered previously unknown ocean processes such as aerobic anoxygenic phototrophy and anammox; new symbiotic associations between bacteria and archaea that are responsible for one of the most important biogeochemical processes in marine sediments—anaerobic methane oxidation; an organism that may be the simplest and smallest form of cellular life—*Nanoarchaeum*; and giant viruses that have transferred entire biosynthetic pathways from their hosts to ensure their replication. What other surprises are out there?

Glossary of Key Terms

ABC transporter ATP-binding cassette transporter; membrane system comprising a substrate-binding protein, an ATP-hydrolyzing protein, and a channel protein spanning the membrane

acetylene reduction assay a sensitive assay for measuring the activity of nitrogenase

active transport energy-dependent transport of substances into and out of the cell

adaptive bleaching hypothesis hypothesis that corals that have expelled zooxanthellae due to stress (bleaching) are recolonized by genetically different types with greater tolerance of the stress conditions

adherence the ability of microbes to stick to surfaces

aerobe an organism that grows in the presence of oxygen

aerobic anoxygenic photosynthesis (AAnP) process of photosynthesis occurring under aerobic conditions in which electron donors such as sulfide or organic matter are used, without evolution of oxygen

algae common name for polyphyletic group of unicellular or multicellular protists usually obtaining nutrition by photosynthesis (may be mixotrophic)

anaerobe an organism that grows in the absence of oxygen; obligate anaerobes may be killed by its presence

anaerobic respiration use of substrate other than oxygen in oxidation of a substrate via an electron transport chain

anammox anaerobic ammonia oxidation

annotation the process of identifying the function of genes in a new genome sequence by comparison with homologous genes of known function

anoxic oxygen-free (<0.1 μM oxygen detectable)

anoxygenic photosynthesis use of light energy to synthesize ATP by cyclic photophosphorylation without the evolution of oxygen

antibiotic a chemical substance (usually low molecular weight) produced naturally by one microorganism that inhibits or kills another microorganism; in common usage this includes chemically synthesized or modified agents

antibody a protein (immunoglobulin) produced by the immune system that binds specifically to an antigen

antigen a substance that elicits an immune response

Archaea a phylogenetic domain of organisms characterized by a relatively undifferentiated simple cell structure, isoprenoid glycerol diether or diglycerol tetraether membrane lipids, archaeal rRNA, complex RNA polymerase, and other distinctive properties

atomic force microscopy instrumentation that can reveal complex surface architectures at the nanometer to micrometer scale

attenuation loss of virulence of a pathogen

autotroph an organism able to grow using CO_2 as the principal source of carbon

Bacteria a phylogenetic domain of organisms characterized by a relatively undifferentiated simple cell structure, diacyl glycerol diester membrane lipids, bacterial rRNA, simple RNA polymerase, and other distinctive properties

bacteriophage a virus that infects bacteria; the term phage is now more widely used to indicate agents that infect members of the *Bacteria* and *Archaea*

bacterioplankton free-floating aquatic bacteria

barophile (piezophile) organism that requires high pressures for growth (usually >400 atm)

barotolerant organism that can grow at high pressures, but is not dependent on them

benthic; benthos the seafloor; organisms living on or in the sediment of the ocean floor

bioaugmentation modification of microbial community composition by addition of specific microbes to improve the rate of bioremediation

biodegradation breakdown of complex organic compounds

biodeterioration damage to natural or fabricated materials through microbial activities

biofilm organized structure of microbial cells, extracellular products, and associated substances formed on surfaces

biofouling colonization of marine surfaces by microbes with successive colonization by algae and animals

biogeochemical cycles movements through the Earth system of key elements essential to life, such as carbon, nitrogen, oxygen, sulfur and phosphorus

bioinformatics the computational storage, retrieval, and analysis of information about biological structure, sequence, and function

biological pump process by which CO_2 at the ocean–atmosphere interface is fixed by photoautotrophs into organic matter, redistributing carbon throughout the oceans and sediments

bioluminescence production of light by living organisms

biomagnification the accumulation and amplification of toxic chemicals at each trophic level so that organisms at the top of the food chain will accumulate high levels of the chemical in their tissues

biomass the quantity of living matter (weight per unit area or volume)

biomimetics design process mimicking processes or principles of assembly found in living organisms

bioremediation biological process to enhance the rate or extent of naturally occurring biodegradation of pollutants

biosphere the global sum of all ecosystems on Earth

biotechnology application of scientific and engineering principles to provide goods and services through mediation of biological agents

bloom the sudden and rapid multiplication of plankton cells

brine water with a very high salinity (e.g. in sea ice)

calcification the deposition of calcium carbonate structures by organisms

Calvin cycle or Calvin–Benson cycle the mechanism by which many autotrophic organisms fix carbon dioxide into organic compounds

Candidatus an intermediate taxonomic category used for *Bacteria* and *Archaea* whose phenotypic characteristics have not been fully documented in accordance with rules defined by the International Committee on Systematics of Prokaryotes

carbonate compensation depth the depth in the ocean below which calcium carbonate dissolves and does not reach the sea floor

carbon cycle the flux of carbon through interconnected reservoirs (atmosphere, terrestrial biosphere, oceans, sediments, and fossil fuels)

carbon fixation incorporation of CO_2 into cellular organic material

carbon sequestration uptake and storage of carbon via biological and geological processes

CARD-FISH fluorescent *in situ* hybridization, in which the signal is amplified by an enzyme-linked reaction

chemocline a boundary layer in ocean water with a strong gradient of a particular chemical

chemokinesis stimulation of microbial movement by a chemical substance

chemolithoautotroph microbe using CO_2 as carbon source, deriving energy and electrons from oxidization of reduced inorganic compounds

chemoorganotrophic heterotroph organism using organic compounds as a source of carbon, energy, and electrons

chemotaxis microbial behavior in which microbes move toward attractant chemicals or away from repellents

chemotherapy the treatment of a disease with antibiotics or synthetic drugs

chitin a long-chain polymer of *N*-acetylglucosamine; a component of fungal cell walls and widely distributed in the marine environment as a component of the exoskeleton of crustaceans and internal structures of cephalopods and other animals

clade a monophyletic group or lineage of organisms which share common inherited characteristics

climate change significant changes in global climate patterns, often synonymous with "global warming"

clone library a genomic library of different DNA sequences that have been inserted into a host (cloning vector) for propagation of multiple copies

cloning [molecular] isolation of a DNA sequence and propagation of multiple copies in a host organism, usually a bacterium, for production of large quantities of DNA for molecular analysis

coccolithophores unicellular marine bloom-forming algae with cell surface covered by calcified plates

community fingerprinting analysis of the genetic sequences in an assemblage of different types of microbes, usually achieved via DGGE of PCR-amplified genes

complement a cascade system of serum proteins in vertebrates that react to antigen–antibody complexes or bacterial surface components to amplify defense responses to a pathogen

conditioning film layer of proteins and polysaccharides that coats surfaces within a short period of immersion in seawater; an essential first step in biofilm formation

confocal laser scanning microscopy microscopic method in which laser light scans the specimen at one level, yielding an image with high contrast and resolution; especially valuable for examining biofilms and biological tissues

consensus (or conserved) sequence a DNA sequence in which particular bases are found at the same position when sequences from different organisms are compared

contig a contiguous length of DNA sequence in which the order of bases is known with high confidence; overlapping segments are used to reconstruct the original genome sequence in shotgun sequencing projects

coral bleaching the loss or impaired function of symbiotic zooxanthellae from corals

Coriolis effect an apparent force arising because of the Earth's rotation; moving objects, winds, and water masses are deflected to the right in the northern hemisphere and to the left in the southern hemisphere

Cyanobacteria a large group of oxygenic photosynthetic *Bacteria*

cyanophage a phage infecting cyanobacteria

deep sea usually, marine waters more than 1000 m deep

denaturing gradient gel electrophoresis (DGGE) a widely used technique for separating amplified DNA sequences from environmental samples

denitrification reduction of nitrate to nitrogen during anaerobic respiration

diatoms unicellular or chain-forming marine and freshwater algal protists; major contributors to primary productivity

diazotroph a microorganism that fixes nitrogen

differentiation the development of modified cell structures and functions arising by sequential gene expression in response to specific stimuli

dimethylsulphide (DMS) volatile compound produced from DMS propionate, a major component of marine algae; DMS has important effects on climatic processes

dinoflagellate unicellular algal protists with two flagella and spinning motion; photosynthetic, phagotrophic or mixotrophic; some are parasites and pathogens

DNA–DNA hybridization method for determining the relatedness of genetic sequences by determining hybridization of single-stranded DNA extracted from two organisms

DNA vaccination direct administration of DNA encoding antigenic proteins into tissue such that the recipient produces an effective immune response

ecology the study of the interrelationships between organisms and their environment

ectoenzyme an enzyme secreted to the outside of the cell; usually located on the cell surface for the breakdown of high-molecular-weight compounds into monomers that can be absorbed by the cell

electron acceptor a substance that accepts electrons during an oxidation–reduction process

electron donor a substance that donates electrons during an oxidation–reduction process

electrophoresis the separation of charged molecules such as nucleic acids or proteins in an electric field

ELISA enzyme-linked immunosorbent assay; an immunological technique for detection of antigens or antibodies

El Niño extended warming of the central and eastern Pacific that leads to a major shift in ocean currents and weather patterns across the Pacific; occurs at irregular intervals of 2–7 years

endemic a disease that is naturally present within a population, involving low numbers of infected individuals

endosymbiont microbe that lives symbiotically within the body of another organism (usually used for intracellular associations)

enrichment culture the use of particular incubation conditions and culture media to encourage the growth of specific microbes from an environment

environmental genomics direct extraction and sequencing of nucleic acids from environmental samples, without the need for isolation or culture of the constituent organisms

epibiotic microbe that lives on the surface of another organism

epidemic an outbreak of disease involving an unusually high number of infected persons

epifluorescence light microscopy (ELM) method for visualizing bacteria, viruses, and other particles on the surface of a membrane after staining with a fluorochrome, which emits light at a particular wavelength after illumination

epizootic an outbreak of epidemic disease in a population of animals

EST expressed sequence tag; a short sequence of a transcribed cDNA sequence

eukaryotic cells with a membrane-bound nucleus and organelles (the spelling eucaryotic is also used)

euphotic zone see *photic zone*

eutrophic environment enriched by high levels of nutrients

exopolymer a high-molecular-weight polymer excreted by cells; important in the formation of biofilms and in aggregation of particles to form marine snow

export production amount of fixed organic matter produced in the photic zone of the oceans which is exported to deeper waters (carbon flux)

fermentation energy-yielding process in which the substrate is oxidized without an exogenous electron acceptor; organic molecules usually serve as both electron donors and acceptors

flow cytometry (FCM) method for quantifying and determining the properties of particles passed in a "single-file" flow through a laser beam and detected by their fluorescent properties; cells or viruses are usually tagged with a fluorochrome, which is often attached via specific oligonucleotides or antibodies

fluorescent *in situ* hybridization (FISH) a technique in which a cell is made fluorescent by use of an oligonucleotide probe attached to a fluorescent dye

Fungi monophyletic group of heterotrophic eukaryotes with absorptive nutrition; unicellular (yeasts) or mycelial

Gaia hypothesis concept that temperature, gaseous composition, and oxidation state of the atmosphere, oceans, and Earth's surface are actively controlled by its living organisms behaving as a system to maintain a stability conducive to the maintenance of life

GC ratio the percentage of guanine + cytosine base pairs in DNA (mol% GC)

gene expression the transcription of a gene into mRNA, followed by translation into a protein

gene probe oligonucleotide sequence used in hybridization methods to detect organisms belonging to a specific group

generation time the time taken for a microbial population to double in size

genome complete complement of genetic information in a cell or virus

genomic fingerprinting method for distinguishing closely related individuals based on small differences in DNA sequences

genomics study of the molecular organization of genomes and gene products using sequence information in coding and noncoding regions

genotype the complete genetic description of an organism

glycocalyx layer of interconnected polysaccharides surrounding bacterial cells; important in cell interactions and biofilm formation

Gram-negative bacterial cells with a complex cell wall composed of a thin layer of peptidoglycan plus an outer membrane containing lipopolysaccharide and lipoprotein; based on staining reaction in the Gram stain

Gram-positive bacterial cells with a simple cell wall composed mainly of peptidoglycan; based on staining reaction in the Gram stain

greenhouse effect natural warming of the atmosphere by absorbance and re-emission of infrared radiation while allowing shortwave radiation to pass; usually refers to the enhanced effect due to elevated levels of gases such as water vapor, CO_2, methane, nitrous oxide, and chlorofluorocarbons

halogen *in situ* hybridization (HISH) a technique employing halogenated oligonucleotide probes used in nanoSIMS for simultaneous phylogenetic identification of specimens and quantification of metabolic activities of single microbial cells in the environment; e.g. the rare element fluorine is used as a label detected by the mass spectrometer, analogous to the visualization of the fluorescent dye in CARD-FISH

halophile a microbe that requires high levels of sodium chloride for growth

harmful algal bloom (HAB) unusual excessive growth of cyanobacterial, microalgal, or macroalgal species resulting in toxin production, mortalities to marine life, foaming, or other nuisance effects

heat-shock response expression of a range of proteins in response to sudden exposure to elevated temperatures or other stressful conditions; the response helps to protect cells from damage

hemolysin a microbial toxin causing lysis of red blood cells

heterocyst a differentiated nitrogen-fixing cell formed by cyanobacteria; nitrogenase enzymes are segregated from the generation of oxygen because they are strongly inhibited by its presence

heterotroph short form of chemoorganotrophic heterotroph

HNLC high nutrient, low chlorophyll; ocean regions characterized by low phytoplankton growth despite ample concentrations of nitrate, and iron is probably the limiting nutrient

homology relatedness of gene sequences sufficient to suggest common phylogenetic ancestry

horizontal (lateral) gene transfer transfer of genes from one independent, mature organism to another; occurs via physical contact (conjugation), transfer of naked DNA (transformation) or via phages (transduction)

hyperthermophile extremely thermophilic bacterium (optimum growth above 80°C)

iron acquisition active mechanism by which ocean microbes and pathogens must obtain iron for cellular processes via secretion of siderophores and/or by surface components

iron hypothesis hypothesis that concentration of free iron in oceans plays a major regulatory role in phytoplankton productivity; most offshore waters contain very low levels since they are remote from land masses and so receive low inputs of iron from terrestrial sources

limiting nutrient nutrient in shortest supply, the concentration of which limits growth and reproduction of particular types of organisms (includes N, P, Si, and Fe)

live attenuated vaccine a virus or bacterium in which virulence has been eliminated; used to stimulate immunity against infection

lysogeny incorporation of a phage genome into a bacterial genome so that it replicates without initiating a lytic cycle unless stimulated to do so by inducing conditions

marine snow particles composed of aggregated cellular detritus, polymers, and living microbes

mesocosm experimental system holding large volumes of seawater to simulate open-water conditions

metagenome; metagenomics the entire genetic complement of all biological entities collected from a particular environment; the science of obtaining the metagenomes (see *environmental genomics*)

metaproteomics the large-scale study of proteins within a microbial community, usually obtained by electrophoretic separation and mass spectrometry

metatranscriptomics the large scale study of gene expression within a microbial community, obtained by sequencing DNA complementary to mRNA

methanogens group of strictly anaerobic *Archaea* that obtain energy by producing methane from CO_2, H_2, acetate, and some other compounds

methanotrophs subset of methylotrophic organisms capable of oxidizing methane to CO_2

methylotrophs group of aerobic *Bacteria* that oxidize organic compounds without carbon–carbon bonds, including methanol, methylamine, and sometimes methane, as a sole source of carbon and energy (cf. methanotrophs)

microalgae microscopic (mostly unicellular) protists, traditionally classified as algae

microarray technology method for determining gene expression by the binding of mRNA or cDNA from cells to an array of oligonucleotides immobilized on a surface

microautoradiography (MAR) technique for the measurement of the uptake of radioactive compounds by individual cells, after exposure of silver grains in a photographic emulsion

microbe in this book, a microbe is defined as any microscopic biological entity, including *Bacteria*, *Archaea*, *Fungi*, protists, and viruses

microbial loop process by which organic matter synthesized by photosynthetic organisms is remineralized by the activity of bacteria and protists, enabling reuse of minerals and CO_2 by primary producers and other heterotrophs

microbial mat complex layered community of microbes on aquatic surfaces, characterized by chemical gradients and associated physiological activities

microorganism in this book, all microbes except viruses which are not cellular

mixed layer uppermost layer of the ocean, mixed by wind; depth varies in different regions, depending on temperature, upwelling, and seasonal effects containing the surface waters

mixotrophy combination of autotrophic and heterotrophic processes

monophyletic a lineage of organisms belonging to the same phylogentic cluster or clade

most probable number (MPN) method for determining microbial population in a liquid based on statistical probability of growth after inoculation of media with various dilutions of the sample

mutualism symbiotic association in which both partners benefit

nanoSIMS a technique of secondary ion mass spectrometry enabling elemental transformations to be monitored at subcellular resolution

nanotechnology construction of materials and functional objects assembled from basic molecular building blocks

neuston the surface film between water and the atmosphere

nitrification oxidation of ammonia to nitrate by certain chemolithotrophic bacteria

nitrogen fixation conversion of atmospheric nitrogen to ammonia, carried out by *Cyanobacteria* and some other microbes

nucleocapsid the nucleic acid and protein of a virus particle

ocean acidification an increase in the concentration of hydrogen ions (fall in pH) due to absorption of CO_2 from the atmosphere

oligonucleotide a nucleic acid molecule containing a small number of nucleotides

oligotrophic environment with very low nutrient levels; also used to describe an organism adapted to such low-nutrient conditions

open reading frame (ORF) arrangement of nucleotides in triplet codons in DNA which does not contain a stop codon; sequences larger than 100 are considered to be potential protein-coding regions

operon a group of genes transcribed into a single mRNA under regulatory control of the same region of the DNA

ortholog a gene found in one organism that closely resembles that in another organism, but separated by speciation

osmoprotectant compatible solute accumulated in the cytoplasm to protect cells from loss of water to the external environment

osmotrophy feeding by absorption of soluble nutrients

OTU operational taxonomic unit; used to distinguish organisms defined on the basis of their SSU rRNA gene sequences (typically, bacteria differing by <3% in sequence identity)

oxic contains oxygen at a concentration high enough to permit aerobic growth and respiration

oxidation–reduction (redox) reaction a paired reaction in which one compound is oxidized by losing electrons or hydrogen atoms, and another compound is reduced by acquiring these electrons or hydrogen atoms

oxygen-minimum zone in common usage, defines the suboxic zone found at intermediate depths (usually 1000–1500 m); more specifically, defines zones at much shallower depths (typically 50–200 m) with extremely low oxygen concentrations (< 5 μM) at their core

pandemic an epidemic disease that spreads to many countries worldwide

pangenome the full complement of genes within a collection of strains in a species of bacteria or archaea in which there is usually a high degree of variation in gene content; it consists of a core genome (genes present in all strains), a dispensable genome (genes shared by two or more stains), and genes unique to individual strains

pathogenicity the ability of a microbe to cause disease in a susceptible host

pathogenicity island an acquired region identified in a bacterial chromosome containing a cluster of genes associated with survival in the host and production of disease

pelagic water column in the open ocean; also used to describe organisms in this habitat

periplasm the space between the cytoplasmic membrane and outer membrane of Gram-negative *Bacteria*

phage a virus infecting members of the *Bacteria* and *Archaea*

phagotrophy feeding by ingestion of particles into vacuoles

phenotype the observable characteristics of a culture organism, such as enzyme activities

photic zone upper layer of ocean water penetrated by light; the term *euphotic zone* may be used to indicate sufficient light of appropriate wavelengths to permit phototrophic metabolism of phytoplankton

photoautotroph organism that grows using light energy, inorganic compound(s) as a source of electrons and CO_2 as a carbon source

photoheterotroph (photoorganotroph) organism that grows using light energy, with organic compounds as its source of electrons and carbon

photophosphorylation the transfer of light energy for synthesis of energy-rich bonds in ATP

photoreactivation DNA repair process for excision of pyrimidine dimers by an enzyme activated by blue light

photosynthesis use of light as a source of energy for the formation of organic compounds from CO_2

phototaxis movement of a microorganism toward light

phototrophy use of light as a source of energy for metabolism

phylogenetic classification a system of classification based on the genetic relatedness and evolutionary history of organisms, rather than similarity of current characteristics

phylotype a unique sequence of a marker gene in microbial community analysis

phytoplankton all free-floating photoautotrophic organisms, including diatoms, dinoflagellates, prymnesiophytes, picoeukaryotes, and other protists, together with cyanobacteria

picoeukaryotes autotrophic and heterotrophic protists in the picoplankton size range

picophytoplankton photoautotrophic plankton microbes in the 0.2–2.0-μm size range

picoplankton plankton microbes in the 0.2–2.0-μm size range (*Bacteria*, *Archaea*, and some flagellates)

piezophile see *barophile*

plankton general term for free-floating microscopic organisms in water

plaque a zone of lysis caused by a virus on a lawn culture of cells

plasmid a double-stranded DNA molecule in bacteria, carrying genes for specialized functions that replicates independently or can be integrated into the chromosome

polymerase chain reaction (PCR) *in vitro* amplification of DNA fragments employing sequence-specific oligonucleotide primers and thermostable polymerases

polyphyletic lineages of organisms belonging to different phylogentic clusters or clades

polysaccharide a chain of sugar molecules linked by glycosidic bonds

primary productivity rate of carbon fixation by autotrophic processes in the oceans; either gross (total biomass) or net (gross productivity less the respiration rate of producing organisms)

prokaryotic cells without a true membrane-bound nucleus found in *Bacteria* and *Archaea* (the spelling procaryotic is also used); although widely used, this book encourages recognition of these groups as separate domains

prophage the genome of a temperate virus when it is integrated into the host genome

proteomics analysis of the complete protein complement of a cell using two-dimensional electrophoresis, mass spectrometry, and other techniques

proteorhodopsin a light-sensitive proton-transport membrane protein that is widely found in various groups of planktonic *Bacteria* and *Archaea*

protists simple eukaryotes, usually unicellular or may be colonial without true tissues

protozoa common name for a polyphyletic group of unicellular protists that lack cell walls usually feeding by phagotrophy (may be mixotrophic)

psychrophilic microbe adapted to growth at low temperatures (grows well around 0°C; temperature maximum usually 15–20°C)

pycnocline a zone having a marked change in density of water as a function of depth

quorum sensing mechanism of regulating gene expression in which bacteria measure their population density by secretion and sensing of signaling molecules; when these reach a certain critical level, the bacterium will express specific genes

radioisotope unstable isotope of an element that decays or disintegrates spontaneously, emitting radiation; compounds labeled with ^{14}C, ^{3}H, ^{32}P, or ^{35}S are commonly used to study metabolic pathways and the rate of reactions

recombinant DNA technology insertion of a gene into a cloning vector such as a plasmid and transfer into another organism to produce a recombinant molecule (genetic engineering)

Redfield ratio the relatively constant proportions maintained between the elements C, N, P, and O taken up during synthesis of cellular material by marine organisms and released by subsequent remineralization

remote sensing collection and interpretation of information about an object without being in physical contact; usually refers to measurement of physical and chemical properties of the oceans via satellite instruments

respiration energy-yielding metabolic process in the substrate is oxidized by transfer in an electron-transport chain to an exogenous terminal electron acceptor such as O_2, nitrate or certain organic compounds

ribotyping identification of microorganisms by analysis of DNA fragments produced by restriction enzyme digestion of 16S rRNA genes

rRNA analysis analysis of the nucleotide sequence of ribosomal RNA molecules or of the genes that encode them together with noncoding spacer regions (this may be designated rDNA); the main method used in phylogenetic classification and diversity studies

scaffold in metagenomics, a portion of the genome sequence reconstructed from end-sequenced whole-genome shotgun clones comprising contigs and gaps

SDS-PAGE sodium dodecyl sulfate polyacrylamide electrophoresis; a technique for separation of proteins

shotgun sequencing random sequencing of DNA from small fragments of a genome followed by assembly to reconstruct the genome sequence

siderophore organic molecule excreted by bacteria, which complexes with iron in the environment and transports it into the cell

signature sequence oligonucleotide sequence (e.g. in rRNA) that characterizes a particular group of organisms and is used to design gene probes

S-layer a paracrystalline layer of protein subunits found on the surface of many bacterial and archaeal cells

solubility pump ocean process producing a vertical gradient of dissolved inorganic carbon due to increased solubility of CO_2 in cold water; deep water with high carbon content is formed in high latitudes and is transported by deep ocean currents, upwelling at low latitudes

Southern blotting technique for transferring denatured DNA fragments from an agarose gel to a nitrocellulose sheet for identification using a hybridization probe

species in microbiology, defined as a collection of strains that share many relatively stable common properties and differ significantly from other collections of strains or forming monophyletic, exclusive groups based on DNA sequence

stable isotope analysis analysis of "heavy" and "light" forms of an element used to distinguish biological from purely geochemical processes; for example by measuring the ratio of $^{13}C:^{12}C$ or $^{34}S:^{32}S$ in organic material

strain [of bacteria] a population descended from an isolate in pure culture

suboxic containing a low level of dissolved oxygen (<20 μM); usually referring to the base of the photic zone or a layer occurring at intermediate depths (usually 1000–1500 m)

sulfate-reducing bacteria (SRB) bacteria able to use sulfate or elemental sulfur as a terminal electron acceptor during anaerobic respiration; usually applied specifically to a group of δ *proteobacteria*, although some *Archaea* can also reduce sulfate

symbiosis living together; close association of two different organisms (usually interpreted as mutualism)

syntrophic association between microbes in which the growth of one or both partners depends on the provision of nutrients or growth factors from the activities of the other

taxon a group of one or more units considered to be a unit because they are phylogenetically related and share common characteristics

taxonomy the process of classifying and naming organisms according to prescribed rules appropriate to different groups

TCA cycle series of metabolic reactions in which a molecule of acetyl CoA is completely oxidized to CO_2, generating precursors for biosynthesis and NADH and $FADH_2$, which are oxidized in the electron transport chain; also known as the Krebs cycle and citric acid cycle

thermocline boundary layer in ocean water separating water of different temperatures

thermophilic organism that grows above 55°C

thiotrophic using oxidation of sulfur or reduced sulfur compounds as an energy source

transposon mutagenesis insertion of a transposon (a type of mobile genetic element) into a gene, resulting in inactivation of that gene and acquisition of the phenotype encoded by the transposon (usually antibiotic resistance)

two-component regulatory system a system comprising a sensor kinase protein in the membrane that becomes phosphorylated in response to an external signal and passes the phosphoryl group (directly or via a cascade mechanism) to a response regulator protein, which usually affects transcription of specific genes (in bacterial chemotaxis there is a direct effect on flagellar motility)

type III secretion system a molecular syringe system used by some bacterial pathogens to secrete effector proteins directly into host cells

ultramicrobacteria term used to describe very small bacteria (as small as 0.2 μm diameter) occurring naturally in seawater or formed by miniaturization under low-nutrient conditions

upwelling upward motion of deep water, driven by effects of winds, temperature, and density, bringing nutrients to the upper layers of the ocean

viable but nonculturable (VBNC) different interpretations are made of this term; in this book, used to describe a physiological state in which it is not possible to culture an organism on media that normally supports its growth, although cells retain indicators of metabolic activity

viral shunt (or loop) the mechanism by which viral lysis of cells results in retention of nutrients for heterotrophic metabolism, lessening transfer to other trophic levels

virion the extracellular particle form of a virus; nucleic acid enclosed in protein, sometimes containing other substances

virome a viral metagenome; the entire complement of all viral genes collected from a particular environment

virulence the degree of pathogenicity of a microorganism as indicated by the production of disease signs

virus a biological entity containing either DNA or RNA that replicates inside host cells

VLP virus-like particle; a structure observed in electron microscopy that has not been proved to contain genetic information

Western blotting technique for transferring protein bands from a polyacrylamide gel to a nitrocellulose sheet for identification using an enzyme-linked antibody

xenobiotic an artificial synthetic compound that does not occur naturally

zoonosis disease generally associated with animals that can be transmitted to humans

zooplankton free-floating small animals and nonphotosynthetic protists

Abbreviations

AAnP	aerobic anoxygenic photosynthesis
Ab	antibody
ABC	ATP-binding cassette
AFLP	amplified fragment length polymorphism
Ag	antigen
AHL	acyl homoserine lactone
AI	autoinducer
ALOHA	a long-term oligotrophic habitat assessment
AMP	adenosine monophosphate
AMO	anaerobic methane oxidation
AO	acridine orange
AOA	ammonia oxidizing archaea
APS	adenosine-5′-phosphosulfate
ASP	amnesic shellfish poisoning
atm	unit of atmospheric pressure [= 0.103 MPa]
ATP	adenosine triphosphate
AUV	autonomous underwater vehicle
BAC	bacterial artificial chromosome
BATS	Bermuda atlantic time series
BBD	black band disease
BBL	benthic boundary layer
BKD	bacterial kidney disease
BPV	baculovirus penaei virus
CARD-FISH	catalyzed reporter deposition fluorescent *in situ* hybridization
cDNA	complementary DNA
CDV	canine distemper virus
CFB	*Cytophaga–Flavobacterium–Bacteroides* group
CFP	ciguatera fish poisoning
CFU	colony-forming units
CLOD	coralline lethal orange disease
CMV	cetacean morbillivirus
CpBV	*Cancer pagarus* bacilliform virus
CTC	5-cyano-2,3-dilotyl tetrazolium chloride
CTD	conductivity, temperature, depth
CZCS	Coastal Zone Color Scanner
DAPI	4′6′-diamido-2-phenylindole
ddNTP	dideoxyribonucleotide triphosphate
DDT	dichloro-diphenyl-trichloroethane
DGGE	denaturing gradient gel electrophoresis
DHA	docosahexaenoic acid
DHAB	deep hypersaline anoxic basin
DIC	dissolved organic carbon
DIN	dissolved inorganic nitrogen
DIP	dissolved inorganic phosphorus

DMS	dimethyl sulfide
DMSP	dimethylsulfoniopropionate
DNA	deoxyribonucleic acid
dNTP	deoxyribonucleotide triphosphate
DOC	dissolved organic carbon
DOM	dissolved organic material
DON	dissolved organic nitrogen
DOP	dissolved organic phosphorus
DSP	diarrhetic shellfish poisoning
EhV	*Emiliania huxleyi* virus
ELISA	enzyme-linked immunosorbent assay
EPS	exopolymeric substances
EST	expressed sequence tag
ETSP	eastern tropical South Pacific
FACS	fluorescence-activated cell sorting (or sorter)
FAME	fatty acid methyl esters
FAT	fluorescent antibody technique
FISH	fluorescence *in situ* hybridization
FITC	fluorescein isothiocyanate
GC or G + C	guanine + cytosine base pair
GCAT	glycerophospholipid:cholesterol acyltransferase
GFP	green fluorescent protein
GOS	Global Ocean Survey
GPA	calcium-binding protein
GTA	gene transfer agent
HAB	harmful algal bloom
HGT	horizontal gene transfer
HNHC	high nutrient, high chlorophyll
HNLC	high nutrient, low chlorophyll
HPLC	high-performance liquid chromatography
HPV	hepatopancreatic parvovirus
HRP	horseradish peroxidase
ICOMM	International Census of Marine Microbes
IHHNV	infectious hypodermal and hematopoietic necrosis virus
IHN (IHNV)	infectious hematopoietic necrosis (virus)
IPN (IPNV)	infectious pancreatic necrosis (virus)
IROMPs	iron-restricted outer-membrane proteins
ISA (ISAV)	infectious salmon anemia (virus)
ITCV	International Committee on Taxonomy of Viruses
ITS	internal transcribed spacer
JCVI	J. Craig Venter Institute
JOD	juvenile oyster disease

kDa	kilodalton
LGT	lateral gene transfer
LNHC	low nutrient, high chlorophyll
LNLC	low nutrient, low chlorophyll
LPS	lipopolysaccharide
LSU	large subunit
M	molar
mAbs	monoclonal antibodies
MABV	marine birnavirus(es)
MALV	marine alveolates
MAR	microautoradiography
Mb	megabase (10^6 nucleotides)
MBV	monodon baculovirus
MCP	major coat protein
MLSA	multilocus sequence analysis
MLST	multilocus sequence typing
MPN	most probable number
mRNA	messenger RNA
MUG	methylumbeliferyl-β-glucuronide
NADPH	nicotinamide adenine dinucleotide phosphate (reduced)
NAG	*N*-acetyl glucosamine
NASA	National Space and Aeronautics Administration
NCDLV	nucleo-cytoplasmic large DNA virus
NMR	nuclear magnetic resonance
NOAA	National Oceanic and Atmospheric Administration
NSP	neurotoxic shellfish poisoning
NTP	nucleotide triphosphate
OA	ocean acidification
OEC	oxygen-evolving complex
OM	outer membrane (of Gram-negative bacteria)
OMZ	oxygen-minimum zone
ONPG	o-nitrophenol-β-galactopyranoside
ORF	open reading frame
OTU	operational taxonomic unit
PAGE	polyacrylamide gel electrophoresis
PAH	polyaromatic hydrocarbon
PCB	polychlorinated biphenyl
PCD	programmed cell death
PCR	polymerase chain reaction
PDV	phocine distemper virus
PFGE	pulsed field gel electrophoresis
PFU	plaque-forming units
PKS	polyketide synthase
PMMoV	pepper mild mottle virus
POM	particulate organic material
PP$_i$	inorganic phosphate
ppm	parts per million
PR	proteorhodopsin
PS	photosystem
PSP	paralytic shellfish poisoning
PUFA	polyunsaturated fatty acid
PyMS	pyrolysis mass spectroscopy
Q-PCR	quantitative (real-time) polymerase chain reaction
QS	quorum sensing
RAPD	random amplified polymorphic DNA
rDNA	DNA encoding ribosomal RNA
RFLP	restriction fragment length polymorphism
RNA	ribonucleic acid
ROD	*Roseovarius* oyster disease
ROV	remotely operated vehicle
rRNA	ribosomal RNA
RT	reverse transcriptase
RTN	rapid tissue necrosis
RuBisCO	ribulose bisphosphate carboxylase
RUBP	ribulose bisphosphate
S	Svedberg unit
SA	surface area
SA/V	surface area:volume ratio
SBP	sphingolipid biosynthesis pathway
SCUBA	self-contained underwater breathing apparatus
SDS-PAGE	sodium dodecyl sulfate polyacrylamide gel electrophoresis
SeaWiFS	Sea-viewing Wide Field-of-view Sensor
SEM	scanning electron microscopy
SIMS	secondary ion mass spectrometry
SMV	spawner mortality virus
SRB	sulfate-reducing bacteria
SSU	small subunit
SWI	sediment–water interface
TBT	tributyl tin
TCA	tricarboxylic acid
TCBS	thiosulfate-citrate-bile-sucrose
TCP	toxin coregulated pili
TDH	thermostable direct hemolysin
TEM	transmission electron microscopy
TGGE	temperature gradient gel electrophoresis
TIGR	The Institute for Genomic Research
Tm	dissociation temperature ("melting point") of DNA
TRFLP	terminal restriction fragment length polymorphism
TRH	thermostable related hemolysin
tRNA	transfer RNA
TSV	Taura syndrome virus
TTSS	type III secretion system
TTX	tetrodotoxin
USEPA	United States Environment Protection Agency
UV	ultraviolet
V	volume
VAI	vibrio autoinducer
VBNC	viable but nonculturable
VHML	*Vibrio harveyi* myovirus like
VHS (VHSV)	viral hemorrhagic septicemia (virus)
VLP	virus-like particle
VPI	*Vibrio* pathogenicity island
WGS	whole genome shotgun sequencing
WHO	World Health Organization
WPD	white plague disease
WS	white syndrome
WSSV	white spot syndrome virus
w/v	weight per volume

Index

Note: page numbers followed by F refer to figures, those followed by T refer to tables, and those followed by B refer to marginal query and information boxes

Color plates

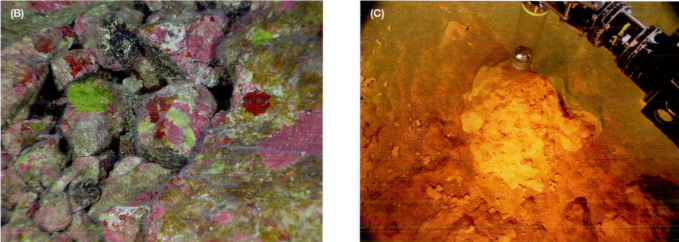

Plate 1.1 Microbial mats at an undersea volcano at Mariana Arc region, Western Pacific Ocean. (A) Boulders covered with a mat of bacteria, which use reduced inorganic compounds from the hydrothermal vent for chemosynthesis. (B) Unusual co-occurrence of chemosynthetic bacterial mats and photosynthetic green and red algae near the top of the East Diamante volcano; sufficient light penetrates for algal growth. (C) Remotely operated vehicle (ROV) sampling of a thick iron-rich bacterial mat found at a diffuse warm vent site near the summit of Northwest Eifuku volcano. (Images courtesy of Pacific Ring of Fire 2004 Expeditions, NOAA Office of Ocean Exploration.)

Plate 2.1 The detection of a bioluminescent bacterial bloom demonstrating a "milky sea" phenomenon in the Arabian Sea, by analysis of archived records of a meteorological satellite. (A–C) show unfiltered images and (D–F) show images after filtering spectral frequencies of the output. The dotted line in (D) represents the track of the ship based on coordinates provided in the captain's report of the observation. (Reprinted from Miller et al. [2005] by permission of the National Academies of Sciences.)

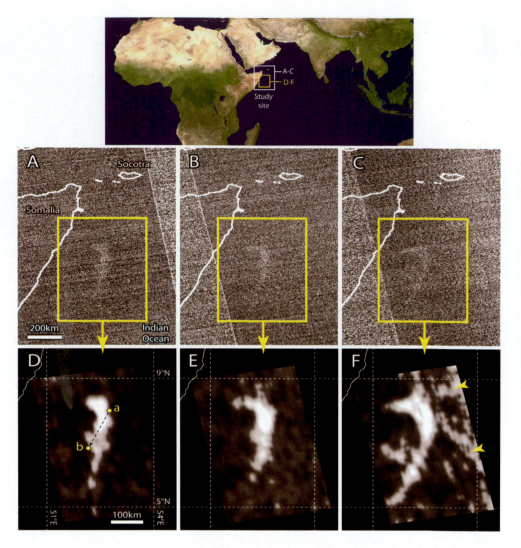

Plate 2.2 The claw of a remotely operated vehicle positions a trap for microbial colonization at a hydrothermal vent field on the Explorer Ridge, North East Pacific. The trap will later be recovered for laboratory study. (Image courtesy of Pacific Ring of Fire 2002 Expedition, NOAA Office of Ocean Exploration.)

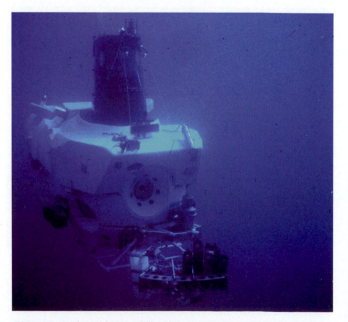

Plate 2.3 The submersible *Alvin* in 1978, 1 year after the discovery of hydrothermal vents. (Image courtesy of NOAA Office of Ocean Exploration.)

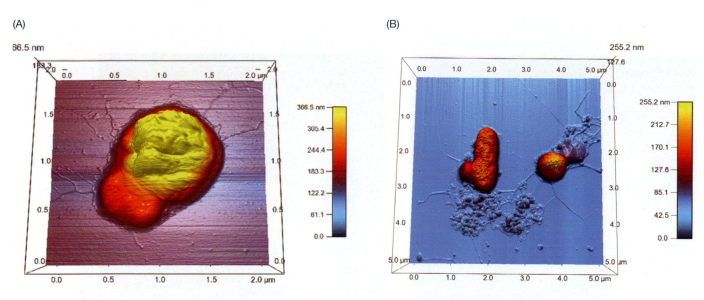

Plate 2.4 Atomic force microscopy of interactions between pelagic bacteria. Scale bar indicates cell height. (A) A heterotrophic bacterium (*red*) closely associated with a larger *Synechococcus* cell (*yellow*). (Reprinted from Malfatti et al. [2010] by permission of Macmillan Publishers Ltd.) (B) Bacteria surrounded by an apparent network of gel and nanometer-sized particles. (Image courtesy of Francesca Malfatti and Farooq Azam, Scripps Institution of Oceanography.)

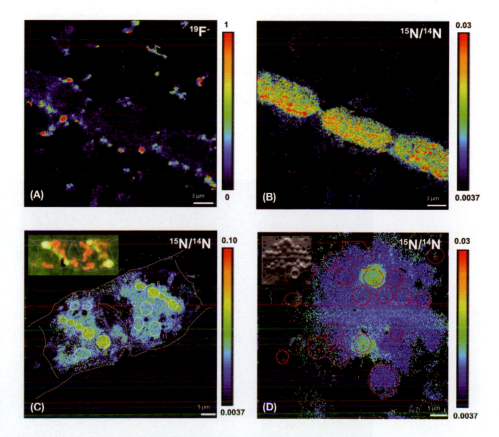

Plate 2.5 Use of NanoSIMS to show nitrogen fixation and transfer. (A) *Cytophaga* bacteria attached to filamentous cyanobacteria, identified by HISH-SIMS. (B) Fixed N is transferred directly to the bacteria. (C) Symbiosis between the diatom *Hemiaulas* and the intracellular cyanobacterium *Richelia*. Inset is the epifluorescent image of cells prior to nanoSIMS analysis. N uptake is localized to the symbionts and also to host cells. (D) A cluster of unicellular *Crocosphaera*, and a filament of *Trichodesmium* incubated in $^{15}N_2$. Note the differential uptake of N into only a few *Crocosphaera* cells. (Images courtesy of Rachel A. Foster, Birgit Adam, Tomas Vagner, Niculina Musat, and Marcel Kuypers, Max Planck Institute for Marine Microbiology, Bremen.)

(A)

(B)

15N 12C/14N 12C

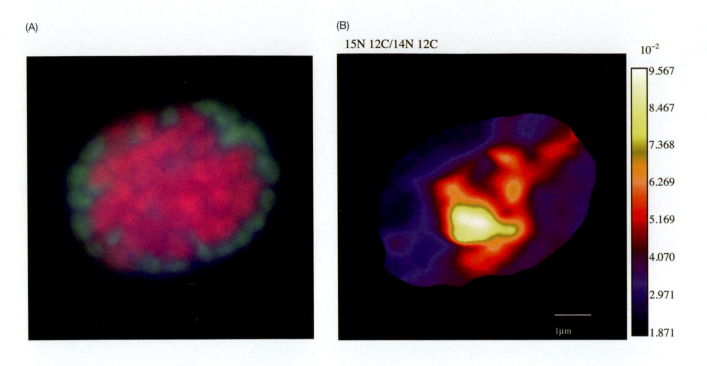

10⁻²

9.567

8.467

7.368

6.269

5.169

4.070

2.971

1.871

1μm

Plate 5.1 Anaerobic oxidation of methane in marine sediments by a metabolic consortium of methane-oxidizing archaea surrounded by sulfate-reducing bacteria (SRB). (A) Confocal micrograph visualized by FISH with probes directed against the archaeal (*red*) and bacterial (*green*) partners. (B) NanoSIMS image showing ¹⁵N incorporation, indicating transfer of fixed nitrogen from the archaeal cells to SRB. (Image courtesy of Anne Dekas and Victoria Orphan, California Institute of Technology.)

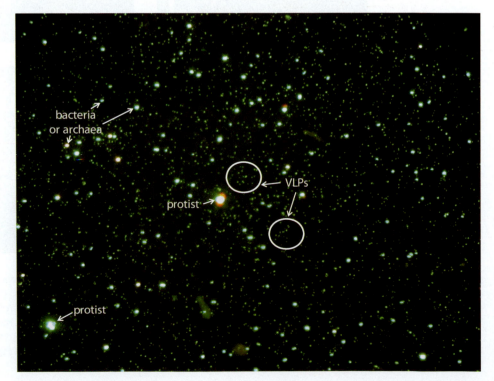

Plate 7.1 Epifluorescence micrograph of a filtered (0.02-μm Anodisc) water sample stained with SYBR green for enumeration of virus-like particles (VLPs). The smallest bright dots are VLPs, the next largest are bacterial or archaeal cells. The two larger cells shown are protists. (Image courtesy of Jed Fuhrman, University of Southern California, using the protocol of Patel et al. [2007].)

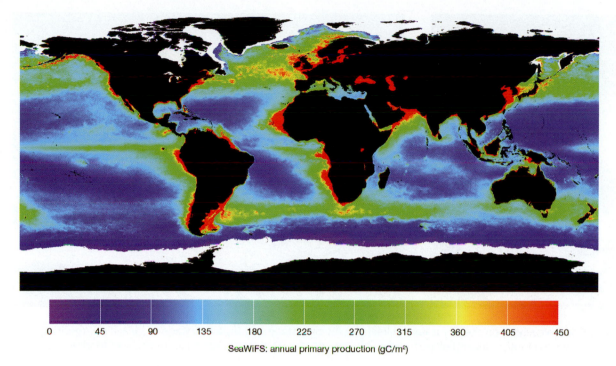

0   45   90   135   180   225   270   315   360   405   450

SeaWiFS: annual primary production (gC/m²)

Plate 8.1 Annual global productivity 1998–1989 (g carbon fixed per m²). Modeled following the method of Behrenfeld and Falkowski (1997), with surface irradiance corrected for cloudiness. (Image courtesy of ICMS Ocean Productivity Study, Rutgers, The State University of New Jersey Institute of Marine and Coastal Sciences.)

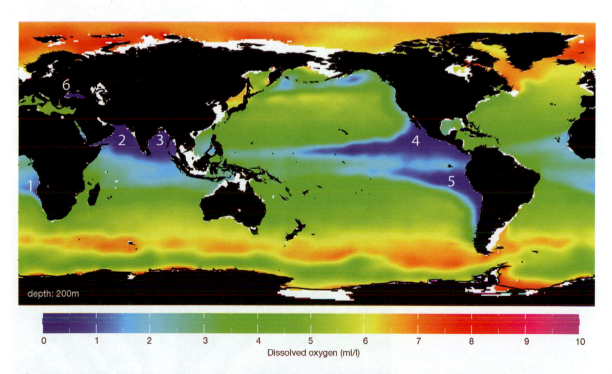

0   1   2   3   4   5   6   7   8   9   10

Dissolved oxygen (ml/l)

Plate 9.1 Location of the major oxygen-minimum zones (OMZs) where very low oxygen minima exist in shallow waters. The scale bar shows the oxygen concentration at 200 m depth. 1, Southwest African continental margin; 2, Arabian Sea; 3, Bay of Bengal; 4, eastern tropical North Pacific; 5, eastern tropical South Pacific; 6, Black Sea. (Image courtesy of NOAA, National Oceanographic Data Center, World Ocean Atlas.)

Plate 10.1 Coral bleaching on the Great Barrier Reef. (A) An early stage and (B) a late stage of bleaching, resulting in the loss of pigmented symbiotic zooxanthellae. (Images courtesy of Mary Wakeford, Australian Institute of Marine Science.)

Plate 10.2 The giant clam *Tridacna maxima* growing on the Great Barrier Reef. The brightly colored mantle is exposed to the light and harbors photosynthetic zooxanthellae. (Image courtesy of NOAA Coral Kingdom Collection.)

Plate 10.3 The giant tubeworm, *Riftia pachyptila*. This colony was photographed at a hydrothermal vent at the East Pacific Rise at 2500 m depth; each tubeworm is about 1 m in length. (Image courtesy of Monika Bright, University of Vienna.)

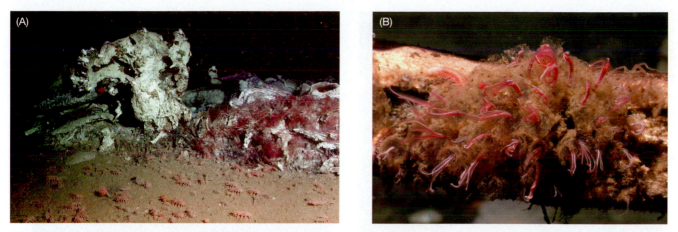

Plate 10.4 Part of the skeleton of a gray whale carcass discovered in 2002, in Monterey Canyon at >3000 m depth. (A) shows the head of the whale and (B) is a close-up of a bone densely covered in *Osedax frankpressi* gutless worms, which penetrate the bones. Symbiotic bacteria within the worm's cells break down collagen, cholesterol, and lipids in the bones. (Copyright Monterey Bay Aquarium Research Institute, reprinted by permission.)

Plate 10.5 (*above*) The "Yeti crab" *Kiwa hirsuta* discovered at hydrothermal vents on the Pacific Antarctic Ridge. The animal is covered in setae that are densely colonized by bacteria belonging to the *Epsilon-* and *Gammaproteobacteria* and *Bacteroidetes* (Goffredi et al., [2008]). (Copyright Ifremer/A. Fifis, reprinted by permission.)

Plate 10.6 The "scaly foot" snail discovered at a hydrothermal vent in the Indian Ocean. (A) Side view. (B) The foot of the snail, showing dense coat of overlapping scales of iron sulfide, colonized by *Epsilonproteobacteria* and sulfate-reducing *Deltaproteobacteria*; these scales are thought to confer protection against predators. (Reprinted from Suzuki et al. [2006] by permission of Elsevier.)

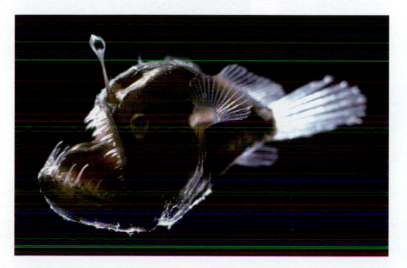

Plate 10.7 (*left*) The angler fish *Melanocetus johnsoni*. The projection from the head (esca) contains bioluminescent symbiotic bacteria. (Image courtesy of Edie Widder, Ocean Research & Conservation Association.)

Plate 11.1 Examples of coral diseases associated with bacterial or fungal infection. (A) Black-band disease on *Diploria strigosa*. (B) Yellow blotch/band disease on *Montastrea* sp. (C) White pox on *Acropora palmata*. (D) White band on *A. palmata*. (E) Aspergillosis on *Gorgonia ventalina*. (F) White syndrome on *Acropora* sp. [Images courtesy of (A) Christina Kellogg, (C, D) Thierry Work and (F) Ginger Garrison, US Geological Survey; (B) James Cervino, Woods Hole Oceanographic Institution; (E) Meir Sussman, Australian Institute of Marine Science.]